# UCLA Symposia on Molecular and Cellular Biology, New Series

*Series Editor,* C. Fred Fox

**RECENT TITLES**

*Volume 50*
**Interferons as Cell Growth Inhibitors and Antitumor Factors,** Robert M. Friedman, Thomas Merigan, and T. Sreevalsan, *Editors*

*Volume 51*
**Molecular Approaches to Developmental Biology,** Richard A. Firtel and Eric H. Davidson, *Editors*

*Volume 52*
**Transcriptional Control Mechanisms,** Daryl Granner, Michael G. Rosenfeld, and Shing Chang, *Editors*

*Volume 53*
**Progress in Bone Marrow Transplantation,** Robert Peter Gale and Richard Champlin, *Editors*

*Volume 54*
**Positive Strand RNA Viruses,** Margo A. Brinton and Roland R. Rueckert, *Editors*

*Volume 55*
**Amino Acids in Health and Disease: New Perspectives,** Seymour Kaufman, *Editor*

*Volume 56*
**Cellular and Molecular Biology of Tumors and Potential Clinical Applications,** John Minna and W. Michael Kuehl, *Editors*

*Volume 57*
**Proteases in Biological Control and Biotechnology,** Dennis D. Cunningham and George L. Long, *Editors*

*Volume 58*
**Growth Factors, Tumor Promoters, and Cancer Genes,** Nancy H. Colburn, Harold L. Moses, and Eric J. Stanbridge, *Editors*

*Volume 59*
**Chronic Lymphocytic Leukemia: Recent Progress and Future Direction,** Robert Peter Gale and Kanti R. Rai, *Editors*

*Volume 60*
**Molecular Paradigms for Eradicating Helminthic Parasites,** Austin J. MacInnis, *Editor*

*Volume 61*
**Recent Advances in Leukemia and Lymphoma,** Robert Peter Gale and David W. Golde, *Editors*

*Volume 62*
**Plant Gene Systems and Their Biology,** Joe L. Key and Lee McIntosh, *Editors*

*Volume 63*
**Plant Membranes: Structure, Function, Biogenesis,** Christopher Leaver and Heven Sze, *Editors*

*Volume 64*
**Bacteria–Host Cell Interactions,** Marcus A. Horwitz, *Editor*

*Volume 65*
**The Pharmacology and Toxicology of Proteins,** John S. Holcenberg and Jeffrey L. Winkelhake, *Editors*

*Volume 66*
**Molecular Biology of Invertebrate Development,** John D. O'Connor, *Editor*

*Volume 67*
**Mechanisms of Control of Gene Expression,** Bryan Cullen, L. Patrick Gage, M.A.Q. Siddiqui, Anna Marie Skalka, and Herbert Weissbach, *Editors*

*Volume 68*
**Protein Purification: Micro to Macro,** Richard Burgess, *Editor*

*Volume 69*
**Protein Structure, Folding, and Design 2,** Dale L. Oxender, *Editor*

*Volume 70*
**Hepadna Viruses,** William Robinson, Katsuro Koike, and Hans Will, *Editors*

*Volume 71*
**Human Retroviruses, Cancer, and AIDS: Approaches to Prevention and Therapy,** Dani Bolognesi, *Editor*

Please contact the publisher for information about previous titles in this series.

# UCLA Symposia Board

**C. Fred Fox**, Ph.D., Director
Professor of Microbiology, University of California, Los Angeles

**Charles J. Arntzen**, Ph.D.
Director, Plant Science and Microbiology
E.I. du Pont de Nemours and Company

**Floyd E. Bloom**, M.D.
Director, Preclinical Neurosciences/
 Endocrinology
Scripps Clinic and Research Institute

**Ralph A. Bradshaw**, Ph.D.
Chairman, Department of Biological Chemistry
University of California, Irvine

**Francis J. Bullock**, M.D.
Vice President, Research
Schering Corporation

**Ronald E. Cape**, Ph.D., M.B.A.
Chairman
Cetus Corporation

**Ralph E. Christoffersen**, Ph.D.
Executive Director of Biotechnology
Upjohn Company

**John Cole**, Ph.D.
Vice President of Research
and Development
Triton Biosciences

**Pedro Cuatrecasas**, M.D.
Vice President of Research
Glaxo, Inc.

**Mark M. Davis**, Ph.D.
Department of Medical Microbiology
Stanford University

**J. Eugene Fox**, Ph.D.
Vice President, Research
and Development
Miles Laboratories

**J. Lawrence Fox**, Ph.D.
Vice President, Biotechnology Research
Abbott Laboratories

**L. Patrick Gage**, Ph.D.
Director of Exploratory Research
Hoffmann-La Roche, Inc.

**Gideon Goldstein**, M.D., Ph.D.
Vice President, Immunology
Ortho Pharmaceutical Corp.

**Ernest G. Jaworski**, Ph.D.
Director of Biological Sciences
Monsanto Corp.

**Irving S. Johnson**, Ph.D.
Vice President of Research
Lilly Research Laboratories

**Paul A. Marks**, M.D.
President
Sloan-Kettering Memorial Institute

**David W. Martin, Jr.**, M.D.
Vice President of Research
Genentech, Inc.

**Hugh O. McDevitt**, M.D.
Professor of Medical Microbiology
Stanford University School of Medicine

**Dale L. Oxender**, Ph.D.
Director, Center for Molecular Genetics
University of Michigan

**Mark L. Pearson**, Ph.D.
Director of Molecular Biology
E.I. du Pont de Nemours and Company

**George Poste**, Ph.D.
Vice President and Director of Research
and Development
Smith, Kline and French Laboratories

**William Rutter**, Ph.D.
Director, Hormone Research Institute
University of California, San Francisco

**George A. Somkuti**, Ph.D.
Eastern Regional Research Center
 USDA-ARS

**Donald F. Steiner**, M.D.
Professor of Biochemistry
University of Chicago

# Growth Factors, Tumor Promoters, and Cancer Genes

# Growth Factors, Tumor Promoters, and Cancer Genes

Proceedings of a Triton Biosciences-UCLA Symposium
Held in Steamboat Springs, Colorado, April 6–13, 1986

## Editors

**Nancy H. Colburn**
National Cancer Institute
Laboratory of Viral Carcinogenesis
Cell Biology Section
Frederick, Maryland

**Harold L. Moses**
Department of Cell Biology
Vanderbilt University School of Medicine
Nashville, Tennessee

**Eric J. Stanbridge**
Department of Microbiology and Molecular Genetics
University of California at Irvine
Irvine, California

Alan R. Liss, Inc. • New York

**Address all Inquiries to the Publisher**
Alan R. Liss, Inc., 41 East 11th Street, New York, NY 10003

**Copyright © 1988 Alan R. Liss, Inc.**

**Printed in the United States of America**

Under the conditions stated below the owner of copyright for this book hereby grants permission to users to make photocopy reproductions of any part or all of its contents for personal or internal organizational use, or for personal or internal use of specific clients. This consent is given on the condition that the copier pay the stated per-copy fee through the Copyright Clearance Center, Incorporated, 27 Congress Street, Salem, MA 01970, as listed in the most current issue of "Permissions to Photocopy" (Publisher's Fee List, distributed by CCC, Inc.), for copying beyond that permitted by sections 107 or 108 of the US Copyright Law. This consent does not extend to other kinds of copying, such as copying for general distribution, for advertising or promotional purposes, for creating new collective works, or for resale.

**Library of Congress Cataloging-in-Publication Data**

UCLA Symposium on Growth Factors, Tumor Promoters, and
   Cancer Genes (1986 : Steamboat Springs, Colo.)
   Growth factors, tumor promoters, and cancer genes.

   (UCLA symposia on molecular and cellular biology ;
new ser., v. 58)
   "UCLA Symposium on Growth Factors, Tumor Promoters,
and Cancer Genes"—Pref.
   Includes bibliographies and index.
   1. Cancer—Genetic aspects—Congresses. 2. Oncogenes—Congresses. 3. Growth promoting substances—Congresses. 4. Carcinogenesis—Congresses. I. Colburn, Nancy H. II. Moses, Harold L. III. Stanbridge, Eric J. IV. Title. V. Series. [DNLM: 1. Carcinogens—congresses. 2. Cell Transformation, Neoplastic—congresses. 3. Gene Expression Regulation—congresses. 4. Growth Substances—congresses. W3 UN17N new ser. v.58 / QZ 202 U175g 1986]
RC268.4.U28   1986      616.99'4071      87-17271
ISBN 0-8451-2657-1

Pages 1–155 of this volume are reprinted from the Journal of Cellular Biochemistry, Volumes 32, 33, and 34. The Journal is the only appropriate literature citation for the articles printed on these pages. The page numbers in the table of contents, contributors list, and index of this volume correspond to the page numbers at the foot of these pages.

The table of contents does not necessarily follow the pattern of the plenary sessions. Instead, it reflects the thrust of the meeting as it evolved from the combination of plenary sessions, poster sessions, and workshops, culminating in the final collection of invited papers, submitted papers, and workshop summaries. The order in which articles appear in this volume does not follow the order of citation in the table of contents. Many of the articles in this volume were published in the Journal of Cellular Biochemistry, and they are reprinted here. These articles appear in the order in which they were accepted for publication and then published in the Journal. They are followed by papers which were submitted solely for publication in the proceedings.

# Contents

**Contributors** . . . . . . . . . . . . . . . . . . . . . . . . . . . . . . . . . . xiii

**Preface**
Nancy H. Colburn, Harold L. Moses, and Eric J. Stanbridge . . . . . . . . . . . . xix

**Acknowledgments** . . . . . . . . . . . . . . . . . . . . . . . . . . . . . . . . xxi

## I. CELL CYCLE AND DIFFERENTIATION

**Proliferative Control in Normal and Neoplastic Cells**
Arthur B. Pardee . . . . . . . . . . . . . . . . . . . . . . . . . . . . . . 157

**Heterologous Regulation of EGF Receptors in Fibroblastic Cells**
Nancy E. Olashaw and W.J. Pledger . . . . . . . . . . . . . . . . . . . . . 141

**Growth-Regulated Genes and Human Leukemias**
Bruno Calabretta, Leszek Kaczmarek, and Renato Baserga . . . . . . . . . . 161

**Differentiation and Its Role in Carcinogenesis and Anticarcinogenesis**
Rodney L. Sparks, David N. Estervig, and Robert E. Scott . . . . . . . . . 171

**Aberrant Regulation of Differentiation in Epidermal Carcinogenesis**
S.H. Yuspa, U. Lichti, J. Strickland, S. Jaken, D. Lowy, J. Harper, D. Roop,
and H. Hennings . . . . . . . . . . . . . . . . . . . . . . . . . . . . . . 183

**Growth and Differentiation Programs of Normal and Transformed Human Bronchial Epithelial Cells**
Tohru Masui, John F. Lechner, George E. Mark III, Andrea M.A. Pfeifer,
Masao Miyashita, George H. Yoakum, James C. Willey, Dean L. Mann, and
Curtis C. Harris . . . . . . . . . . . . . . . . . . . . . . . . . . . . . 191

**Analysis of Murine Homeo Box Genes and Their Expression During Development**
Stephan D. Voss, Anamaris M. Colberg-Poley, and Peter Gruss . . . . . . . . 203

## II. SIGNAL TRANSDUCTION

**Induction of c-*fos* Is Mediated by Diverse Biochemical Pathways**
Tom Curran and James I. Morgan . . . . . . . . . . . . . . . . . . . . . . 215

**Role of Protein Kinase C in Cell Surface Signal Transduction**
Ushio Kikkawa . . . . . . . . . . . . . . . . . . . . . . . . . . . . . . . 223

**Growth Factors Modify the Epidermal Growth Factor Receptor Through Multiple Pathways**
BethAnn Friedman and Marsha Rich Rosner . . . . . . . . . . . . . . . . . . 89

**Characterization of the Epidermal Growth Factor- and Vanadate-Activated Calcium Influx in A431 Cells**
Ian G. Macara and George M. Gray . . . . . . . . . . . . . . . . . . . . . 233

**Vaccinia Virus and the EGF Receptor: A Portal for Infectivity?**
Y. Vivienne Marsh and Deborah A. Eppstein . . . . . . . . . . . . . . . . . 149

## Contents

**Oxidant Tumor Promoters**
Peter A. Cerutti . . . . . . . . . . . . . . . . . . . . . . . . . . . 239

**Signals and Sequences Involved in the Ultraviolet- and 12-*O*-tetradecanoyl-phorbol-13-acetate (TPA)-Dependent Induction of Genes**
Peter Herrlich, Carsten Jonat, Hans Jobst Rahmsdorf, Peter Angel, Alois Haslinger, Masayoshi Imagawa, and Michael Karin . . . . . . . . . . . . . . . 249

**Tyrosyl and Phosphatidylinositol Kinases of Human Erythrocyte Membranes**
Mark R. Vossler, Anna Coco, Bimmie T. Strausser, Christine Zaricznyj, and
Ian G. Macara . . . . . . . . . . . . . . . . . . . . . . . . . . . 57

### III. GENES INVOLVED IN MULTISTAGE CARCINOGENESIS

**Growth Factors, Oncogenes, and Multistage Carcinogenesis**
I. Bernard Weinstein . . . . . . . . . . . . . . . . . . . . . . . . 45

**Oncogene Mutation and Amplification During Initiation and Progression Stages of Mouse Skin Carcinogenesis**
M. Quintanilla, K. Brown, F. Fee, S. Young, and A. Balmain . . . . . . . . . . . 257

**In Vivo and In Vitro Expression Pattern of Genes Activated During Multistage Carcinogenesis in the Mouse Skin**
Peter Krieg, Karl Melber, Gerhard Furstenberger, and G. Tim Bowden . . . . . . . 267

**Tumorigenic Transformation of Human Teratocarcinoma Cells by Activated *ras* Oncogene but Not the Homologous Photo-Oncogene**
Michael A. Tainsky . . . . . . . . . . . . . . . . . . . . . . . . . 277

### IV. GENETIC DETERMINANTS OF SUSCEPTIBILITY TO GROWTH FACTORS AND TUMOR PROMOTERS

**Genes That Cooperate With Tumor Promoters in Transformation**
Nancy H. Colburn and Bonita M. Smith . . . . . . . . . . . . . . . . . 127

**Molecular Cloning of a Tumor Promoter-Inducible mRNA Found in JB6 Mouse Epidermal Cells: Induction Is Stable at High, but not at Low, Cell Densities**
James H. Smith and David T. Denhardt . . . . . . . . . . . . . . . . . 101

**Use of Cell Variants to Study the Molecular and Cellular Determinants of Tumor Promotion**
H. Yamasaki, M. Hollstein, E. Hamel, L. Giroldi, E. Rivedal, T. Sanner, and
T. Kakunaga . . . . . . . . . . . . . . . . . . . . . . . . . . . 283

**Mitogen-Specific Nonproliferative Variants of Swiss 3T3 Cells**
Harvey R. Herschman . . . . . . . . . . . . . . . . . . . . . . . . 295

**Workshop: "The Use of Genetic Variants to Find Genes for Growth and Transformation"**
H. Herschman, M. Gottesman, and L. Cantley . . . . . . . . . . . . . . . 311

### V. ONCOGENES AND THEIR PRODUCTS

**Transformation by the v-*fms* Oncogene Product: An Analog of the CSF-1 Receptor**
Carl W. Rettenmier, Suzanne Jackowski, Charles O. Rock, Martine F. Roussel, and
Charles J. Sherr . . . . . . . . . . . . . . . . . . . . . . . . . . 27

**c-fos Proto-Oncogene Expression Is Necessary for Normal Growth of Mouse 3T3 Cells**
J.T. Holt and A.W. Nienhuis . . . . . . . . . . . . . . . . . . . . . 313

**Structure and Function of p21 *ras* Proteins: Biochemical, Immunochemical, and Site-Directed Mutagenesis Studies**
    Thomas Y. Shih, David J. Clanton, Seisuke Hattori, Linda S. Ulsh, and Zhang-qun Chen . . . . . . . . . . . . . . . . . . . . . . . . . . . . . 321

**Trans-Activation by the Adenovirus E1A Gene Product**
    Joseph R. Nevins, Imre Kovesdi, and Ronald Reichel . . . . . . . . . . . . . . . 333

**Mutations in the E1a Gene of Type 5 Adenovirus Result in Oncogenic Transformation of Fischer Rat Embryo Cells**
    Gregory J. Duigou, Lee E. Babiss, Wen-Shing Liaw, Stephen G. Zimmer, Harold S. Ginsberg, and Paul B. Fisher . . . . . . . . . . . . . . . . . . . . 35

**Transcriptionally Active Domains in the 5′ Flanking Sequence of Human c-*myc***
    Bruce Whitelaw, Neil M. Wilkie, Katherine A. Jones, James T. Kadonaga, Robert Tjian, and Jas C. Lang . . . . . . . . . . . . . . . . . . . . . . . . 337

## VI. TUMOR SUPPRESSOR GENES

**Tumor Suppressor Genes**
    Ruth Sager . . . . . . . . . . . . . . . . . . . . . . . . . . . . . . . . . . 353

**Oncogene and Chemical-Induced Neoplastic Progression: Role of Tumor Suppression**
    J. Carl Barrett, Minoru Koi, Tona M. Gilmer, and Mitsuo Oshimura . . . . . . . 359

**The Ha-ras-Induced Transformed Phenotype of Rat-1 Cells can be Suppressed in Hybrids With Rat Embryonic Fibroblasts**
    Klaus Willecke, Sabine Griegel, Wolfgang Martin, Otto Traub, and Reinhold Schäfer . . . . . . . . . . . . . . . . . . . . . . . . . . . . . . . 111

**Cellular Signal Transduction and the Reversal of Malignancy**
    Arthur H. Lockwood, Suzanne K. Murphy, Steven Borislow, Adam Lazarus, and Maryanne Pendergast . . . . . . . . . . . . . . . . . . . . . . . . . . . . . 69

**Inhibition of HeLa Cell Growth Following Transfection With Genomic DNA From Human Embryo Fibroblasts**
    Raji Padmanabhan, Tazuko Howard, and Bruce H. Howard . . . . . . . . . . . 369

**Suppression of Tumorigenicity in Somatic Cell Hybrids Does not Involve Quantitative Changes in Transcription of Cellular Ha-ras, Ki-ras, myc, and fos Oncogenes**
    R. Schäfer, S. Geisse, and K. Willecke . . . . . . . . . . . . . . . . . . . . . 119

## VII. TRANSFORMING GROWTH FACTORS

**Transforming Growth Factor-α: Structure and Biological Activities**
    Rik Derynck . . . . . . . . . . . . . . . . . . . . . . . . . . . . . . . . . 1

**Transforming Growth Factors and Control of Neoplastic Cell Growth**
    Jorma Keski-Oja, Edward B. Leof, Russette M. Lyons, Robert J. Coffey, Jr., and Harold L. Moses . . . . . . . . . . . . . . . . . . . . . . . . . . . . . . . 13

**Index** . . . . . . . . . . . . . . . . . . . . . . . . . . . . . . . . . . . . . 377

# Contributors

**Peter Angel,** Kernforschungszentrum Karlsruhe, Institute of Genetics and Toxicology, and Institute of Genetics, University of Karlsruhe, D-7500 Karlsruhe 1, Federal Republic of Germany [249]

**Lee E. Babiss,** Rockefeller University, New York, NY 10021 [35]

**A. Balmain,** The Beatson Institute for Cancer Research, Bearsden, Glasgow G61 1BD, Scotland [257]

**J. Carl Barrett,** Environmental Carcinogenesis Group, Laboratory of Pulmonary Pathobiology, National Institute of Environmental Health Sciences, Research Triangle Park, NC 27709 [359]

**Renato Baserga,** Department of Pathology and Fels Research Institute, Temple University Medical School, Philadelphia, PA 19140 [161]

**Steven Borislow,** Department of Pediatrics, Albert Einstein Medical Center and Temple University School of Medicine, Philadelphia, PA 19141 [69]

**G. Tim Bowden,** Radiation Oncology Department, University of Arizona, Health Sciences Center, Tucson, AZ 85724 [267]

**K. Brown,** The Beatson Institute for Cancer Research, Bearsden, Glasgow G61 1BD, Scotland [257]

**Bruno Calabretta,** Department of Pathology and Fels Research Institute, Temple University Medical School, Philadelphia, PA 19140 [161]

**L. Cantley,** Department of Physiology, Tufts University School of Medicine, Boston, MA 02111 [311]

**Peter A. Cerutti,** Department of Carcinogenesis, Swiss Institute for Experimental Cancer Research, 1066 Epalinges s/Lausanne, Switzerland [239]

**Zhang-qun Chen,** Laboratory of Molecular Oncology, Division of Cancer Etiology, National Cancer Institute, National Institutes of Health, Frederick, MD 21701 [321]

**David J. Clanton,** Laboratory of Molecular Oncology, Division of Cancer Etiology, National Cancer Institute, National Institutes of Health, Frederick, MD 21701 [321]

**Anna Coco,** Department of Biophysics, School of Medicine and Dentistry, University of Rochester, Rochester, NY 14642 [57]

**Robert J. Coffey, Jr.,** Departments of Cell Biology, Pathology and Medicine, Vanderbilt University School of Medicine, Nashville, TN 37232 [13]

**Anamaris M. Colberg-Poley,** Zentrum für Molekulare Biologie der Universität Heidelberg, D-6900 Heidelberg, Federal Republic of Germany; present address: Central Research and Development, Experimental Station Building 328, E.I. DuPont de Nemours and Co., Inc., Wilmington, DE 19898 [203]

**Nancy H. Colburn,** National Cancer Institute, Laboratory of Viral Carcinogenesis, Cell Biology Section, Frederick, MD 21701 [xix,127]

**Tom Curran,** Department of Molecular Oncology, Roche Institute of Molecular Biology, Roche Research Center, Nutley, NJ 07110 [215]

**David T. Denhardt,** Cancer Research Laboratory, University of Western Ontario, London, Ontario N6A 5B7 [101]

The number in brackets is the opening page number of the contributor's article.

## Contributors

**Rik Derynck,** Department of Molecular Biology, Genentech, Inc., South San Francisco, CA 94080 [1]

**Gregory J. Duigou,** Department of Microbiology, Comprehensive Cancer Center Columbia University, College of Physicians and Surgeons, New York, NY 10032 [35]

**Deborah A. Eppstein,** Institute of Bio-Organic Chemistry, Syntex Research, Palo Alto, CA 94304 [149]

**David N. Estervig,** Section of Experimental Pathology, Department of Biochemistry and Molecular Biology, Mayo Clinic/Foundation, Rochester, MN 55905 [171]

**F. Fee,** The Beatson Institute for Cancer Research, Bearsden, Glasgow G61 1BD, Scotland [257]

**Paul B. Fisher,** Department of Microbiology, Comprehensive Cancer Center Columbia University, College of Physicians and Surgeons, New York, NY 10032 [35]

**BethAnn Friedman,** Massachusetts Institute of Technology, Cambridge, MA 02139 [89]

**Gerhard Furstenberger,** Institutes for Biochemistry, German Cancer Research Center, D-6900 Heidelberg, Federal Republic of Germany [267]

**S. Geisse,** Institut für Zellbiologie (Tumorforschung), Universität Essen (GH), D-4300 Essen 1, Federal Republic of Germany [119]

**Tona M. Gilmer,** Environmental Carcinogenesis Group, Laboratory of Pulmonary Pathobiology, National Institute of Environmental Health Sciences, Research Triangle Park, NC 27709 [359]

**Harold S. Ginsberg,** Department of Microbiology, Comprehensive Cancer Center Columbia University, College of Physicians and Surgeons, New York, NY 10032 [35]

**L. Giroldi,** International Agency for Research on Cancer, 69372 Lyon, France [283]

**M. Gottesman,** Laboratory of Molecular Biology, National Cancer Institute, National Institutes of Health, Bethesda, MD 20892 [311]

**George M. Gray,** Division of Toxicology, Department of Biophysics, University of Rochester Medical Center, Rochester, NY 14642 [233]

**Sabine Griegel,** Institut für Zellbiologie, Universität Essen, Hufelandstrasse 55, 4300 Essen 1, Federal Republic of Germany [111]

**Peter Gruss,** Zentrum für Molekulare Biologie der Universität Heidelberg, D-6900 Heidelberg, Federal Republic of Germany [203]

**E. Hamel,** International Agency for Research on Cancer, 69372 Lyon, France [283]

**J. Harper,** Laboratories of Cellular Carcinogenesis and Tumor Promotion, National Cancer Institute, Bethesda, MD 20892 [183]

**Curtis C. Harris,** Laboratory of Human Carcinogenesis, Division of Cancer Etiology, National Cancer Institute, Bethesda, MD 20892 [191]

**Alois Haslinger,** Division of Pharmacology, School of Medicine, M-036, University of California, San Diego, La Jolla, CA 92093 [249]

**Seisuke Hattori,** Laboratory of Molecular Oncology, Division of Cancer Etiology, National Cancer Institute, National Institutes of Health, Frederick, MD 21701; present address: Department of Pure and Applied Sciences, University of Tokyo, Tokyo 153, Japan [321]

**H. Hennings,** Laboratories of Cellular Carcinogenesis and Tumor Promotion, National Cancer Institute, Bethesda, MD 20892 [183]

**Peter Herrlich,** Kernforschungszentrum Karlsruhe, Institute of Genetics and Toxicology, and Institute of Genetics, University of Karlsruhe, D-7500 Karlsruhe 1, Federal Republic of Germany [249]

**Harvey R. Herschman,** Department of Biological Chemistry and Laboratory of Biomedical and Environmental Sciences, Center for the Health Sciences, University of California, Los Angeles, CA 90024 [295,311]

**M. Hollstein,** International Agency for Research on Cancer, 69372 Lyon, France [283]

**J.T. Holt,** Clinical Hematology Branch, National Heart, Lung, and Blood Institute, National Institutes of Health, Bethesda, MD 20892 [313]

**Bruce H. Howard,** Laboratory of Molecular Biology, Division of Cancer Biology and Diagnosis, National Cancer Institute, Bethesda, MD 20892 [369]

**Tazuko Howard,** Laboratory of Molecular Biology, Division of Cancer Biology and Diagnosis, National Cancer Institute, Bethesda, MD 20892 [369]

**Masayoshi Imagawa,** Division of Pharmacology, School of Medicine, M-036, University of California, San Diego, La Jolla, CA 92093 [249]

**Suzanne Jackowski,** Department of Biochemistry, St. Jude Children's Research Hospital, Memphis, TN 38105 [27]

**S. Jaken,** Center for Drugs and Biologics, FDA, Division of Virology, National Institutes of Health, Bethesda, MD 20892 [183]

**Carsten Jonat,** Kernforschungszentrum Karlsruhe, Institute of Genetics and Toxicology, and Institute of Genetics, University of Karlsruhe, D-7500 Karlsruhe 1, Federal Republic of Germany [249]

**Katherine A. Jones,** Department of Biochemistry, University of California, Berkeley, CA 94720 [337]

**Leszek Kaczmarek,** Department of Pathology and Fels Research Institute, Temple University Medical School, Philadelphia, PA 19140 [161]

**James T. Kadonaga,** Department of Biochemistry, University of California, Berkeley, CA 94720 [337]

**T. Kakunaga,** Osaka University, Osaka 565, Japan [283]

**Michael Karin,** Division of Pharmacology, School of Medicine, M-036, University of California, San Diego, La Jolla, CA 92093 [249]

**Jorma Keski-Oja,** Departments of Cell Biology, Pathology and Medicine, Vanderbilt University School of Medicine, Nashville, TN 37232 [13]

**Ushio Kikkawa,** Department of Biochemistry, Kobe University School of Medicine, Kobe 650, Japan [223]

**Minoru Koi,** Environmental Carcinogenesis Group, Laboratory of Pulmonary Pathobiology, National Institute of Environmental Health Sciences, Research Triangle Park, NC 27709 [359]

**Imre Kovesdi,** Howard Hughes Medical Institute, The Rockefeller University, New York, NY 10021 [333]

**Peter Krieg,** Institutes for Virus Research, German Cancer Research Center, D-6900 Heidelberg, Federal Republic of Germany [267]

**Jas C. Lang,** Beatson Institute for Cancer Research, Bearsden, Glasgow, Scotland [337]

**Adam Lazarus,** Department of Pediatrics, Albert Einstein Medical Center and Temple University School of Medicine, Philadelphia, PA 19141 [69]

**John F. Lechner,** Laboratory of Human Carcinogenesis, Division of Cancer Etiology, National Cancer Institute, Bethesda, MD 20892 [191]

**Edward B. Leof,** Departments of Cell Biology, Pathology and Medicine, Vanderbilt University School of Medicine, Nashville, TN 37232 [13]

**Wen-Shing Liaw,** Department of Microbiology, Comprehensive Cancer Center Columbia University, College of Physicians and Surgeons, New York, NY 10032 [35]

**U. Lichti,** Laboratories of Cellular Carcinogenesis and Tumor Promotion, National Cancer Institute, Bethesda, MD 20892 [183]

**Arthur H. Lockwood,** Department of Pediatrics, Albert Einstein Medical Center and Temple University School of Medicine, Philadelphia, PA 19141 [69]

**D. Lowy,** Laboratory of Cellular Oncology, National Cancer Institute, Bethesda, MD 20892 [183]

## Contributors

**Russette M. Lyons,** Departments of Cell Biology, Pathology and Medicine, Vanderbilt University School of Medicine, Nashville, TN 37232 [13]

**Ian G. Macara,** Division of Toxicology, Department of Biophysics, University of Rochester Medical Center, Rochester, NY 14642 [57, 233]

**Dean L. Mann,** Laboratory of Human Carcinogenesis, Division of Cancer Etiology, National Cancer Institute, Bethesda, MD 20892 [191]

**George E. Mark III,** Laboratory of Human Carcinogenesis, Division of Cancer Etiology, National Cancer Institute, Bethesda, MA 20892 [191]

**Y. Vivienne Marsh,** Institute of Bio-Organic Chemistry, Syntex Research, Palo Alto, CA 94304 [149]

**Wolfgang Martin,** Institut für Zellbiologie, Universität Essen, Hufelandstrasse 55, 4300 Essen 1, Federal Republic of Germany [111]

**Tohru Masui,** Laboratory of Human Carcinogenesis, Division of Cancer Etiology, National Cancer Institute, Bethesda, MD 20892 [191]

**Karl Melber,** Institutes for Virus Research, German Cancer Research Center, D-6900 Heidelberg, Federal Republic of Germany [267]

**Masao Miyashita,** Laboratory of Human Carcinogenesis, Division of Cancer Etiology, National Cancer Institute, Bethesda, MD 20892 [191]

**James I. Morgan,** Department of Neurosciences, Roche Institute of Molecular Biology, Roche Research Center, Nutley, NJ 07110 [215]

**Harold L. Moses,** Department of Cell Biology, Vanderbilt University School of Medicine, Nashville, TN 37232 [xix,13]

**Suzanne K. Murphy,** Department of Pediatrics, Albert Einstein Medical Center and Temple University School of Medicine, Philadelphia, PA 19141; present address: The Philadelphia College of Pharmacy and Science, Philadelphia, PA 19104 [69]

**Joseph R. Nevins,** Howard Hughes Medical Institute, The Rockefeller University, New York, NY 10021 [333]

**A.W. Nienhuis,** Clinical Hematology Branch, National Heart, Lung, and Blood Institute, National Institutes of Health, Bethesda, MD 20892 [313]

**Nancy E. Olashaw,** Department of Cell Biology, Vanderbilt University School of Medicine, Nashville, TN 37232 [141]

**Mitsuo Oshimura,** Environmental Carcinogenesis Group, Laboratory of Pulmonary Pathobiology, National Institute of Environmental Health Sciences, Research Triangle Park, NC 27709 [359]

**Raji Padmanabhan,** Laboratory of Molecular Biology, Division of Cancer Biology and Diagnosis, National Cancer Institute, Bethesda, MD 20892 [369]

**Arthur B. Pardee,** Dana-Farber Cancer Institute, Boston, MA 02115 [157]

**Maryanne Pendergast,** Department of Pediatrics, Albert Einstein Medical Center and Temple University School of Medicine, Philadelphia, PA 19141 [69]

**Andrea M.A. Pfeifer,** Laboratory of Human Carcinogenesis, Division of Cancer Etiology, National Cancer Institute, Bethesda, MD 20892 [191]

**W.J. Pledger,** Department of Cell Biology, Vanderbilt University School of Medicine, Nashville, TN 37232 [141]

**M. Quintanilla,** The Beatson Institute for Cancer Research, Bearsden, Glasgow G61 1BD, Scotland [257]

**Hans Jobst Rahmsdorf,** Kernforschungszentrum Karlsruhe, Institute of Genetics and Toxicology, and Institute of Genetics, University of Karlsruhe, D-7500 Karlsruhe 1, Federal Republic of Germany [249]

**Ronald Reichel,** Howard Hughes Medical Institute, The Rockefeller University, New York, NY 10021 [333]

**Carl W. Rettenmier,** Department of Tumor Cell Biology, St. Jude Children's Research Hospital, Memphis, TN 38105 [27]

**E. Rivedal,** Norwegian Radium Hospital, Oslo 3, Norway [283]

**Charles O. Rock,** Department of Biochemistry, St. Jude Children's Research Hospital, Memphis, TN 38105 [27]

**D. Roop,** Laboratories of Cellular Carcinogenesis and Tumor Promotion, National Cancer Institute, Bethesda, MD 20892 [183]

**Marsha Rich Rosner,** Massachusetts Institute of Technology, Cambridge, MA 02139 [89]

**Martine F. Roussel,** Department of Tumor Cell Biology, St. Jude Children's Research Hospital, Memphis, TN 38105 [27]

**Ruth Sager,** Division of Cancer Genetics, Dana-Farber Cancer Institute, Boston, MA 02115 [353]

**T. Sanner,** Norwegian Radium Hospital, Oslo 3, Norway [283]

**Reinhold Schäfer,** Institut für Zellbiologie, Universität Essen, 4300 Essen 1, Federal Republic of Germany; present address: Ludwig Institute for Cancer Research, Bern Branch, Inselspital, 3010 Bern, Switzerland [111,119]

**Robert E. Scott,** Section of Experimental Pathology, Department of Biochemistry and Molecular Biology, Mayo Clinic/Foundation, Rochester, MN 55905 [171]

**Charles J. Sherr,** Department of Tumor Cell Biology, St. Jude Children's Research Hospital, Memphis, TN 38105 [27]

**Thomas Y. Shih,** Laboratory of Molecular Oncology, Division of Cancer Etiology, National Cancer Institute, National Institutes of Health, Frederick, MD 21701 [321]

**Bonita M. Smith,** National Cancer Institute, Laboratory of Viral Carcinogenesis, Cell Biology Section, Frederick, MD 21701 [127]

**James H. Smith,** Cancer Research Laboratory, University of Western Ontario, London, Ontario N6A 5B7 [101]

**Rodney L. Sparks,** Section of Experimental Pathology, Department of Biochemistry and Molecular Biology, Mayo Clinic/Foundation, Rochester, MN 55905 [171]

**Eric J. Stanbridge,** Department of Microbiology and Molecular Genetics, University of California at Irvine, Irvine, CA 92717 [xix]

**Bimmie T. Strausser,** Department of Biophysics, School of Medicine and Dentistry, University of Rochester, Rochester, NY 14642 [57]

**J. Strickland,** Laboratories of Cellular Carcinogenesis and Tumor Promotion, National Cancer Institute, Bethesda, MD 20892 [183]

**Michael A. Tainsky,** Department of Tumor Biology, M.D. Anderson Hospital and Tumor Institute, Houston, TX 77030 [277]

**Robert Tjian,** Department of Biochemistry, University of California, Berkeley, CA 94720 [337]

**Otto Traub,** Institut für Zellbiologie, Universität Essen, 4300 Essen 1, Federal Republic of Germany [111]

**Linda S. Ulsh,** Laboratory of Molecular Oncology, Division of Cancer Etiology, National Cancer Institute, National Institutes of Health, Frederick, MD 21701 [321]

**Stephan D. Voss,** Zentrum für Molekulare Biologie der Universität Heidelberg, D-6900 Heidelberg, Federal Republic of Germany; present address: Department of Human Oncology, University of Wisconsin, Madison, WI 53792 [203]

**Mark R. Vossler,** Department of Biophysics, School of Medicine and Dentistry, University of Rochester, Rochester, NY 14642 [57]

**I. Bernard Weinstein,** Comprehensive Cancer Center and Institute of Cancer Research, Department of Medicine and School of Public Health, Columbia University, New York, NY 10032 [45]

**Bruce Whitelaw,** Beatson Institute for Cancer Research, Bearsden, Glasgow, Scotland [337]

**Neil M. Wilkie,** Beatson Institute for Cancer Research, Bearsden, Glasgow, Scotland [337]

**Klaus Willecke,** Institut für Zellbiologie, Universität Essen, 4300 Essen 1, Federal Republic of Germany [111,119]

**James C. Willey,** Laboratory of Human Carcinogenesis, Division of Cancer Etiology, National Cancer Institute, Bethesda, MD 20892 [191]

**H. Yamasaki,** International Agency for Research on Cancer, 69372 Lyon, France [283]

**George H. Yoakum,** Laboratory of Human Carcinogenesis, Division of Cancer Etiology, National Cancer Institute, Bethesda, MD 20892 [191]

**S. Young,** The Beatson Institute for Cancer Research, Bearsden, Glasgow G61 1BD, Scotland [257]

**S.H. Yuspa,** Laboratories of Cellular Carcinogenesis and Tumor Promotion, National Cancer Institute, Bethesda, MD 20892 [183]

**Christine Zaricznyj,** Department of Biophysics, School of Medicine and Dentistry, University of Rochester, Rochester, NY 14642 [57]

**Stephen G. Zimmer,** Department of Medical Microbiology and Immunology, University of Kentucky Medical Center, Lexington, KY 40536 [35]

# Preface

Although the somatic mutation hypothesis of carcinogenesis was put forth many years ago, it is only recently that the question of what genes are involved in cancer induction has been productively addressed. Similarly, it was clear long ago that carcinogenesis is a multistage process that includes at least some steps that are irreversibly traversed; the understanding that many if not all of these stages may involve expression of specific genes has emerged only recently. That growth factors and tumor promoters can modulate the expression of such cancer-related genes has become clear, as has the knowledge that a number of oncogene products represent forms of growth factors or their receptors. Finally, new knowledge of the signals (or second messengers) that connect receptor occupancy to induction of gene expression has made it possible to understand pathways that may be causally related to cancer induction. These advances facilitate new more rational strategies for the prevention or treatment of cancer.

What are the implications of the new knowledge of the connection between growth factors, tumor promoters, and cancer genes? Can we understand the genetic bases for normal or abnormal regulation of growth and differentiation? What are the signals by which growth factors and tumor promoters transmit messages from cell surface receptors to the nucleus for altered gene expression? Can we identify genes that cooperate to bring about the multistage process of carcinogenesis, or that operate to determine susceptibility to the action of growth factors or tumor promoters? What has been learned about oncogene products and their functions? How do antagonists of carcinogenesis such as beta-transforming growth factors or the products of tumor suppressor genes work? To explore approaches and answers to these questions, a Symposium on Growth Factors, Tumor Promoters, and Cancer Genes was convened in Steamboat Springs, Colorado, April 6-13, 1986. The program brought together leaders in the fields of oncogenes and their products, growth factors, protein kinases, lipid kinases, gene regulation, transcriptional signals, and genetics of multistage carcinogenesis. The current interest in both genes and signal transduction involved in cancer, the distinction of the invited speakers, the clarity and stimulation provided by the talks, and the logical flow of concepts addressed by the program attracted a large and diverse but focused audience. The resulting forum was characterized by lively discussion and enthusiastic participation from start to finish. The present volume offers a representative sampling of the proceedings.

The conference organizers are grateful to Robin Yeaton for her untiring and competent efforts in assembling the program and to Hank Harwood for his help in procuring funds.

**Nancy H. Colburn**
**Harold L. Moses**
**Eric J. Stanbridge**

# Acknowledgments

We thank Triton Biosciences, Inc., for their generous sponsorship of this meeting. We also wish to acknowledge additional financial support from Genetics Institute, Syntex Research and Hoffman-La Roche, Inc.

# Transforming Growth Factor-α: Structure and Biological Activities

## Rik Derynck

*Department of Molecular Biology, Genentech, Inc., South San Francisco, California 94080*

Two types of growth factors have been termed transforming growth factors (TGFs). One of these, TGF-α, is related to epidermal growth factor (EGF) and binds to the EGF receptor, while the other one, TGF-β, is a structurally unrelated protein with a distinct receptor. The initial observation that led to the identification of TGF-α was that some retrovirally transformed fibroblasts displayed a strongly reduced number of EGF binding sites at their surface [1]. It was subsequently shown that these cells release an EGF-like factor that is able to bind to the EGF receptors, which then become unavailable for binding of an externally added ligand. This EGF receptor binding factor was first isolated from murine-sarcoma-virus-transformed fibroblast cultures and was therefore initially called sarcoma growth factor [2]. Subsequent examination of a variety of cell sources showed that this factor was made by many more transformed cells but not by adult normal cells in culture [3–6].

Sarcoma growth factor preparations are able to induce profound morphological changes in rat fibroblasts when added to the medium. These changes result in a phenotype similar to that of virally transformed cells. Removal of these growth factor preparations results in a reversion of the cellular phenotype back toward the normal. It was also shown that these preparations enable normal anchorage-dependent rat fibroblasts to grow in soft agar. However, when these anchorage-independent soft agar colonies are selected and subsequently plated in the absence of these growth factor preparations, they grow again as normal contact-inhibited fibroblasts [2,6]. The fact that preparations of this factor were able to convert the normal rat kidney (NRK) cells into phenotypically transformed cells and the synthesis of this factor by several different transformed cells led to the name transforming growth factor.

Initially it was assumed that sarcoma or transforming growth factor was a single peptide [2]. Extensive biochemical purification and characterization showed later that the preparations consisted of the structurally unrelated peptides TGF-α and -β. While the binding to the EGF receptor is solely due to the presence of TGF-α, the profound morphological changes observed with the rat fibroblasts are due to the cooperative effect of TGF-α and -β [7]. TGF-β by itself will not induce any colony formation of

Received April 21, 1986; revised and accepted July 28, 1986.

© 1986 Alan R. Liss, Inc.

NRK cells in soft agar. Pure TGF-α preparations will only have a minimal effect in this assay system but can elicit the formation of few relatively small colonies, which may be ascribed to the synergistic activity of low levels of active TGF-β in the bovine serum used. In contrast, the simultaneous presence of both TGF-α and -β will result in the acquisition of the transformed phenotype by normal rat kidney cells as shown by the appearance of a high number of large colonies in the soft agar assay. It should be stressed that the need for both growth factors in order to promote phenotypic transformation is dependent upon the cell system used. Both TGF-α and -β are indeed needed in the NRK system, but in many other cell systems they do not have a cooperative effect on proliferation or transformation. In some systems TGF-α (or EGF) and -β could even function as antagonists [8].

## TGF-α and Its Precursor

TGFs-α have been detected in culture supernatants and extracts from several transformed rodent and human cells [2,3,9]. These TGFs-α, which all bind to the EGF receptor, display upon gel filtration a heterogeneity in apparent molecular weights ranging from a 6-kd species secreted by several tumor cell lines [9] to the 34-kd TGF-α species detected in the urine of cancer patients. The low molecular weight TGF-α species has been purified to homogeneity from several cell sources [10]. Subsequent amino acid sequencing led to the establishment of the complete amino acid sequence of the 50-amino-acid-long rat TGF-α [11]. These data on the structure of TGF-α have now been confirmed and extended by cDNA analysis. The sequence of a human TGF-α cDNA derived from a renal cell carcinoma cDNA library indicates that the 50-amino-acid TGF-α is synthesized as a larger precursor [12], as has now been confirmed by the sequence analysis of rat TGF-α cDNAs [13].

Human TGF-α is encoded by a 4.5–4.8-kb mRNA. cDNA clones derived from either a renal cell carcinoma or a fibrosarcoma cell line reveal that the 50-amino-acid TGF-α is initially translated as an internal part of a 160-amino-acid precursor from which it is derived after proteolytic cleavage (Figs. 1 and 2.) The initiator ATG is followed by a short hydrophobic sequence between positions 8 and 18, suggesting the presence of an amino-terminal signal sequence. Comparison with other signal sequences suggests that the cleavage by the signal peptidase could occur following the Ala at position 19, the Cys at position 20, or the Ala at position 22, although no experimental proof for this is available. The Asn-Ser-Thr triplet at positions 25–27 could possibly be a site for N-glycosylation.

Comparison of the precursor sequence with the available direct amino acid sequencing data of rat, mouse, and human TGF-α [9,11] starts with the N-terminus Val-Val at positions 40–41 and ends at the Leu-Ala dipeptide (amino acids 88–89; Fig. 1). In order to generate the 50-amino-acid TGF-α, proteolytic cleavage of the precursor must occur at both the amino and carboxy terminus between an alanine residue and valine dipeptide. This Ala-Val-Val trimer is located within the sequence Val-Ala-*Ala-Val-Val* at the amino terminus of the 50-amino-acid TGF-α and within the similar sequence *Ala-Val-val*-Ala-Ala at its carboxyl end. Proteolytic processing of a precursor protein by a protease with such specificity has not been described for any other polypeptide. In contrast, cleavage of precursors for polypeptide hormones often takes place at dibasic residues. An extremely hydrophobic domain begins nine residues downstream of the carboxy terminus of the 50-amino-acid TGF-α. This region, which consists almost exclusively of isoleucines, leucines, and valines, is 23

```
                                                                              20
Human:  Met Val Pro Ser Ala Gly Gln Leu Ala Leu Phe Ala Leu Gly Ile Val Leu Ala Ala Cys
Rat:                    Ala                         Leu                 Leu Val     Val
                                                                              40
Human:  Gln Ala Leu Glu Asn Ser Thr Ser Pro Leu Ser Ala Asp Pro Pro Val Ala Ala Ala VAL
Rat:                            Pro             ---         Ser
                                                                              60
Human:  VAL SER HIS PHE ASN ASP CYS PRO ASP SER HIS THR GLN PHE CYS PHE HIS GLY THR CYS
Rat:                    LYS                                 TYR
                                                                              80
Human:  ARG PHE LEU VAL GLN GLU ASP LYS PRO ALA CYS VAL CYS HIS SER GLY TYR VAL GLY ALA
Rat:                            GLU                                                 VAL
                                                                             100
Human:  ARG CYS GLU HIS ALA ASP LEU LEU ALA Val Val Ala Ala Ser Gln Lys Lys Gln Ala Ile
Rat:                                                                        ----------
                                                                             120
Human:  Thr Ala Leu Val Val Val Ser Ile Val Ala Leu Ala Val Leu Ile Ile Thr Cys Val Leu
Rat:    -------------------------------------------------------------------------------
                                                                             140
Human:  Ile His Cys Cys Gln Val Arg Lys His Cys Glu Trp Cys Arg Ala Leu Ile Cys Arg His
Rat:    -----                                                       Val
                                                                             160
Human:  Glu Lys Pro Ser Ala Leu Leu Lys Gly Arg Thr Ala Cys Cys His Ser Glu Thr Val Val
Rat:
```

Fig. 1. Comparison of the amino acid sequences for the human and rat TGF-α precursors. Only the residues different from the human sequences are indicated for the rat precursor. The sequence of the 50-amino-acid fully processed TGF-α is shown in capitals and is overlined. The 23-amino-acid hydrophobic sequence that may function as a transmembrane region is underlined with a dashed line.

amino acids long (residues 99–121; Fig. 1) and is flanked by paired basic amino acids at positions 96–97 and 127–128. The length of the hydrophobic domain and the basic character of the flanking amino acids is characteristic of transmembrane domains of membrane proteins. This would thus suggest that the TGF-α precursor, following the removal of the $NH_2$-terminal signal peptide, would be inserted into the membrane (Fig. 2). Subsequently, the Ala-Val-Val-specific protease, which is likely located at the external side of the membrane, would then cleave the external segment of the precursor and in this way release the 50-amino-acid TGF-α. It may be conceivable that not all cells have this protease. This could then result in the continuous anchorage of the TGF-α precursor in the membrane and the absence of TGF-α in the medium. The sequence downstream of the 23-amino-acid hydrophobic domain is very rich in cysteines. Indeed, seven cysteines are present in the 37 amino acids at the C-terminus of the precursor. If the TGF-α precursor constitutes a transmembrane protein, then this C-terminal portion would remain at the cytoplasmic side of the membrane and probably not undergo any disulfide bond formation in this protein domain.

The rat TGF-α precursor, deduced from cDNA cloning [13], is 159 amino acids long, ie, one residue shorter than the human counterpart, and has all the features described above for the human precursor (Fig. 1). Comparison of the deduced rat and human precursor sequences indicates a very strong homology. Only four differences are observed in the 50-amino-acid-long TGF-α sequence. The precursor sequence downstream of the TGF-α peptide is even more conserved and displays only conservative differences between both species. This striking conservation of the sequence, which encompasses the putative transmembrane region and the potentially cytoplasmic segment, suggests a potentially important biological function for the

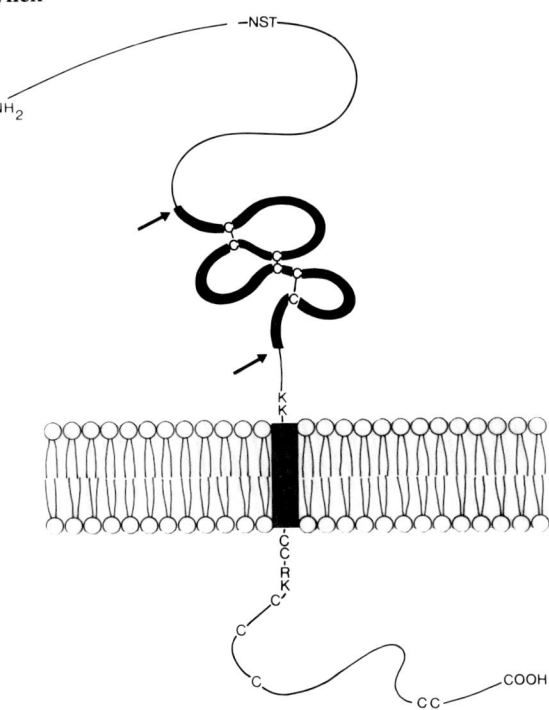

Fig. 2. Depiction of a hypothetical model of the TGF-α precursor as a transmembrane protein. The NH$_2$-terminal signal sequence is shown as already cleaved from the precursor. The 50-amino-acid TGF-α with its three proposed cysteine (C)-disulfide bridges is shown as a heavy line, flanked by the proteolytic cleavage sites (arrows). The boxed transmembrane region is flanked at each side by two basic amino acids (KK and RK). The carboxy-terminal cytoplasmic domain shown below the membrane is rich in cysteines (C).

C-terminus. However, it is as yet unclear what the physiological role of this peptide would be. It could be proposed that the intracellular segment is cleaved from the transmembrane region and thus exerts a biological function as a separate entity. However, cleavage at the Arg-Lys residues that flank the putative transmembrane region at the cytoplasmic side is unlikely, since the processing protease that cleaves at the dibastic peptide is presumably located at the external side of the membrane. Alternatively, the cytoplasmic peptide would remain covalently attached to the transmembrane region, and it could then be suggested that it plays a role in some type of signal transduction, possibly in a way similar to a receptor. The high number of Cys-residues would then probably have biological significance. However, it is not known if the TGF-α precursor, perhaps in its unprocessed form, can function as a receptor molecule as has been proposed for the much larger EGF precursor. Whatever the function of the C-terminal segment of the precursor may be, it is important to recognize that the synthesis and secretion of TGF-α goes together with the synthesis of the C-terminal precursor segment, which may have separate activities in the physiology of the cell.

As mentioned above, several larger forms of TGF-α besides the 50-amino-acid species can be detected in the medium of transformed cells. Genomic hybridizations have not revealed the existence of more than one TGF-α gene [12], which suggests that the larger forms may be derived from the same gene and thus from the same precursor. The fact that several larger TGF-α species isolated from retrovirally transformed cells react with a polyclonal antiserum to the C-terminus of the 50-amino-acid TGF-α supports this view [14]. Proteolytic processing at different sites of the precursor, eg, at the dibasic residues, or the lack of processing could then result in the generation of various other forms. The larger forms could possibly be due to some form of aggregation to other proteins or to a type of dimerization or oligomerization.

## Structural Homology With EGF

TGF-α and EGF have an unambiguous sequence homology [11,12]. Twenty-one of the 50 residues of human TGF-α, including all six cysteines, are found in corresponding positions in the human EGF sequence, while rat TGF-α and human EGF have 17 amino acids in common. However, there is a major difference in isoelectric point, ie, 6.8 for rat TGF-α and 4.5 for mouse EGF. The positions of the three disulfide bonds in murine EGF have been determined [15]. It is likely that the same disulfide bridges exist in TGF-α, since both peptides bind to the same receptor. This would imply the presence of disulfide bridges between the first and third, second and fourth, and fifth and sixth cysteines (Fig. 2). The homology between TGF-α and EGF is most concentrated in the third disulfide-bounded loop of the peptides, suggesting that this region could be most important in the binding to the receptor. This suggestion may be strengthened by the fact that synthetic peptides corresponding to only this loop are able to bind to the receptor, albeit with a much lower affinity [16].

A 140-amino-acid-long polypeptide encoded by vaccinia virus [17] contains a sequence that appears to be closely related to EGF and TGF-α. The conservation of all six cysteines in this polypeptide segment suggest that post-translational processing could result in the release of a peptide, which, owing to a disulfide bond configuration similar to that of EGF and TGF-α, could bind to the EGF/TGF-α receptor. Experimental evidence shows that vaccinia-virus-infected cells do indeed release an EGF receptor binding protein, which has been named vaccinia virus growth factor (VVGF). Biochemical characterization indicates that VVGF is 77 amino acids long and is glycosylated, in contrast to EGF or TGF-α. N-terminal sequencing of VVGF shows that the first cysteine is preceded by a 25-residue-long segment, which is much longer than the corresponding sequence in TGF-α and EGF [18].

All three peptides that bind to the same receptor and presumably have a homologous disulfide bridge formation, EGF, TGF-α, and VVGF, are initially synthesized as larger precursors. The human TGF-α [12] and VVGF [17,18] precursors are 160 and 140 amino acids long, while the murine EGF precursor is 1,217 residues long [19,20]. The corresponding EGF/TGF-α-like sequences in all three precursors are flanked by sequences that also show an unambiguous sequence homology. In all three precursors, the EGF/TGF-α-like sequence is closely followed at a similar distance by a hydrophobic domain, which could act as a transmembrane region. However, the sites of proteolytic cleavage, which results in the release of both peptides, are certainly dissimilar in sequence, thus indicating a different processing mechanism. No clear sequence homology is present in the C-terminal regions, which supposedly are located at the cytoplasmic side of the membrane.

## Cellular Sources of TGF-α Synthesis

TGF-α activity was initially detected in culture supernatants of rodent fibroblasts transformed with Moloney or Kirsten murine sarcoma viruses [3–6]. An initial survey of various transformed cell lines indicated that EGF receptor binding activity could be found predominantly in the medium of retrovirally transformed cells, but to a much lesser extent in cells transformed with DNA viruses or chemical carcinogens [6]. However, it has been reported that transformation by SV40 [21] or polyoma [22] will induce TGF-α secretion. In the latter case, transfection of rat cells with the DNA segment coding for middle T is sufficient to induce both the transformed phenotype and TGF-α production. The close correlation between TGF-α synthesis and transformation is also illustrated by experiments with rat cells transformed with a Kirsten murine sarcoma virus, which secretes TGF-α only when phenotypic transformation occurs at the permissive temperature [4].

The reported secretion of TGF-α by retrovirally transformed fibroblasts and the apparent lack of it by normal fibroblasts in cultures led us to examine a large variety of human tumor cell lines and surgically removed tumors for the presence of TGF-α mRNA [23]. This study showed that TGF-α mRNA could not be detected in any of ten tumor cell lines of hematopoietic origin. On the other hand, it was detectable in many solid tumors or cell lines derived from such origin. While TGF-α mRNA is present in several sarcomas or sarcoma cell lines, its occurrence is biased toward carcinomas and tumors of neuroectodermal origin. TGF-α mRNA is most consistently synthesized in renal carcinomas and in squamous carcinomas, irrespective of their location, but can also be frequently found in many mammary carcinomas and in tumors of neuronal origin. It is possible that TGF-α is also consistently synthesized by some other tumor types, but the low number of samples of a particular tumor type precludes generalizations. The occurrence of TGF-α mRNA in such a large variety of solid tumors suggests that the synthesis of TGFα may play a biological role in malignant transformation and tumor development in vivo. The synthesis of TGF-α by these tumors could then explain the presence of a high molecular weight EGF receptor binding factor that reacts with specific anti-TGF-α antibodies in the urine of some cancer patients and not in the urine of normal controls [24,25].

The initial observations suggested that secretion of TGF-α is tightly linked with malignant transformation. Accordingly, TGF-α could not be detected in medium from normal cells in vitro and is not known to be made in normal fully developed tissues. However, it should be stressed that only a very limited number of observations have been reported, and that as yet it cannot be excluded that TGF-α may play a role in the normal physiology of the adult organism. Recent evidence from specific antibody-based detection and Northern hybridizations have indicated that TGF-α is synthesized during early fetal development [25,13]. The TGF-α expression in the murine fetus appears to peak around day 9 and quickly levels off, so that there is no detectable TGF-α birth (day 21). This indicates that TGF-α may function as a normal embryonic version of a family of EGF-related growth factors. The expression of the gene may be reinitiated during the process of malignant transformation and tumor development, indicating that TGF-α is an oncodevelopmental antigen.

## Role of TGF-α in Malignant Transformation

It has now been convincingly illustrated that the induction of anchorage independence by the original TGF preparations in normal rat fibroblasts was due to the cooperative activity of TGF-α and TGF-β [7]. As these soft agar assays are usually performed in the presence of serum that contains a variety of growth factors, it cannot be excluded that still other factors may cooperate. TGF-α by itself can exert mitogenic activities but may not be very effective in inducing anchorage independence. It is, however, important to recognize that both in vitro and in vivo the cells are continuously exposed to an environment of growth factors. Therefore, a change in expression levels of one particular factor may in a cooperative fashion trigger major changes in the behavior and the phenotype of particular cells. It is thus conceivable that initiation of TGF-α synthesis may trigger or contribute to phenotypic transformation owing to cooperativity with TGF-β or other factors. It has been postulated that during the transformation process, TGFs exert their action via an autocrine mechanism, whereby they help sustain the transformed character of the same cells from which they are secreted [26,27]. In the case of TGF-α, this would be due to an interaction of the growth factor with the EGF receptor, which would then induce a down-regulation of the ligand-receptor complex and induce subsequent physiological changes. Such an autocrine mechanism could explain the initial observation that retrovirally transformed cells have a lower number of EGF binding sites at their surface as a result of endogenous TGF-α secretion [1]. While TGF-α synthesis could contribute to malignant transformation through an autocrine mechanism, it could also exert some activities on other cell populations via a paracrine mechanism.

Much attention has been focused in the last few years on whether the secretion of a growth factor can induce malignant transformation via an autocrine mechanism. One of the best studied systems to date is the woolly monkey simian sarcoma virus (SSV), which contains the v-*sis* gene, which is highly homologous to the c-*sis* gene coding for the B chain of platelet-derived growth factor [28,29]. Experimental evidence indicates that the binding of the v-*sis* gene product to the receptor for platelet-derived growth factor is responsible for the SSV transformation via an autocrine mechanism [30,31]. This is in agreement with the experimentally induced phenotypic transformation that is due to the overexpression of the human cellular homologue, the c-*sis* gene [32]. An autocrine mechanism has also been invoked for the phenotypic changes triggered by superinfection of v-*myb*-transformed chicken myeloblasts with retroviruses carrying src-related oncogenes. This superinfection induced growth factor independence in these cells, which are otherwise dependent for their growth on the presence of a specific myelomonocytic growth factor [33]. In another hematopoietic system, transfection of recombinant retroviral vectors that express granulocyte-macrophage colony stimulating factor (GM-CSF) into a hematopoietic precursor cell line resulted in the acquisition of malignant characteristics by these cells. This was confirmed by the ability of these GM-CSF producing cells to develop leukemias in mice [34]. As illustrated by these three examples, cellular endogenous production of a growth factor can induce malignant transformation. Overproduction of TGF-α could thus possibly be involved in transformation via a similar autocrine mechanism. The fact that TGF-α mRNA is most consistently produced in squamous carcinomas, which all contain relatively high levels of EGF receptor mRNA, would be in agreement with this hypothesis. However, it has not yet been reported whether constitutive expression of TGF-α is sufficient to induce malignant transformation via an autocrine mechanism. Also, while TGF-α expression is quite common in solid tumors and tumor cell lines, it is still debatable if the autocrine mechanism of transformation by TGF-α or by any growth factors will induce or contribute to the development of human malignancies in vivo.

## Biological Activities of TGF-α: Comparison With EGF

Studies of the biological activities of TGF-α have been very much hampered by the low availability of sufficient pure TGF-α from transformed fibroblasts. The need to carry out the biological experiments with TGF-α devoid of any other biologically active peptides is exemplified by the fact that the initial TGF preparations contained both TGF-α and -β, which resulted in the induction of anchorage independence, in contrast to results with homogeneously pure TGF-α and EGF. Most experiments with natural TGF-α have used rat TGF-α from Snyder-Theilen feline-sarcoma-virus-transformed rat fibroblasts [10,11]. The determination of the complete sequence of the 50-amino-acid rat TGF-α [11] has enabled the direct synthesis of larger amounts of rat TGF-α by solid phase techniques [35]. The isolation of a human TGF-α cDNA has also led to the synthesis and purification of relatively large amounts of human TGF-α from properly engineered *Escherichia coli* [12].

It is generally accepted that the biological actions of TGF-α, like the actions of other polypeptide hormones, are mediated through the binding to specific cell surface receptors. Earlier studies have indicated that TGF-α can interact not only with the EGF receptor but also with a 60-kd membrane component that does not bind EGF. It was proposed that this 60-kd protein was a putative TGF-α receptor species, which may mediate TGF-α effects that are directly involved in the induction of the transformed phenotype [36]. However, the induction of anchorage independence can be neutralized by blocking antibodies raised against the EGF receptor [37], which make the identity of the 60-kd protein as a specific TGF-α receptor unlikely. The binding of TGF-α to the EGF receptor makes it possible to quantify TGF-α on the basis of a generally used radioreceptor assay in which TGF-α competes with $^{125}$I-EGF for receptor binding. Comparison between murine EGF and the 50-amino-acid TGF-α secreted by transformed rat fibroblasts has indicated that both ligands exhibit a remarkably similar mode of interaction with the EGF receptor and that both peptides compete for receptor binding with the same potency and to the same extent [38]. However, the 50-amino-acid recombinant human TGF-α purified from *E. coli* cultures and subsequently refolded appears to be only about half as potent in EGF receptor binding as murine EGF (M. Winkler, personal communication). Comparison of binding characteristics has revealed that natural rat TGF-α requires a stringent pH optimum for receptor binding in contrast to EGF. Exposure of placental membranes, which contain EGF receptors, to several lectins will modify the binding of EGF or TGF-α in a parallel way. Continued exposure of A431 cells to either TGF-α or EGF induces down-regulation of the receptors according to similar kinetics [38]. Interaction of TGF-α to the EGF receptor will also mimic the action of EGF to activate a receptor-associated kinase [39]. While our current knowledge thus indicates that TGF-α may exert its activities through the EGF receptor, it is not known how EGF receptors with different stages of affinity will behave in vivo with respect to their binding properties for TGF-α or EGF.

As sufficient quantities of TGF-α are becoming available, EGF and TGF-α are being compared for their biological activities. It is now well established that the full induction of anchorage independence of normal rat kidney fibroblasts by the earlier sarcoma growth factor or transforming growth factor preparations is due to the cooperative effect of both TGF-α and -β [7]. Pure TGF-α and EGF are equally potent in these soft agar colony formation assays in the presence of TGF-α, and both growth factors appear therefore to be interchangeable. The absence of exogenously added TGF-α results in a highly depressed efficiency of induction of anchorage independence, but again the concentration-dependent response curves for EGF and TGF-α are superimposable [12]. It is assumed that the low but still significant responses seen in the soft agar assay with either of these peptides is due to the presence of TGF-$\beta$ in the serum.

One of the first biological activities established for EGF was that it is able to induce precocious eyelid opening in newborn mice. Injection of EGF into newborn mice accelerates eyelid opening in a concentration-dependent way, from 12 days in the absence of exogenous growth factor to 8 days at the maximal EGF concentration. Comparison of human TGF-$\alpha$ with murine or human EGF indicates that also in this assay both growth factors induce similar responses [40]. In addition to the effects on eyelid opening, both EGF and TGF-$\alpha$ also induce other changes in the somatic development of the mouse. They induce accelerated tooth eruption, retard the growth rate, and inhibit hair growth. While it has been reported that TGF-$\alpha$ and EGF do not differ significantly in these activities [41], it is important to evaluate these data with caution, since the group treated with EGF consisted only of two or four animals, depending upon the experimental parameter examined. Also, the control animals were injected with EGF only at a single concentration that presumably triggers maximal responses. Because of these limitations, it is impossible to derive relevant conclusions concerning the relative responses and especially the relative potencies of both growth factors. In addition, the photographic evidence of the effects on hair growth and morphology would suggest that TGF-$\alpha$ exerts more pronounced effects on the hair morphology and on hair follicle development [41].

Cell ruffling is a very early response of cells in culture to administration of various growth factors. Both TGF-$\alpha$ and EGF are able to induce rapid and transient ruffling responses in sparsely cultured cells. At lower doses, the magnitude and duration of the responses to either factor is similar, but at high doses the maximal responses for both parameters are higher with TGF-$\alpha$ than with EGF. Pretreatment of the cells with TGF-$\beta$ greatly enhances the ruffling response to TGF-$\alpha$ but will antagonize the EGF-induced ruffling [42]. TGF-$\alpha$ and EGF have also been compared for their ability to induce proliferation of human epidermal cells. Also, in this cell culture system differences in activity between EGF and TGF-$\alpha$ are apparent. TGF-$\alpha$ elicits a greater effect in inducing the formation of epidermal cell colonies than does EGF (Barrandon and Green, personal communication).

The biological activities of EGF and TGF-$\alpha$ have been compared in several other systems, which also revealed differential responses to both factors. The ability of a factor to induce release of calcium ions in a bone organ culture is often studied in two established systems. Either murine calvaria or fetal rat long bones prelabeled with $^{45}Ca^{2+}$ are incubated in the presence of the factor. An increase of $Ca^{2+}$ release owing to the presence of the growth factor is a measure of bone resorption and could bear relevance to hypercalcemia in vivo. Both EGF and TGF-$\alpha$ are able to induce $Ca^{2+}$ release in the calvaria system, but TGF-$\alpha$ is about three- to ten-fold more potent than EGF and induces a response at concentrations as low as 0.5 ng/ml. The difference is more striking in the fetal rat long bone system, since TGF-$\alpha$ induces a pronounced $Ca^{2+}$ release in a dose-dependent manner, while EGF does not trigger any statistically significant effect [43,44]. Studies on cultured cells indicate that this induction of bone resorption by TGF-$\alpha$ may be due to an inhibition of osteoblast activity as measured by the effects on collagen synthesis and by an activation of the osteoclast population [44]. In a recently developed in vitro system, TGF-$\alpha$ was about 10- to 100-fold more potent than EGF in stimulating the proliferation of osteoclast-precursor cells [45]. Hypercalcemia in vivo is often observed in conjunction with advanced stages of malignancies. It can therefore be speculated that the strong potency of TGF-α to induce bone resorption in vitro may have significance in vivo. This hypothesis may be enforced since TGF-α mRNA is produced most consistently in squamous, renal, and mammary carcinomas and in melanomas [23], which often induce malignancy-associated hypercalcemia [46]. It should be emphasized, however, that other factors have been implicated in hypercalcemia and that parathyroid hormone-like polypeptides released by several tumor cells can also induce bone resorption [46]. It is therefore likely that a single factor may not be responsible for all cases of malignancy-associated hypercalcemia, and it is also possible that several factors in conjunction with each other may trigger hypercalcemia in vivo. If TGF-α released by the tumor cells indeed induces hypercalcemia in vivo, then these activities are exerted via a paracrine or even an endocrine mode.

TGF-α has also been tested for its effect on angiogenesis. Different quantities of TGF-α or EGF were absorbed to blue Sepharose and subcutaneously implanted in the hamster cheek pouch. This in vivo system allows the monitoring of the extent of neovascularization that is due to the formation of capillaries that migrate toward the site of implantation. EGF is a relatively poor inducer of angiogenesis, but TGF-α is able to induce neovascularization at low concentrations that are without any effect in the case of EGF. However, both factors were equally efficient in inducing their mitogenic effects on cultured endothelial cells and on several other cells in vitro [47]. Thus, the differential angiogenic response cannot be explained by simple differences in mitogenicity on the capillary endothelial cells. It is well recognized that tumor cells secrete angiogenic factors and that tumor-derived angiogenesis is crucial to tumor development. It is possible that TGF-α, which is apparently synthesized by many tumors, could play a role in the induction of neovascularization in the tumor, could play a role in the induction of neovascularization in the tumor. This would be in agreement with the lack of TGF-α synthesis by cell lines derived from hematopoietic tumors, which do not require neovascularization. It is important, however, to recognize that several tumors have been shown to synthesize other potent angiogenic polypeptides, which could either alone or in combination with TGF-α induce neovascularization in the tumor. It is not yet known to what extent TGF-α or these other factors will induce angiogenesis in the different tumor types in vivo. The production of TGF-α by the tumor may thus be important to tumor development in at least two ways. It could act as a stimulator of tumor cell proliferation in an autocrine way, but it could also contribute to the induction of tumor-derived angiogenesis using a paracrine mechanism. The latter activity of TGF-α could also be physiologically important during early fetal development [48,49].

EGF exerts a potent activity on vascular tissue, which results in an increase in regional arterial blood flow in a variety of vascular beds. TGF-α and EGF displayed an equal potency in this system, but the maximal response obtained for TGF-α was much higher than with EGF. In addition, prior exposure of the vascular tissue to TGF-α markedly desensitized the arterial system to EGF but not to TGF-α. The synthesis of TGF-α by tumor cells and the fact that TGF-α does not cause desensitization to its own action may suggest that it could play a persistent role in the local vascular hyperdynamic state associated with malignancy (M.D. Hollenberg, personal communication).

The results briefly discussed above indicate that TGF-α and EGF behave differently in several biological systems. In many cases, TGF-α is much more potent than EGF and seems to behave as a superagonist. In other systems, the responses elicited by both factors are very similar, if not identical. In any case, it is important that TGF-α and EGF should not be considered as mere analogues because they bind to the same receptor and cannot be discriminated in their biological effects in some assays. It could be argued that some quantitative differences in response by TGF-α or EGF may be due to differential stability in the assay systems or to differential aggregation with binding proteins. However, such an explanation does not agree with the data obtained in cell culture or in the in vivo eyelid opening assay, nor could it account for differences in maximal effects or for the qualitatively different responses observed for both growth factors. It is unclear how both TGF-α and EGF, which bind to the same receptor, can trigger differential responses. It has been proposed that there are EGF receptors with high and low affinity for EGF, and it may be possible that TGF-α and EGF will exhibit differences in their binding to these. One could also postulate that there are differences in behavior of the ligand-receptor complex during internalization. It is also possible that receptor binding of EGF or TGF-α will trigger different effects in some specialized cell types such as osteoblasts or endothelial cells in contrast to, for example, fibroblasts. Detailed studies on receptor-ligand interactions and subsequently triggered physiological events will therefore be needed to explain the differential activities of TGF-α and EGF.

## ACKNOWLEDGMENTS

I acknowledge all colleagues at Genentech who contributed to the determination of the human TGF-α cDNA sequence and its expression in bacteria: E. Chen, D. Eaton, D. Goeddel, and M. Winkler. I thank M.D. Hollenberg (University of Calgary) and H. Green (Harvard University) for allowing the communication of data prior to submission of a manuscript. J. Arch is acknowledged for her skillful assistance in the preparation of this manuscript.

## REFERENCES

1. Todaro GJ, De Larco JE, Cohen S: Nature 264:26–31, 1976.
2. De Larco J, Todaro GJ: Proc Natl Acad Sci USA 75:4001–4005, 1978.
3. Todaro GJ, Fryling C, De Larco JE: Proc Natl Acad Sci USA 77:5258–5262, 1980.
4. Ozanne B, Fulton RJ, Kaplan PL: J Cell Physiol 105:163–180, 1980.
5. Roberts AB, Lamb LC, Newton DL, Sporn MB, De Larco JE, Todaro GJ: Proc Natl Acad Sci USA 77:3494–3498, 1980.
6. Todaro GJ, Lee DC, Webb NR, Rose TM, Brown, JP: In Feramisco J, Ozanne B, Stiles C (eds): "Cancer Cells 3." Cold Spring Harbor, NY: Cold Spring Harbor Laboratories, 1985, pp 51–58.
7. Anzano MA, Roberts AB, Smith JM, Sporn MB, De Larco JE: Proc Natl Acad Sci USA 80:6264–6268, 1983.
8. Roberts AB, Anzano MA, Wakefield LM, Roche NS, Stern DF, Sporn MB: Proc Natl Acad Sci USA 82:119–123, 1985.
9. Marquardt H, Hunkapiller MW, Hood LE, Twardzik DR, De Larco JE, Stephenson JR, Todaro GJ: Proc Natl Acad Sci USA 80:4684–4688, 1983.
10. Massague JJ: J Biol Chem 258:13606–13613, 1983.
11. Marquardt H, Hunkapiller MW, Hood LE, Todaro GJ: Science 223:1079–1082, 1984.
12. Derynck R, Roberts AB, Winkler ME, Chen EY, Goeddel DV: Cell 38:287–297, 1984.
13. Lee DC, Rose TM, Webb NR, Todaro GJ: Nature 313:489–491, 1985.
14. Linsley D, Hargreaves W, Twardzik D, Todaro GJ: Proc Natl Acad Sci USA 82:356–360, 1985.
15. Savage CR Jr, Hash JH, Cohen S: J Biol Chem 248:7669–7672, 1973.

16. Nestor JJ Jr, Newman SR, De Lustro B, Todaro GJ, Schreiber AB: Biochem Biophys Res Commun 129:226–232, 1985.
17. Venkatesan S, Gershowitz A, Moss B: J Virol 49:637–646, 1982.
18. Stroobant P, Rice AP, Gullick WJ, Cheng DJ, Kertz IM, Waterfield MD: Cell 42:383–393, 1985.
19. Gray A, Dull TJ, Ullrich A: Nature 303:722–725, 1983.
20. Scott J, Urdea M, Quiroga M, Sanchez-Pescador R, Fong N, Selby M, Rutter WJ, Bell GI: Science 221:236–240, 1983.
21. Kaplan DL, Topp WC, Ozanne B: Virology 108:484–490, 1981.
22. Kaplan PL, Ozzane B: Virology 12:372–380, 1982.
23. Derynck R, Goeddel DV, Ullrich A, Gutterman JU, Williams R, Bringman TS, Berger WH: Submitted.
24. Sherwin SA, Twardzik DR, Bohn WH, Cockley KD, Todaro GJ: Cancer Res 43:403–407, 1983.
25. Twardzik DR, Kimball ES, Sherwin SA, Ranchalis JE, Todaro GJ: Cancer Res 45:1934–1939, 1985.
26. Sporn MB, Todaro GJ, N Engl J Med 303:878–880, 1980.
27. Sporn MB, Roberts AB: Nature 313:745–747, 1985.
28. Waterfield MD, Scrace GT, Whittle N, Stroobant P, Johnsson A, Wasteson A, Westermark B, Heldin C-H, Huang JS, Deuel TF: Nature 304:35–39, 1983.
29. Doolittle RF, Hunkapiller MW, Hood LE, Devare SG, Robbins KC, Aaronson SA, Antoniades HN: Science 221:275–277, 1983.
30. Huang JS, Huang SS, Deuel TF: Cell 39:79–87, 1984.
31. Johnsson A, Betsholtz C, Heldin CH, Westermark B: Nature 317:438–440, 1985.
32. Clarke MF, Westin E, Schmidt D, Josephs SF, Ratner L, Wong-Staal F, Gallo RC, Reste MS: Nature 308:464–467, 1984.
33. Adkins B, Leutz A, Graf T: Cell 39:439–445, 1984.
34. Lang RA, Metcalf D, Gough NM, Dunn AR, Gonda TJ: Cell 43:531–542, 1985.
35. Tam JP, Marquardt H, Rosberger DF, Wong TW, Todaro GJ: Nature 309:376–378, 1983.
36. Massague J, Czech MP, Iwata K, De Larco JE, Todaro GJ: Proc Natl Acad Sci USA 79:6822–6826, 1982.
37. Carpenter G, Stoscheck CM, Preston YA, De Larco JE: Proc Natl Acad Sci USA 80:5627–5630, 1983.
38. Massague J: J Biol Chem 258:13614–13620, 1983.
39. Pike LJ, Marquardt H, Todaro GJ, Gallis B, Casnellic JE, Bornstein PE, Krebs G: J Biol Chem 257:14628–14631, 1983.
40. Smith JM, Sporn MB, Roberts AB, Derynck R, Winkler ME, Gregory H: Nature 315:515–516, 1985.
41. Tam JP: Science 229:673–675, 1985.
42. Myrdal S: J Cell Biol 101:2440, 1985.
43. Stern PH, Krieger NS, Nissenson RA, Williams RD, Winkler ME, Derynck R, Strewler GJ: J Clin Invest 76:2016–2019, 1985.
44. Ibbotson KJ, Harrod J, Gowen M, D'Souza S, Winkler ME, Derynck R, Mundy GR: Proc Natl Acad Sci USA 83:2228–2232, 1986.
45. Takahashi N, McDonald BR, Hon J, Winkler ME, Derynck R, Mundy GR, Roodman GD: J Clin Invest (in press).
46. Mundy GR, Ibbotson KJ, D'Souza SM: J Clin Invest 76:391–394, 1985.
47. Schrieber AB, Winkler ME, Derynck R: Science 232:1250–1253, 1986.
48. Twardzik DR: Cancer Res 45:5413–5416, 1985.
49. Lee DC, Rochford RM, Todaro GJ, Villareal LP: Mol Cel Biol 5:3644–3646, 1985.

# Transforming Growth Factors and Control of Neoplastic Cell Growth

### Jorma Keski-Oja, Edward B. Leof, Russette M. Lyons, Robert J. Coffey, Jr., and Harold L. Moses

*Departments of Cell Biology, Pathology and Medicine, Vanderbilt University School of Medicine, Nashville, Tennessee 37232*

    Transforming growth factors (TGFs) are peptides that affect the growth and phenotype of cultured cells and bring about in nonmalignant fibroblastic cells phenotypic properties that resemble those of malignant cells. Two types of TGFs have been well characterized. One of these, TGFα, is related to epidermal growth factor (EGF) and binds to the EGF receptor, whereas the other, TGFβ, is not structurally or functionally related to TGFα or EGF and mediates its effects via distinct receptors.

    TGFβ is produced by a variety of normal and malignant cells. Depending upon the assay system employed, TGFβ has both growth-inhibitory and growth-stimulating properties. Many of the mitogenic effects of TGFβ are probably an indirect result of the activation of certain growth factor genes in the target cell. The ubiquitous nature of the TGFβ receptor and the production of TGFβ in a latent form by most cultured cells suggests that the differing cellular responses to TGFβ are regulated either by events involved in the activation of the factor or by postreceptor mechanisms. The combined effects of TGFβ with other growth factors or inhibitors evidently play a central role in the control of normal and malignant cellular growth as well as in cell differentiation and morphogenesis. Since transforming growth factor as a concept has partially proven misleading and insufficient, there is a need to find a new nomenclature for these regulators of cellular growth and differentiation.

**Key words:** transforming growth factors, TGFβ, oncogene activation, growth stimulation, growth inhibition, neoplastic growth, cancer cell

    Transforming growth factor-β (TGFβ) was discovered as a growth-stimulatory molecule in two laboratories concurrently [1,2]. Its major effect was the ability to induced soft agar growth of nontumorigenic fibroblastic cells. It was originally thought that TGFs would be found only in malignant cells since similar effects had earlier been ascribed to sarcoma growth factors (SGFs) that were found in medium conditioned by murine sarcoma virus-transformed fibroblasts [3]. It was later found

---

Received June 9, 1986; revised and accepted September 15, 1986.

© 1987 Alan R. Liss, Inc.

that SGF is composed of an epidermal growth factor (EGF)-like growth factor (TGFα, see below) and TGFβ [4].

Recent observations, however, have demonstrated that TGFs are found in normal cells and tissues, and that other growth factors such as platelet-derived growth factor (PDGF) can stimulate anchorage-independent growth of nonmalignant cells in the presence of serum [5,6]. The ability of cells to grow in soft agar probably results from the synergistic action of intracellular factors such as activated oncogenes [5,7] and external ones such as polypeptide growth factors including PDGF, EGF, TGFα, and TGFβ [8] and their extracellular matrices [9].

Recent studies have demonstrated that TGFβ can function as a growth stimulator only for fibroblastic cells. A model has been proposed that suggests that its effect is indirect and is mediated via the induction of the *sis*-proto-oncogene that codes for the B chain of PDGF [7]. On the other hand, TGFβ is a potent growth inhibitor of several types of normal epithelial cells and several types of malignant cells [10]. TGFβ, or a very similar factor, had been described earlier as a growth inhibitor (GI) for certain epithelial cell types [11,12] before its growth stimulatory effects were discovered. TGFβ can thus function both as a stimulator and as an inhibitor of growth. These properties, when modified by other growth factors, may be essential in the regulation of both physiological and malignant growth. In addition, novel anchorage-independent growth modifying peptides have been purified and characterized from several different sources during the past few years [13-18]. Elucidation of their molecular properties will add to our understanding of the regulation of the malignant phenotype.

## TGFα—RELATIONSHIPS TO EGF AND TGFβ

TGFα was originally observed in the conditioned medium of murine sarcoma virus-transformed cells as EGF-competing activity that was associated with anchorage-independent growth of fibroblastic rat kidney NRK cells [3]. The growth factor was termed sarcoma growth factor (SGF), and later it was shown that SGF was composed of TGFα and TGFβ molecules [4]. Examination of a variety of cell lines indicated that TGFα was produced by different malignant cells including those that were transformed by viruses but not by normal adult cells [19-21]. TGFα is possibly an embryonic form of EGF, which is inappropriately expressed in various malignancies.

The addition of TGFα to NRK cells resulted in the formation of small colonies of cells without exogenous TGFβ; however, the simultaneous presence of both TGFα and TGFβ resulted in enhanced soft agar colony formation [2]. Which TGF(s) is (are) required for growth in soft agar depends upon the cell assay system employed. AKR-2B, Balb/c-3T3, human foreskin fibroblasts, or EGF receptorless NR6 cells only require the addition of TGFβ to the serum supplemented medium for induction of soft agar growth [10]; there is no EGF or TGFα requirement.

Radioreceptor assays have shown that EGF and TGFα compete similarly for binding to cell surface EGF receptors [22-24]. Amino acid sequencing of rat, mouse, and human TGFαs demonstrated that the protein is highly conserved [25, 26, cf.27], which indicates that the protein is phylogenetically important [28-30].

The human TGFα gene has been mapped to chromosome 2 (2p13) close to the breakage point in Burkitt's lymphoma [31,32], which suggests that a similar activation mechanism may operate in certain malignancies. Molecular cloning of TGFα indicated that human TGFα is encoded by a 4.5–4.8-kb mRNA. Analyses of the cDNA clones indicated that the 50-amino acid TGFα is translated as a part of a precursor molecule of 160 amino acids. A very hydrophobic region of the TGFα precursor is located nine residues downstream of the COOH terminus of the 50-amino acid TGFα molecule. The structure of this region is characteristic of transmembrane regions of membrane proteins, and it is possible that TGFα is secreted from cells followed by a proteolytic cleavage at the cell membrane [cf. 27].

In addition to the TGFα molecule, several larger proteins with TGFα activity can be detected in the conditioned medium of transformed cells [cf.27]. These larger peptides are evidently derived from the same gene and are presumably from the same precursor [23]. Variations in the proteolytic processing may explain the different molecular sizes of TGFα activity.

Other molecules that are structurally related to EGF and TGFα and that contain EGF-like sequences include the EGF precursor molecule [33,34], the vaccinia virus growth factor (VVGF) [35–37], and the low density lipoprotein (LDL)-receptor [38,39]. Several serine proteinases including the plasminogen activators and blood coagulation factors contain EGF-like domains whose functions have not as yet been elucidated [40].

## PURIFICATION AND MOLECULAR CHARACTERIZATION OF TGFβ

TGFβ (see Table I) is widely distributed in different tissues [cf 41–43] and has been purified from placenta [44,45], kidneys [46], and platelets [47,48] as well as from cultured cells [1,2,11; see 41–43]. Platelets have been the most common source for purification because of their high TGFβ content [48]. It has been estimated that platelets contain threefold greater amounts of TGFβ than PDGF. The intact molecule from all sources has a molecular weight of 25,000 and is composed of two apparently identical subunits of 12,500. The reduced, dissociated subunit is biologically inactive [48].

Derynck and coworkers [49] have cloned the gene for TGFβ from a human genomic library and from cDNA libraries derived from human term placenta and from the human fibrosarcoma line HT-1080. The amino acid sequence deduced from

**TABLE I. Properties of TGFβ***

| | |
|---|---|
| Molecule | 25,000-dalton disulfide-linked homodimer (112 amino acids each chain) [48] |
| | Cleaved from a larger precursor protein [49] |
| Chromosomal location | 19q (subbands q13.1–q13.3) [51] |
| Sources | Cultured cells [1–3], platelets [47,48], placenta [44,45], kidney [46] |
| Activation | Dissociation from a binding protein or cleavage from a precursor protein [43,80,81] |
| Receptor | High molecular weight dimeric glycoprotein ($M_r$ 560,000) [61,62] |
| | Abundant in all cells [59] |
| | No enzymatic functions known |
| Homologies | Inhibin [52]; Activins [53,54] |
| | Müllerian-inhibiting substance [55] |
| Identities | Growth inhibitor (GI) [12] |
| | Cartilage-inducing factor-A (CIF-A) [57] |
| | Differentiation inhibitor of BRL cells [58] |

*The reference nos. of some representative publications are included.

sequencing of overlapping cDNA fragments suggests a subunit of 112 amino acids. Amino acid sequence analysis of reduced human platelet-derived TGFβ further confirmed that the two chains were identical [49]. These studies furthermore suggested a precursor encoded in the 391-residue open reading frame where each subunit is encoded by residues 280–391. The murine TGFβ gene has recently been cloned, and the cDNA sequence has been determined [50]. The COOH-terminal precursor cDNA sequence representing the TGFβ coding region is identical in murine and human clones except for one amino acid at position 354 (serine in murine, alanine in human TGFβ). This high degree of evolutionary conservation suggests that most regions of the TGFβ molecule are necessary for biological activity and that TGFβ probably plays an essential role in normal growth and development. The human TGFβ gene has been localized to the long arm of chromosome 19 and to chromosome 7 in the mouse, which share four homologous loci [51].

Analysis of the TGFβ sequence indicated homology with the β-chains ($\beta_A$ and $\beta_B$ of inhibin [52]. Inhibin is a potent inhibitor of follicle-stimulating hormone (FSH) secretion, and is produced by both the ovaries and the testis. It is composed of two different polypeptide chains ($\alpha$ and $\beta_A$ or $\beta_B$) and it is structurally related to certain glycoprotein hormones of the pituitary and placenta. Interestingly, the β-subunits can form homodimers and heterodimers, which have effects opposite to those of inhibin ("activins" [53,54]). In addition, a structural homology between Müllerian-inhibiting substance ($M_r$ 150,000) and TGFβ has also been found [55]. These molecules appear to form a group of regulators of growth and differentiation with multiple functions [see 56]. In addition, biological and polypeptide analyses have shown that cartilage-inducing factor-A (CIF-A) [57] and the differentiation inhibitor of Buffalo rat liver (BRL)-cells [58] are actually TGFβ.

## TGFβ RECEPTOR

TGFβ, unlike TGFα, has its own specific cell membrane receptors that like the TGFβ molecule itself, are ubiquitous. Specific binding of $^{125}$I-TGFβ to various mesenchymal and epithelial cells in primary and secondary cultures and continuous cell lines, both normal and neoplastic, has been reported [59]. The development of radioreceptor assays for TGFβ has allowed the quantitation of dissociation constants (25–140 pM) and receptor number per cell (10,000–40,000) [59,60]. The receptors bind TGFβ from different species equally well (see Fig. 1) suggesting that these growth factor-receptor systems are highly conserved. The TGFβ receptor is different from either the EGF or platelet-derived growth factor (PDGF) receptors, which function as tyrosine-specific protein kinases. Affinity labeling of the receptor in mouse cells has identified both multimeric complexes, and a $M_r$ 565,000 complex that is apparently a dimeric glycoprotein. The receptor dissociates in the presence of disulfide reducing agents into two subunits of $M_r$ 280,000–290,000 [61]. Two types of TGFβ-receptors have recently been proposed on the basis of cross-linking studies [62]. Thus far no kinase or other enzymatic activity has been reported for the TGFβ receptor.

TGFβ enhances the affinity of the EGF-receptors of NRK cells [63,64], and the activation of the EGF-receptor appears to be essential for the action of TGFβ in these cells [65]. The need for insulin-like growth factors in TGFβ-induced cell transformation has also been proposed [66].

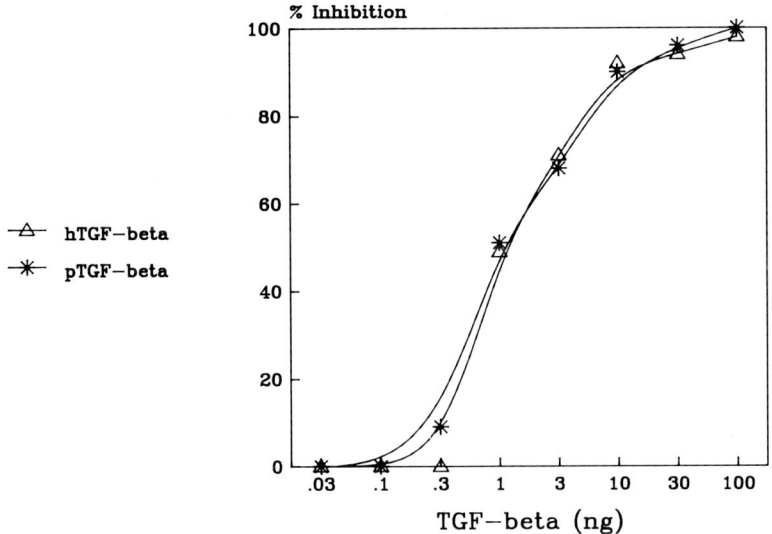

Fig. 1. Inhibition of the binding of human TGFβ (hTGFβ) to cellular receptors by human and porcine TGFβ (pTGFβ). The radioreceptor assays were carried out using $^{125}$I-labeled hTGFβ and 84A cells [59]. The inhibition of binding with pTGFβ occurs at nanogram concentrations as with hTGFβ [59] which indicates that their affinities for the cellular receptors are of the same magnitude.

## MECHANISM OF GROWTH STIMULATION BY TGFβ

Cell stimulation by TGFβ brings about alterations in the functions of the cell membrane. The uptake of glucose and amino acids into the cells is enhanced [65,66]. However, the relationships of these membrane changes to growth stimulation have not been elucidated.

### Anchorage-Independent Growth

TGFβ stimulates soft agar growth of murine AKR-2B and 3T3 cells, rat NRK cells, and secondary cultures of human foreskin fibroblasts [1,2,10]. All of the cells stimulated to proliferate by TGFβ are mesenchymal in origin and are either fibroblasts or fibroblast-like cells. The requirement of additional growth factors in medium already containing serum for soft agar growth has led to significant confusion in the interpretation of the results [see 10]. Studies of growth factor requirements for growth in soft agar have, in general, demonstrated that the same growth factors that are required for cell growth as substratum-attached layers, in addition to TGFβ, are required for the support of proliferation of anchorage-independent growth [67,68].

Recently it has been shown that fibronectin plays a role in the soft agar growth of NRK (49F) cells [9]. Both fibronectin and procollagen production were enhanced by TGFβ, and fibronectin, at microgram concentrations, enhanced the soft agar growth of the indicator cells. Synthetic peptide inhibitors of fibronectin binding to its cell surface receptors inhibited the colony-forming effect of TGFβ, suggesting that the formation of pericellular matrix structures may have been important in the actions of TGFβ [9]. Recently it has also been reported that retinoic acid, in combination with insulin and EGF or PDGF, can induce soft agar growth of NRK cells [69]. Whether this takes place via the stimulation of extracellular matrix formation is not known. The multiple known biological effects of TGFβ (Table II) may take place via

**TABLE II. Biological Effects of TGFβ***

Inhibits the growth of most normal, especially epithelial cells, and several malignant cells [10,72,74]
Acts as a mitogen for mesenchymal cells (activates the c-*sis* oncogene) [1,2,7,70]
Stimulates anchorage-independent growth of nonmalignant fibroblastic cells (frequently in concert with other growth factors) [1,2]
Inhibits adipogenic differentiation of 3T3 cells [105,106] and stimulates terminal differentiation of bronchial epithelial cells [108]
Enhances wound healing [87,88]
Induces fibroblast chemotaxis [97]
Regulates the plasminogen activator activity of cultured cells; induces endothelial type plasminogen activator inhibitors [91,92]
Enhances the production of connective tissue components [9,87]
Acts as a modifier of different immunological responses [98]

*The reference nos. of some representative publications are included.

the induction of multiple genes that act synergistically to support the cell's ability to grow in soft agar.

## Mitogenicity of TGFβ

TGFβ stimulates DNA synthesis in quiescent substratum-attached cultures of mouse AKR-2B cells without other added growth factors in a completely defined medium, but with delayed kinetics relative to stimulation with other growth factors [70]. Stimulation with EGF and insulin, PDGF, fibroblast growth factor (FGF) or serum resulted in a 12–14-hr lag phase before the onset of DNA synthesis, which then peaked at 20–25 hr following stimulation. Cultures stimulated with TGFβ exhibited a lag phase that was prolonged to 24 hr with a peak of DNA synthesis between 30 and 35 hr. In an effort to determine why TGFβ stimulated DNA synthesis with such unusual, delayed kinetics relative to stimulation with other growth factors, the possibility was examined that TGFβ was acting as an indirect mitogen through induction of synthesis of endogenous growth factors. It was found that TGFβ stimulation of quiescent cultures of AKR-2B cells resulted in an early induction of c-*sis* mRNA [7] (see Fig. 2). The rise in c-*sis* mRNA was followed by a corresponding increase of a PDGF-like protein in the culture medium. In addition, PDGF-regulated genes (c-*fos* and c-*myc*) were stimulated by TGFβ with delayed kinetics relative to that seen with direct PDGF stimulation. The data suggest that the mitogenicity of TGFβ for adherent cells is mediated by the induction of c-*sis* (PDGF) with the subsequent autocrine stimulation of c-*fos*, c-*myc*, other PDGF inducible genes, and DNA synthesis. Although this model of indirect mitogenicity might explain the results of TGFβ stimulation of adherent mesenchymal cells, it does not account for the growth inhibitory actions of TGFβ (see below).

## INHIBITION OF CELL PROLIFERATION BY TGFβ

Recently, Tucker et al [12] have shown that TGFβ and the growth inhibitor (GI) that Holley and coworkers [11] isolated from BSC-1 cell conditioned media are similar, if not identical, molecules. Under conditions in which TGFβ is stimulatory for fibroblastic AKR-2B cells, there is an inhibition of the early S phase induced by EGF and insulin or PDGF [12,70]. In addition, GI purified from medium conditioned by BSC-1 cells [71] and TGFβ purified from human platelets have almost identical

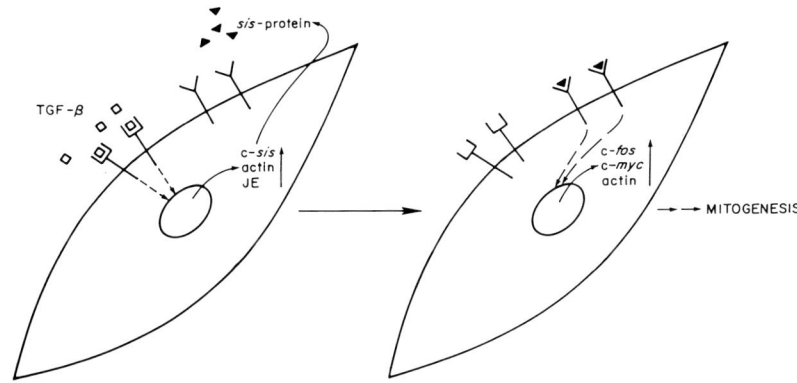

Fig. 2. Proto-oncogene activation by TGFβ. The binding of TGFβ to its cell surface receptors is followed by the activation of certain cellular proto-oncogenes [7]. Among the first activated genes are c-*sis*, actin, and JE genes [96]. The *sis*-protein presumably activates the c-*fos* and c-*myc* genes which then leads to DNA synthesis.

biological activities in stimulating growth of AKR-2B cells in soft agar and inhibiting DNA synthesis in BSC-1 and CCL-64 epithelial mink lung cells. GI was also able to compete for binding with $^{125}$I-labeled TGFβ to membrane receptors nearly as effectively as the native platelet-derived TGFβ [59]. Both the GI and TGFβ have apparent molecular weights of 25,000 and migrate as a single polypeptide of 12,500 in sodium dodecyl sulfate (SDS)-polyacrylamide gels under reducing conditions [48,71]. In addition, our recent immunological analyses have indicated antigenic cross-reactivity (unpublished).

TGFβ is inhibitory for spontaneous soft agar growth of several human carcinoma cell lines [10,72]. No epithelial cell type, either neoplastic or non-neoplastic, has been demonstrated to be stimulated to proliferate by TGFβ. The epithelial cells or carcinoma cell lines that have been tested so far were either inhibited or showed no response to TGFβ under usual cell culture conditions [10,73]. TGFβ is a potent inhibitor of growth of secondary cultures of human foreskin keratinocytes [10,74]. The keratinocytes were reversibly inhibited in the G1 phase of the cell cycle by TGFβ. TGFβ is also a potent inhibitor of EGF-induced stimulation of DNA synthesis in primary cultures of rat hepatocytes [75,76], and it inhibits the growth of primary cultures of human megakaryocytic and erythroid precursors [77]. TGFβ inhibits effectively also fibroblast growth factor (FGF)-stimulated endothelial cell proliferation [78,79].

The mechanisms by which TGFβ inhibits cell proliferation are largely unknown. It is possible that TGFβ is primarily an inhibitor for all cell types and that stimulation of fibroblastic cells is fortuitous through the induction of c-*sis* and autocrine activity by the PDGF-like *sis*-protein, which is the direct mitogen.

## POTENTIAL ROLES OF TGFβ IN NEOPLASIA AND OTHER DISEASE STATES

We used mouse embryo-derived cell culture model systems for neoplastic transformation (AKR-2B and C3H/10T½ cell lines), to demonstrate that the chemically transformed derivatives of these cell lines both produced and responded to

TGFβ [1,10]. Although the parent cell lines released as much TGFβ into serum-free conditioned medium as their chemically transformed derivatives, the TGFβ released by both the parent cells and the transformed derivatives was in an inactive form that was irreversibly activated by acid treatment.

These studies are in agreement with those of Lawrence et al [80] who demonstrated that many cell types release TGFβ in an inactive form. The inactive form of TGFβ released by cells in culture appears to be in a higher molecular weight form than the active molecule, perhaps reflecting an association with a binding protein [43,81]. The physiological mechanism of TGFβ activation is not known. Proteolytic cleavage from a larger precursor or dissociation from a binding protein appear to be the most plausible explanations (see Fig. 3).

**Cellular Responsiveness to TGFβ**

The major change observed in the chemically transformed cells relative to their parent cell lines was the development of a markedly increased sensitivity to TGFβ stimulation of growth in soft agar [10]. The possible mechanisms of the enhanced responsiveness of these cells to TGFβ were examined. Studies on the TGFβ receptor revealed very slightly reduced numbers of receptors on the chemically transformed cells relative to the parent cell lines with no detectable change in affinity. This suggested that a postreceptor mechanism was responsible for the increased TGFβ sensitivity observed in the chemically transformed AKR-2B cells. Using the C3H/10T½ cells, which are completely unresponsive to TGFβ with respect to stimulation of growth in soft agar, transfection was carried out with a mouse c-*myc* gene linked to an SV40 promotor and/or with an activated H-*ras* gene, both of which were cotransfected with the dominant neomycin-resistance marker (E.B. Leof and H.L. Moses, unpublished observations). The c-*myc* gene-transfected cells became highly responsive to stimulation of growth in soft agar by TGFβ, which suggested that c-*myc* expression, at least in part, controlled cellular responsiveness to TGFβ. The H-*ras* gene-transfected cells demonstrated marked morphologic transformation in

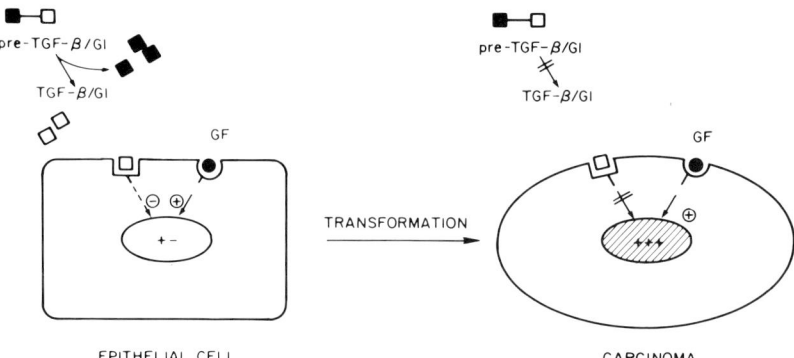

Fig. 3. Escape from TGFβ growth inhibition in carcinoma cells. During the growth of epithelial cells they receive both stimulatory and inhibitory signals from their cell membrane growth factor receptors. An alteration in the cell's ability to activate latent TGFβ or an alteration in the function of the receptor or postreceptor mechanisms may sensitize the cells in a pathological way to normal growth stimulatory signals. This may result in malignant growth [74]. Abbreviations: pre-TGFβ, the inactive precursor of TGFβ (possibly TGFβ bound to a binding protein); GI, growth inhibitor (structurally and immunologically identical with TGFβ).

culture and grew spontaneously in soft agar in the absence of TGFβ. The growth of the H-*ras* gene-transfected cells was only slightly enhanced by TGFβ. These data demonstrated that transfection of C3H/10T½ cells with an activated H-*ras* gene induced a similar phenotype to that induced by TGFβ, but without the requirement for added TGFβ. This suggests that $p21^{H\text{-}ras}$ may enhance autocrine stimulation by endogenous TGFβ or may be involved in the transduction of the TGFβ signal. However, divergent results of similar transfection experiments and TGFβ responsiveness using Fisher rat 3T3 cells have recently been reported [82].

## Epithelial Cell Transformation

The potential role of TGFβ in neoplastic transformation of epithelial and other nonfibroblastic cell types may be entirely different from that involved in fibroblastic cells. We have observed that a squamous carcinoma cell line has lost the inhibitory response to TGFβ characteristically exhibited by normal keratinocytes [74]. Loss of TGFβ-effected growth inhibition in epithelial cells could result in an enhanced proliferative potential, which would produce the same effect as the activation of a stimulatory response (see Fig. 3).

The growth of many normal epithelial cells in primary or secondary culture, including human keratinocytes [83] and mammary [84] and bronchial [85] epithelial cells, is inhibited by serum or by the addition of pure platelet-derived TGFβ [10,74]. Although serum contains relatively large quantities of platelet-derived TGFβ [47], most of the TGFβ in serum is in an inactive form [43,80,81]. It is probable that the TGFβ in serum is responsible for the inhibitory effect of serum on epithelial cells [86]. It is of interest to note that many carcinoma cell lines grow well in serum, but are inhibited in their growth by active TGFβ [10].

These observations suggest that certain epithelial cells can activate the TGFβ present in serum and that at least some carcinoma cells have lost that capability. Since many cells, including epithelial cells [74], produce TGFβ in an inactive form, have receptors for TGFβ, and are capable of responding to active TGFβ, a major regulatory step in TGFβ action is probably at the level of activation of the inactive TGFβ precursor. If this is the case, the loss of the ability to activate TGFβ in cells that are normally inhibited by this molecule could lead to a growth advantage (Fig. 3).

## Effects on Connective Tissue and Wound Healing

Recent studies by Roberts and coworkers [87] demonstrated that TGFβ induced a marked desmoplastic reaction (induction of angiogenesis and connective tissue formation) when injected into mice. Since TGFβ is abundant in platelets, its role in wound healing was soon appreciated [88]. Another growth factor of platelets, PDGF, enhances wound healing by increasing the proliferation of fibroblasts and the formation of granulation tissue [89]. These and other studies indicated that TGFβ might be involved in wound healing [90] and further suggested that the TGFβ released by carcinoma cells could contribute to the stromal cell proliferation necessary for the formation of large tumors. The data further suggested that TGFβ could play a major role in many disease states involving fibroblastic proliferation and collagen deposition. Potential mechanisms for augmented connective tissue formation are that TGFβ both enhances the expression of fibronectin and procollagen in cells [9] and regulates pericellular proteolysis by inducing the production of endothelial-type plasminogen activator inhibitors [91,92]. TGFβ-like factors that enhance the production of proteinase inhibitors are produced by 8387 human fibrosarcoma cells [93]. These kinds of factors may operate in the regulation of both normal and pathological proteolysis.

An interesting field of research with connections to wound healing is atherosclerosis. According to the response-to-injury hypotheses of Ross and coworkers [cf 94] an injury to the endothelium exposes the subendothelium at sites of turbulent blood flow; platelets become attached at these sites, aggregate and release PDGF. Given that TGF$\beta$ is also released from platelets [90] it may act directly or indirectly (via the induction of c-*sis*) in the pathogenesis of atherosclerosis. PDGF and TGF$\beta$ may participate in the formation of the atheroma plaques at least by stimulating the growth of smooth muscle cells [see 94]. PDGF enhances the transcription of the actin genes [95], and similar direct effects have recently been described for TGF$\beta$ [96]. The activation of the actin genes is probably an indication of the enhanced motility of cells and may explain in part the chemotactic effects of these growth factors [see 89,97].

## Immunological Aspects

A potential role for TGF$\beta$ in the modulation of the immune response has recently been suggested. Mitogenic treatment of human T lymphocytes results in accumulation of TGF$\beta$ mRNA [98], and TGF$\beta$ appears to act as an antagonist of interleukin-2. In addition, Mizel et al [99] have shown the production of immunosuppressive factors by murine sarcoma virus (MSV)-transformed cells, and TGF$\beta$ inhibits the production of IgG and IgM [see 98]. These results suggest that TGF$\beta$ is capable of modifying immunological responses and might participate in the pathogenesis of different immunological diseases.

## Effects on Bone Resorption

It has been suggested that growth factors such as EGF, TGF$\alpha$, TGF$\beta$ and PDGF regulate bone resorption in animals [100,101]. This is possibly a cause of the hypercalcemia associated with many cancers [102]. The most important growth factor regulating bone resorption appears to be TGF$\alpha$, and it has been recently reported that it is more effective in this regard that EGF, indicating for the first time a difference in action for these closely related growth factors [103]. Interestingly, the tumor necrosis factors (TNFs) also have bone-resorbing activity, and may act in concert with TGF$\alpha$ in vivo [104].

## TGF$\beta$ IN THE REGULATION OF DIFFERENTIATION

TGF$\beta$ appears to play a role in certain steps of cell differentiation. TGF$\beta$ inhibits insulin- and dexamethasone-induced differentiation of mouse 3T3 cells to adipocytes [105,106], and it also controls myogenesis [107]. On the other hand, it has been reported that it can enhance terminal differentiation of bronchial epithelial cells to squamous cells [108]. One of the functions of TGF$\beta$ appears to be to control the differentiation of BRL cells [58]. When the effects of SGFs (TGF$\alpha$ + TGF$\beta$) on the developing tooth germ were studied in organ culture, it was found that they prevented morphogenesis and stimulated connective tissue formation and neovascularization [109]. The inhibition of tooth germ morphogenesis was probably due to the TGF$\beta$ inhibition of epithelial cell growth.

Available evidence thus suggests that TGF$\beta$ may, in concert with other growth factors, participate in various events of cell differentiation and morphogenesis and thus also have a function as a developmental protein.

## CONCLUDING REMARKS

It is obvious that TGFβ has a multifaceted role in growth regulation. Both stimulatory and inhibitory properties have been reported. The ubiquitous nature and extreme evolutionary conservation of this molecule suggest that TGFβ plays a central role in mediating the cellular response to a variety of environmental stimuli. Characterization of the subsequent molecular response will undoubtedly expand our understanding of normal growth control as well as of cellular transformation.

## ACKNOWLEDGMENTS

We thank Ms. Rebecca Koransky and Ms. Dot Blue for assistance in the preparation of the manuscript. Our original research has been supported by PHS grants CA 42749 and CA 42750 awarded by the National Cancer Institute, DHHS.

## REFERENCES

1. Moses HL, Branum EB, Proper JA, Robinson RA: Cancer Res 41:2842–2848, 1981
2. Roberts AB, Anzano MA, Lamb LC, Smith JM, Sporn MB: Proc Natl Acad Sci USA 78:5339–5343, 1981.
3. DeLarco JE, Todaro GJ: Proc Natl Acad Sci USA 75:4001–4005, 1978.
4. Anzano MA, Roberts AB, Smith JM, Sporn MB, DeLarco JE: Proc Natl Acad Sci USA 80:6264–6268, 1983.
5. Kaplan PL, Ozanne B: Cell 33:931–938, 1983.
6. Anzano MA, Roberts AB, Sporn MB: J Cell Physiol 126:312–318, 1986.
7. Leof EB, Proper JA, Goustin AS, Shipley GD, DiCorleto PE, Moses HL: Proc Natl Acad Sci USA, 83:2453–2457, 1986.
8. Assoian RK, Grotendorst GR, Miller DM, Sporn MB: Nature 309:804–806, 1984.
9. Ignotz RA, Massaque J: J Biol Chem 261:4337–4345, 1986.
10. Moses HL, Tucker RF, Leof EB, Coffey RJ Jr, Halper J, Shipley GD: In Feramisco J, Ozanne B, Stiles C (eds): "Cancer Cells 3." Cold Spring Harbor, NY: Cold Spring Harbor Press, 1985, pp 65–71.
11. Holley RW, Armour R, Baldwin JH: Proc Natl Acad Sci USA 75:864–1866, 1978.
12. Tucker RF, Shipley GD, Moses HL, Holley RW: Science 226:705–707, 1984.
13. Hirai R, Yamaoka K, Mitsui H: Cancer Res 43:5742–5746, 1983.
14. Yamaoka K, Hirai R, Tsugita A, Mitsui H: J Cell Physiol 119:307–314, 1984.
15. Halper J, Moses HL: Cancer Res 43:1972–1979, 1983.
16. Iwata KK, Fryling CM, Knott WB, Todaro GJ: Cancer Res 45:2689–2694, 1985.
17. Fryling CM, Iwata KK, Johnson PA, Knott WB, Todaro GJ: Cancer Res 45:1695–1699, 1985.
18. DeLarco JE, Pigott DA, Lazarus JA: Proc Natl Acad Sci USA 82:5015–5019, 1985.
19. Todaro GJ, Fryling C, DeLarco JE: Proc Natl Acad Sci USA 77:5258–5262, 1980.
20. Roberts AB, Lamb LC, Newton DL, Sporn MB, DeLarco JE, Todaro GJ: Proc Natl Acad Sci USA 77:3494–3498, 1980.
21. Ozanne B, Fulton RJ, Kaplan PL: J Cell Physiol 105:163–180, 1980.
22. Massague J: J Biol Chem 258:13614–13620, 1983.
23. Derynck R, Roberts AB, Winkler ME, Chen EY, Goeddel DV: Cell 38:287–297, 1984.
24. Tam, JP, Marquardt H, Rosberger DF, Wong TW, Todaro GJ: Nature 309:376–378, 1984.
25. Marquardt H, Hunkapiller MW, Hood LE, Twardzik DR, DeLarco JE, Stephenson JR, Todaro GJ: Proc Natl Acad Sci USA 80:4684–4688, 1983.
26. Marquardt H, Hunkapiller MW, Hood LE, Todaro GJ: Science 223:1079–1082, 1984.
27. Derynck R: J Cell Biochem (in press), 1986.
28. Twardzik DR: Cancer Res 45:5413–5416, 1985.
29. Lee DC, Rochford R, Todaro GJ, Villarreal LP: Mol Cell Biol 5:3644–3646, 1985.
30. Tam JP: Science 229:673–675, 1985.

31. Brissenden JE, Derynck R, Francke U: Cancer Res 45:5593–5597, 1985.
32. Tricoli JV, Nakai H, Byers MG, Rall LB, Bell GI, Shows TB: Cytogenet Cell Genet 42:94–98, 1986.
33. Scott J, Urdea M, Quiroga M, Sanchez-Pescador R, Fong N, Selby M, Rutter WJ, Bell GI: Science 221:236–240, 1983.
34. Gray A, Dull TJ, Ullrich A: Nature 303:711–715, 1983.
35. Venkatesan S, Gershowitz A, Moss B: J Virol 44:637–646, 1982.
36. Strobant P, Rice AP, Gullick WJ, Cheng DJ, Kerr IM, Waterfield MD: Cell 42:383–393, 1985.
37. Twardzik DR, Brown JP, Ranchalis JE, Todaro GJ, Moss B: Proc Natl Acad Sci USA 82:5300–530, 1985.
38. Russell DW, Schneider WJ, Yamamoto T, Luskey KL, Brown MS, Goldstein JL: Cell 37:577–585, 1984.
39. Yamamoto T, Davis G, Brown MS, Schneider WJ, Casey HL, Goldstein JL, Russell DW: Cell 39:27–38, 1984.
40. Patthy L: Cell 41:657–663, 1985.
41. Goustin AS, Leof EB, Shipley GD, Moses HL: Cancer Res 46:1015–1029, 1986.
42. Sporn MB, Roberts AB: Nature 313:745–747, 1985.
43. Moses HL, Shipley GD, Leof EB, Halper J, Coffey RJ Jr, Tucker RF: In Leffert, HL, Boynton AL (eds): "Control of Animal Cell Proliferation." New York: Academic Press, 1986 (in press).
44. Stromberg K, Pigott DA, Ranchalis JE, Twardzik DR: Biochem Biophys Res Commun 106:354–361, 1982.
45. Roberts AB, Anzano MA, Meyers CA, Wideman J, Blacher R, Pan Y-CE, Stein S, Lehrman SR, Smith LC, Lamb LC, Sporn MB: Biochemistry 22:5692–5698, 1983.
46. Frolik CA, Dart LL, Meyers CA, Smith DM, Sporn MB: Proc Natl Acad Sci USA 80:3676–3680, 1983.
47. Childs C, Proper JA, Tucker RF, Moses HL: Proc Natl Acad Sci USA 79:5312–5316, 1982.
48. Assoian RK, Komoriya A, Meyers CA, Miller DM, Sporn MB: J Biol Chem 258:7155–7160, 1983.
49. Derynck R, Jarrett JA, Chen EY, Eaton DH, Bell JR, Assoian RK, Roberts AB, Sporn MB, Goeddel DV: Nature 316:701–705, 1985.
50. Derynck R, Jarrett JA, Chen EY, Goeddel DV: J Biol Chem 261:4377–4379, 1986.
51. Fujii D, Brissenden JE, Derynck R, Francke U: Somatic Cell Mol Genet 12:281–288, 1986.
52. Mason AJ, Hayflick JS, Ling N, Esch F, Ueno N, Ying S-Y, Guillemin R, Niall H, Seeburg PH: Nature 318:659–663, 1985.
53. Vale W, Rivier J, Vaughan J, McClintock R, Corrigan A, Woo W, Karr D, Spiess J: Nature 321:776–779, 1986.
54. Ling N, Ying S, Ueno N, Shimasaki S, Esch F, Hotta M, Guillemin R: Nature 321:779–782, 1986.
55. Cate RL, Mattaliano RJ, Hession C, Tizard R, Farber NM, Cheung A, Ninfa EG, Frey AZ, Gash DJ, Chow EP, Risher RA, Bertonis JM, Torres G, Wallner BP, Ramachandran KL, Ragin RC, Manganaro TF, MacLaughlin DT, Donahoe PK: Cell 45:685–698, 1986.
56. Keski-Oja J, Moses HL: Med Biol (in press), 1987.
57. Seyeden SM, Thompson AY, Bentz H, Rosen DM, McPherson JM, Conti A, Siegel NR, Galluppi GR, Piez KA: J Biol Chem 261:5693–5695, 1986.
58. Florini JR, Roberts AB, Ewton D, Falen SL, Lancers KC, Sporn MB: J Biol Chem (submitted), 1986.
59. Tucker RF, Branum EL, Shipley GD, Ryan RJ, Moses L: Proc Natl Acad Sci USA 81:6757–6761, 1984.
60. Frolik CA, Wakefield LM, Smith DM, Sporn MB: J Biol Chem 259:10995–11000, 1984.
61. Massague J: J Biol Chem 260:7059–7066, 1985.
62. Cheifetz S, Like B, Massague J: J Biol Chem 261:9972–9978, 1986.
63. Assoian RK, Frolik CA, Roberts AB, Miller DM, Sporn B: Cell 36:35–41, 1984.
64. Massague J: J Cell Biol 100:1508–1514, 1985.
65. Inman WH, Colowick SP: Proc Natl Acad Sci USA 82:1346–1349, 1985.
66. Boerner P, Resnick RJ, Racker E: Proc Natl Acad Sci USA 82:1350–1353, 1985.
67. Massague J, Kelly B, Mottola C: J Biol Chem 260:4551-4554, 1085.
68. Rizzino A: In Vitro Cell Dev Biol 20:815–822, 1984

69. van Zoelen EJJ, van Oostwaard TMJ, deLaat SW: J Biol Chem 261:5003–5009, 1986.
70. Shipley GD, Tucker RF, Moses HL: Proc Natl Acad Sci USA 82: 4147–4151, 1985.
71. Holley RW, Armour R, Baldwin JH, Greenfield S: Cell Biol Int Rep 7:525–526, 1983.
72. Roberts AB, Anzano MA, Wakefield LM, Roche NS, Stern F, Sporn MB: Proc Natl Acad Sci USA 82:119–123, 1985.
73. Coffey RJ Jr. Shipley GD, Moses HL: Cancer Res 46:1164–1169, 1986.
74. Shipley GD, Pittelkow MR, Wille JJ Jr, Scott RE, Moses HL: Cancer Res 46:2068–2071, 1986.
75. Nakamura T, Tomita Y, Hirai R, Yamaoka K, Kaji K, Ichihara A: Biochem Biophys Res Commun 133:1042–4050, 1985.
76. Carr BI, Hayashi I, Branum EL, Moses HL: Cancer Res 46:2330–2334, 1986.
77. Solberg LA Jr, Tucker RF, Jenkins RB, Moses HL: J Cell Biochem 10C:187, 1986.
78. Frater-Schröder M, Müller G, Birchmeier W, Böhlen P: Biochem Biophys Res Commun 137:295–302, 1986.
79. Baird A, Durkin T: Biochem Biophys Res Commun 138:476–482, 1986.
80. Lawrence DA, Pircher R, Kryceve-Martinerie C, Jullien P: J Cell Physiol 121:184–188, 1984.
81. Lawrence DA, Pircher R, Jullien P: Biochem Biophys Res Commun 133:1026–1034, 1985.
82. Stern DF, Roberts AB, Roche NS, Sporn MB, Weinberg RA: Mol Cell Biol 6:870–877, 1986.
83. Boyce ST, Ham RG: J Invest Dermatol 81:33s–40s, 1983.
84. Hammond SL, Ham RG, Stampfer MR: Proc Natl Acad Sci USA 81:5435–5439, 1984.
85. Lechner JF, McClendon IA, LaVeck MA, Schamsuddin AM, Harris CC: Cancer Res 43:5915–5921, 1983.
86. Masui T, Wakefield LM, Lechner JF, LaVeck MA, Sporn MB, Harris CC: Proc Natl Acad Sci USA 83:2438–2442, 1986.
87. Roberts AB, Sporn MB, Assoian RK, Smith JM, Roche NS, Wakefield LM, Heine UI, Liotta LA, Falanga V, Kehrl JH, Fauci AS: Proc Natl Acad Sci USA 83:4167–4171, 1986.
88. Sporn MB, Roberts AB, Shull JH, Smith JM, Ward JM, Sodek J: Science 219:1329–1331, 1983.
89. Grotendorst GR, Martin GR, Pencev D, Sodek J, Harvey AK: J Clin Invest 76:2323–2329, 1985.
90. Assoian RK, Sporn MB: J Cell Biol 102:1217–1223, 1986.
91. Laiho M, Saksela O, Keski-Oja J: Exp Cell Res 164:399–407, 1986.
92. Laiho M, Saksela O, Andreasen PA, Keski-Oja J; J Cell Biol 103:(in press), 1986.
93. Saksela O, Laiho M, Keski-Oja J: Cancer Res 45:2314–2319, 1985.
94. Ross R: N Engl J Med 314:488–500, 1986.
95. Greenberg ME, Ziff EB: Nature 311:433–438, 1984.
96. Leof EB, Proper JA, Getz MJ, Moses HL: J Cell Physiol 127:83–88, 1986.
97. Postlethwaite AE, Keski-Oja J, Moses HL, Kang AH: J Exp Med (in press), 1987.
98. Kehrl JH, Wakefield LM, Roabertgs AB, Jakowlew S, Alvarez-Mon M, Derynck R, Sporn MB, Fauci AS: J Exp Med 163:1037–1050, 1986.
99. Mizel SB, DeLarco JE, Todaro GJ, Farrar WL, Hilfiker ML: Proc Natl Acad Sci USA 77:2205–2208, 1980.
100. Ibbotson KJ, Ng KW, Osborne CK, Niall M, Martin TJ, Mundy GR: Science 221:1292–1294, 1983.
101. Tashjian AHJ, Voelkel EF, Lazzaro M, Singer FR, Roberts AB, Derynck R, Winkler ME, Levine L: Proc Natl Acad Sci USA 82:4535–4538, 1985.
102. Mundy GR, Ibbotson KJ, D'Souza SM, Simpson EL, Jacobs W, Martin TJ: N Engl J Med 310:1718–1727, 1984.
103. Ibbotson KJ, Harrod J, Gowen M, D'Souza S, Smith DD, Winkler ME, Derynck R, Mundy GR: Proc Natl Acad Sci USA 83:2228–2232, 1986.
104. Bertolini DR, Nedwin GE, Bringman TS, Smith DD, Mundy GR: Nature 319:516–518, 1986.
105. Ignotz RA, Massague J: Proc Natl Acad Sci USA 82:8530–8534, 1985.
106. Sparks RL, Scott RE: Exp Cell Res 165:345–352, 1986.
107. Massaque J, Chiefetz S, Endo T, Nadal-Ginard B: submitted 1986.
108. Masui T, Wakefield LM. Lechner JF, LaVeck MA, Sporn MB, Harris CC: Proc Natl Acad Sci USA 83:2438–2442, 1986.
109. Thesleff I, Ekblom P, Keski-Oja J: Cancer Res 43:5902–5909, 1983.

# Transformation by the v-*fms* Oncogene Product: An Analog of the CSF-1 Receptor

Carl W. Rettenmier, Suzanne Jackowski, Charles O. Rock, Martine F. Roussel, and Charles J. Sherr

*Departments of Tumor Cell Biology (C.W.R., M.F.R., C.J.S.) and Biochemistry (S.J., C.O.R.), St. Jude Children's Research Hospital, Memphis, Tennessee 38105*

The product of the c-*fms* proto-oncogene is related to, and possibly identical with, the receptor for the macrophage colony-stimulating factor, M-CSF (CSF-1). Unlike the product of the v-*erb*B oncogene, which is a truncated version of the EGF receptor, the glycoprotein encoded by the v-*fms* oncogene retains an intact extracellular ligand-binding domain so that cells transformed by v-*fms* express CSF-1 receptors at their surface. Although fibroblasts susceptible to transformation by v-*fms* generally produce CSF-1, v-*fms*-mediated transformation does not depend on an exogenous source of the growth factor, and neutralizing antibodies to CSF-1 do not affect the transformed phenotype. An alteration of the v-*fms* gene product at its extreme carboxyl-terminus represents the major structural difference between it and the c-*fms*-coded glycoprotein and may affect the tyrosine kinase activity of the v-*fms*-coded receptor. Consistent with this interpretation, tyrosine phosphorylation of the v-*fms* products in membranes was observed in the absence of CSF-1 and was not enhanced by addition of the murine growth factor. Cells transformed by v-*fms* have a constitutively elevated specific activity of a guanine nucleotide-dependent, phosphatidylinositol-4,5-diphosphate-specific phospholipase C. We speculate that the tyrosine kinase activity of the v-*fms*/c-*fms* gene products may be coupled to this phospholipase C, possibly through a G regulatory protein, thereby increasing phosphatidylinositol turnover and generating the intracellular second messengers diacylglycerol and inositol triphosphate.

Key words: growth factors, tyrosine-specific protein kinase, phospholipase C, second messengers

## THE v-*fms* ONCOGENE PRODUCTS

Transforming oncogenes of retroviruses (v-*onc* genes) are generated by recombination between replicating viral vectors and cellular proto-oncogenes (c-*onc* genes) that are presumed to play pivotal roles in controlling normal growth and differentiation [1]. Structural alterations in v-*onc* genes and their inappropriate expression owing

Received April 30, 1986; revised and accepted July 28, 1986.

© 1987 Alan R. Liss, Inc.

to retroviral transduction are responsible for the ability of the recombined genes to cause neoplastic tumors in animals and to transform cultured cells. The study of viral transforming genes and their proto-oncogene progenitors has therefore begun to provide insights into the mechanisms by which normal cells are transformed into their malignant counterparts.

The v-*fms* oncogene of the Susan McDonough strain of feline sarcoma virus (SM-FeSV) was isolated from a spontaneously arising fibrosarcoma of a domestic cat [2]. The virus is able to transform fibroblast cell lines from several mammalian species [3], and, similarly, the molecularly cloned SM-FeSV provirus is biologically active in transforming mouse NIH/3T3 fibroblasts after DNA-mediated transfection [4]. Nucleotide sequencing of the SM-FeSV genome revealed that v-*fms* coding sequences were transduced into the viral *gag* gene in the same reading frame [5]. The primary translation product was predicted to be a *gag-fms* fusion polyprotein consisting of 536 amino-terminal *gag*-coded residues and 975 carboxyl-terminal v-*fms*-coded amino acids. Addition of carbohydrate to the polypeptide chain yields a 180-kilodalton (kDa) glycoprotein ($gP180^{gag-fms}$) containing asparagine (N)-linked oligosaccharide chains. Some of the $gP180^{gag-fms}$ molecules are cotranslationally cleaved at a site near the *gag-fms* junction to generate two distinct products: an unglycosylated 55-kDa fragment that is precipitated by antibodies to *gag*-coded determinants and a 120-kDa v-*fms*-coded glycoprotein ($gp120^{v-fms}$) [6,7]. Both $gP180^{gag-fms}$ and $gp120^{v-fms}$ contain oligosaccharides of the high-mannose type typical of immature glycoproteins. Some $gp120^{v-fms}$ molecules are transported through the Golgi compartment, where their N-linked oligosaccharides are processed to complex carhbohydrate chains, and on to the plasma membrane, where they are detected as glycoproteins of greater apparent molecular weight ($gp140^{v-fms}$) [8,9]. Once expressed at the cell surface, the $gp140^{v-fms}$ molecules become associated with clathrin-coated pits and are returned in endocytic vesicles to the interior of the cell, where they are probably degraded [10].

The location of hydrophobic segments within $gP180^{gag-fms}$ suggested a model for the transmembrane orientation of the v-*fms*-coded glycoprotein during translocation [11,12]. Vectorial transfer of newly synthesized molecules into the lumen of the endoplasmic reticulum (ER) proceeds until a segment of 26 hydrophobic amino acids is encountered near the middle of the v-*fms*-coded sequence. This hydrophobic peptide was predicted to stop translocation and to anchor the v-*fms*-coded glycoprotein in the membrane with its amino-terminal portion in the lumen of the ER and its carboxyl-terminal 406 amino acids in the cytoplasm. Canonical sites for addition of N-linked oligosaccharides (Asn-X-Ser/Thr) are clustered in the v-*fms*-coded sequence on the amino-terminal side of the transmembrane segment where they are accessible to glycosyl transferases within the ER cisternae. Intracellular transport via membrane vesicles would be expected to orient $gp140^{v-fms}$ at the cell surface with its glycosylated amino-terminal domain outside the cell and its carboxyl-terminal domain at the cytoplasmic face of the plasma membrane. This transmembrane orientation of the v-*fms*-coded glycoproteins was confirmed experimentally [12].

The cytoplasmic carboxyl-terminal domain of the v-*fms* gene product shares amino acid sequence homology with the products of other oncogenes that code for tyrosine-specific protein kinases [5]. When the v-*fms*-coded glycoproteins are precipitated with appropriate antisera and incubated with radiolabeled ATP in the presence of $Mn^{++}$ ions, an associated kinase activity transfers the $\gamma$-phosphate from the nucleotide onto tyrosyl residues of the oncogene product itself or to exogenously added protein substrates [13]. A series of v-*fms* mutants encoding products that lack the associated kinase activity do not transform cells, suggesting that this enzymatic function is required for transformation. However, unlike prototypic members of the tyrosine kinase family of retroviral oncogenes, the v-*fms* products themselves are not extensively phosphorylated on tyrosine residues in vivo, and cells transformed by SM-FeSV do not exhibit an increase of the phosphotyrosine detected in total cellular proteins.

Intracellular transport of the v-*fms* oncogene product positions the carboxyl-terminal kinase domain of $gp140^{v-fms}$ at the inner surface of the plasma membrane. This subcellular localization is shared by the products of several other oncogenes [14–19], suggesting that the intracellular targets for transforming signals generated by these proteins are at or near the plasma membrane. Indeed, transformation is blocked by mutations in the polypeptide chain [20] or by inhibitors of glycosylational processing [21] that prevent cell surface expression of the v-*fms* gene product. The topological features of the v-*fms*-coded glycoprotein are similar to those of a family of cell surface receptors for polypeptide growth factors that exhibit a ligand-stimulated tyrosine-specific protein kinase activity. The glycosylated amino-terminal portion is analogous to the extracellular ligand-binding domain of a receptor linked by the transmembrane segment to an intracellular signal-transducing function (the kinase domain). A relationship between oncogenes and growth factors or their receptors is also evident from the findings that the v-*sis* oncogene product is homologous to a chain of the platelet-derived growth factor (PDGF) [22,23], and the v-*erb* B oncogene product represents a truncated form of the receptor for epidermal growth factor (EGF) [24].

## THE c-*fms* PROTO-ONCOGENE PRODUCT AND CSF-1 RECEPTOR

Although the properties of the v-*fms* product strongly suggested that it was derived from a proto-oncogene coding for a cell surface receptor, the identity of that receptor was unknown. To deduce the function of the c-*fms* proto-oncogene product, expression of the c-*fms* gene was assayed in different tissues of the domestic cat. High levels of a polyadenylated c-*fms* RNA about 4 kilobases in length were detected in spleen, whereas significantly lower levels of transcripts were found in liver, lymph nodes, bone marrow, and brain [25]. When monoclonal antibodies to v-*fms*-coded epitopes [8] and the immune complex kinase reaction were used, c-*fms*-coded glycoproteins of 130 kDa and 170 kDa were detected in cat splenocytes and were phosphorylated on tyrosine residues by an associated enzymatic activity. The patterns of glycosylation of the c-*fms*-coded products were analogous to those of the glycoproteins encoded by v-*fms*, suggesting that feline $gp130^{c-fms}$ is a precursor of the mature cell surface form, $gp170^{c-fms}$ [25]. Fluorescence-activated flow cytometry performed with the same monoclonal antibodies demonstrated that expression of the cat c-*fms* gene product was primarily restricted to macrophages [26]. Although these monoclonal antibodies to feline v-*fms*-coded epitopes do not react with c-*fms*-coded glycoproteins in species other than the domestic cat, antisera to a recombinant v-*fms*-coded polypeptide expressed in bacteria precipitated c-*fms*-coded molecules from mouse [26] and man [27,28]. Again, the expression of the c-*fms* gene product in somatic tissues from these other species was primarily restricted to cells of the mononuclear phagocyte series, including peripheral blood monocytes and tissue macrophages.

Since the only known growth factor specific in its action for mononuclear phagocytes is the macrophage colony-stimulating factor CSF-1 (M-CSF) [29], a possible relationship between the c-*fms* gene product and CSF-1 receptor was investigated. Purified CSF-1 from mouse L cells was found to bind with high affinity to the murine c-*fms* gene product, and, in membrane preparations, it enhanced tyrosine-specific phosphorylation of the c-*fms*-coded glycoprotein [26]. The viral oncogene product, $gp140^{v\text{-}fms}$, was also found to specifically bind murine CSF-1, so that cells transformed by SM-FeSV acquire CSF-1 binding sites on their surface [30]. These results suggest that the c-*fms* proto-oncogene is closely related to, and possibly identical with, the gene encoding the receptor for CSF-1. However, because the amino acid sequence of the purified CSF-1 receptor has not yet been determined, there is as yet no formal proof that these are identical genes.

## MECHANISMS OF TRANSFORMATION

An implication of these findings is that the mechanism of viral transformation may involve the transduction of a competent receptor gene into cells that synthesize the corresponding growth factor. Indeed, CSF-1 is normally produced by mesenchymal cells in the bone marrow [29] and also by fibroblast cell lines susceptible to transformation by SM-FeSV [30]. SM-FeSV-mediated transformation does not depend on an exogenous source of CSF-1, and neutralizing antibodies to CSF-1 do not affect the transformed phenotype [30]. It remains formally possible that an interaction between the colony-stimulating factor and the v-*fms*-coded receptor might occur intracellularly within the secretory compartment. However, it seems more likely that rearrangement of the v-*fms* gene product at its extreme carboxyl-terminus [31] or more subtle mutations at other positions within the polypeptide affect the kinase activity of the viral-coded receptor so that the enzyme either acts constitutively, has enhanced activity in the presence of ligand, or phosphorylates substrates not recognized by its normal cellular counterpart.

It is intriguing that deletions of the long arm of human chromosome 5 at bands q33.2–q34, the site of the c-*fms* locus [32,33], are associated with a variety of hematopoietic disorders including refractory anemia, myelodysplastic syndromes, and therapy-related acute nonlymphocytic leukemias (ANNL) [34–37]. This raises the possibility that rearrangements at or near the c-*fms* locus may contribute in some way to the altered hematopoiesis seen in these disorders. Deletion of c-*fms* has been demonstrated both in cases of the 5q− refractory anemia syndrome and in ANNL [38,39], although the locations of the precise breakpoints in the long arm of chromosome 5 remain unknown. Moreover, the genes coding for both CSF-1 [39] and for the granulocyte-macrophage colony-stimulating factor (GM-CSF) [40,41] map to a neighboring region of chromosome 5q implying that the observed deletions may affect any of these genes or, perhaps, additional genes regulating hematopoiesis that cluster within this region. Although rearrangements of c-*fms*, GM-CSF, and the CSF-1 genes have not yet been documented in ANNL, a search for such alterations now seems warranted.

Although the CSF-1 receptor has been defined through its role in hematopoiesis, c-*fms* transcripts are also detected at high levels in placental trophoblasts and in malignant choriocarcinoma cell lines [42]. As predicted, the human c-*fms* gene product was detected in several independently derived human choriocarcinoma cell lines and was identical in its biochemical properties with the c-*fms*-coded glycoprotein precipitated from monocytes and macrophages [27,28]. Thus, the CSF-1 receptor may play some additional role in placental development that differs from its function in adult tissues.

## GUANINE-NUCLEOTIDE-DEPENDENT PHOSPHOLIPASE C IN v-*fms* TRANSFORMANTS

The identity of the intracellular target(s) of the CSF-1 receptor kinase and the v-*fms*-coded transforming glycoprotein is unknown, but the generation of phosphatidylinositol(PtdIns)-derived second messengers may be physiologically important. The catalytic component of this signaling system is a hormone-responsive, PtdIns 4,5-diphosphate (PtdIns-$P_2$) phospholipase C that hydrolyzes PtdIns-$P_2$, giving rise to inositol 1,4,5-triphosphate (Ins(1,4,5)-$P_3$) and diacylglycerol. The liberated Ins(1,4,5)-$P_3$ mobilizes calcium from intracellular stores, thereby activating a host of calcium-regulated functions, and the diacylglycerol stimulates the phosphorylation of a number of proteins via the activation of protein kinase C [43–45]. These two intracellular effector molecules are capable of eliciting a wide range of biological responses, and their production is stimulated following the exposure of cells to many growth-promoting polypeptide hormones [46–58].

Enhanced PtdIns turnover is also associated with cell transformation induced by a variety of oncogenes and chemical agents [59–67]. Our recent observations that cells transformed by the v-*fms* or v-*fes* oncogenes have elevated specific activities of guanine-nucleotide-dependent PtdIns-$P_2$ phospholipase C suggest that an increase in this enzymatic activity may be responsible for the higher rates of PtdIns turnover exhibited by these cells [67]. The levels of PtdIns turnover and membrane-associated PtdIns-$P_2$ phospholipase C activity were higher in v-*fes* transformants than in v-*fms* transformants, correlating with the relative activities of these transforming proteins as tyrosine-specific protein kinases in vivo [13,68]. It is therefore tempting to speculate that the tyrosine kinase activities of these oncogene products are responsible for the increased PtdIns-$P_2$ phospholipase C activity. The tyrosine kinases may directly activate the phospholipase C or may alter the properties of an intermediate transducing protein (G-protein) that mediates the interaction of receptors and catalytic units. The removal of an endogenous phospholipase C inhibitor could also account for these results.

If the v-*fms*-coded glycoprotein activates PtdIns metabolism, we would anticipate that CSF-1 would also fall within the class of polypeptide growth factors that mediate a rapid PtdIns response. In Balb/c 3T3 fibroblasts, growth factors act synergistically in stimulating DNA synthesis and have been classified according to their biological effects by "order of addition" experiments. For example, PDGF does not need to be continually present during the cell cycle, but exerts its action in the GO/G1 transition and renders 3T3 fibroblasts "competent" to begin traversing the cell cycle [69]. PDGF-primed cells require a second hormone (ie, EGF or insulin) to initiate a round of division, and these latter "progression" factors must be present throughout the G1 phase of the cell cycle to exert their effects [70]. In contrast to PDGF, EGF and insulin apparently do not stimulate PtdIns hydrolysis in Balb/c 3T3 fibroblasts [71]. Thus, competence factors in these cells are associated with the activation of PtdIns metabolism, whereas progression factors do not elicit a PtdIns response. Other fibroblast cell lines do not obey these rules [72,73], and whether the action of CSF-1 can be classified by an analogous scheme remains unclear. CSF-1 stimulates mouse bone marrow-derived macrophages to enter S phase and appears to be necessary throughout the duration of G1 [74]. However, recent experiments demonstrating that the human c-*fms* gene product and the mouse PDGF receptor are closely related to one another [75] suggest that the PDGF and the CSF-1 receptor kinases might phosphorylate a similar class of physiologic target molecules.

## ACKNOWLEDGMENTS

We acknowledge grants CA-38187, GM-28035, and GM-34496 from the National Institutes of Health and the American Lebanese Syrian Associated Charities of St. Jude Children's Research Hospital. C.W.R. was supported by Biomedical Research Support grant RR-05584-21 from the NIH to St. Jude Hospital.

## REFERENCES

1. Bishop JM: Cell 42:23, 1985.
2. McDonough SK, Larsen S, Brodey RS, Stock ND, Hardy Jr WD: Cancer Res 31:953, 1971.
3. Sarma PS, Sharar AL, McDonough, S: Proc Soc Exp Biol Med 140:1365, 1972.
4. Donner L, Fedele LA, Garon CF, Anderson SJ, Sherr CJ: J Virol 41:489, 1982.
5. Hampe A, Gobet M, Sherr CJ, Galibert F: Proc Natl Acad Sci USA 81:85, 1984.
6. Barbacid M, Lauver AV, Devare SG: J Virol 33:196, 1980.
7. Ruscetti SK, Turek LP, Sherr CJ: J Virol 35:259, 1980.
8. Anderson SJ, Furth M, Wolff L, Ruscetti SK, Sherr CJ: J Virol 44:696, 1982.
9. Anderson SJ, Gonda MA, Rettenmier CW, Sherr CJ: J Virol 51:730, 1984.
10. Manger R, Najita L, Nichols EJ, Hakomori S-I, Rohrschneider L: Cell 39:327, 1984.
11. Sherr CJ, Anderson SJ, Rettenmier CW, Roussel MF: In Vande Woude GF, Levine AJ, Topp WC, Watson JD (eds): "Cancer Cells 2/Oncogenes and Viral Genes." Cold Spring Harbor, NY: Cold Spring Harbor Laboratory Press, 1984, pp 329–338.
12. Rettenmier CW, Roussel MF, Quinn CO, Kitchingman GR, Look AT, Sherr CJ: Cell 40:971, 1985.
13. Barbacid M, Lauver AV: J Virol 40:812, 1981.
14. Willingham MC, Jay G, Pastan I: Cell 18:125, 1979.
15. Courtneidge S, Levinson AD, Bishop JM: Proc Natl Acad Sci USA 77:3783, 1980.
16. Krueger JG, Wang E, Goldberg AR: Virology 101:25, 1980.
17. Willingham MC, Pastan I, Shih TY, Scolnick EM: Cell 19:1005, 1980.
18. Hayman MJ, Ramsay GM, Savin K, Kitchener G, Graf T, Beug H: Cell 32:579, 1983.
19. Privalsky ML, Sealy L, Bishop JM, McGrath JP, Levinson AD: Cell 32:1257, 1983.
20. Roussel MF, Rettenmier CW, Look AT, Sherr CJ: Mol Cell Biol 4:1999, 1984.
21. Nichols EJ, Manger R, Hakomori S, Herscovics A, Rohrschneider LR: Mol Cell Biol 5:3467, 1985.
22. Waterfield MD, Scrace GT, Whittle N, Stroobant P, Johnsson A, Wasteson A, Westermark B, Heldin C-H, Huang JS, Deuel TF: Nature 304:35, 1983.
23. Doolittle RF, Hunkapiller MW, Hood LE, Devare SG, Robbins KC, Aaronson SA, Antoniades HN: Science 221:275, 1983.
24. Downward J, Yarden Y, Mayes E, Scrace G, Totty N, Stockwell P, Ullrich A, Schlessinger J, Waterfield MD: Nature 307:521, 1984.
25. Rettenmier CW, Chen JH, Roussel MF, Sherr CJ: Science 228:320, 1985.
26. Sherr CJ, Rettenmier CW, Sacca R, Roussel MF, Look AT, Stanley ER: Cell 41:665, 1985.
27. Woolford J, Rothwell V, Rohrschneider L: Mol Cell Biol 5:3458, 1985.
28. Rettenmier CW, Sacca R, Furman WL, Roussel MF, Holt JT, Nienhuis AW, Stanley ER, Sherr CJ: J Clin Invest 77:1740, 1986.

29. Stanley ER, Guilbert LJ, Tushinski RJ, Bartelmez SH: J Cell Biochem 21:151, 1983.
30. Sacca R, Stanley ER, Sherr CJ, Rettenmier CW: Proc Natl Acad Sci USA 83:3331, 1986.
31. Coussens L, Van Beveren C, Smith D, Chen E, Mitchell RL, Isacke CM, Verma IM, Ullrich A: Nature 320:277, 1986.
32. Groffen J, Heisterkamp N, Spurr N, Dana S, Wasmuth JJ, Stephenson JR: Nucleic Acids Res 11:6331, 1983.
33. Roussel MF, Sherr CJ, Barker PE, Ruddle FH: J Virol 48:770, 1983.
34. Van den Berghe H, Cassiman J-J, David G, Fryns J-P, Michaux J-L, Sokal G: Nature 251:437, 1974.
35. Van den Berghe H, David G, Michaux J-L, Sokal G, Verwilghen R: Blood 48:624, 1976.
36. Rowley JD, Golomb HM, Vardiman JW: Blood 58:759, 1981.
37. Wisniewski LP, Hirschhorn K: Am J Hematol 15:295, 1983.
38. Nienhuis AW, Bunn HF, Turner PH, Gopal TV, Nash WG, O'Brien SJ, Sherr CJ: Cell 42:421, 1985.
39. Le Beau MM, Pattenati MJ, Lemons RS, Diaz MO, Westbrook CA, Larson RA, Sherr CJ, Rowley JD: Cold Spring Harbor Symp Quant Biol 51:(in press).
40. Huebner K, Isobe M, Croce CM, Golde DW, Kaufman SE, Gasson JC: Science 230:1282, 1985.
41. Le Beau MM, Westbrook CA, Diaz MO, Larson RA, Rowley JD, Gasson JC, Golde DW, Sherr CJ: Science 231:984, 1986.
42. Müller R, Tremblay JM, Adamson ED, Verma IM: Nature 304:454, 1983.
43. Berridge MJ: Biochem J 220:345, 1984.
44. Fain JN: Vitam Horm 41:117, 1984.
45. Hokin LE: Annu Rev Biochem 54:205, 1985.
46. Rozengurt E, Legg A, Pettican P: Proc Natl Acad Sci USA 76:1284, 1979.
47. Dicker P, Rozengurt E: Nature 287:607, 1980.
48. Glenn KC, Carney DH, Fenton JW, Cunningham DD: J Biol Chem 255:6609, 1980.
49. Habenicht AJR, Glomset JA, King WC, Nist C, Mitchell CD, Ross R: J Biol Chem 256:12329, 1981.
50. Creba JA, Downes CP, Hawkins PT, Brewster G, Michell RH, Kirk CJ: Biochem J 212:733, 1983.
51. Berridge MJ, Heslop JP, Irvine RF, Brown KD: Biochem J 222:195, 1984.
52. Carney DH, Scott DL, Gordon EA, LaBelle EF: Cell 42:479, 1985.
53. Chu S-HW, Hoban CJ, Owen AJ, Geyer RP: J Cell Physiol 124:391, 1985.
54. Farrar WL, Anderson WB: Nature 315:233, 1985.
55. Farrar WL, Thomas TP, Anderson WB: Nature 315:235, 1985.
56. Kaibuchi K, Tsuda T, Kikuchi A, Tanimoto T, Yamashita T, Takai Y: J Biol Chem 261:1187, 1986.
57. Tsuda T, Kaibuchi K, Kawahara Y, Fukuzaki H, Takai Y: FEBS Lett 191:205, 1985.
58. VanObberghen-Schilling E, Chambard JC, Paris S, L'Allemain G, Pouyssegur J: EMBO J 4:2927, 1985.
59. Marggraf WD, Diringer H, Koch MA, Anderer FA: Z Physiol Chem 353:1761, 1972.
60. Koch MA, Diringer H: Biochem Biophys Res Commun 55:305, 1973.
61. Diringer H, Friis RR: Cancer Res 37:2979, 1977.
62. Sugimoto Y, Whitman M, Cantley LC, Erikson RL: Proc Natl Acad Sci USA 81:2117, 1984.
63. Macara IG, Marinetti GV, Balduzzi PC: Proc Natl Acad Sci USA 81:2728, 1984.
64. Fry MJ, Gebhardt A, Parker PJ, Foulkes JG: EMBO J 4:3173, 1985.
65. Kubota Y, Inoue H, Yoshioka T: Biochim Biophys Acta 875:1, 1986.
66. Fleischman LF, Chahwala SB, Cantley L: Science 231:407, 1986.
67. Jackowski S, Rettenmier CW, Sherr CJ, Rock CO: J Biol Chem 261:4978, 1986.
68. Reynolds FH, Van de Ven WJM, Blomberg J, Stephenson JR: J Virol 38:1084, 1981.
69. Pledger WJ, Stiles CD, Antoniades HN, Scher CD: Proc Natl Acad Sci USA 75:2839, 1978.
70. Stiles CD, Capone GT, Scher CD, Antoniades HN, Van Wyk JJ, Pledger WJ: Proc Natl Acad Sci USA 76:1279, 1979.
71. Besterman JM, Watson SP, Cuatrecasas P: J Biol Chem 261:723, 1986.
72. Westermark B, Heldin CH: J Cell Physiol 124:43, 1985.
73. Liboi E, Pelosi E, Testa U, Peschle C, Rossi GB: Mol Cell Biol 6:2275, 1986.
74. Tushinski RJ, Stanley ER: J Cell Physiol 122:221, 1985.
75. Yarden Y, Escobedo JA, Kuang WJ, Yang-Feng TL, Daniel TO, Tremble PM, Chen EY, Ando ME, Harkins RN, Francke U, Fried VA, Ullrich A, Williams LT: Nature 323:226, 1986.

# Mutations in the E1a Gene of Type 5 Adenovirus Result in Oncogenic Transformation of Fischer Rat Embryo Cells

Gregory J. Duigou, Lee E. Babiss, Wen-Shing Liaw, Stephen G. Zimmer, Harold S. Ginsberg, and Paul B. Fisher

*Department of Microbiology (G.J.D., W-S.L., H.S.G., P.B.F.), and Department of Urology (G.J.D., P.B.F.), Comprehensive Cancer Center Columbia University, College of Physicians and Surgeons, New York, New York 10032, Rockefeller University, New York, New York 10021 (L.E.B.), Department of Medical Microbiology and Immunology, University of Kentucky Medical Center, Lexington, Kentucky 40536 (S.G.Z.)*

Transformation of a specific clone of Fischer rat embryo (CREF) cells with wild-type 5 adenovirus (Ad5) or the E1a plus E1b transforming gene regions of Ad5 results in epithelioid transformants that grow efficiently in agar but that do not induce tumors when inoculated into nude mice or syngeneic Fischer rats. In contrast, CREF cells transformed by a host-range Ad5 mutant, H5hr1, which contains a single base-pair deletion of nucleotide 1055 in E1a resulting in a 28-kd protein (calculated) in place of the wild-type 51-kd acidic protein, display a cold-sensitive transformation phenotype and an incomplete fibroblastic morphology but surprisingly do induce tumors in nude mice and syngeneic rats. Tumors develop in both types of animals following injection of CREF cells transformed by other cold-sensitive Ad5 E1a mutants (H5dl101 and H5in106), which contain alterations in their 13S mRNA and consequently truncated 289AA proteins. CREF cells transformed with only the E1a gene (0–4.5 m.u.) from H5hr1 or H5dl101 also produce tumors in these animals. To directly determine the role of the 13S E1a encoded 289AA protein and the 12S E1a encoded 243AA protein in initiating an oncogenic phenotype in adenovirus-transformed CREF cells, we generated transformed cell lines following infection with the Ad2 mutant pm975, which synthesizes the 289AA E1a protein but not the 243AA protein, and the Ad5 mutant H5dl520 and the Ad2 mutant H2dl1500, which do not produce the 289AA E1a protein but synthesize the normal 243AA E1a protein. All three types of mutant adenovirus-transformed CREF cells induced tumors in nude mice and syngeneic rats. Tumor formation by these mutant adenovirus-transformed CREF cells was not associated with changes in the arrangement of integrated adenovirus DNA or in the expression of adenovirus early genes. These results indicate, therefore, that oncogenic transformation of CREF cells can occur in the presence of a wild-type

Received May 20, 1986; accepted July 10, 1986.

© 1987 Alan R. Liss, Inc.

13S E1a protein or a wild-type 12S E1a protein when either protein is present alone, but does not occur when both wild-type E1a proteins are present.

**Key words: integration of Ad5 DNA, primary transcription of Ad5 genes, CREF cells, cell transformation, Ad5 mutant genes, tumorigenicity, mutated Ad5 E1a gene**

Although it is well established that viral and cellular oncogenes can independently or cooperatively induce cell transformation in vitro and that certain transformed cell populations can induce tumors when injected into susceptible animals, the underlying molecular mechanism(s) by which different transforming genes modulate the cellular phenotype have not been defined [1,2]. In the case of type 2 or type 5 adenovirus, complete morphological transformation of both primary and established cell cultures requires the expression of the early region 1a (E1a) and 1b (E1b) viral transcriptional units [3–11]. When a wild-type E1a gene is transferred into primary baby rat kidney cells it can immortalize these cells [12–14], but the cells acquire an incomplete transformed phenotype; ie, cells display a fibroblastic as opposed to an epithelioid morphology, and transformed cells are nontumorigenic [12]. In contrast, when primary baby rat kidney cells are cotransfected with a combination of E1a and the Ha-ras (T24) oncogene, complete oncogenic transformation results [13,14]. The ability of E1a to immortalize cells in culture and to cooperate with the Ha-ras oncogene in inducing oncogenic transformation is not unique to E1a since this function has been found with several other genes, including v-myc [15], c-myc [15], n-myc [16,17], polyoma large T-antigen [13,14], and the cellular tumor antigen p53 [18–20]. These studies have led to the concept that transformation of cultured primary cells requires at least two cooperating gene functions, ie, those responsible for immortalization and those responsible for transformation [14,15,21,22].

Recent studies have demonstrated that the biological phenotype induced by a specific oncogene depends not only on the level of its expression [23] or the expression of other cellular genes [24] but also on the target cell in which it is inserted [14,25,26]. For example, although a mutated or a normal Ha-ras gene transcriptionally activated by a long terminal repeat does not induce morphological transformation of primary rodent cells, these genes can immortalize primary cells [23]. Similarly, when c-myc or p53-cDNA genes are linked to strong viral promoters, they can induce oncogenic transformation of cells without dramatically altering cellular morphology [18,25,27]. In the case of adenovirus E1a, different effects have been observed depending on the target cell, including the induction of morphological transformation [7,9,11,28], cooperativity with Ha-ras in inducing transformation [13,14,29] (Duigou, Liaw, Babiss, and Fisher, unpublished data), and tumorigenic conversion [25]. In the established REF52 rat cell line, the combination of E1a and T24 Ha-ras genes was required to convert these cells to a tumorigenic phenotype [26], while in the 3T3 cell line a wild-type Ad5 E1a gene was sufficient to induce oncogenic conversion of these transformed cells [25]. In both REF52 and 3T3 cells insertion of E1a was not associated with dramatic alterations in cellular morphology, whereas in CREF [9] and the established 3Y1 [7] rat cell lines E1a induced morphologically transformed foci, but clones generated from these foci were not tumorigenic [26]. In contrast, when the Ad5 E1a 289 or 243AA protein is expressed alone in 3T3 cells [28] or when the levels of Ad2 E1a proteins are increased by substituting the strong promoter from the mouse metallothionein gene in place of the Ad2 E1a promoter [30], morphological changes are induced in NIH 3T3 cells. These studies demonstrate that the E1a gene of Ad5 can do more than just immortalize cells, and they emphasize the importance of defining appropriate target cells and endpoints for analyzing the immortalization and/or transformation (morphological and/or oncogenic) potentials of specific genes or sets of genes.

In order to facilitate analysis of the mechanism by which adenovirus genes induce transformation, we have isolated and characterized a specific clone of Fischer rat embryo cells, called CREF [31,32], in which complete adenovirus virions [8–11,26,31–33] as well as transfected adenovirus transforming genes transform at high frequencies [9,11,34]. The CREF cell line has now been utilized for adenovirus transformation studies in several laboratories and has been shown to be valuable in defining the functions of specific areas of the E1a and E1b genes of adenovirus in regulating both the initiation and the maintenance of the transformed phenotype [10,11,33,35,36]. The CREF cell line, although immortal, exhibits many of the properties of normal rat embryo cells, including a high degree of contact inhibition, an inability to form macroscopic colonies when grown in agar suspension, and the inability to form tumors when inoculated into nude mice or syngeneic Fischer rats [26,32]. In the present study, we investigated the role of the 13S E1a encoded 289AA protein and the 12S E1a encoded 243AA protein in regulating the expression of the tumorigenic phenotype in transformed CREF cells.

## MATERIALS AND METHODS
### Cell Cultures

CREF cells [31,32] and the transformed CREF clones described in Table I were grown in Dulbecco's modified Eagle's medium (DMEM) containing 5% fetal bovine serum. The isolation and subsequent cloning of the cell lines used in these studies (Table I) have been described previously [8,9,26,32]. The Ad2 and Ad5 mutants and the plasmids containing various Ad5 transforming genes used in these studies have been described previously [8–11].

### Tumor Induction in Nude Mice and Syngeneic Fischer Rats

The ability of cloned Ad5-transformed CREF cells to form tumors in 4-week-old Balb/c nude mice and 6–9-week-old immunocompetent Fischer 344 rats was determined by injecting $2 \times 10^6$ cells per animal. Monolayer cultures of tumor cells were established by trypsinization of excised tumors, and these are referred to as nmt (nude mouse tumor) or imm (immunocompetent rat tumor) cells.

### DNA Analysis

High molecular weight DNA was isolated from the adenovirus-transformed CREF cell lines as previously described [37]. The presence of viral sequences within each cellular genome was determined by DNA filter hybridization analysis as previously described [31,38–40].

### RNA Analysis

The in vitro nuclear run-off assay described by Hofer and Darnell [41] and Weber et al [42] was used to label nuclei from adenovirus-transformed CREF cells. The transcription assay utilized approximately $10^7$ nuclei, which were incubated in the presence of 200 $\mu$Ci of $[^{32}P]$-UTP (3,000 $\mu$Ci/mmol) for 15 min at 30°C. Nuclear

**TABLE I. Tumorigenic Properties of Wild-Type and Mutant Type 5 Adenovirus-Transformed CREF Cells**

| Cell line | Transforming agent[a] | Tumorigenicity[b] | |
|---|---|---|---|
| | | Nude mice | Fischer rats |
| CREF[d] | None | 0/9 | 0/6 |
| wt3A[d] | H5wt | 1/9 (90)[c] | 0/6 |
| c2[d] | H5wt 0–15.5 m.u. (E1a +E1b) | 0/3 | 0/3 |
| d2[d] | H5wt 0–15.5 m.u. (E1a + E1b) | 0/6 | 0/6 |
| hr1A2[d] | H5hr1 | 6/6 (18) | 6/6 (42) |
| g8t[d] | H5hr1 0–15.5 m.u. (E1a + E1b) | 9/9 (44) | 0/6 |
| S2 | H5dl101 0–15.5 m.u. (E1a + E1b) | 3/3 (24) | 2/2 (91) |
| d3t[d] | H5hr1 0–4.5 m.u. (E1a) | 6/6 (14) | 3/3 (95) |
| 01[d] | H5dl101 0–4.5 m.u. (E1a) | 6/6 (27) | 6/6 (157) |
| 975-6 | pm975 (No E1a 12S) | 2/3 (78) | 0/3 |
| 975-8 | pm975 (No E1a 12S) | 3/3 (30) | 2/2 (113) |
| 520-1 | H5dl520 (No E1a 13S) | 3/3 (34) | 0/3 |
| 520-2 | H5dl520 (No E1a 13S) | 1/3 (39) | 0/3 |
| 520-4 | H5dl520 (No E1a 13S) | 3/3 (44) | 3/3 (105) |
| 1500-2 | H2dl1500 (No E1a 13S) | 3/3 (76) | 2/2 (87) |
| 1500-6 | H2dl1500 (No E1a 13S) | 3/3 (25) | 2/2 (128) |
| 1500-10 | H2dl1500 (No E1a 13S) | 3/3 (11) | 2/2 (71) |
| 1500-11 | H2dl1500 (No E1a 13S) | 3/3 (19) | 2/2 (108) |

[a]H5wt: wild-type 5 adenovirus; H5wt 0–15.5: transfected CREF clone containing E1a and E1b (0–15.5 m.u.) of H5wt; H5hr1: host range mutant of Ad5 (contains a 1bp deletion of nucleotide 1055); H5hr1 0–15.5: transfected CREF clone containing E1a and E1b region of H5hr1; H5dl101 0–15.5: transfected CREF clone containing E1a and E1b region from H5dl101 virus (contains one 5bp deletion of nucleotides 1008–1012); H5hr1 0–4.5: transfected CREF clone containing E1a region of H5hr1; H5dl101 0–4.5: transfected CREF clone containing E1a region of H5dl101; pm975: host range mutant of Ad2 that lacks 12S E1a mRNA and the 243AA protein encoded by this mRNA; H5dl520: host range mutant of Ad5 that lacks a 13S E1a mRNA and the 289AA protein encoded by this mRNA; H2dl1500: host range mutant of Ad2 that lacks a 13S E1a mRNA and the 289AA protein encoded by this mRNA.
[b]Four-week-old Balb/C nude mice or 6–8-week-old Fischer 344 rats were injected with $2 \times 10^6$ cells. Results indicate No. of animals with tumors/ No. of animals injected, and bracketed No. indicates the average latency time in days, ie, first appearance of a palpable tumor. Nude mice were observed for tumor formation for $\geq 180$ days, and syngeneic rats were observed for tumor formation for $\geq 270$ days.
[c]This tumor developed late, was very small, and with further time it regressed.
[d]Tumorgenicity studies on these cell lines have been reprinted in [26].

RNA was isolated by the guanidinium isothiocyanate method of Ulrich et al [43], and prior to hybridization it was cleaved by treatment with 0.2 N NaOH for 15 min on ice [44]. The preparation of filters containing "dots" of denatured DNA was as described by Kafatos et al [45].

## RESULTS

CREF cells transformed by complete Ad5 virions or the E1a and E1b genes from wild-type Ad5 failed to produce tumors when injected into either nude mice or syngeneic Fischer rats [26] (Table I). In contrast, CREF cells transformed by the host-range cold-sensitive mutant H5hr1 [8,46–49], or by similar mutants such as

H5dl101 and H5in106 [10], induced tumors in both nude mice and syngeneic rats [26] (Table I). These data suggest that the ability of these mutant-virus-transformed CREF cells to induce tumors in animals results from the defined mutation in the unique region of the E1a gene, which results in an altered 13S mRNA and consequently a truncated E1a 51-kd acidic polypeptide [47]. These results also indicate that tumorigenic transformation of CREF cells does not require E1b genetic information, since CREF cells transformed by only the E1a gene from H5hr1 or H5dl101 form tumors in both types of animals [26] (Table I).

To determine directly the role of the two E1a proteins, the 289AA (encoded by the 13S E1a mRNA) and the 243AA (encoded by the 12S E1a mRNA) protein, in regulating expression of the tumorigenic phenotype of CREF cells we have utilized the Ad2 mutant pm975, which produces only the 289AA E1a protein [10] and the Ad5 mutant H5dl520 [11] and the Ad2 mutant H2dl1500 [10], which produce only the 243AA E1a protein. Specific foci of CREF cells transformed by pm975 exhibited an epithelioid morphology similar to wild-type Ad5-transformed CREF cells but induced tumors in both nude mice and syngeneic Fischer rats (Table I). CREF cells transformed by H5dl520 and H2dl1500 displayed a cold-sensitive transformation phenotype, as did H5hr1-, H5dl101-, and H5in106-transformed CREF cells [8,9,48], and were tumorigenic in both nude mice and syngeneic rats (Table I). In the case of H5dl520-transformed CREF cells, cell cultures established from two different transformed foci, 520-1 and 520-2, induced tumors in nude mice (100% tumor incidence for 520-1 and 33% tumor incidence for 520-2) but did not form tumors in syngeneic rats. In contrast, the 520-4 transformed cell line produced tumors (100% of animals) in both types of animals. A different situation was observed in H2dl1500-transformed CREF cells in which all four transformed lines produced tumors in 100% of the nude mice and Fischer rats (Table I). Differences in latency time were observed in the formation of tumors in nude mice (11–44-day averages) and syngeneic rats (71–128-day averages) inoculated with the two types of mutant-transformed CREF cells. No direct correlation was observed between the latency period for tumor formation in nude mice and either the ability to form tumors or the latency time for tumor development in Fischer rats; eg, 520-1 formed tumors in three of three injected nude mice with an average latency period of 34 days but did not form tumors in rats, whereas 1500-2 formed tumors in three of three injected nude mice with an average latency period of 76 days and formed tumors in two of two injected rats with an average latency period of 87 days.

The process of tumor formation in both nude mice and syngeneic Fischer rats by the H5hr1-transformed CREF clone A2 was not associated with major alterations in the pattern of integration of viral DNA sequences, the number of integrated viral genomes, or the transcription rate of early viral gene regions, including E1a, E1b, E2a, E3, or E4 [26]. Similarly, the induction of tumors in nude mice and syngeneic rats by pm975-, H5dl520-, and H2dl1500-transformed cells also did not result in the rearrangement of integrated adenovirus DNA sequences (Fig. 1 and 2, and unpublished data). Subclones of H2dl1500-transformed CREF cells derived from the same original transformed focus; ie, 10-1 and 10-2, and 11-1, 11-2, 11-3, and 11-4, displayed similar patterns of DNA integration but differed in their tumorigenic potential (unpublished data).

To determine if tumor formation by H2dl1500 was associated with an alteration in the expression of viral genes other than E1a and E1b, the rate of transcription of

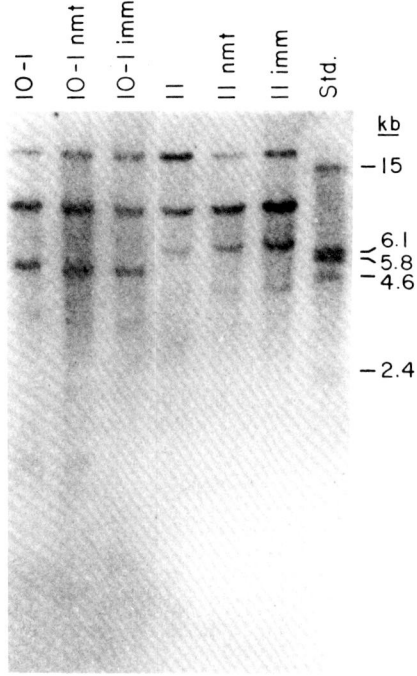

Fig. 1. Analysis of viral DNA in mutant adenovirus-transformed CREF cell lines. Cellular DNA (10 μg) was digested with XbaI, size-fractionated through 0.6% agarose, transferred to nitrocellulose filters, and probed using Ad2 DNA labeled with $^{32}$P by nick-translation. 10-1 represents DNA of a cloned cell line isolated from an expanded focus of H2dl1500-transformed CREF cells. 11 contains DNA of a cell line derived from a focus of H2dl1500-transformed CREF cells. nmt and imm denote cell lines derived from tumors produced after injection of the different transformants into nude mice (nmt) or immunocompetent Fischer rats (imm). Std. contains $10^{-5}$ μg of XhoI-digested Ad5 DNA and is included as a size standard.

early viral genes was analyzed by labeling nascent RNAs from isolated nuclei with [$^{32}$P]-UTP and hybridizing to appropriate dots of DNA immobilized on nitrocellulose filters [26,41–45]. This procedure is useful for determining differential rates of RNA transcription initiation [41] since chain initiation in isolated nuclei is an inefficient process, whereas chain elongation of previously initiated RNA by RNA polymerase II occurs faithfully. By applying similar amounts of total labeled RNA (isolated from approximately equivalent numbers of cell nuclei) from each nuclear sample, the hybridization assays permitted a determination of the rate of transcription of the defined genes in each cell type. As previously found with the H5hr1-transformed A2 CREF clone, tumor formation by H2dl1500-transformed CREF cells followed by reestablishment in culture did not result in dramatic alterations in the rate of transcription of early adenovirus genes (Fig. 3). The two- to threefold decrease in the transcription rate of the E1a and E1b genes observed for the clone 11 and 11-imm cells compared to 11-nmt cells was not observed in another experiment. This conclusion is further supported by the increased hybridization signal for actin in the 11-nmt cell line.

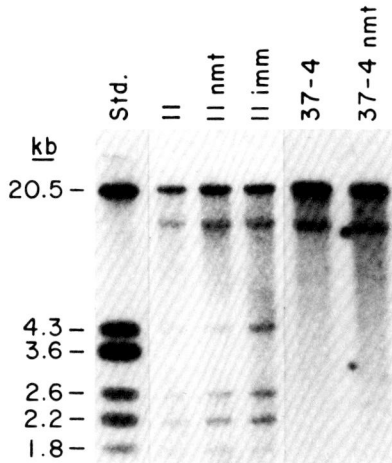

Fig. 2. Analysis of viral DNA in E1a 13S minus Ad2- and Ad5-transformed CREF cell lines. Cellular DNA (10 μg) was digested with EcoRI, size-fractionated through 0.6% agarose, transferred to nitrocellulose filters, and analyzed using $^{32}$P-labeled Ad2 DNA. 11 contains DNA of a cell line derived from a focus of H2dl1500-transformed CREF cells. 37-4 represents DNA from a cell line expanded from a focus of H5dl520-transformed CREF cells. nmt and imm denote DNA from cell lines derived from tumors produced after injection of transformed CREF cells in nude mice (nmt) or immunocompetent Fischer rats (imm). Std. contains $10^{-5}$ μg of EcoRI-digested Ad2 DNA and is included as a size standard.

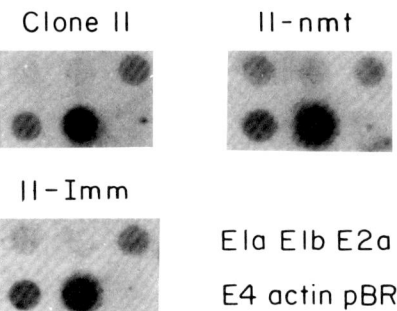

Fig. 3. Dot hybridization of in vitro labeled RNA isolated from cell lines clone 11, 11-nmt, and 11-imm. Nuclear RNA was labeled in vitro with $^{32}$P-UTP as described in the text, and RNA containing $4 \times 10^6$ cpm was hybridized to a dot containing 7 μg of E1a (0–4.5 m.u.), E1b (7.9–9.6 m.u.), E2a (60.1–63 m.u.), E4 (93–100 m.u.), chicken actin cDNA, and pBR322 (pBR) DNA.

## DISCUSSION

This study demonstrates that the 243AA protein encoded by the Ad5 E1a 12S mRNA, when present alone or in combination with a truncated form of the 289AA protein encoded by a mutated E1a-derived 13S mRNA, in concert with or without other Ad5 genes is capable of eliciting a tumorigenic phenotype in CREF cells. In addition, tumorigenic CREF cells were also generated following transformation by the Ad2 mutant virus pm975 [10], which expresses the E1a 289AA protein but not the E1a 243AA protein. Tumorigenicity was not associated with major changes in the integration profile of Ad5 DNA sequences or in the transcription rates of several early viral and cellular genes. In contrast, when both wild-type E1a proteins, alone or in combination with E1b and other Ad5 gene products, are expressed in CREF cells, transformed clones do not produce tumors in either nude mice or syngeneic Fischer rats.

The mechanism is not known by which both wild-type E1a proteins repress the oncogenic phenotype in CREF cells, while alterations in the types of E1a protein(s) in CREF cells result in the tumorigenic conversion of these cells. Possible insights into this mechanism come from the observations that the E1a proteins exert pleiotropic responses in cells [14,28,47,50-53] and the 13S and 12S gene products, when present separately, may differ in their ability to elicit specific responses in target cells [52,53,57-62]. For example, it has been suggested that the 289AA E1a protein is more effective than the 243AA protein in transcriptionally activating other early Ad5 genes [47,50-53], such as E2, and nonviral genes [54-58], and that it is more active in inducing cell cycle progression [60]. In contrast, both the 13S and 12S gene products are similarly effective in immortalizing rodent cells [14], cooperating with the Ha-ras oncogene in inducing transformation in primary rodent cells [14], stimulating cellular DNA synthesis [10], and transcriptionally activating other adenovirus genes when microinjected as purified proteins into cells [61]. The diversity of E1a effects in cells is emphasized by the observations by means of DNA transfection techniques that E1a proteins can also inhibit transcription from the Ad2 E1a, SV4O, and polyomavirus enhancers [62,63], and the E2A late promoter [64]. The 13S and 12S gene products are similarly effective in suppressing transcription from the SV40 early promoter [63]. No studies have yet addressed the question of regulation of the expression of the 289AA and 243AA proteins when the genes coding for these proteins are expressed singly by transformed cells.

Recent studies by Schrier et al [65] and Bernards et al [66] demonstrate that 1) Ad12-transformed baby rat kidney cells lack at least two cellular proteins that are expressed in nononcogenic Ad5-transformed cells and in untransformed cells, and one of these proteins is encoded by the rat class I major histocompatibility complex; and 2) the product of the 13S mRNA of Ad12 may allow cells to escape T-cell immunity and may therefore be the mechanism by which these transformed cells induce tumors in syngeneic animals. Similarly, the importance of natural killer cells and macrophages in determining the tumorigenic potential of various DNA-virus transformed hamster and rat cells has been emphasized [67-72]. Based on these studies, a possible mechanism by which various CREF transformants that contain mutated E1a genes induce tumors in animals may involve the suppression of specific cellular surface antigens that mediate tumor cell rejection. Since transformed cells expressing both E1a proteins fail to induce tumors, the present results suggest that the combination of wild-type Ad5 E1a proteins may be unable to suppress the cellular genes effecting tumor rejection, whereas expression or overexpression [30] of either the 289AA or the 243AA protein alone may result in changes in cellular gene expression resulting in cell surface alterations mediating tumor formation. Support for this hypothesis comes from studies indicating that CREF cells transformed by the Ad2 mutant pm975 [10], which produces the 13S E1a-encoded 289AA protein and not the 12S E1a-encoded 243AA protein, as well as CREF cells transformed by the Ad2 mutant H2dl1500 [10] or the Ad5 mutant H5dl520 [11], which produce the 12S E1a-encoded 243AA protein and not the 13S E1a-encoded 289AA protein, are capable of inducing tumors in both nude mice and syngeneic rats (Table I). The present challenge, therefore, is to determine the nature of the cellular alterations induced by various mutations in the E1a gene of Ad5 and Ad2 that result in expression of the tumorigenic phenotype by CREF cells.

## ACKNOWLEDGMENTS

This work was supported by National Cancer Institute grants CA-35675 (P.B.F.), CA-33434 (S.G.Z.), and AI-12052 (H.S.G.). Dr. Lee E. Babiss was supported by a Damon Runyon-Walter Winchell Postdoctoral Fellowship (DRG-794). We thank A.J. Berk for providing the viral mutants pm975 and H2dl1500 and N.C. Jones for the Ad5 mutant H5dl520. We also thank Ms. Ellen Reynolds for assistance in the preparation of this manuscript and Mrs. C. Bruser for her technical expertise in the transformed cell tumorigenicity studies.

## REFERENCES

1. Bishop JM: Annu Rev Biochem 52:310, 1983.
2. Bishop JM: Cell 42:23, 1985.
3. Gallimore PH, Sharp PA, Sambrook J: J Mol Biol 89:49, 1974.
4. Graham FL, Abrahams PJ, Mulder C, Heijneker HL, Warnaar SO, deVries FAJ, Fiers W, van der Eb AJ: Cold Spring Harbor Symp Quant Biol 39:637, 1974.
5. van der Eb AJ, Mulder C, Graham FL, Houweling A: Gene 2:115, 1977.
6. Jones N, Shenk T: Cell 17:683, 1979.
7. Shiroki K, Maruyama K, Saito I, Fukui Y, Shimojo H: Virology 138:1048, 1981.
8. Babiss LE, Ginsberg HS, Fisher PB: Proc Natl Acad Sci USA 80:1352, 1983.
9. Babiss LE, Fisher PB, Ginsberg HS: J Virol 49:731, 1984.
10. Montell C, Courtois G, Eng C, Berk AJ: Cell 36:951, 1984.
11. Haley KP, Overhauser J, Babiss LE, Ginsberg HS, Jones NC: Proc Natl Acad Sci USA 81:5734, 1984.
12. Houweling A, van der Elsen PJ, van der Eb AJ: Virology 105:537, 1980.
13. Ruley HE: Nature 304:602, 1983.
14. Zerler B, Moran B, Maruyama K, Moomaw J, Grodzicker T, Ruley HE: Mol Cell Biol 6:887, 1986.
15. Land H, Parada L, Weinberg RA: Nature 304:596, 1983.
16. Schwab M, Varmus HE, Bishop JM: Nature 316:160, 1985.
17. Yancopoulos GD, Nisen PD, Tesfaye A, Kohl NE, Goldfarb MP, Alt FW: Proc Natl Acad Sci USA 82:5455, 1985.
18. Eliyahu D, Raz A, Gruss P, Givol D, Oren M: Nature 312:646, 1984.
19. Parada L, Land H, Weinberg RA, Wolf D, Rotter V: Nature 312:649, 1984.
20. Jenkins JR, Rudge K, Currie G: Nature 312:651, 1984.
21. Land H, Parada LF, Weinberg RA: Science 222:771, 1983.
22. Newbold RF, Overell RW: Nature 304:648, 1983.
23. Spandidos DA, Wilkie NM: Nature 310:469, 1984.
24. Babiss LE, Guernsey DL, Fisher PB: Cancer Res 45:6017, 1985.
25. Kelekar A, Cole MD: Mol Cell Biol 6:7, 1986.
26. Babiss LE, Liaw WS, Zimmer SG, Godman GC, Ginsberg HS, Fisher PB: Proc Natl Acad Sci USA 83:2167, 1986.
27. Keath EJ, Caimi PG, Cole MD: Cell 39:339, 1984.
28. Roberts BE, Miller JS, Kimelman D, Cepko CL, Lemischka IR, Mulligan RC: J Virol 56:404, 1985.
29. Franza BR, Jr, Maruyama K, Garrels JI, Ruley HE: Cell 44:409, 1986.
30. Senear AW, Lewis JB: Mol Cell Biol 6:1253, 1986.
31. Fisher PB, Mufson RA, Weinstein IB, Little JB: Carcinogenesis 2:183, 1981.
32. Fisher PB, Babiss LE, Weinstein IB, Ginsberg HS: Proc Natl Acad Sci USA 79:3527, 1982.

33. Carter TH, Nicolas JC, Young CSH, Fisher PB: Virology 117:519, 1982.
34. Babiss LE, Fisher PB, Ginsberg HS: J Virol 52:389, 1984.
35. Winsberg G, Shenk T: EMBO J 3:1907, 1984.
36. Hurwitz DR, Chinnadurai G: Proc Natl Acad Sci USA 82:163, 1985.
37. Dorsch-Hasler K, Fisher PB, Weinstein IB, Ginsberg HS: J Virol 34:305, 1980.
38. Rigby PWJ, Dieckmann M, Rhodes C, Berg P: J Mol Biol 113:237, 1977.
39. Southern EM: J Mol Biol 98:503, 1975.
40. Wahl GM, Stern M, Stark G: Proc Natl Acad Sci USA 76:3683, 1979.
41. Hofer E, Darnell JE, Jr: Cell 23:585, 1981.
42. Weber J, Jelinek W, Darnell JE, Jr: Cell 10:611, 1977.
43. Ulrich A, Shine H, Chirgwin J, Picket R, Tischer E, Rutter WJ, Goodman HM: Science 196:1313, 1977.
44. Nevins JR: Methods Enzymol 65:768, 1980.
45. Kafatos FC, Jones CW, Efstratiadis A: Nucleic Acids Res 7:1541, 1979.
46. Graham FL, Harrison TJ, Williams JF: Virology 86:10, 1978.
47. Ricciardi RP, Jones RL, Cepko CL, Sharp PA, Roberts BE: Proc Natl Acad Sci USA 78:6121, 1981.
48. Ho YS, Galos R, Williams J: Virology 122:109, 1982.
49. Hermo H Jr, Babiss LE, Liaw WS, Pinto IM, McDonald RJ, Fisher PB: In Rein R (ed): "Molecular Basis of Cancer: Part A: Macromolecular Structure, Carcinogens and Oncogenes." New York: Alan R Liss, Inc., 1985, pp 489–512.
50. Berk AJ, Lee F, Harrison T, Williams J, Sharp PA: Cell 17:935, 1979.
51. Jones N, Shenk T: Proc Natl Acad Sci USA 76:3665, 1979.
52. Carlock LR, Jones NC: J Virol 40:657, 1981.
53. Montell C, Fisher EF, Caruthers MH, Berk AJ: Nature 295:380, 1982.
54. Treisman R, Green MR, Maniatis T: Proc Natl Acad Sci USA 80:7428, 1983.
55. Kao H-T, Nevins JR: Mol Cell Biol 3:2058, 1983.
56. Green MR, Treisman R, Maniatis T: Cell 35:137, 1983.
57. Svensson C, Akusjarvi G: EMBO J 3:789, 1984.
58. Stein R, Ziff EB: Mol Cell Biol 4:2792, 1984.
59. Gaynor RB, Hillman D, Berk AJ: Proc Natl Acad Sci USA 81:1193, 1984.
60. Bellett AJD, Li P, David ET, Mackey EJ, Braithwaite AW, Cutt JR: Mol Cell Biol 5:1933, 1985.
61. Ferguson B, Krippl B, Andrisani O, Jones N, Westphal H, Rosenberg M: Mol Cell Biol 5:2653, 1985.
62. Borrelli E, Hen R, Chambon P: Nature 312:608, 1984.
63. Velcih A, Ziff E: Cell 40:705, 1985.
64. Rossini M: Virology 131:49, 1983.
65. Schrier PI, Bernards R, Vaessen RTMJ, Houweling A, van der Eb AJ: Nature 305:771, 1983.
66. Bernards R, Schrier PH, Houweling A, Bos JL, van der Eb AJ, Zijlstra M, Melief CJM: Nature 305:776, 1983.
67. Raska K, Jr, Dougherty J, Gallimore PH: Virology 117:530, 1982.
68. Raska K, Jr, Gallimore PH: Virology 123:8, 1982.
69. Cook JL, Hibbs JB, Lewis AM, Jr: Int J Cancer 30:795, 1982.
70. Sheil JM, Gallimore PH, Zimmer SG, Sopori ML: Immunology 132:1578, 1984.
71. Lewis AM, Jr, Cook JL: Science 227:15, 1985.
72. Sawada Y, Fohring B, Shenk TE, Raska K, Jr: Virology 147:413, 1985.

# Growth Factors, Oncogenes, and Multistage Carcinogenesis

### I. Bernard Weinstein

*Comprehensive Cancer Center and Institute of Cancer Research, Department of Medicine and School of Public Health, Columbia University, New York, New York 10032*

> This paper presents evidence that the full repertoire of cellular genes involved in the carcinogenic process is several times larger than that of the known list of proto-oncogenes. Furthermore, this repertoire includes genes whose normal function is related to growth stimulation, as well as genes whose normal function is to inhibit growth or induce terminal differentiation. Multistage carcinogenesis probably results from a complex series of changes in both categories of genes. Despite this complexity, carcinogenesis can be conceived in terms of disturbances in biochemical functions that normally control the expression or function of growth factors, receptors, and pathways of signal transduction. Several protein kinases play a central role in the process of signal transduction. Our laboratory has recently isolated cDNA clones for the enzyme protein kinase C (PKC). These clones should be useful for clarifying the role of PKC in growth control and tumor promotion. Finally, the existence of genes whose normal function is to inhibit cell growth provides a rationale for new strategies of cancer prevention and treatment.

**Key words:** protein kinases, protein kinase C, interferon, phorbol esters, tumor promotion, signal transduction, growth factors, oncogenes, multistage carcinogenesis

The development of tumors in humans and in a number of experimental animal models occurs through an extremely complex multistep process, which can occupy over one-half of the life span of the organism. These, and other characteristics of the carcinogenic process, predict that multiple cellular genes and multiple mechanisms are involved in the conversion of normal cells to fully malignant tumor cells [1,2]. It is obvious that chemical carcinogens, in contrast to viruses, cannot introduce new genetic information into target cells. They must, therefore, call upon and distort the function of normally present cellular genes. These considerations predicted, in a sense, the existence of "proto-oncogenes." Until recently, however, it was not clear how to identify the specific genes involved. Recent studies of the acute transforming retroviruses and DNA transfection procedures have now directly identified at least 30 proto-oncogenes [3,4] and have thus revolutionized our approach to the genetic basis of multistage carcinogenesis. The increasing evidence that these proto-oncogenes

Received July 15, 1986; accepted October 13, 1986.

© 1987 Alan R. Liss, Inc.

normally code for proteins that play a role in various stages of the action of growth factors [3-5] also provides an exciting unitary theme to the underlying physiologic mechanisms.

This symposium on "Growth Factors, Tumor Promoters, and Cancer Genes" will reveal recent advances in the areas of oncogene and growth factor research. In this introductory talk I would like to broaden the scope of the current paradigm by speculating about the number of growth factors and proto-oncogenes that normally exist, and how this diversity complicates our understanding of the carcinogenic process. I will then present recent findings from our laboratory on the cloning and sequence of the gene(s) for protein kinase C (PKC), an enzyme that plays a central role in signal transduction. Finally, I will consider the subject of genes and protein factors that inhibit growth, the relevance of negative control mechanisms to carcinogenesis, and the implications of negative control with respect to the design of new strategies of cancer therapy. The latter theme will, I trust, be of particular interest to those attending the symposium "Interferons as Cell Growth Inhibitors and Antitumor Factors."

## SIGNAL TRANSDUCTION

It is apparent that in a multicellular organism the behavior of individual cells must be highly coordinated with that of others. This coordination is accomplished, in part, by the endocrine system, ie, the transmission of specific hormones between tissues. It is now apparent that another level of coordination is accomplished within tissues by paracrine mechanisms, ie, mechanisms involving growth factors and differentiation factors that are synthesized within tissues and that act at short ranges on neighboring cells [5,6]. At least 20 of these factors have already been identified, and each year several new ones are discovered [5]. Most of these factors are polypeptides, although it seems likely that certain prostaglandins, and other arachidonic acid derivatives, play analogous roles [10]. Figure 1 displays in schematic form how some of these extracellular signal molecules are perceived by cellular receptors, which are often located at the cell surface, and how the occupancy of these receptors leads to a cascade of signal transduction through the cytoplasm and eventually into the nucleus, thus altering patterns of gene expression. Figure 1 also emphasizes the central role that a series of protein kinase enzymes plays in this process of signal transduction. A major theme that has emerged is that the proto-oncogenes represent a subset of genes that normally code for components in these pathways of signal transduction. Alterations in the structure and function of these proto-oncogenes can convert them to "activated" oncogenes, which cause aberrations in signal transduction and thus disrupt normal growth, differentiation, and intercellular coordination.

There is now evidence that with platelet-derived growth factor (PDGF), epidermal growth factor (EGF), insulin, and certain lymphokines, receptor occupancy leads to activation of a tyrosine kinase domain present in the cytoplasmic portion of the receptor [5,7]. Another mechanism of signal transduction is exemplified by the beta-adrenergic system in which occupancy of the receptor by the agonist leads to activation of the enzyme adenylate cyclase, which is coupled to the receptor through G regulatory proteins [8]. The resulting increase in cytoplasmic cAMP then activates protein kinase A, a serine and threonine kinase [9]. The role of this pathway in growth control is not clear at the present time. It is possible that certain prostaglandins

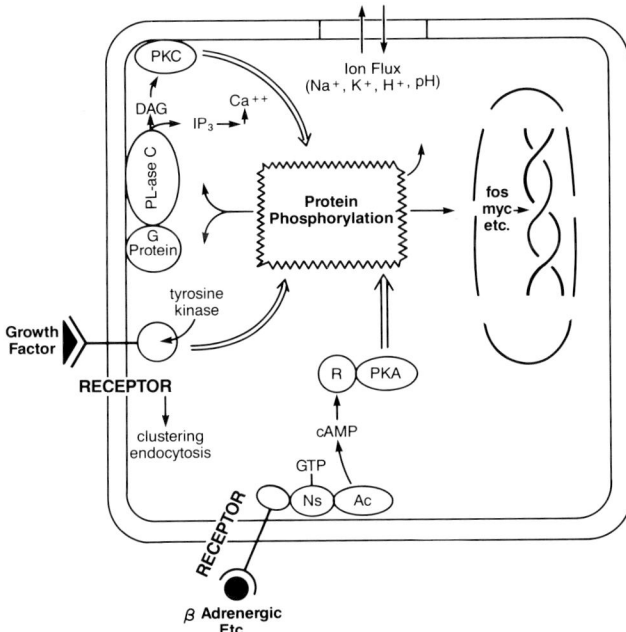

Fig. 1. A schematic diagram of a cell showing various pathways of membrane-associated receptors and signal transduction. The growth factor pathway applies to EGF, PDGF, and insulin. The beta-adrenergic pathway involves coupling via a G regulatory protein (Ns) to adenyl cyclase (Ac), and cyclic AMP (cAMP) binding to the regulatory subunit (R) of protein kinase A (PKA). Various agonists can activate phospholipase C (PL-ase C), presumably via a G protein, leading to the release of diacylglycerol (DAG) and inositol 1,4,5-triphosphate (IP$_3$). DAG activates protein kinase (PKC) and IP$_3$ causes the release of Ca$^{++}$ from the endoplasmic reticulum. These events lead to a cascade of protein phosphorylation that alters the functions of membrane-associated receptors, ion channels, and cytoplasmic proteins. Signals (undefined) also enter the nucleus to induce the expression of various genes including c-fos and c-myc (see also Table I).

may mediate their effects via adenylate cyclase-coupled receptors [10], but this requires further study. Since in some cell systems increases in cellular cAMP can induce reversion of the transformed phenotype [11,12], the adenylate cyclase pathway may exert negative regulation (ie, inhibition) of growth. I shall return to the theme of negative regulation later. A third pathway of signal transduction involves the turnover of phosphatidylinositol and the activation of a phospholipid and Ca$^{2+}$-dependent serine and threonine protein kinase, designated protein kinase C (PKC) (for review, see 14). It would appear that PKC plays a central role in a variety of membrane-related signal transduction events [14]. This is because several agonists lead to the activation of a phospholipase C activity that hydrolyzes phosphatidylinositol 4,5 diphosphate to diacylglycerol (DAG) and inositol 1,4,5 triphosphate (IP$_3$) [13]. DAG then activates PKC [14], and IP$_3$ binds to receptors present on the endoplasmic reticulum (ER), causing the release of Ca$^{2+}$ from storage sites in the ER [13,15]. The resulting increase in cytoplasmic Ca$^{2+}$ then activates several calmodulin-dependent enzymes (protein kinases, phosphatases, phosphodiesterases), and also produces effects on the cytoskeleton. The fact that the tumor promoter 12-0-tetradecanoyl phorbol-13-acetate (TPA), and related tumor promoters, apparently act in place of DAG, and thus usurp the function of PKC [14,16] provides a satisfying unity between the action of tumor promoters and the current conceptual framework of growth control.

The protein targets that become phosphorylated by the above-described protein kinases include receptors and membrane-associated ion channels [14]. Thus, there occurs a cascade of receptor-receptor cross-talk and a highly pleiotropic series of biochemical events. A major gap in our current knowledge is the mechanism by which signals are ultimately conveyed to the nucleus and how they act at the level of the chromatin. Rapid progress is, however, being made in elucidating the genes whose expression is increased in fibroblasts undergoing a mitogenic response to growth factors or tumor promoters [17]. This mitogenic program is described in Table I. Our laboratory has recently obtained cDNA clones for genes whose expression is induced or repressed by the tumor promoter TPA [18]. The gene that is repressed is, I think, of particular interest because it is very likely that the fine tuning of growth control is achieved through both increases and decreases in the expression of specific genes.

## SPECULATIONS ABOUT THE NUMBER OF PROTO-ONCOGENES

It is of importance to consider, within the above model of cellular control mechanisms, the number of proto-oncogenes that exist in higher organisms, since this may define the magnitude and complexity of our endeavor to understand the evolution of the cancer cell at the genetic level.

I will define a "proto-oncogene" as a gene that normally plays a major role in the control of growth and/or differentiation and thus a gene that has the capacity, when mutated or "activated," to become an "oncogene," ie, a gene that contributes to the abnormal behavior of tumor cells. The current number of known oncogenes totals about 30 [3,4]. These were originally discovered in the acute transforming retroviruses, or in rodent or human tumors by the DNA transfection procedure. These methods continue to reveal new oncogenes at a rate of a few per year, so it is clear that the full repertoire has not been revealed.

Based on theoretical considerations, I would predict that the genomes of higher organisms contain several hundred proto-oncogenes. Vertebrates contain at least thirty cell types (neurons, glia, hepatocytes, renal cells, bronchial cells, mammary cells, lymphocytes, etc) [19]. It is likely that the growth of each of these cell types is under the control of several growth factors, since we know, for example, that lymphocytes are controlled by at least three lymphokines [5]. It is also clear that several factors are required for the growth and maturation of myeloid progenitors [5,20]. Thus multi CSF (IL-3) stimulates proliferation of granulocyte and macrophage lineages, M-CSF stimulates cells committed to the macrophage lineage, and G-CSF stimulates cells

**TABLE I. The Mitogenic Program***

1. Altered ion flux and increased cellular pH
2. Increased transport of glucose and other nutrients
3. Increased turnover of phospholipids (activation of phospholipases $A_2$ and C)
4. Increased activity of tyrosine kinases, protein kinase C, and protein kinase G
5. Increased mRNA for *fos*, *myc*, *actin*, ornithine decarboxylase, MRP, and various cDNA clones

*Inducers of this program in specific cell types include TPA, EGF, PDGF, serum. For review see [17].

committed to the granulocyte lineage. Erythropoetin plays a role in erythroepoesis, and thrombopoetin plays a role in platelet formation. Thus, at least six distinct factors are involved in controlling the growth and maturation of myeloid elements. It seems likely, therefore, that multiple paracrine factors are also involved in controlling the growth and maturation of specific cellular elements in other tissues. Indeed, there is evidence for the existence of a series of glia-promoting polypeptide factors (GPFs), which differ in their specificities for oligodendrocytes and astrocytes [21].

It is not unreasonable, therefore, to assume that at least three growth factors are involved in controlling the growth and differentiation of each cell type. Each growth factor, in turn occupies a specific receptor, which might also be considered a proto-oncogene. For example, the proto-oncogene *fms* apparently encodes the CSF-1 receptor [20]. It is also reasonable to assume that the pathway of signal transduction through the cytoplasm, for each receptor, is mediated by at least three additional gene products. Finally, I assume that in the nucleus there exist at least three specific transcription factors that control the expression of the responsive gene(s). Taken together, this suggests that normal cells contain about 360 proto-oncogenes.

Other considerations also suggest that the list of proto-oncogenes is large. 1) Of the approximately 50,000 genes per mammalian cell it seems reasonable that nature has committed at least 1% of these genes (ie, 500) to the control of growth. 2) New growth factors and differentiation factors continue to be discovered. 3) There is increasing evidence that some of the known proto-oncogenes belong to families; the *myc* family includes *c-myc, N-myc,* and *L-myc* [22]; the EGF receptor gene has a homolog, the *c-neu* or *c-erbB-2* gene [23]; and the *ras* family includes *H-ras, K-ras, N-ras,* and others [4,24]. Studies with the *myc* family suggest that individual members display tissue-specific expression and tissue-specific activation during carcinogenesis [22]. If all of the proto-oncogenes belong to families of at least three per family, this alone would almost triple the number of known proto-oncogenes. 4) Genes that code for receptors or components of the signal transduction pathways for some of the more conventional hormones, such as the glucocorticoids, estrogens, androgens, progesterones, prolactin, and prostaglandins, can also be considered proto-oncogenes, since aberrations in these genes could contribute to the autonomous growth of tumors. In this context it is of interest that the viral oncogene *erb A* shares homology with steroid receptors [25].

Later I will discuss the evidence that there also exists a repertoire of growth inhibitor genes whose normal function is to inhibit growth and/or induce terminal differentiation. Mutations in these genes could also lead to disturbances in growth control. Furthermore, the mammalian genome contains thousands of copies of long terminal repeat (LTR)-like sequences, and alterations in the function or state of integration of these genes could also produce disturbances in growth control [2].

The evidence presented above suggesting that at least several hundred genes may play a role in growth control and differentiation, and that this multitude of genes provides a large repertoire of potential oncogenes, might cast pessimism on our ability to identify all of the oncogenes that contribute to the phenotypes of human tumors. There are, however, at least three reasons to be optimistic. The first is that, despite the apparently large repertoire of proto-oncogenes, it would appear that certain members of this repertoire (ie, *ras* and *myc*) become activated preferentially or are strongly selected for during carcinogenesis [4,26]. This might reflect the ease with which activation can occur. For example, a single base substitution in the 12th or 61st codon of a *ras* proto-oncogene will cause its activation [26], and increased expression of *c-myc* or *N-myc* will activate these genes. In contrast activation of *c-src*, which thus far has not been found to occur in spontaneous rodent or human tumors, might require multiple changes, since its structure differs in several respects from *v-src* [3,4]. A second reason for optimism is that, although there may be a few hundred cellular proto-oncogenes, it is likely that they can be divided into a few categories (perhaps four or five) in terms of their structural homologies and mechanisms of action [3,4]. This greatly simplifies the identification of new proto-oncogenes and our understanding of the mechanisms by which they contribute to the tumor cell phenotype. It also gives rise to the hope that research in this area will provide general strategies of therapy, ie, therapy tailored to a class of oncogenes rather than to each specific oncogene.

## STUDIES ON PROTEIN KINASE C (PKC)

I would now like to discuss recent data from our laboratory related to PKC. Because of the central role of PKC in signal transduction, growth control, and tumor promotion [14,16], our laboratory has recently studied the function of this enzyme and cloned the related DNA sequences. Figure 2 presents a hypothetical diagram of PKC emphasizing the fact that the enzyme has two domains: an active site, containing an ATP binding site and the region to which protein substrates bind, and a regulatory domain whose activity is controlled by lipid, $Ca^{2+}$, and DAG or by lipid and TPA. We hypothesize that the usual function of the regulatory domain is to inactivate the enzyme, by "closing" the catalytic site, and that the binding of appropriate factors to

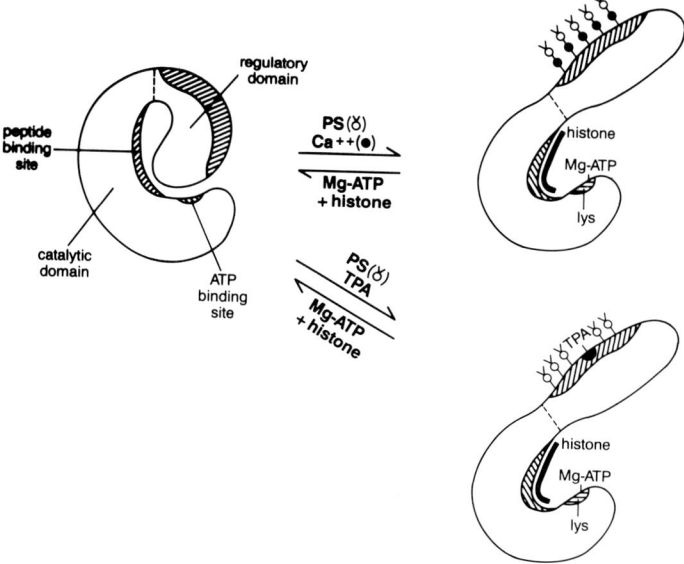

Fig. 2. A hypothetical model of protein kinase C emphasizing a catalytic domain that contains the ATP and peptide substrate binding sites and a regulatory domain that binds phosphatidylserine (PS), and $Ca^{++}$, or PS and TPA. We postulate that binding of PS plus $Ca^{++}$ or PS plus TPA to the regulatory domain induces a conformational change in the enzyme that "opens" the catalytic domain, thus activating the phosphorylation of a protein substrate, for example, histone H1.

the regulatory domain induces a conformational change that "opens" the catalytic site and thus activates enzyme function. Consistent with this scheme is evidence that limited proteolysis of the enzyme yields a fragment of about 66 kD that is active in the absence of lipid and cofactors [14]. In addition, there exist inhibitors of PKC that appear to act preferentially on the regulatory domain or the catalytic domain [16]. The development of pharmacologic agents that specifically inhibit PKC could provide a novel and nonmutagenic strategy of cancer chemoprevention and cancer chemotherapy.

Molecular studies on PKC would be tremendously enhanced by having available the cloned genes for this enzyme. We have, therefore, purified the enzyme from rat brain by ammonium sulphate precipitation and a series of column purification steps. As a final step in the purification we allowed the partially purified enzyme to undergo autophosphorylation and then purified the $^{32}$P-labeled protein by gel electrophoresis. The homogenous protein obtained from the gel (82 kD) was then reduced, carboxymethylated, and cleaved with the protease endolyse C to yield a series of polypeptides. These were separated by high-pressure liquid chromatography, and the amino acid sequences of a few of these polypeptides were then determined by microsequencing techniques. A peptide, designated P2, has the sequence Ser-Val-Asp-Trp Trp-Ala-Phe-Gly-Val-Leu-Leu-Tyr-Glu-Met-Leu-Ala-Gly-Gln. It was particularly useful since it is 18 amino acids long and contains two adjacent tryptophan residues. A 53-base pair oligonucleotide probe whose sequence corresponds to the predicted coding sequence for P2 (based on mammalian codon usage frequencies and codon degeneracy) was then synthesized and used to probe a rat brain cDNA library. Several homologous clones have been obtained, and we are now determining their complete nucleotide sequences [27].

The sequences that we have obtained display rather striking homologies to other protein kinase genes. Thus the sequence of the P2 peptide shares strong homology with a region present in the catalytic domain of protein kinase A. We have also identified an ATP binding site sequence, as well as several other sequences, that are homologous to sequences found in several protein kinases. As expected, in each of these regions the PKC clones show more homology to the serine and threonine kinases than to the tyrosine kinases. In a recent review [7], Dr. Tony Hunter has stressed the evolutionary relationship between several protein kinases. Our studies with PKC extend this theme. In several protein kinases, the catalytic domain is located in the carboxyterminal end of the molecule, whereas the regulatory domain is at the amino terminal end or is a separate protein subunit. Thus it would appear that during evolution the amino terminal end of these proteins has diverged, and in some cases has become a separate polypeptide chain, so as to provide regulation of protein kinase activities by diverse agonists. Depending on the particular kinase, the regulatory domain is responsive to cAMP in the case of protein kinase A, to cGMP in the case of protein kinase G, to $Ca^{2+}$ in the case of the myosin light-chain kinase (which is regulated by calmodulin), and to EGF in the case of the EGF receptor. We plan studies to determine whether deletion of the regulatory domain of PKC by recombinant DNA methods will, following transfection into mammalian cells, produce a protein that is autonomous and, therefore, cause disturbances in growth control. These studies will test the possibility that PKC can function as a proto-oncogene during carcinogenesis.

We have used our cDNA PKC clones from rat brain as probes to analyze the poly $A^+$ RNAs obtained from rat brain, heart, and liver, using Northern blot hybridizations. The highest levels of PKC mRNA were found in brain, with much lower levels in heart and liver. This reflects the relative levels of PKC activity observed in these tissues [14]. It is of interest that one of our PKC clones hybridizes to a single RNA species that is 7.5 kb, whereas a second clone hybridizes to two RNA species, one of 3.5 kb and the other of about 9 kb. We are currently comparing the nucleotide sequences of these two clones and find that there are subtle differences. These results suggest that there may be more than one gene for PKC and multiple forms of PKC enzyme, but this requires further study. Consistent with this possibility is the fact that our purified PKC protein displays a doublet profile on gel electrophoresis, and other laboratories have also obtained evidence for multiple species of PKC [28,29] (T. Hunter et al, personal communication). If subsequent studies verify the existence of more than one gene for PKC, then different forms of the enzyme could account for some of the tissue-specific and pleiotropic effects of this enzyme.

As I mentioned earlier, a major gap in our knowledge of the pathway of signal transduction is the question, how are signals conveyed from the cytoplasm to the nucleus? In the case of PKC it seems unlikely that the enzyme itself is translocated to the nucleus because attempts to demonstrate significant levels of the enzyme in the nucleus, using either TPA tagged with a fluorescent dansyl residue [30] or immunofluorescence with PKC antibody [31], have been negative. Another unsolved problem is how, during the process of tumor promotion, the imprint of a tumor promoter such as TPA eventually becomes fixed, so that cells remain abnormal even when the tumor promoter is no longer applied. If tumor promoters such as TPA act simply through activation of PKC, then one would expect cells to revert to their previous state once the promoter is removed.

We have found that when mouse [32] or rat fibroblast [33] cell lines are transfected with an activated H-*ras* oncogene and are grown in the presence of TPA, the tumor promoter markedly enhances the yield of transformed foci. This finding has been confirmed and extended to early passage rodent cells [34]. Since the cells obtained from these foci remain transformed even in the absence of TPA, this system may be useful for analyzing the mechanism underlying stable effects of TPA. A recent study [35] indicating that PKC selectively phosphorylates the DNA methyltransferase enzyme raises the intriguing possibility that the imprinting may be produced via an alteration in patterns of DNA methylation that influence gene expression. This would assume that PKC or a catalytically active fraction of the enzyme does enter the nucleus, or that the methyltransferase is phosphorylated by PKC in the cytoplasm before the methyltransferase enters the nucleus. Alternatively, prolonged activation of PKC might activate pathways related to the production of activated forms of oxygen [36] or alter cellular levels of poly ADP ribose [37], and thus produce stable effects on DNA or chromatin structure.

## GROWTH INHIBITOR GENES AND NEW STRATEGIES OF CANCER THERAPY

Finally, I want to discuss the likelihood that, in addition to genes that code for stimulatory growth factors (such as EGF and PDGF) and activated oncogenes that cause aberrant cell growth, there exists a reciprocal set of genes whose products inhibit cell growth and/or induce cells to undergo terminal differentiation. I shall refer to these genes as "growth inhibitor genes" although there is evidence that the products of specific genes can also induce terminal differentiation [38]. A priori, it seems likely that nature employs both growth stimulator and growth inhibitor genes to achieve the complex control that is required for normal growth, development, and differentiation. In the case of growth stimulator genes (or proto-oncogenes), mutations that result in activated oncogenes exert a dominant effect on growth. On the other hand, mutations that *inactivate* growth inhibitor or differentiation genes (by, for example, deletions) would lead to growth stimulation, and/or failure of terminal differentiation. Such mutations would be recessive since it would be necessary to inactivate the function of both alleles for tumors to occur, and replacement of the normal gene product would correct the defect. Tumors could result from either mechanism or, as seems most likely in terms of a multistep-multigene model of carcinogenesis, from a distortion in the net balance in function of both types of genes. Cell-cell hybridization studies lend support to the latter formulation [38,39].

Genetic studies provide strong evidence for the existence of genes that inhibit growth and modulate differentiation. In *Drosophila*, recessive mutations in at least 24 different genes can produce tumors of various types [for review, see 40]. It is of interest that all of these genes normally play a role in development. When the mutant allele is homozygous it is often lethal, because development is arrested at a specific stage. Thus, in flies homozygous for the mutation *giant larvae* there is an arrest in the development of the presumptive optic centers of the larval brain so that ganglion precursor cells continue to proliferate, eventually giving rise to neuroblastomas. The hereditary forms of human retinoblastoma and Wilm's tumor provide evidence for similar genes in humans [41,42]. Table II lists other evidence for growth inhibitor genes. Other authors [41–46] have emphasized this theme and refer to this category of genes as suppressor genes or "anti-oncogenes" [41–46], although it is not clear that these genes act simply by directly inhibiting the function of proto-oncogenes.

Except for the *Drosophila* gene *giant larvae* [47], none of these growth inhibitor genes has been cloned. A major challenge is to isolate these genes and elucidate their mechanisms of action. Until this is achieved, studies on proto-oncogenes and oncogenes will provide only a portion of the full repertoire of genes that underlie the carcinogenic process. Within the context of the signal transduction model displayed in Figure 1, what might be the biochemical functions of the proteins encoded by these putative growth inhibitor genes? I suggest that these might include: 1) protein phosphatases; 2) protein kinases that have effects that oppose those of PKC and the tyrosine protein kinases, for example, protein kinase A; 3) phospholipase inhibitors, for example, lipocortin or related inhibitors that inhibit specific phospholipases; 4) transcription control factors that suppress rather than enhance the transcription of specific genes involved in cell proliferation, for example, oncogenes; 5) translation control factors that inhibit the translation of specific mRNAs; and 6) genes that induce

**TABLE II. Evidence for Growth Inhibitor Genes ("Anti-Oncogenes")***

1. Fusion of normal with malignant cells suppresses malignancy
2. Certain hereditary human tumors (Retinoblastoma and Wilm's tumor) are associated with chromosomal *deletions* and loss of both alleles in the tumor
3. In *Drosophila* there exist recessive mutations in over 20 loci that predispose to developmental abnormalities and malignancy
4. In fish hybrids the absence of a gene that controls melanocyte differentiation causes melanomas

*For review see [41–46].

terminal differentiation by mechanisms that are not well understood at the present time. Presumably growth inhibitor factors, such as TGF-beta or specific interferons (see below), act through one or more of these biochemical mechanisms. Some of these factors may be lipids rather than polypeptides since there is evidence that specific prostaglandins can inhibit the growth of tumor cells or induce differentiation [10]. TGF-beta provides an example of a factor that can either inhibit or stimulate cell growth depending on the cell type [5]. Obviously, decreased production of a growth inhibitor or aberrations in its receptor could lead to abnormal cell proliferation and/or failure of normal development and differentiation.

The existence of growth inhibitory genes suggests, of course, new strategies for cancer prevention and treatment, by employing agents that mimic the action of such genes or actually induce their expression. There already exist several agents that appear to have such effects since they can, in appropriate cell systems, inhibit the growth of tumor cells and/or induce such cells to differentiate. These agents include: 1) retinoids, 2) vitamin D derivatives, 3) glucocorticoids, 4) dimethylsulfoxide (DMSO) and hexamethylene bis acetamide (HMBA), 5) cyclic AMP, 6) butyrate, 7) TPA and mezerein, 8) TGF-beta, and 9) specific interferons [for review see 5,38,48]. The growth inhibitory effects of the latter two substances will be discussed in considerable detail at these meetings.

My colleagues and I have been intrigued with the ability of the interferons to inhibit the growth and induce the differentiation of certain human tumor cell lines [48]. We have also found that these effects are markedly enhanced by the combination of interferon with either retinoids or mezerein [48]. It has been demonstrated that the growth suppression induced by treatment with interferon is associated with decreased expression of ornithine decarboxylase (ODC), *c-myc*, and *c-fos*, as well as increased expression of the 2′5′-oligoA synthetase [49]. Curiously, these events are then followed by increased production by these leukemic cells of an endogenous interferon. It is possible, therefore, that the normal role of specific interferons is to inhibit cell growth and/or modulate cellular differentiation. In this sense, specific interferon genes might normally function as "anti-oncogenes."

Obviously much more work remains to be done, but I am confident that the exciting papers that will be presented during the coming week at the parallel conferences on "Growth Factors, Tumor Promoters, and Cancer Genes" and "Interferons as Cell Growth Inhibitors and Antitumor Factors" will contribute insights into the multigenic basis of cancer as well as suggest new strategies for cancer prevention and treatment.

## ACKNOWLEDGMENTS

The author is indebted to several colleagues in his laboratory who carried out the studies described in this paper, in particular, Catherine O'Brian, Gerard Housey, Paul Kirschmeier, Mark Johnson, and Wendy Hsiao. Janusz Wideman played a valuable role in obtaining the amino acid sequence data on PKC. These studies were supported by NIH grant CA 02656, the Alma Toorock Memorial for Cancer Research and the National Foundation for Cancer Research.

## REFERENCES

1. Weinstein IB: J Supramol Struct Cell Biochem 17:99–120, 1981.
2. Weinstein IB, Gattoni-Celli S, Kirschmeier P, Lambert M, Hsiao W, Backer J, Jeffrey A: In Levine A, Vande Woude G, Watson JD, Topp WC (eds): "Cancer Cells 1, The Transformed Phenotype." Cold Spring Harbor, New York: Cold Spring Harbor Laboratory, 1984, pp 229–237.
3. Bishop JM: Cell 42:23–38, 1985.
4. Vande Woude GF, Levine AJ, Topp WC, Watson JD (eds): "The Cancer Cell. II, Oncogenes and Viral Genes." Cold Spring Harbor NY: Cold Spring Harbor Laboratory, 1984.
5. Goustin AS, Leof EB, Shipley GD, Moses HL: Cancer Res 46:1015–1029, 1986.
6. Sporn MB, Roberts AB: Nature 313:745–747, 1985.
7. Hunter T, Cooper JA: Annu Rev Biochem 54:897–930, 1985.
8. Gilman AG: Cell 36:577–579, 1984.
9. Nairn AC, Hemmings HC, Greengard P: Annu Rev Biochem 54:931–976, 1985.
10. Powles TJ, Bockman RS, Honn KV: Ramwell P (eds): "Prostaglandins and Cancer: First International Conference." New York: Alan R. Liss, Inc., 1982.
11. Ashall F, Puck TT: Proc Natl Acad Sci USA 82:5145–5149, 1984.
12. Sisskin E, Weinstein IB: J Cell Physiol 102:141–154, 1980.
13. Berridge MJ: Sci Am 253:142–152, 1985.
14. Nishizuka Y: J Natl Cancer Inst 76:363–370, 1986.
15. Spat A, Bradford PG, McKinney JS, Rubin RP, Putney JW Jr, Nature 391:514–516, 1986.
16. O'Brian CA, Liskamp RM, Arcoleo JP, Hsiao WL-W, Housey GM, Weinstein IB: In Poste PG, Crooke ST (eds): "New Insights into Cell and Membrane Transport Processes." New York: Plenum Publishing Co., 1986.
17. Bishop R, Martinez R, Weber MJ, Blackshear PJ, Beatty S, Lim R, Herschman HR: Mol Cell Biol 5:2231–2237, 1985.
18. Johnson M, Housey GM, Kirschmeier P, Weinstein IB: J Cell Biochem [Suppl] 10C(Abstracts):133, 1986.
19. Robbins SL, Cotran RS: "Pathologic Basis of Disease," 3rd Edition. Philadelphia: W.B. Sanders Co., 1985.
20. LeBeau MM, Westbrook CA, Diaz MO, Larson RA, Rowley JD, Gasson JC, Golde DW, Sherr CJ: Science 231:984–987, 1986.
21. Giulian D, Allen RL, Baker TJ, Tomorura Y: J Cell Biol 102:803–811, 1986.
22. Nau MN, Brooks BJ, Battey J, Sausville E, Gazdar AF, Kirsch IR, McBride OW, Bertness V, Hollis GF, Minna JD: Nature 318:69–73, 1985.
23. Cousseus L, Yang-Feng TL, Liao YC, Chen E, Gray A, McGrath J, Seeburg PH, Liberman TA, Schlessinger J, Francke U, Levinson A, Ullrich A: Science 230:1132–1139, 1985.
24. Madaule P, Axel R: Cell 41:31–40, 1985.
25. Bishop JM: Nature 321:112–113, 1986.
26. Sukumar S, Notario V, Martin-Zanca D, Barbacid M: Nature 306:658–661, 1983.
27. Housey GM, O'Brian CA, Johnson MD, Kirschmeier PT, Roth JE, Weinstein IB: J Cell Biochem [Suppl] 10C (Abstracts):132, 1986.
28. Kikkawa U, Go M, Koumoto J, Nishizuka Y: Biochem Biophys Res Commun 135:636–643, 1986.
29. Kiley SC, Jaken S: J Cell Biochem [Suppl] 10C(Abstracts):200, 1986.
30. Liskamp RMJ, Brothman AR, Arcoleo JP, Miller OJ, Weinstein IB: Biochem Biophys Res Commun 131:920–027, 1985.
31. Kikkawa U, Nishizuka Y: J Cell Biochem [Suppl] 10C(Abstracts):107, 1986.
32. Hsiao W-LW, Gattoni-Celli S, Weinstein IB: Science 226:552–555, 1984.
33. Hsiao W-LW, Wu T, Weinstein IB: Mol Cell Biol 6:1943–1950, 1986.
34. Dotto GP, Parada LF, Weinberg RA: Nature 381:472–475, 1985.
35. DePaoli-Roach A, Roach PJ, Zucker KE, Smith SS: FEBS Lett 197:149–153, 1986.
36. Cerutti PA: Science 227:375–381, 1985.
37. Singh N, Leduc Y, Poirier G, Cerutti P: Carcinogenesis 6:1489–1494, 1985.
38. Sachs L: Sci Am 254:30–37, 1986.
39. Stanbridge EJ: Cancer Surveys 3:335–350, 1984.
40. Gateff E: Adv Cancer Res 37:33–74, 1982.
41. Cavenee WK, Hansen MF, Norderskjold M, Kock E, Maumenee I, Squire JA, Phillips RA, Gallie BL: Science 228:501–503, 1985.

42. Koufos A, Hansen MF, Lampkin BC: Nature 309:170–172, 1984.
43. Green AR, Wyke JA: Lancet 2:475–477, 1985.
44. Comings DE: Proc Natl Acad Sci USA 70:3324–3328, 1973.
45. Knudson AG: Cancer Res 45:1437–43, 1985.
46. Sager R: Cancer Res 46:1573–1580, 1986.
47. Mechler B, McGinnis W, Gehring W: EMBO J 4:1551–1557, 1985.
48. Fisher PB, Hermo H Jr, Pestka S, Weinstein IB: In Bagnara J, Klaus SN, Paul E, Schartle M (eds): "Pigment Cell 1985, Biological and Clinical Aspects of Pigmentation." Tokyo:University of Tokyo Press, 1985, pp 325–332.
49. Einat M, Resnitzky D, Kimchi A: Proc Natl Acad Sci USA 82:7608–7612, 1985.

## NOTE ADDED IN PROOF

Since the preparation of this manuscript four other groups have also reported the isolation and analysis of cDNAs encoding protein kinase C (PKC) (Ono Y, et al: FEBS Lett 203:111–115, 1986; Knopf JL, et al: Cell 46:491–502, 1986; Parker PJ: Science 233:853–858, 1986; Coussens L: Science 233:859–866, 1986). Their results and additional data from our laboratory (Housey MD, et al: Proc Natl Acad Sci USA, in press) provide convincing evidence that PKC belongs to a multigene family. On the subject of recessive genes involved in cancer, a recent publication describes a human DNA segment that appears to correspond to the gene that predisposes to retinoblastoma and osteosarcoma (Friend SH, et al: Nature 323:643–645, 1986).

# Tyrosyl and Phosphatidylinositol Kinases of Human Erythrocyte Membranes

Mark R. Vossler, Anna Coco, Bimmie T. Strausser, Christine Zaricznyj, and Ian G. Macara

Department of Biophysics, School of Medicine and Dentistry, University of Rochester, Rochester, New York 14642

The tyrosyl kinase and phosphatidylinositol (PI) kinase activities of human red cells have been partially purified and characterized. Although the PI kinase required detergent for solubilization, the major tyrosyl kinase of the red cell could be extracted by high salt. A very small residual activity remained associated with the membranes, however, that was solubilized with the PI kinase and copurified through an ammonium sulfate precipitation and diethylaminoethyl (DEAE) ion-exchange step gradient elution. However, the two activities were found to differ with respect to their apparent $K_m$s for ATP and $Mg^{2+}$; they showed different half-lives for temperature inactivation, possessed different relative activities in the presence of $Mn^{2+}$ and $Ca^{2+}$, and were separable by elution from a DEAE-Trisacryl ion exchange column using a linear NaCl gradient. The kinetic parameters of the membrane-associated tyrosyl kinase differed from those of the salt-extracted enzyme. PI kinase was not activated by pretreatment with the tyrosyl kinase p68$^{v\text{-}ros}$ or by addition of the phosphotyrosyl phosphatase inhibitor, vanadate, to intact membranes, and was not competitively inhibited by the tyrosyl kinase substrate poly(Glu$_4$,Tyr). We conclude that the human red cell phosphatidylinositol and tyrosyl kinases are distinct and separable activities, and that at least two separable tyrosyl kinases are present in human erythrocytes.

Key words: tyrosine kinase, erythrocyte, vanadate, phosphatidylinositol

The proteins encoded by a number of oncogenes of retroviral origin have been shown to possess a tyrosyl-specific protein kinase activity (for recent review, see [1]). Their normal counterparts encoded by the homologous proto-oncogenes can also phosphorylate tyrosyl residues [2]. Two members of this group, p68$^{v\text{-}gag\text{-}ros}$ and p60$^{v\text{-}src}$, appeared to be associated also with a phosphatidylinositol (PI) kinase activity [3,4] and the polyoma middle T antigen has been reported to activate a PI kinase associated with cellular p60$^{src}$ [5]. The relationship between tyrosyl kinases and PI kinase activity remains obscure, however. A residual PI kinase activity present in highly purified epidermal growth factor (EGF) receptor has recently been separated

Received April 15, 1986; revised and accepted June 30, 1986.

© 1987 Alan R. Liss, Inc.

from the intrinsic EGF binding and tyrosyl kinase activities [6], and the major tyrosyl kinase activities of two transformed cell lines, RS-1 and LSTRA, have been shown to be kinetically and immunologically distinct from the total cell PI kinase activities [7]. Moreover, we have recently distinguished $p68^{v\text{-}gag\text{-}ros}$ from its associated PI kinase by a *ros*-specific antibody that inhibits only the tyrosyl kinase activity in anti-*gag* immunoprecipitates from UR2-transformed cells [8]. Nonetheless, phosphatidylinositol phosphorylation remains of possible importance in the regulation of cell growth because it is required for the generation of inositol trisphosphate and diacylglycerol C [9]. These two products are second messengers that increase cytosolic $Ca^{2+}$ and protein kinase C activity, respectively. Protein kinase C acts as the receptor for mitogenic tumor promoters such as phorbol esters [10], and we have proposed [3,11] that cellular transformation might in part be a consequence of the constitutive activation of protein kinase C.

The evidence against any association of PI kinases and tyrosyl kinase activities remains somewhat ambiguous, because the receptor that was studied [6] is not associated in vivo with the activation of PI turnover [11] and the kinase activities of the transformed cell lines were not purified.

To resolve this issue, we have partially purified and characterized an "authentic" PI kinase from the membranes of human erythrocytes. Erythrocytes were chosen because they contain high concentrations of the phosphorylated phosphatidylinositols [12] and have been previously shown also to possess unusually high tyrosyl kinase activity [13]. Their use also avoids the problems of interpretation associated with cells containing multiple types of membrane.

## MATERIALS AND METHODS
### Materials

Recently outdated packed red blood cells were provided by the American Red Cross. Triton X-100 was from Packard, ammonium sulfate was from Schwartz-Mann, DEAE-Trisacryl was from LKB, and poly(Glu$_4$,Tyr), angiotensin II, phospholipids, protease inhibitors, NP-40, and ATP were from Sigma. $\gamma$-$^{32}$P-ATP was from New England Nuclear. Cell lysates from UR2-infected chick embryo fibroblasts and anti-*gag* antiserum were provided by P. Balduzzi (Rochester, NY). Vanadate stock solutions were routinely boiled and allowed to stand for several days before use to allow decomposition of polymeric species.

### Purification

Ghosts were prepared from human red cells by lysis in hypotonic solution (5 mM Na phosphate, ph 8.0, 0.5 mM ethylenediamine tetraacetic acid [EDTA], 0.5 mM EGTA, 1 mM dithiothrietol [DTT], plus 1 µg/ml pepstatin and 20 µg/ml phenylmethyl sulfonyl fluoride as protease inhibitors) and washed by repeated centrifugation and suspension [14]. The resulting membranes were solubilized by incubation with buffer A (10 mM Na phosphate, pH 7.4, containing 0.5 mM EDTA, 1 mM DTT and 1% Triton X-100). Insoluble cytoskeletal elements were removed by centrifugation. Sodium phosphate (pH 7.4) was then added to bring the concentration to 50 mM, followed by addition of solid ammonium sulfate to 2 M. The suspension was centrifuged (20,000g for 30 min), and the protein-detergent complex, which had floated to the surface, was collected by removal of the subnatant. The precipitate was dissolved and dialysed against buffer B (as buffer A but with 0.05% detergent). DEAE-Trisacryl, previously equilibrated with buffer B, was stirred into the dialyzed material and incubated for 10 min. The resin was washed with three volumes of buffer B and a single-step elution was performed using three volumes of buffer B plus 200 mM NaCl. The eluant was concentrated by ultrafiltration with an Amicon YM100 filter (100,000 dalton cut-off) and stored at $-80°C$ after freezing in liquid nitrogen.

## Assays

PI kinase activity was determined by a modification of the procedure of Buckley [15], using 1 mM $\gamma$-$^{32}$P-ATP (10 $\mu$Ci/$\mu$mol) in 25 mM Hepes-NaOH, pH 7.4, 10 mM $MgCl_2$, 10 mM dithiothrietol, 0.5% Triton X-100 plus 0.5 mM phosphatidylinositol, and terminated by addition of 5% (w/v) trichloracetic acid (0°C) and 100 $\mu$g of carrier bovine serum albumin. The precipitate was centrifuged, and the pellet was rinsed with 0.5 ml of ice-cold water. The pellet was dissolved by addition of 200 $\mu$l of chloroform/methanol/HCl (2:1:0.03 N). A volume of 50 $\mu$l was spotted onto a Baker Si250 thin layer chromatography (t.l.c.) plate and developed in chloroform/methanol/ammonium hydroxide (9:7:4 N). The $^{32}$P-phosphatidylinositol-4-phosphate was located by autoradiography, scraped off the plate, and counted for $^{32}$P.

Tyrosyl kinase activity was routinely assayed using as substrate 1 mg of poly($Glu_4$,Tyr) per ml [16] in 25 mM Hepes-NaOH, pH 7.4, 10 mM $MgCl_2$. The reaction was terminated by addition of electrophoresis sample buffer. An aliquot was electrophoresed on a 10% sodium dodecyl sulfate (SDS) acrylamide gel, stained with Coomassie blue, dried, and either autoradiographed or directly counted by excising each lane between molecular weight markers of 27K and 55K daltons and counting for $^{32}$P.

Protein concentrations were determined by a modification of the method of Lowry [17].

## RESULTS

### Assay and Purification of Phosphatidylinositol and Tyrosyl Kinase Activities

The PI kinase assay used in the present study proved considerably more rapid than those described previously [eg, 12,15] because of the absence of two-phase extraction steps common to earlier methods. Carrier protein was necessary to ensure complete precipitation of the phospholipid. Using solubilized membranes assayed under standard conditions, the only $^{32}$P-labeled spots detectable above the origin on t.l.c. were phosphatidylinositol-4-phosphate and lysophosphatidylinositol-4-phosphate, as judged by comparison with authentic marker lipids. The time course for phosphorylation was linear for at least 60 min using solubilized material. As reported previously [18], red cell membranes also possess tyrosyl-specific protein kinase activity. Under standard assay conditions the reaction was linear for 5 min when using membranes and for more than 30 min when using solubilized material.

Most of the red cell tyrosyl kinase activity was found to be membrane bound, but was released by treatment with a high salt concentration (Table I). The $K_{1/2}$ for release was about 150 mM NaCl (Fig. 1). The soluble tyrosyl kinase activity was stable for several months after freezing in liquid nitrogen and storage at $-80°C$. The $K_m$ for ATP was 9.7 $\mu$M. Addition of 0.5% Triton X-100 to the assay buffer increased the $K_m$ slightly, to about 20 $\mu$M. The PI kinase activity, on the other hand, together

**TABLE I. Partial Purification of Phosphatidylinositol and Tyrosyl Kinase Activities From Human Erythrocytes***

| Purification step | Total protein (mg) | PI kinase activity (nmol mg$^{-1}$ min$^{-1}$) | Fold purification | Percent yield | Tyrosyl kinase activity (nmol mg$^{-1}$ min$^{-1}$) |
|---|---|---|---|---|---|
| Total cell lysate | 33903 | 0.02 | 1 | 100 | 0.016 |
| Membranes | 661 | 0.66 | 44 | 64.3 | 0.87 |
| Salt wash (250 mM NaCl) | N.D. | <0.01 | — | — | 0.85 |
| Triton X-100 supernatant | 366 | 1.27 | 64 | 68.6 | 0.004 |
| Ammonium sulfate precipitation | 33.3 | 3.73 | 187 | 18.3 | 0.026 |
| DEAE step-elution | 3.6 | 22.2 | 1110 | 11.8 | 0.280 |

*Activities were purified from one unit of recently-outdated human red cells, as described in "Materials and Methods." PI kinase was assayed using 1 mM ATP ($\gamma$-$^{22}$-P-ATP, 50 $\mu$Ci/$\mu$mol) and 0.5 mM PI; tyrosol kinase was assayed using 0.2 mM ATP ($\gamma$-$^{32}$-P-ATP, 250 $\mu$Ci/$\mu$mol) and 0.2 mg/ml poly (Glu$_4$, Tyr) as substrate. This purification was typical of six similar preparations, providing a mean PI kinase activity after the DEAE step of 23.8 $\pm$ 5.6 nmol/mg/min ($\pm$ 1 SD).

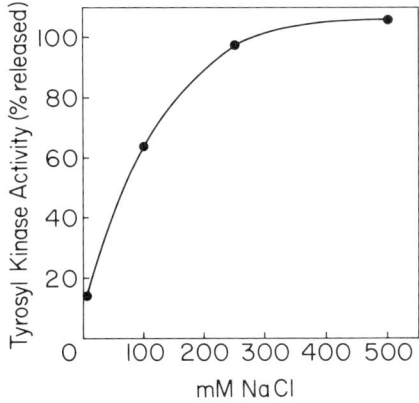

Fig. 1. NaCl-dependent elution of tyrosyl kinase from human red cell membranes. Membranes were diluted with NaCl to the concentrations indicated, incubated on ice for 5 min, then centrifuged in a microfuge for 10 min (4°C). Supernatants were adjusted to the same final NaCl concentration and assayed as described in "Materials and Methods."

with a small fraction of the tyrosyl kinase activity, required the presence of a nonionic detergent such as Triton X-100 for solubilization. These two activities were precipitated by 40% ammonium sulfate (w/v) and were retained by a DEAE anion exchanger at pH 7.8 at low ionic strength and eluted together by 200 mM NaCl. The salt-extractable tyrosyl kinase, however, was not retained by DEAE cellulose at pH 7.8 (data not shown). Typical purification results are shown in Table I. In six similar purifications, the mean PI kinase activity of the DEAE eluant was 23.8 $\pm$ 5.6 nmol/mg/min, representing about a 1,000-fold purification from the initial cell lysate. The membrane tyrosyl kinase activity was about two orders of magnitude lower than this, under standard assay conditions. No PI-phosphate kinase or diacylglycerol kinase activity was detectable.

The material obtained by batch elution from the DEAE-Trisacryl was used for further kinetic analysis.

## Characterization of Erythrocyte PI and Tyrosyl Kinase Activities (Table II)

To determine whether a single enzyme is responsible for the two detergent-solubilized kinase activities found in the erythrocyte membrane, we examined their kinetic characteristics with respect to ATP and $M^{2+}$ concentration and to thermal inactivation. The effect of varying ATP is shown in Figure 2. Curves were fit to the data by nonlinear least-squares using the standard Mento-Michaelis equation. The apparent $K_m$s for PI and tyrosyl kinase were 0.14 mM and 0.34 mM, respectively. That these values are significantly different can be judged from the dashed curves of Figure 2, generated by reversing the respective $K_m$ values. The $K_m$ for the detergent-solubilized tyrosyl kinase is also significantly different from that of the salt-extracted enzyme (0.02 mM) when measured under very similar conditions.

The effect of varying $Mg^{2+}$ concentration is shown in Figure 3. Free $Mg^{2+}$ concentrations were calculated assuming a pKa for ATP of 6.97 and a $pK(Mg \cdot ATP^{4-})$ of 4.49. Again the apparent $K_m$s differ significantly, being 0.63 mM and 7.2 mM for the PI and tyrosyl kinases, respectively. We also found that addition of 10 mM $Ca^{2+}$ inhibited the PI kinase activity by 98% but the tyrosyl kinase activity by only 75%. $Ca^{2+}$ alone was unable to support substrate phosphorylation by either activity. $Mn^{2+}$ (2 mM) supported 30% of the PI kinase activity observed with 10 mM $Mg^{2+}$, possibly as a result of contaminating $Mg^{2+}$. Under the same conditions, tyrosyl kinase activity was undetectably low.

Inactivation rates at two different temperatures were also compared. At 37°C the rate of PI kinase inactivation ($t_{1/2}$ of 4.8 hr) was significantly slower than that of tyrosyl kinase ($t_{1/2}$ of 2.8 hr). At 50°C, in the absence of the phospholipid, inactivation was rapid and apparently first order, with $t_{1/2}$s of 2.5 min and 5.1 min for the PI and tyrosyl kinase, respectively.

To determine if the substrates of the two kinases were competitive, PI kinase activity was examined in the presence of poly($Glu_4$,Tyr) at 0–4 mg/ml, plus an

TABLE II. Comparison of Phosphatidylinositol and Tyrosyl Kinase Activities of Human Erythrocyte Membranes*

| Enzyme | $K_m$ for ATP | $K_m$ for $Mg^{2+}$ | Preferred cation | $T_{1/2}$ for inactivation | |
|---|---|---|---|---|---|
| | | | | 37°C | 50°C |
| Salt-extracted tyrosyl kinase | 0.01-0.02 mM[a] | 10mM[c] | $Mn^{2+}$[b] | N.D. | N.D. |
| Detergent-solubilized tyrosyl kinase | 0.34 mM | 7.2 mM | $Mg^{2+}$ | 2.8 hr | 5.1 min |
| Detergent-solubilized phosphatidylinositol kinase | 0.14 mM | 0.63 mM | $Mg^{2+}$ | 4.8 hr | 2.5 min |

*Parameters were determined as described in "Materials and Methods."
[a]The apparent $K_m$ was increased by inclusion of 0.5% Triton X-100 in the assay buffer. Mohamed and Steck [19] reported a $K_m$ of 2.5 μM in the absence of detergent.
[b]Mohamed and Steck [19].
[c]Phan-Dinh-Tuy et al. [18].

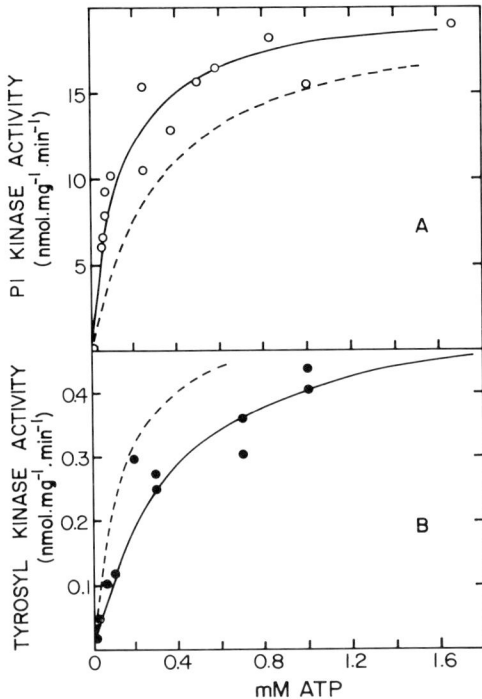

Fig. 2. Effect of varying ATP concentrations on PI (**A**) and tyrosyl (**B**) kinase activities. Assays were performed on DEAE-purified material (Table I) as described in "Materials and Methods." Curves were fit to the data by an iterative nonlinear least-squares procedure, using the standard Menton-Michaelis equation. The best-fit parameters were: (A) $K_m = 0.14$ mM, $V_{max} = 0.49$ mol mg$^{-1}$ min$^{-1}$. Dashed lines show fits using the same values for $V_{max}$ but opposite values for the $K_m$ (0.34 in A and 0.14 in B).

additional 5 mM MgCl$_2$ to avoid artifactual effects of complexation by the polyanionic substrate. No inhibition of the PI phosphorylation was detected (data not shown).

These results all indicate that the human red cell membrane contains distinct tyrosyl and PI kinases. Moreover, the kinetic data for the detergent-solubilized tyrosyl kinase are significantly different from those recently reported for the salt-extractable enzyme [18,19], suggesting that at least two different tyrosyl kinases are present in red cell membranes.

## Separation of Kinase Activities by Ion-Exchange Chromatography

In view of the above results, we further examined the copurification of the two detergent-solubilized kinase activities by using a DEAE-Trisacryl column loaded with protein from the ammonium sulfate precipitation step and eluted with a linear NaCl gradient. Results are in Figure 4. As expected, both enzyme activities were retained by the column and eluted at 100–200 mM NaCl. However, the PI kinase eluted three fractions ahead of the tyrosyl kinase (at about 130 and 170 mM NaCl, respectively). Only one major peak of each kinase was detected. A similar separation was obtained on two different preparations. Therefore, the two kinase activities represent different enzymes with nonoverlapping substrate specificities.

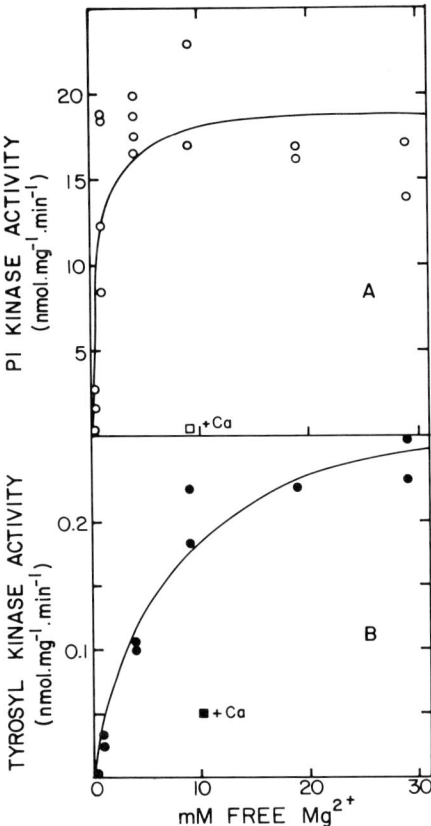

Fig. 3. Effect of varying $Mg^{2+}$ concentration on PI (**A**) and tyrosyl (**B**) kinase activities. In one instance 10 mM $CaCl_2$ was also added (□, ■). Curves were fit to the data using an iterative nonlinear least-squares procedure with the standard Menton-Michaelis equation. The $K_{1/2}$s obtained were 0.63 (A) and 7.2 mM (B).

## Effect of Tyrosyl Kinases on PI Kinase Activity

To investigate whether tyrosyl phosphorylation of the PI kinase might activate the enzyme, we preincubated the partially purified PI kinase in the presence of 100 µM cold ATP at 30°C for 30 min with $p68^{v\text{-}gag\text{-}ros}$, which had been immunoprecipitated from UR2-infected chick embryo fibroblasts by anti-gag antiserum as in [3]. The immunoprecipitate was then removed by centrifugation, and the supernatant PI kinase activity was compared with a mock-treated control. No significant difference in phospholipid phosphorylation was detected between the two samples. The $p68^{v\text{-}gag\text{-}ros}$ was able to phosphorylate poly(Glu$_4$,Tyr), but no phosphorylated bands were detected when a sample of the preincubated PI kinase was analysed by electrophoresis on a 10% SDS gel. These results suggest that the PI kinase is not a good substrate for the $p68^{v\text{-}gag\text{-}ros}$ tyrosyl kinase.

It remained possible, however, that the PI kinase might be phosphorylated and activated specifically by the erythrocyte tyrosyl kinase under certain conditions. We have therefore studied the effect of added vanadate on PI and tyrosyl phosphorylations in intact erythrocyte membranes. Vanadate is a potent inhibitor of phosphotyrosyl

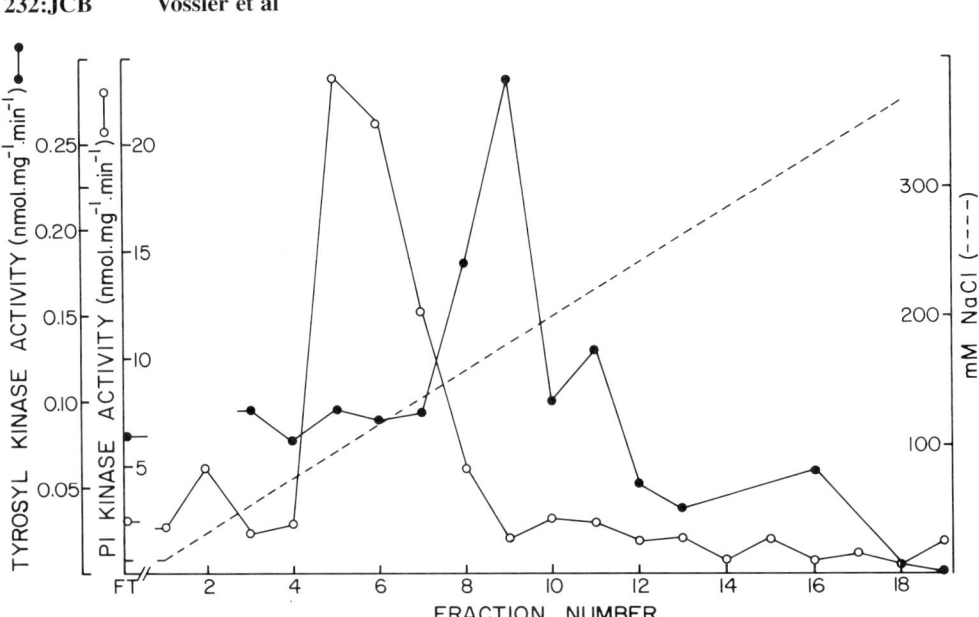

Fig. 4. Separation of PI (○) and tyrosyl kinase (●) activities by DEAE-Trisacryl ion-exchange chromatography. The column was equilibrated with starting buffer (10 mM Na phosphate pH 7.4, 1 mM DTT, 0.5 mM EDTA, 0.05% NP-40) and loaded with ammonium sulfate-precipitated material dialyzed extensively against starting buffer. The column was then eluted with a 0–500 mM linear NaCl gradient in starting buffer. Fractions were assayed for PI, tyrosyl kinase activities, and protein concentration as described in "Materials and Methods."

phosphatases [20] and might be expected to increase the level of tyrosyl phosphorylation of proteins in the membranes. As can be seen from Figure 5A, addition of 20 μM ammonium vanadate more than doubled the incorporation of $^{32}P$ into poly(Glu$_4$,Tyr) and decreased by half the release of $^{32}P$-phosphate catalyzed by the membranes (Fig. 5B). However, vanadate had little effect on the initial rate of phosphate incorporation into PI (Fig. 5C), suggesting that increasing the phosphotyrosyl content of the membranes does not significantly activate the PI kinase. A small and variable stimulation of the total level of PI-$^{32}P$-phosphate formation was frequently observed in these experiments, however. This effect was not a result of an inhibition of PI-phosphate phosphatase activity, which was very low in these membrane preparations and was unaffected by vanadate (rate of PI-$^{32}P$ breakdown was 0.24%·min$^{-1}$ at 37°C). Nor was it likely to have been a result of a variable pre-existing phosphorylation state of the PI kinase, since pretreatment with calf intestinal alkaline phosphatase, which can efficiently dephosphorylate phosphotyrosyl residues, did not significantly effect the PI kinase activity (data not shown). The most likely explanation is that the small stimulation was caused by inhibition of membrane ATPases by the vanadate, resulting in the maintenance of a slightly higher ATP concentration over the course of the experiment (Fig. 5B). Whatever the cause, it appears unlikely that tyrosyl phosphorylation can activate the red cell PI kinase.

## DISCUSSION

We have partially purified the phosphatidylinositol and tyrosyl kinase activities present in human erythrocyte membranes and have shown them to be kinetically

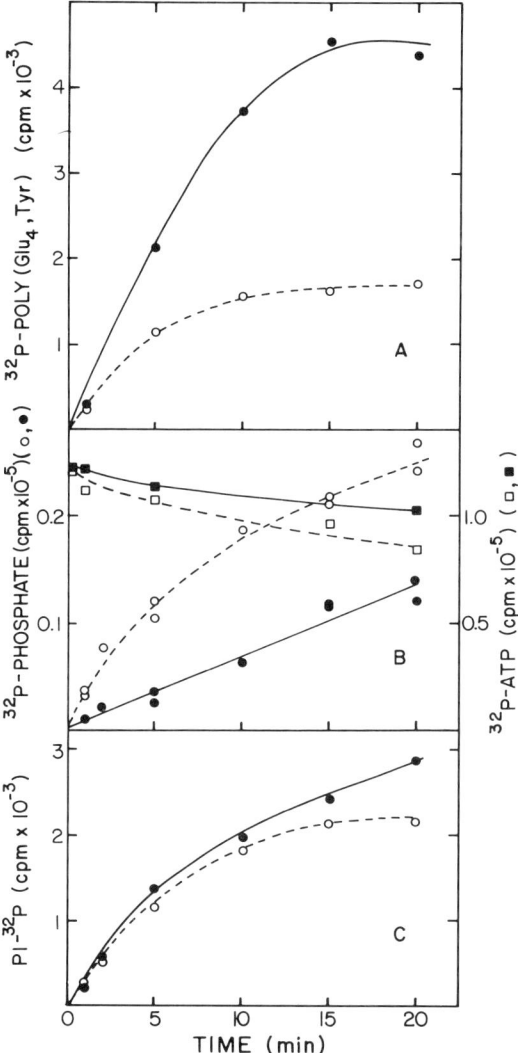

Fig. 5. Effect of vanadate on tyrosyl kinase activity (**A**), phosphate production and ATP utilization (**B**), and PI kinase activity (**C**) by erythrocyte membranes. Membranes were prepared as in [14]. Incubations were performed at 37°C in the presence (●, ■) or absence (○, □) of 20 $\mu$M sodium vanadate. Tyrosyl kinase activity was measured using poly(Glu$_4$,Tyr) as substrate, and PI kinase activity was measured using endogenous PI as substrate. Assays were performed in the absence of detergent. Remaining $^{32}$P-phosphate was precipitated after stopping the reaction, as in [21]. Precipitates were centrifuged, and the pellet was washed three times in the precipitating solution and counted for $^{32}$P (○, ●). A sample of the first supernatant, containing the $\gamma$-$^{32}$P-ATP, was spotted onto a PEI-cellulose t.l.c. plate and chromatographed using 0.75 M Tris, 0.45 HCl, and 0.5 M LiCl as solvent, as in [22]. Following autoradiography the ATP spot was cut out and counted for $^{32}$P (□, ■). Mean recovery of counts was 86.6 ± 8.3% (ATP + phosphate).

distinct and chromatographically separable activities. We have found that at least two tyrosyl kinases are present in the membranes, the principal one of which is a peripheral membrane protein released by high salt, as has recently been reported also by others [19], plus a minor component that copurifies through several steps with the PI kinase. Manipulations to try and phosphorylate the PI kinase did not increase its activity. We conclude that the kinase activities reside on separate polypeptides and that they do not interact in any obvious fashion. Nonetheless the two detergent-solubilized enzymes bear some similarities to one another: they both act as integral membrane proteins and copurify through a number of steps; and they show different but similar rates of heat inactivation and similar divalent cation preferences. Moreover, McDonald et al [7] have recently reported that a monoclonal antibody against p60$^{src}$ appears to be able to recognize a PI kinase from certain cell types. Whether the two classes of kinase show significant homology at the sequence level, however, must await the complete purification of the PI kinase.

The membrane-bound, detergent-solubilized tyrosyl kinase appears to be distinct from the salt-extractable tyrosyl kinase described in this report and reported earlier by Mohamed and Steck [19]. The kinetic properties and characteristics on purification of a tyrosyl kinase from human red cells described by Pha-Dinh-Tuy et al [18] make it likely that it is identical with the salt-extracted enzyme even though detergent was used for the initial solubilization from membranes. It is likely, therefore, that there are two distinct tyrosyl kinases in human red cells. The $K_{1/2}$s of these two tyrosyl kinase activities for $Mg^{2+}$ are similar, but the salt-extracted kinase appears to prefer $Mn^{2+}$ to $Mg^{2+}$ under the assay conditions of Mohamed and Steck [19], and the apparent $K_m$ for ATP is significantly different (20 $\mu$M as compared to 340 $\mu$M for the detergent-solubilized enzyme in the present study). Moreover, the salt-extracted enzyme appears to be positively charged at physiological pH since it binds to a cation exchanger [18] and to the anionic portion of band 3 [19], whereas the detergent-solubilized kinase is not retained by cation exchangers (unpublished observation) and requires approximately 150 mM NaCl for elution from DEAE. The functions of these kinases in the enucleated red cell and their relationship to other known tyrosyl kinases associated with growth factor receptors and oncogene proteins remain obscure.

## ACKNOWLEDGMENTS

This work was supported by grant CA-38888 from the National Institutes of Health.

## REFERENCES

1. Hunter T, Cooper JA: Adv Cyclic Nucleotide Protein Phosphorylation Res 17:443, 1984.
2. Bishop JM, Varmus H: In Weiss R, Teich N, Varmus H, Coffin J (eds): "RNA Tumor Viruses." Cold Spring Harbor, NY: Cold Spring Harbor Laboratory, 1982, pp 999–1108.
3. Macara IG, Marinetti GV, Balduzzi PC: Proc Natl Acad Sci USA 81:2728, 1984.
4. Sugimoto Y, Whitman M, Cantley LC, Erikson RL: Proc Natl Acad Sci USA 81:2117, 1984.
5. Whitman M, Kaplan DR, Schaffhauser B, Cantley L, Roberts TM: Nature 315:239, 1985.
6. Thompson DM, Cocket C, Chambaz EM, Gill G: J Biol Chem 260:8824, 1985.
7. MacDonald ML, Kuerzel EA, Glomset JA, Krebs EG: Proc Natl Acad Sci USA 82:3993, 1985.
8. Balduzzi PC, Chovav M, Christensen JR, Macara IG: J Virol 60:765, 1986.

9. Michell RH, Kirk CJ, Jones LM, Downes C, Creba JA: Philos Trans R Soc Lond [Biol] 296:123, 1981.
10. Nishizuka Y: Nature 308:693, 1984.
11. Macara IG: Am J Physiol 248:C3, 1985.
12. Ferrell Jr JE, Huestis WH: J Cell Biol 98:1992, 1984.
13. Phan-Dinh-Tuy F, Henry J, Rosenfeld C, Kahn A: Nature 305:435, 1983.
14. Steck TL, Kant JA: Methods Enzymol 31:172, 1974.
15. Buckley JT: Biochim Biophys Acta 498:1, 1977.
16. Braun S, Raymond WE, Racker E: J Biol Chem 259:2051, 1984.
17. Wong C-S, Smith RL: Anal Biochem 63:414, 1975.
18. Phan-Dinh-Tuy F, Henry J, Kahn A: Biochem Biophys Res Commun 126:304, 1985.
19. Mohamed AH, Steck TL: J Biol Chem 261:2804, 1986.
20. Swarup G, Cohen S, Garbers DL: Biochem Biophys Res Commun 107:1104, 1982.
21. Bochner BR, Ames BN: Anal Biochem 122:100, 1982.
22. Bochner BR, Ames BN: J Biol Chem 257:9759, 1982.

# Cellular Signal Transduction and the Reversal of Malignancy

Arthur H. Lockwood, Suzanne K. Murphy, Steven Borislow, Adam Lazarus, and Maryanne Pendergast

*Department of Pediatrics, Albert Einstein Medical Center and Temple University School of Medicine, Philadelphia, Pennsylvania 19141*

Animal cells contain only a few defined molecular systems that transduce hormonal and growth signals from the external environment to the intracellular milieu to regulate cellular growth and differentiation. Among the most ubiquitous of these "second messenger" pathways are those utilizing cyclic AMP and phosphatidylinositide turnover. The former activates protein kinase A, while the latter leads to the activation of protein kinase C and mobilization of intracellular calcium. Lesions induced by oncogenes in signal transduction systems may be responsible for the cancerous transformation of cells. In many tumor cell lines, including some transformed by the *ras* and *sis* oncogenes, activation of protein kinase A by elevation of cyclic AMP or activation of protein kinase C by addition of phorbol esters can restore many normal aspects of growth and morphology. Such "reverse transformation" is accompanied by the phosphorylation of unique cellular proteins and alterations in the phosphoinositide cycle. Molecular mechanisms by which activation of signal transduction systems can attenuate the malignant phenotype are considered in the context of cellular growth and differentiation.

Key words: protein kinase C, protein kinase A, phosphatidylinositides, sis oncogenes, reverse transformation by cAMP, ras oncogenes, phorbol esters, reverse transformation, cyclic AMP, signal transduction, cancer

The molecular mechanisms by which hormonal signals are transduced by intracellular second messenger systems to regulate cell metabolism and growth is a subject of growing interest for cancer biologists. Perturbation of second messenger systems and of the hormones and receptors that modulate them may be responsible for the oncogenic transformation of normal cells [1].

Evidence for this concept originated from studies concerning the behavior of the cyclic AMP (cAMP) signaling system as a second messenger in cancer cells. This

---

Suzanne K. Murphy's present address is The Philadelphia College of Pharmacy and Science, Philadelphia, PA 19104.

Received June 26, 1986; revised and accepted October 29, 1986.

© 1987 Alan R. Liss, Inc.

model for studying the control of cell transformation is called "cAMP-mediated reverse transformation" [2,3].

The oncogenic transformation of fibroblasts by tumor viruses, oncogenes, or chemical carcinogens results in the alteration of cell growth and morphology [4]. Transformed fibroblasts grow to high saturation densities, escape from contact inhibition and substrate dependence, and have reduced serum requirements [5,6]. Morphological changes include a more rounded or spindle-like shape, altered membrane topography, loss of tight focal adhesions with the substratum, a disorganized culture morphology, and a reduction in directed motility [2]. Transformation is also frequently accompanied by a reduction in the intracellular concentration of cAMP and attenuation of the activity of the enzyme adenylate cyclase, which synthesizes cAMP.

In many such oncogenic, transformed cell lines, elevation of intracellular cAMP (for example, by addition of membrane-permeable cAMP analogs to the culture) can reverse the pleiotypic effects of transformation [7,8]. This reverse transformation by cAMP results in the rapid acquisition of normal cell growth and morphology [9]. Cells become oriented and polarized (Fig. 1). Lamellar cytoplasm and stress fibers increase, as does adhesiveness to the substratum. Density and contact inhibition of growth are restored [10]. There is an increase in cytoskeletal organization: the cytoplasmic microtubule network expands [11], microfilament bundles assemble, and there is a redistribution of cytoplasmic myosin into these bundles [12]. Thus, treatment of transformed cells with cAMP not only affects cell growth but also induces an assembly of the cytoskeletal structures that organize the cytoplasm and govern cell morphology. The ability of cAMP to restore so many normal properties to cancer cells implies that one major pathway of oncogenic transformation is the disruption of the cellular mechanisms that regulate cAMP levels.

How does cAMP exert its varied effects on the growth and morphology of the transformed cell? The only known mode of action of cAMP in higher eukaryotes is the activation of cAMP-dependent protein kinases [13-15]. As best defined in those hormone-responsive tissues where cAMP serves as a second messenger, the activation of a cAMP-dependent protein kinase (PKA) results in the phosphorylation of specific target proteins. Phosphorylation alters the activity of the proteins, and, often through a cascade, the hormonal response is evoked.

Genetic studies of tumor cells confirm that cAMP-mediated reverse transformation is also orchestrated by protein kinases. Mutants of S49 lymphoma cells defective in PKA no longer show growth inhibition by cAMP [16,17], and mutants of Chinese hamster ovary (CHO) cells defective in PKA lack a morphological and growth response to cAMP [18]. An attractive explanation of the broad spectrum of phenotypic changes induced in transformed cells by cAMP is that a cAMP-dependent protein kinase system acts pleiotypically to phosphorylate—and thereby alter the activity of—a variety of proteins important in cytoskeletal organization and cell proliferation.

Which cellular proteins are phosphorylated by the cAMP-dependent protein kinase system? Until recently, no identification of specific proteins whose phosphorylation state is altered during reverse transformation had been made. Likely candidates would be structural proteins of the cytoskeleton, the enzymes or regulatory proteins of other major second messenger systems perturbed by transforming agents, the proteins encoded by oncogenes, and the cellular proteins involved in growth regulation and cytoskeletal organization that are themselves modulated by oncogene expression.

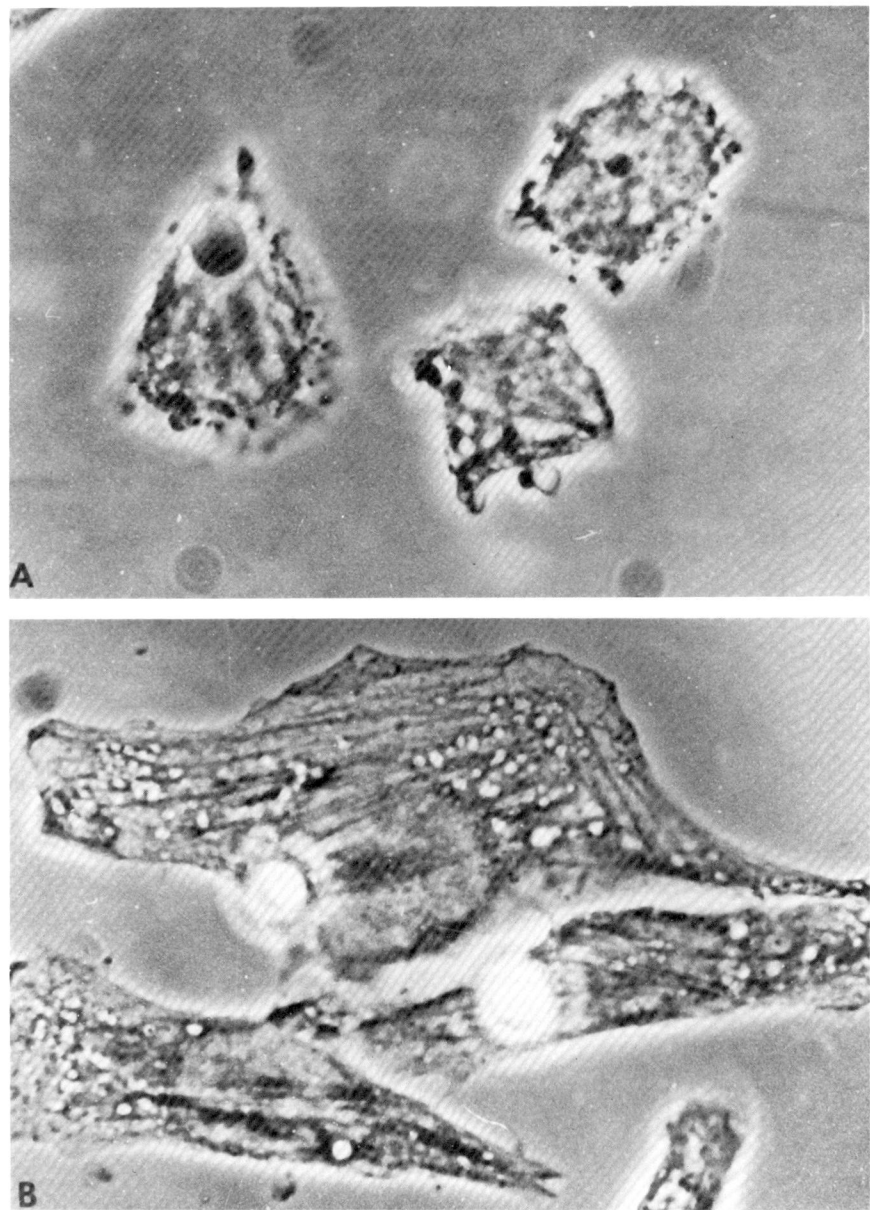

Fig. 1. Effect of db-cAMP on the morphology of CHO cells. **A:** No treatment. **B:** db-cAMP (1 mM) for 18 hr. Phase contrast micrographs.

Research in the authors' laboratory has identified a number of proteins, including a 20,000-dalton myosin light chain, unidentified 45,000- and 50,000-dalton proteins, and a 155,000-dalton protein possibly bound to microtubules, as proteins whose phosphorylation is altered during reverse transformation [3,19]. However, a specific link between the phosphorylation of these species and the phenotypic effects of cAMP on tumor cells remains to be established.

The protein products of a class of retroviral oncogenes are membrane/cytoskeleton-bound protein kinases with the unique ability to phosphorylate tyrosine residues [20,21]. Transformation often results in an elevation of total cellular phosphotyrosine and a moderate increase in the levels of phosphotyrosine in specific proteins. However, despite intensive research in several laboratories, it has so far proved difficult to establish a cause-and-effect relationship between cAMP-dependent or tyrosine phosphorylation of specific cellular proteins and modulation of the transformed phenotype [1,22]. Recently, we and others have begun to study the involvement of another major second messenger system, phosphoinositide turnover, in the events of reverse transformation. Increased turnover of a minor membrane phospholipid, the bis-phosphorylated form of phosphatidylinositol (PI), phosphatidylinositol 4,5-bisphosphate ($PIP_2$), is one of the earliest events following stimulation of cell proliferation by growth factors [23–25]. Phosphoinositide turnover is also increased in Rous sarcoma virus-transformed cells [26,27]. Hydrolysis of $PIP_2$ by a specific phosphodiesterase releases two products: 1,2-diacylglycerol (DAG) and inositol 1,4,5-trisphosphate ($IP_3$). Both function as second messengers [28–30]. $IP_3$ appears to function as a second messenger for the mobilization of calcium [31,32], while DAG is an endogenous activator of the phospholipid- and $Ca^{+2}$-dependent protein kinase C (PKC) [30,33].

Protein kinase C has been implicated in cell proliferation. It also appears to be the major receptor for, and mediator of the action of, phorbol esters, the potent tumor promoters [33,34]. These compounds mimic many of the parameters of transformation in cell culture [70]. Recently, it was shown that retroviral tyrosine kinase such as the *src* and *ros* enzymes are associated in vitro with phosphoinositide kinases and can regulate phosphoinositide metabolism in vivo [27,35]. Hence, unrestricted activation of protein kinase C, either by binding of phorbol esters or by overproduction of inositol phospholipids catalyzed or stimulated by oncogene-encoded tyrosine kinases, may play a major role in regulating the transformed phenotype. This second messenger pathway is thus a likely target for modulation by cAMP and cAMP-dependent phosphorylation.

## PHOSPHOINOSITIDE TURNOVER DURING REVERSE TRANSFORMATION

Our studies demonstrate that phosphoinositide metabolism is strongly affected during cAMP-mediated reverse transformation of CHO cells. They further demonstrate a dramatic influence on the morphologic phenotype of cells during reverse transformation by phorbol 12-myristate 13-acetate (PMA), the most potent known tumor promoter, and diacylglycerol (DAG), both direct activators of protein kinase C. Further, we have found major synergistic effects of PMA, DAG, and cAMP, both on phosphoinositide metabolism and on the phosphorylation of specific CHO cell proteins.

CHO cells were labeled with $^{32}PO_4$ before and after exposure to dibutyryl-cAMP (db-cAMP). Phospholipids were extracted and resolved by thin-layer chromatography [27,36]. Radioautography of the chromatograms demonstrated a substantial increase in levels of the phosphoinositides PI, PIP, and $PIP_2$ (Fig. 2). The correlation of these phosphoinositides and changes with time of exposure of the cells to db-cAMP is shown in Figure 3. Upon removal of cAMP, phospholipid concentrations rapidly return to those found in untreated CHO cells. The dependence of this

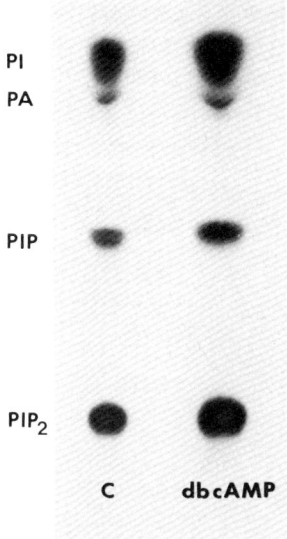

Fig. 2. Changes in phosphoinositide and phosphatidic-acid levels induced in CHO cells by db-cAMP. CHO cells were incubated at 37°C for 18 hr in the absence or presence of db-cAMP (1 mM) and testosterone propionate (15 μM). During the final 90 min the cells were radiolabeled using carrier-free $H_3\,^{32}PO_4$ (0.1 mCi/ml). Cells were removed from the culture dish, and the lipids were extracted, resolved, and analyzed by autoradiography as described [27,36].

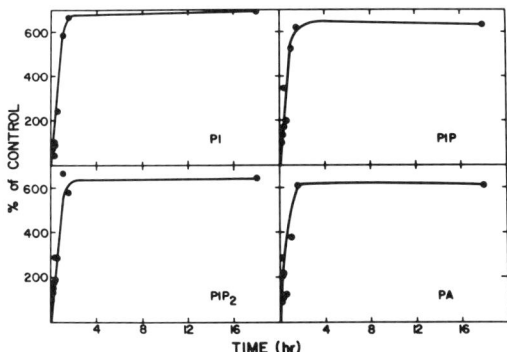

Fig. 3. Time course of db-cAMP-stimulated phosphorylation of the phosphoinositides and phosphatidic acid. CHO cells were exposed to db-cAMP (1 mM) and testosterone propionate (15 μM) at 37°C for the times indicated. Cells were radiolabeled with $H_3\,^{32}PO_4$ (0.1 mCi/ml) for 90 min prior to lipid extraction. The phosphorylated lipids were separated by thin-layer chromatography and analyzed as described [27,36]. Results are calculated as a percentage of control (no treatment) phospholipid. Actual cpm recovered: PI, 1,580; PIP, 5,630; $PIP_2$, 10,560.

effect on db-cAMP concentration shows an $ED_{50}$ of approximately 0.2 mM db-cAMP, a concentration comparable to that required for morphologic reversion. Other agents, such as cholera toxin (a cyclase activator), which elevate intracellular cAMP, induce similar increases in phosphoinositide levels (Table I).

The most straightforward explanation of these results is a cAMP-mediated inhibition of inositol phospholipid breakdown to diacylglycerol and inositol phosphates, perhaps by inactivation of the $PIP_2$ phosphodiesterase, although an effect on the lipid kinase is also possible.

**TABLE I. Effects of Agents That Elevate or Mimic cAMP on Phosphoinositide Levels in CHO Cells***

| Addition | Percentage of control | |
|---|---|---|
| | PIP | $PIP_2$ |
| None | 100 | 100 |
| db-cAMP | 305 | 330 |
| Cholera toxin | 141 | 176 |
| 8-bromo-cAMP | 163 | 155 |
| Theophylline | 370 | 460 |

*CHO cells were incubated at 37°C for 18 hr with the indicated agent: db-cAMP (1 mM); cholera toxin (1 μg/ml); 8-bromo-cAMP (0.5 mM); theophylline (1 mM). Lipids were extracted, separated by thin-layer chromatography, and quantitated as described [27,36]. Values were calculated as a percent of control (no addition) phospholipid. 100% levels of phospholipids were PIP, 1,141 cpm; $PIP_2$, 1,743 cpm.

**TABLE II. Phosphoinositide Levels in CHO Cells Treated With cAMP and Phorbol Ester***

| Phospholipid | PMA (ng/ml) (% control) | | |
|---|---|---|---|
| | 0 | 1 | 10 |
| $PIP_2$ | 100 | 315 | 418 |
| PIP | 100 | 247 | 352 |
| PI | 100 | 170 | 219 |

*CHO cells were treated with 1 mM db-cAMP for 18 hr in the presence of PMA as indicated. Phospholipids were quantitated as described [27,36]. Control values represent phospholipid levels after 18 hr exposure to db-cAMP alone. 100% levels (cpm) of phospholipids were $PIP_2$, 1,564; PIP, 794; PI, 331. In the absence of db-cAMP, levels of phospholipids (% control) with 10 ng PMA were $PIP_2$, 113; PIP, 88; PI, 92.

Activation of protein kinase C by phorbol esters can also affect phosphoinositide metabolism. There is a synergistic effect of PMA and cAMP on PIP and $PIP_2$ levels during reverse transformation. There are large increases in phosphoinositides when PMA is present during reverse transformation (Table II).

One implication of the experiments described above is that cAMP, by altering diacylglycerol production, may change the activity of protein kinase C. Conversely, to examine whether the activity of protein kinase C affects the induction of reverse transformation by cAMP, we determined whether exogenous diacylglycerol or PMA—both protein kinase C activators—could modulate cAMP-induced changes in protein phosphorylation, cell morphology, and growth.

## SYNERGISTIC PROTEIN PHOSPHORYLATION BY cAMP, PHORBOL ESTERS, AND DIACYLGLYCEROL

We examined the effects of activation of PKC by tumor promoters of diacylglycerol on cellular protein phosphorylation during cAMP-mediated reverse transformation. An analysis by SDS-polyacrylamide gel electrophoresis is shown in Figure 4. Cyclic AMP induced the phosphorylation of proteins of 45,000 and 50,000 daltons and the dephosphorylation of a protein of 28,000 daltons. Surprisingly, the phorbol ester PMA induced similar phosphorylation changes. When cells were exposed simultaneously to cAMP and PMA there was a striking synergistic phosphorylation of both the 50,000- and 45,000-dalton proteins and a complete dephosphorylation of the 28,000-dalton protein. Exactly the same effect was observed when diacylglycerol, the natural PKC activator, replaced PMA (Fig. 5). These observations represent one of the largest reported changes in the phosphorylation of specific proteins induced in intact cells. Extraction of CHO cells with nonionic detergents in a buffer that preserves the cytoskeleton demonstrates that the 45,000- and 50,000-dalton phosphoproteins are present mainly in the cytosol/membrane fraction. To analyze further the basis for the synergism between cAMP and phorbol esters, we carried out one-dimensional peptide mapping of the $^{32}$P-labeled proteins. Figure 6 shows the phosphopeptides generated from the 50,000-dalton protein. While cAMP or PMA treatment alone evokes phosphorylation of both common and unique peptides, treatment with both together results in the appearance of two new highly phosphorylated peptides not phosphorylated with either compound alone (shown by arrows in Fig. 6). This observation suggests either that cAMP and PMA synergism results from activation of a third protein kinase or, alternatively, that it results from a conforma-

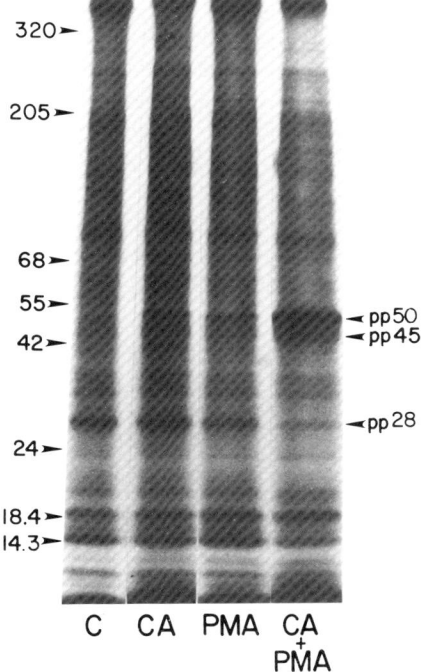

Fig. 4. Phosphoprotein distribution in CHO cells treated with cAMP and PMA. Metabolic labeling of cells with $^{32}$PO$_4$ and analysis of phosphoproteins on 10% SDS-PAGE was as described previously [66]. C, No treatment; CA, 0.75 mM db-cAMP, 15 μM testosterone; PMA, 200 ng PMA; CA+PMA, 0.75 mM db-cAMP, 15 μM testosterone, 200 ng PMA. Numbers and arrows on left-hand side refer to molecular weight standards (kilodaltons); those on the right indicate molecular weights of proteins whose phosphorylation is dramatically altered by CA+PMA.

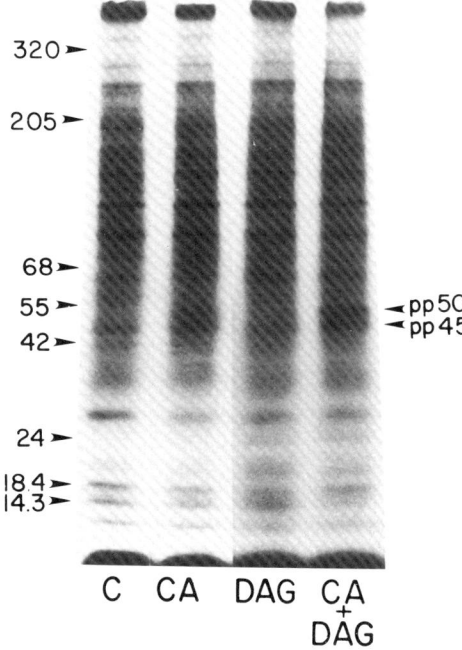

Fig. 5. Phosphoprotein distribution in CHO cells treated with cAMP and DAG. Analysis was on 10% SDS-PAGE [66]. C, No treatment; CA, 0.75 mM db-cAMP, 15 μM testosterone; DAG, 400 μg DAG; CA+DAG, 0.75 mM db-cAMP, 15 μm testosterone, 400 ng DAG.

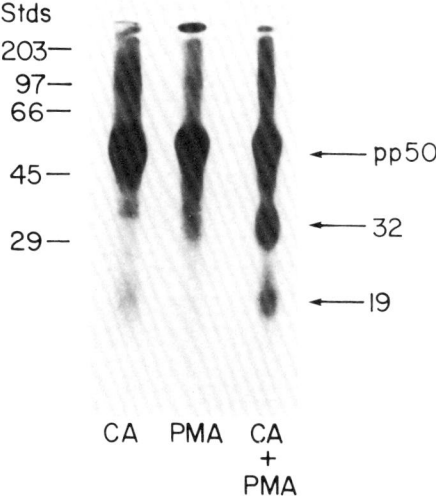

Fig. 6. Phosphopeptide map of pp50. After $^{32}PO_4$ labeling and sodium dodecyl sulfate-polyacrylamide gel electrophoresis (SDS-PAGE) as described in Figure 4, the pp50 band was excised from the gel and subjected to one-dimensional peptide mapping with V8 protease (0.025 μg/well) on a 15% SDS-PAGE gel.

tional change in the substrate protein that makes additional phosphorylation sites accessible to PKA and/or PKC.

## PHORBOL ESTERS AND DAG EXAGGERATE MORPHOLOGIC REVERSION

The influence of phorbol esters and DAG on the cAMP-induced morphological phenotype of CHO cells is shown in Figure 7. At low concentrations neither compound added alone changes cell form (Fig. 7a,c,e). The significant observation is that at these low concentrations both PMA and diacylglycerol dramatically alter the cAMP-induced morphologic phenotype. cAMP alone causes CHO cells to resume a

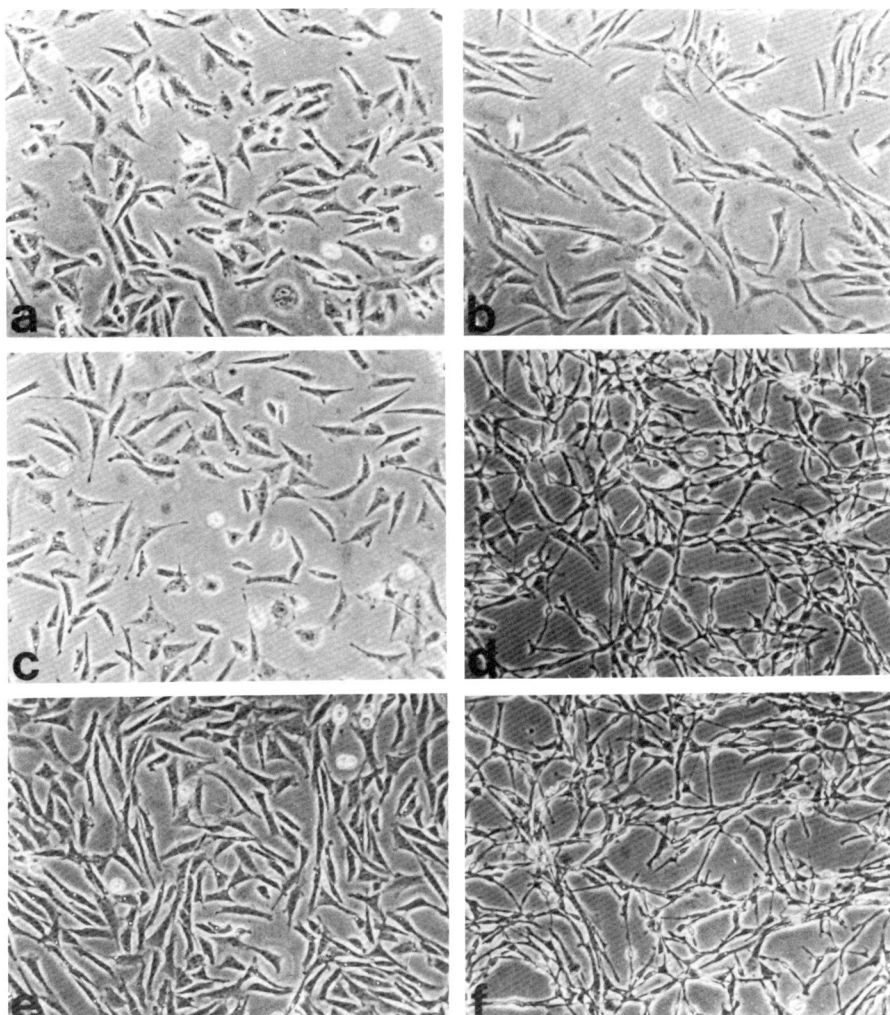

Fig. 7. Effect of phorbol esters and diacylglycerol on the morphology of CHO cells during reverse transformation. CHO cells were incubated for 18 hr at 37°C with the indicated agent. **a:** No treatment. **b:** db-cAMP (0.1 mM) + testosterone propionate (15 μM). **c:** PMA (0.5 ng/ml). **d:** PMA (0.5 ng/ml) + db-cAMP (1 mM) + testosterone propionate (15 μM). **e:** Diacylglycerol (10 μg/ml). **f:** Diacylglycerol (10 μg/ml + db-cAMP (1 mM) + testosterone propionate (15 μM).

normal fibroblastic morphology (Fig. 7b). Cells become flattened and elongated with a tranquil membrane and many actomyosin bundles (stress fibers). In the presence of either PMA (Fig. 7d) or diacylglycerol (Fig. 7f), cAMP is unable to induce cell spreading and flattening; the cell body remains rounded, and each cell extends long bilateral neurite-like processes. It is possible that continued activation of protein kinase C during reverse transformation alters the cAMP-induced assembly of actomyosin bundles. It also potentiates the cAMP-induced microtubule assembly that results in cell elongation and process extension.

## PHORBOL ESTERS AND cAMP COORDINATELY INHIBIT CHO CELL GROWTH

Activation of protein kinase C by phorbol esters strongly potentiates cAMP-induced growth inhibition. As shown in Figure 8, cAMP alone reduces both growth rate and saturation density of CHO cells. PMA alone has little effect on growth. However, a dramatic inhibition of growth occurs when PMA is added simultaneously with cAMP. Cells were completely growth inhibited over many days. Trypan blue exclusion and flow cytofluorometry showed that cells were still viable after PMA and cAMP treatment. Furthermore, the synergistic inhibition of growth by PMA and cAMP was completely reversible upon removal of the drugs.

The synergistic inhibition of tumor cell growth by PMA and cAMP was confirmed by testing a variety of other agents that elevate intracellular cAMP. These included 8-Br-cAMP, theophylline, cholera toxin, and isobutylmethylxanthine (IBMX). Both DAG and phorbol esters showed synergism with each compound for morphological reversion and growth inhibition. [$^3$H] Thymidine incorporation into DNA confirmed a rapid and complete inhibition of DNA replication by PMA and cAMP (95% inhibition after 25 hr) under conditions where PMA alone had no effect and cAMP was 50% inhibitory.

The preceding results suggested that PMA and cAMP inhibit CHO cell growth at a specific phase in the cell cycle. We used computerized laser flow cytospectrofluorometry to determine that cAMP imposes a $G_1$-specific block and that there is a very vigorous potentiation of this block by PMA. Similar results were obtained with diacylglycerol.

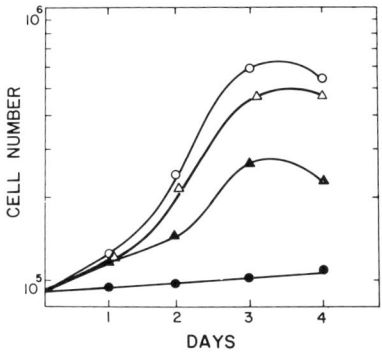

Fig. 8. Effect of PMA and cAMP on the growth of CHO cells. No addition (○——○); 0.5 mM db-cAMP (▲——▲); 0.5 mM cAMP and 20 ng/ml PMA (●——●); 20 ng PMA (△——△).

There is mounting evidence that phorbol esters can also induce differentiation to a normal phenotype in other transformed cell lines [37]. PMA has been shown to inhibit cell growth and induce differentiation of HL-60 human promyelocytic leukemia cells [38,39] and human K562 myeloid leukemia cells [40]. In fact, diacylglycerol mimics the PMA induction of HL-60 cell differentiation [41].

These results demonstrate that tumor-promoting phorbol esters, usually conceived of as mitogens and tumor promoters, can act synergistically with cAMP to reverse the transformed phenotype of cancer cells. Hence, phorbol esters must now also be conceived of as differentiating agents, which may, in some cases, actually function as antitumor drugs.

Taken together, the above observations suggest an intimate interaction in tumor cells between the signal transduction system modulated by cAMP and protein kinase A and that controlled by phosphoinositides and protein kinase C.

Cyclic AMP-mediated reverse transformation of tumor cells thus provides a unique model for dissecting these interactions and for examining the role of phosphoinositide metabolism, phorbol esters, and specific protein phosphorylation in the reversal of the transformed phenotype.

## CYCLIC AMP REVERSES THE TRANSFORMATION OF NIH3T3 CELLS BY THE HUMAN H-ras ONCOGENE

Although cell lines such as CHO provide well characterized models for the study of reverse transformation, one drawback is that the transforming principle in most of these lines has not been defined. To isolate molecular pathways of reverse transformation, it would be valuable to have a cyclic AMP-responsive cell line rendered tumorigenic by a defined oncogene. We have found that the NIH3T3 cell lines transformed by the H-*ras* oncogene is an excellent model.

Human *ras* oncogenes are associated with many aspects of the transformation of normal tissue cells to cancer cells [42]. The proteins encoded by the *ras* genes are 21-kilodalton (kDa) polypeptides called p21 [43]. They are cytoplasmic proteins that are bound, in large part, to the plasma membrane [44]. Molecular cloning and biochemical characterization of the human *ras* proteins revealed that they are able to bind and hydrolyze guanosine triphosphate (GTP) in a manner analogous to protein synthesis factors [45,46], $\alpha$-tubulin [47,48], and the G proteins, which transduce signals between cell-surface receptors and adenylate cyclase [49,50]. Oncogenic mutations in the human H-*ras* gene result in proteins defective in GTP hydrolysis. Such a defect may be the basis for the tumorigenic activation of the *ras* genes [51-55]. Genetic and biochemical studies in yeast have also implicated the *ras* proteins as direct or indirect modulators of cAMP synthesis [56–59]. Although the yeast studies implied that p21 can activate adenylate cyclase, our experience with the response of mammalian cancer cells to cAMP suggested that a negative regulation of cAMP levels by such oncogene proteins is also possible.

We have, in fact, found that an increase in intracellular cAMP can reverse most of the morphological and growth changes associated with the malignant phenotype induced in NIH3T3 cells upon transfection with the human H-*ras* oncogene or its overexpressed cellular progenitor ([60] and manuscript submitted). NIH3T3 clones transfected either with the oncogenic form of the human H-*ras* (T-24 bladder carcinoma) gene carrying a glycine-to-valine substitution at position 12 or with the proto-oncogene linked to a hyperactive simian virus 40 (SV40) promoter were exposed to a variety of agents that elevate intracellular cAMP. These included db-cAMP, 8-Br-cAMP, cholera toxin, IBMX, and theophylline. In each case the *ras*-transformed lines responded identically. There was a rapid alteration in cell morphology (Fig. 9). The

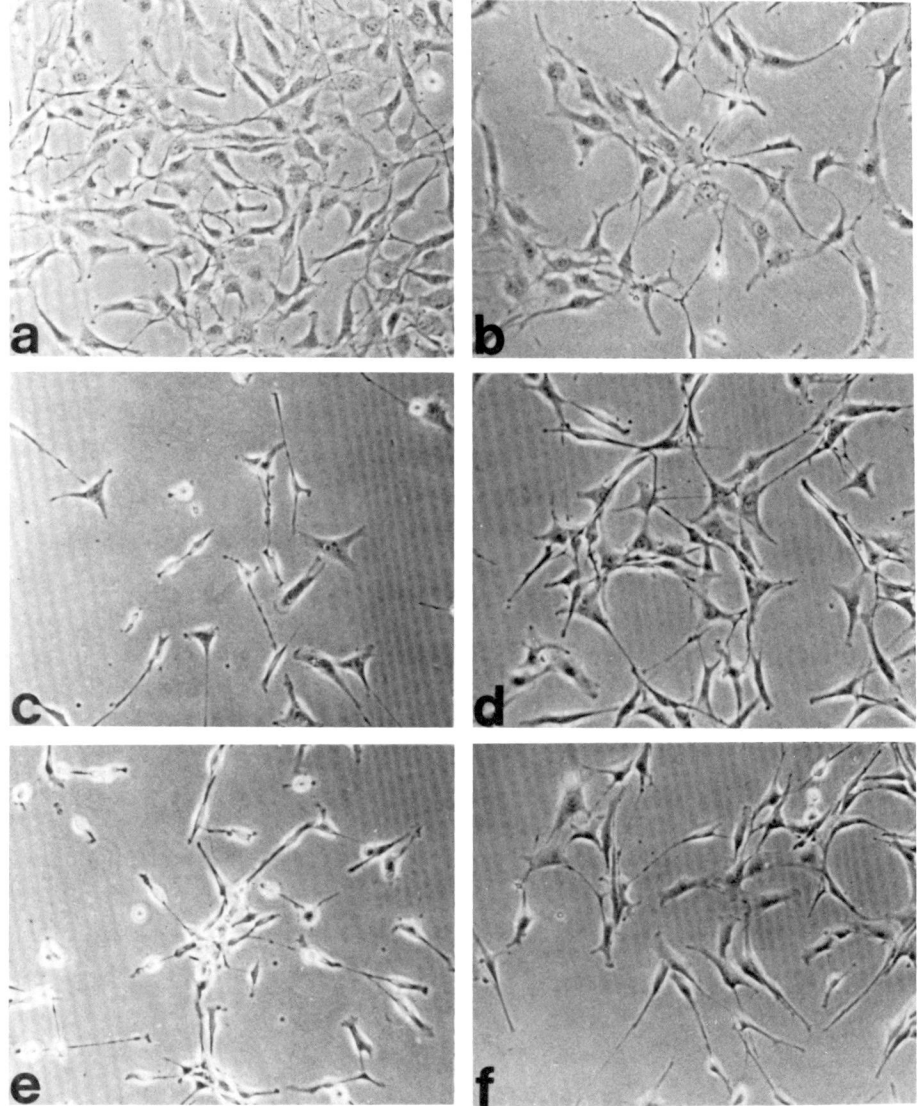

Fig. 9. Effect of elevated intracellular cyclic-AMP on the morphology of human H-*ras*-transformed NIH3T3 cells growing at low density. No addition. **a,c,e:** 0.5 mM 8-Br-cAMP and 50 mM IBMX. **b,d,f:** NIH3T3 (a,b); H-*ras* (T-24) clone 6. c,d: C-H-*ras* clone 2 (e,f). Human H-*ras*-transformed lines were established by transfection of NIH3T3 cells with plasmid cDNA encoding either the human C-H-*ras* gene or the T-24 H-*ras* mutant oncogene (carrying a glycine-to-valine mutation at position 12). The normal C-H-*ras* gene was designed to be overexpressed by insertion between an SV40 early promoter and SV40 termination and polyadenylation signals (S. Yokoyama et al, unpublished). Transformed foci were cloned using established methods.

tumor cells became less refractile, less rounded, had more tranquil membrane, were less adhesive, and spread more on the substratum. Stress fibers and actomyosin bundles were re-established. The culture morphology at confluence became typical of normal fibroblasts (Fig. 10). Contact inhibition of growth was restored (Fig. 10). As shown in Figure 11, for 3T3 cells carrying the oncogenic *ras* gene, there was a major reduction in log phase growth rate, and the saturation density was reduced to or was below that of untransformed 3T3 cells. As is typical of all tumor cell lines responsive to cAMP, the parent 3T3 cells were also partially growth inhibited by cAMP (Fig. 10). Another *ras* 3T3 line, which expresses 25–50 times the normal level of the *ras* proto-oncogene, also showed cAMP-mediated reverse transformation of morphology (Figs. 9, 10) and growth (not shown). Flow cytofluorometric analysis showed that growth arrest occurred in the $G_1$ phase of the cell cycle.

Thus, elevation of intracellular cAMP can reverse many of the in vitro parameters of oncogenicity associated with expression of the human H-*ras* oncogene. Although attempts to demonstrate directly an effect of *ras* proteins on mammalian adenylate cyclase activity have so far been unsuccessful [61], the concept of p21 *ras* oncogene products as signal transduction proteins that modulate second messenger pathways is, as a result of the studies described here, highly attractive.

The literature relating normal and malignant growth to cellular cAMP levels is often discordant. In certain cell types, cessation of growth is associated with increased cAMP, while in other cells cAMP may be mitogenic [62]. This is not surprising in view of the ubiquity of the cAMP pathway as a second messenger system in virtually

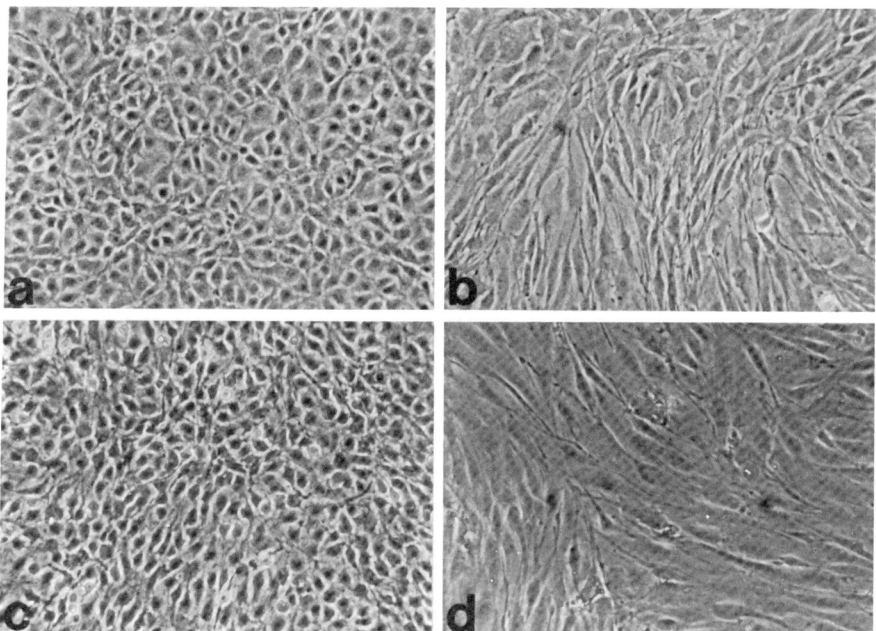

Fig. 10. Effect of elevated intracellular cyclic-AMP on the morphology of human H-*ras*-transformed NIH3T3 cells at confluence. **a:** NIH3T3 H-*ras* (T-24) clone 6 at confluence. **b:** NIH3T3 H-*ras* (T-24) clone 6 at confluence in the presence of 0.5 mM 8-Br-cAMP and 50 $\mu$M IBMX. **c:** NIH3T3 C-H-*ras* clone 2 at confluence. **d:** NIH C-H-*ras* clone 2 at confluence in the presence of 0.5 mM 8-Br-cAMP and 50 $\mu$M IBMX.

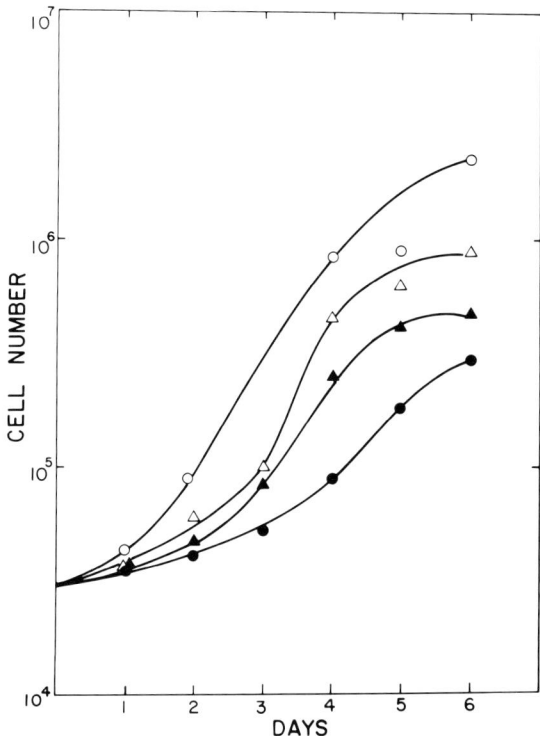

Fig. 11. Effect of cyclic AMP on the growth kinetics of H-*ras* (T-24) oncogene-transformed NIH3T3 cells. Human H-*ras* (T-24) clone 6 cells: No addition (○——○); plus 0.5 mM 8-Br-cAMP (●——●). NIH3T3 cells: No addition (△——△); plus 0.5 mM with 8-Br-cAMP (▲——▲).

all tissues. One would expect that, depending upon the nature of the hormone receptor and final effector targets, elevation of cAMP could serve as either a positive or a negative regulator of growth and malignant transformation.

In NIH3T3 cells, cAMP levels increase as cells become confluent and division ceases [63]. Our preliminary analysis of the human H-*ras* NIH3T3 transformants indicates a reduction in cAMP levels compared to untransformed NIH3T3 cells, particularly at confluence (unpublished data). Such regulation of cAMP levels by H-*ras* expression is supported by early studies that demonstrate a fall in intracellular cAMP in cells transformed by the Harvey sarcoma virus, whose oncogene is v-H-*ras* [64]. Hence an attractive hypothesis is that *ras* oncogene expression acts upon the cAMP pathway most strongly at confluence (or at least upon cell-cell contact), when cAMP levels would normally rise. The resulting decrease in intracellular cAMP would lead to escape from contact inhibition of growth and the loss of normal cell and culture morphology, which is typical of untransformed cells. Artificial elevation of cAMP, as described here, would reverse the effects of *ras* expression at confluence and restore normal saturation density and fibroblastic morphology to the transformed cells.

The only known mode of action of cAMP in eukaryotes is the activation of cAMP-dependent protein kinases, which results in the phosphorylation of specific target proteins [65]. We have previously shown that cAMP-mediated reverse transformation of CHO tumor cells is accompanied by the phosphorylation of unique cytoskeletal and cytoplasmic proteins [66,67]. It is likely that a similar mechanism governs cAMP-mediated reverse transformation of *ras*-transformed NIH3T3 cells. Our initial results indicate that specific protein phosphorylation in response to cAMP does, in fact, occur (Lockwood et al, submitted).

The H-*ras* transformed cells show increased phosphorylation of proteins of molecular weight (MW) 80,000 and 50,000 and decreased phosphorylation of a protein of MW 97,000. Exposure of the H-*ras* cells to cAMP results in enhanced phosphorylation of proteins with MWs of 45,000, 50,000, and 80,000 daltons. Characterization of these proteins offers the promise of identifying the molecular sites of H-*ras* modulation of growth and morphology.

We conclude that, in some mammalian cells, the human H-*ras* p21 oncogene protein must function to reduce cellular cAMP levels. Such regulation could be via direct interaction with components of the adenylate cyclase or cAMP phosphodiesterase system or, indirectly, by modification of another signal transduction pathway such as the phosphatidylinositol cycle [68]. We have, in fact, recently demonstrated an association of phosphatidylinositide turnover with cAMP-mediated reverse transformation [69]. We aso infer that the human c-H-*ras* gene is part of a regulatory system that modulates cAMP levels in normal cells during growth and development.

## COMMON PROTEIN PHOSPHORYLATION INDUCED BY GROWTH FACTORS, cAMP, AND PHORBOL ESTERS

Recent studies in the authors' laboratories indicate that NIH3T3 cells transformed by the cloned *sis* oncogene also undergo cAMP-mediated reverse transformation of growth and morphology (manuscript submitted). The *sis* oncogene encodes a chain of platelet-derived growth factor (PDGF) [71]. PDGF 13 known to stimulate $PIP_2$ breakdown, inositol triphosphate formation, and calcium mobilization in 3T3 cells [68,71]. This effect of cAMP indicates a link between growth factor action and cAMP-mediated protein phosphorylation in transformed cells. Indeed, the patterns of cellular protein phosphorylation induced by PDGF, phorbol esters, and cAMP suggest an intriguing connection. As described earlier in this review, cAMP and PMA show synergistic phosphorylation of proteins in CHO cells with molecular weights of 45,000 and 50,000 daltons and dephosphorylation of a 28,000-dalton protein. In *ras* transformed 3T3 cells, cAMP induces phosphorylation of 50,000- and 80,000-dalton proteins; this phosphorylation is stimulated by PMA. In quiescent 3T3 fibroblasts both PMA and PDGF rapidly stimulate phosphorylation of an 80,000-dalton protein [72]. Both growth factors and PMA stimulate phosphorylation of tyrosine residues of a 42,000-dalton protein in chick embryo fibroblasts [73]. In addition to these common phosphorylations, both cAMP and PMA induce unique phosphoprotein changes [3,66,74]. Hence all of these systems may function, in part, via the phosphorylation of a common set of endogenous cellular substrates. In many cases, phorbol esters decrease the affinity of cell surface receptors for growth factors or hormones including epidermal growth factor [75,76], insulin [77], and transferrin [78]. This is accompanied by, and may be a consequence of, protein kinase C-mediated receptor phosphorylation [79–82]. Although cAMP-mediated receptor phosphorylation has not yet been demonstrated, it would seem a likely event in view of the above studies.

These observations suggest biochemical pathways by which signal transduction systems can interact and by which the pleiotropic biological effects of cAMP, phorbol esters, and oncogenes can be reconciled.

## REVERSAL OF MALIGNANCY BY ACTIVATION OF SIGNAL TRANSDUCTION PATHWAYS—POSSIBLE MECHANISMS

The diverse biological effects of the protein kinase A and protein kinase C signal transduction pathways on cellular differentiation and oncogenic transformation may have underlying biochemical mechanisms in common. It is unlikely in cancer cells that all the cellular mechanisms controlling growth and differentiation have gone awry. Rather, it is more probable that oncogene expression has selectively modified specific second messenger pathways and that malignant cells retain the biochemical and genetic capacity for differentiation.

If this is so, then the potential exists for overcoming the specific malignant lesion, for example by restoration of normal cellular cAMP levels or direct activation of protein kinase C with phorbol esters. The presence of only a few recognized signal transduction systems in all cell types suggests that in each cell type, at specific stages of development, the cAMP/protein kinase A and phosphoinositide/protein kinase C systems can function as components of cellular pathways programmed for specific aspects of growth or differentiation. Depending on the cell type and the stage of maturation at which an oncogene suborns normal function, the biological result will differ. For example, interference with cAMP production might lead to uncontrolled growth in certain cell types such as fibroblasts, yet have little effect on or even induce differentiation in others such as certain endothelial lines.

Restoration of cAMP could then potentially have diverse effects depending on the nature of the transformed cell. Similarly, in the hematopoietic cell lineage, phospatidylinositide turnover and protein kinase C activation might be controlled by normal agents such as colony-stimulating factors. Oncogene-mediated interference with the function of these factors, by alteration in their receptors in the coupling of receptors to the phospholipid cycle, could result in leukemogenesis and defective hematopoietic differentiation. In a manner directly analogous to cAMP action, phorbol esters, by activation of protein kinase C, could circumvent the polyphosphoinositide lesion and induce differentiation of malignant cells.

The studies reviewed here make it obvious that there are intimate molecular interactions between signal transduction pathways and oncogenes in cancer cells. Possible sites at which these systems interact are indicated in Figure 12. Cyclic AMP, by activation of protein kinase A, might induce phosphorylation alterations in components of the phosphoinositide pathway such as the lipid kinases or the $PIP_2$ phosphodiesterase (phospholipase C). The result would be altered activity of protein kinase C. PKA might also directly phosphorylate, and thereby regulate, growth factor receptors such as that for the *sis* oncogene product (PDGF).

A protein kinase cascade involving PKA might potentially regulate expression of the *ras* oncogene at the level of the genome as well as by cytoplasmic protein phosphorylation. Protein kinase C, activated by phosphoinositide turnover or phorbol esters, may phosphorylate protein components of the cytoskeleton and the cAMP pathway to influence growth and differentiation, depending on the nature of the transformed cell. At least in the cancer cell lines reviewed here, these two major signal transduction systems are able to cooperate to inhibit cancer cell growth. Current evidence indicates that both PKA and PKC can phosphorylate a unique and common set of cellular proteins. Identification of these substrates and their function may lead to an understanding of the biochemical mechanisms by which cAMP and protein

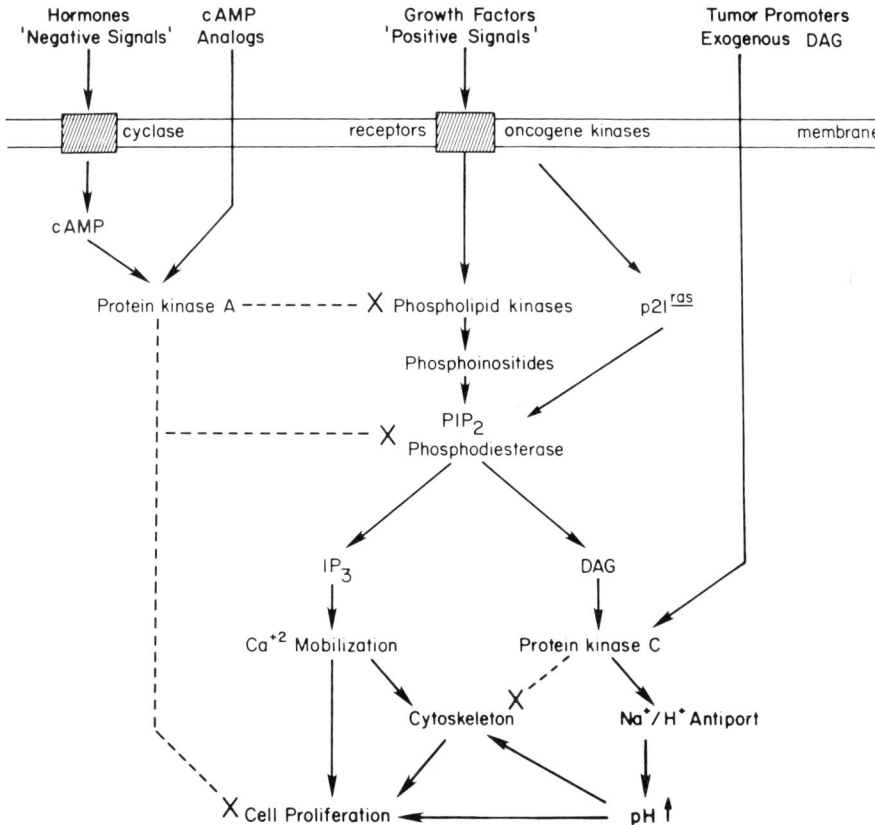

Fig. 12. Model of interactions between cellular signal transduction systems for the regulation of growth and morphology.

kinase C can attenuate oncogene function and reverse malignancy in a variety of cancer cells.

## ACKNOWLEDGMENTS

Studies in the authors' laboratory were supported by grants from the National Institute of Health (CA-39232), the American Cancer Society, and the Leukemia Society of America (Scholar award to Arthur H. Lockwood). We thank Dr. Peter Mamunes for helpful comments, and Esther Morak for preparation of the manuscript.

## REFERENCES

1. Macara IG: Am J Physiol 248:C3, 1985.
2. Pastan I, Willingham M: Nature 274:654, 1978.
3. Lockwood AH, et al: In "Cold Spring Harbor Symposium," Vol. 46. Cold Spring Harbor, NY: Cold Spring Harbor Laboratory Press, 1982, pp 909–919.
4. Temin H, Rubin H: Virology 6:669, 1958.
5. Hanafusa J: In Feankel DH, Wagoner RR (eds): "Comprehensive Virology," Vol. 1. New York: Plenum Press, 1977, pp 401–467.

6. Tooze J (ed): "The Molecular Biology of Tumor Viruses," 2nd ed. Cold Spring Harbor, NY: Cold Spring Harbor Laboratory Press, 1984.
7. Johnson GS, et al: Proc Natl Acad Sci USA 68(2):425, 1971.
8. Guerrant RL, et al: Infect Immun 10(2):320, 1974.
9. Puck TT, et al: Proc Natl Acad Sci USA 69(7):1943, 1972.
10. Hsie AE, et al: Proc Natl Acad Sci USA 68(7):1648, 1971.
11. Brinkley BR, Cartwright J Jr: Ann NY Acad Sci 253:428, 1975.
12. Bloom G, Lockwood AH: Exp Cell Res 129:31, 1980a.
13. Walsh DA, et al: J Biol Chem 243:3763, 1968.
14. Kuo JR, Greengard P: Proc Natl Acad Sci USA 64:1349, 1969.
15. Weller M: "Protein Phosphorylation." London: Pion Ltd, 1979, pp 76–80.
16. Bourne HR, et al: Science 187:750, 1975.
17. Coffino P, et al: J Cell Physiol 85:603, 1975a.
18. LeCam A, et al: J Biol Chem 256(2):933, 1981.
19. Bloom G, Lockwood AH: J Supramol Struct 14:241, 1980b.
20. Erickson RL, et al: Proc Natl Acad Sci USA 76:6260, 1979.
21. Hunter T, Sefton BM: Proc Natl Acad Sci USA 77(13):1311, 1980
22. Hunter T, Sefton BM: Annu Rev Biochem 54:897, 1985.
23. Habenich AJR, et al: J Biol Chem 256:12329, 1981.
24. Sawyer ST, Cohen S: Biochemistry 20:6280, 1981.
25. Farese RV, et al: J Biol Chem 257:4042, 1982.
26. Diringer H, Friis RR: Cancer Res 37:2979, 1977.
27. Sugimoto Y, et al: Proc Natl Acad Sci USA 81:2117, 1984.
28. Michell RH, et al: Philos Trans R Soc Lond [Biol] 296:123, 1981.
29. Berridge MJ, Irvine RF: Nature 312:315, 1984.
30. Nishizuka Y: Nature 308:693, 1984.
31. Biden TJ, et al: Biochem J 223:467, 1984.
32. Joseph SK, et al: J Biol Chem 259:3077, 1984.
33. Takai Y, et al: Biochem Biophys Res Commun 91:1218, 1979.
34. Ashendel CL, et al: Cancer Res 43:4333, 1983.
35. Macara IG, et al: Proc Natl Acad Sci USA 81:2728, 1984.
36. Gonzalez-Sastre F, Folch-Pi J: J Lipid Res 9:532, 1968.
37. Vanderbark GR, Niedel JE: J Natl Cancer Inst 73:1013, 1984.
38. Huberman E, Callahan MF: Proc Natl Acad Sci USA 76:1293, 1979.
39. Rovera G, et al: Science 204:868, 1979.
40. Lozzio CB, et al: Cancer Res 36:4657, 1976.
41. Ebeling JG, et al: Proc Natl Acad Sci USA 82:815, 1985.
42. Shih TY, Weeks MO: Cancer Invest 2:109, 1984.
43. Gibbs JB, et al: TIBS 350, 1985.
44. Ellis RW, et al: In Klein G (ed): "Advances in Virol Oncology." New York: Rover, 1982, p 107.
45. Lockwood AH, et al: Proc Natl Acad Sci USA 68:3122, 1971.
46. Maitra U, et al: Annu Rev Biochem 51:869, 1982.
47. Jacobs M: In Robert K, Hyams JS (eds): "Microtubules." New York: Academic Press, 1979, p 225.
48. Lockwood AH, et al: Fed Proc 34:520, 1975.
49. Ross EM, Gilman AG: Annu Rev Biochem 49:558, 1980.
50. Hurley JB, et al: Science 226:860, 1984.
51. Sweet RW, et al: Nature 311:273, 1984.
52. Manne V, et al: Proc Natl Acad Sci USA 82:376, 1985.
53. McGrath JP, et al: Nature 310:644, 1984.
54. Finkel T, et al: Cell 41:763, 1985.
55. Gibbs JB, et al: Proc Natl Acad Sci USA 81:5704, 1984.
56. Brock G, et al: Cell 41:763, 1985.
57. Kataoka T, et al: Cell 37:437, 1984.
58. Tatchell K, et al: Nature 309:523, 1984.
59. Toda T, et al: Cell 40:27, 1985.
60. Lockwood AH, et al: Fed Proc 45:1708, 1986.

61. Beckner SK, et al: Nature 317:71, 1985.
62. Rozengurt E, et al: Adv Enzyme Regul 19:61, 1981.
63. Pastan I, et al: Annu Rev Biochem 44:491, 1975.
64. Carchman RA, et al: Cell 1:59, 1974.
65. Krebs EG, Beavo JA: Annu Rev Biochem 48:923, 1979.
66. Bloom G, Lockwood AH: J Supramol Struct 14:241, 1980.
67. Pendergast M, Lockwood AH: J Cell Biol 101:248a, 1985.
68. Berridge MJ: J Biochem (Tokyo) 220:345, 1984.
69. Murphy SK, Lockwood AH: J Cell Biol 103:302a (1986).
70. Assoian CL: Nature 309:804, 1984.
71. Berridge MJ, Irvine RF: Nature 312:315, 1984.
72. Rozengurt E, Rodriguez-Pena M, Smith K: Proc Natl Acad Sci USA 80:7244, 1983.
73. Gilmore T, Martin GS: Nature 306:487, 1983.
74. Laszlo A, Radkek, Chin S, Bissell MJ: Proc Natl Acad Sci USA 78:6241, 1981.
75. Shoyar M, DeLarco J, Todaro GJ: Nature 279:387, 1979.
76. Magun BE, Matrisian LM, Bowden GT: J Biol Chem 255:6373, 1980.
77. Grunberger G, Gordon P: Am J Physiol 243:E319, 1982.
78. Rovera G, Ferrero D, Pagliardi GL: Ann NY Acad Sci 397:211, 1982.
79. Jacobs S, Sahyoun NE, Saltiel AR, Cuatrecasas P: Proc Natl Acad Sci USA 80:6211, 1983.
80. May WS, Jacobs S, Cuatrecasas P: Proc Natl Acad Sci USA 81:2016, 1984.
81. Cochet C, et al: J Biol Chem 259:2553, 1984.
82. Iwashita S, Fox CF: J Biol Chem 259:2559, 1984.

# Growth Factors Modify the Epidermal Growth Factor Receptor Through Multiple Pathways

## BethAnn Friedman and Marsha Rich Rosner

Massachusetts Institute of Technology, Cambridge, Massachusetts 02139

Previous results have shown that tumor promoters modify the properties of the epidermal growth factor (EGF) receptor through the activation of protein kinase C. Diacylglycerol-generating factors such as platelet-derived growth factor (PDGF) and p28$^{sis}$ should activate protein kinase C and alter EGF receptor properties in a similar manner. To test directly the involvement of protein kinase C in the action of media from v-sis–transformed cells on the EGF receptor, Swiss 3T3 cells were first extensively treated with various concentrations of the tumor-promoter phorbol dibutyrate (PDBu) This treatement reduced levels of active protein kinase C in the cells, making them less responsive to subsequent rechallenge with the tumor promoter. The results demonstrate that there are at least two components to the action of media from v-sis transformed cells on EGF binding: a labile factor that confers protein kinase C independence and a stable factor that appears to be dependent on protein kinase C. The action of the first factor cannot be mimicked by transforming growth factor-$\beta$ or EGF in either the presence or absence of PDGF. The action of the second factor is similar to that of PDGF. These findings indicate that heterologous regulation of the EGF receptor can occur through both protein kinase C-dependent and -independent pathways.

**Key words: regulation, multiple pathways, EGF receptor**

In previous work, we [1,2] and others [3–6] have shown that tumor promoters modulate the action of the epidermal growth factor (EGF) receptor. Three effects are observed in human cell lines: 1) reduction of EGF binding to the apparent high affinity EGF receptor, 2) decrease in EGF-stimulated tyrosine kinase activity as monitored by EGF receptor tyrosine phosphorylation, and 3) increase in overall EGF receptor phosphorylation at serine and threonine residues. Since these effects appear to be mediated by protein kinase C, growth factors that can generate diacylglycerol, the endogenous activator of protein kinase C [7], should modify the EGF receptor in a similar manner. Platelet-derived growth factor (PDGF), which generates diacylglycerol through phosphatidylinositol turnover, falls into this category [8–11]. Simian

Received June 6, 1986; revised and accepted December 15, 1986.

© 1987 Alan R. Liss, Inc.

sarcoma virus (v-*sis*)-transformed NRK cells contain p28$^{sis}$, a PDGF-like protein that is the product of the v-*sis* oncogene. To investigate directly the role of protein kinase C in mediating changes in EGF-receptor properties, we determined whether conditioned medium from v-*sis*-transformed cells was capable of modifying EGF binding to the EGF receptor in protein kinase C-depleted Swiss 3T3 cells.

The results indicate that 1) there is an alternative pathway to protein kinase C for heterologous regulation of the EGF receptor; 2) this alternative pathway must arise, at least in part, from a component other than PDGF-like factors; and 3) since medium from v-*sis*–transformed cells does not act on cellular regulatory pathways solely through the PDGF receptor, some of the growth-promoting and -transforming properties ascribed to the product of the v-*sis* gene may in fact be due to the action of other factors.

## MATERIALS AND METHODS
### Materials

Mouse EGF (Collaborative Research, Inc., Lexington, MA) was iodinated by the chloramine T method to a specific activity of approximately 6 Ci/μmol using NA-$^{125}$I (Amersham, Arlington Heights, IL). Phorbol diterpene esters were purchased from Sigma (St. Louis, MO). Medium from v-*sis*–transformed NRK cells was obtained from the laboratory of C. Stiles. It was concentrated using a Millipore PT6C series membrane (MW cutoff of 10,000 Da) from serum-free DME/F12 (1:1) that had been conditioned by confluent cultures of v-*sis*–transformed NRK cells for 2 days. One unit of activity is defined as the amount of medium that induces 50% DNA synthesis in a quiescent monolayer of BALB/c-3T3 cells. Purified porcine PDGF was obtained from BRL and shown to be free of TGF-$\beta$ by R. Assoian.

### Cultures

Swiss 3T3 cells were grown in a gassed (5.5% $CO_2$), humidified incubator in Dulbecco's modified Eagle's medium (DME) supplemented with 10% heat-inactivated fetal calf serum. When the cells reached confluence, the medium was removed and replaced with DME containing 0.5% bovine serum albumin (BSA) for 48 hr. To generate cells having different levels of protein kinase C, cultures were treated for an additional 48 hr with 0, 10, 100, or 1,000 ng/ml phorbol dibutyrate (PDBu) in DME/0.5% BSA.

### Quantitation of $^{125}$I-EGF Binding to High Affinity EGF Receptors

Confluent, quiescent cells, grown on 24 well dishes, were treated with various concentrations of PDBu for 48 hr as described above. The cultures were then washed three times with binding medium over a period of 2 hr at 37°C. This medium was removed and appropriate agents added in a total volume of 0.2 ml binding medium for the times indicated in the figure legends. Each variable was tested in triplicate. Cells were then placed on ice and washed with binding medium. $^{125}$I-EGF (0.05–0.1 nM) in binding medium was added for 4–6 hr at 4°C. This concentration of $^{125}$I-EGF binds primarily to high affinity EGF receptors, as demonstrated by the 80% reduction in cpm bound resulting from 37°C PDBu treatment of cells that had not been depleted of protein kinase C. Cells were washed, lysed, and quantitated for specific $^{125}$I-EGF binding as described. Data were normalized to the amount of specific $^{125}$I-EGF binding to cells that were treated with binding medium alone. Since treatment with PDBu elicits a round of mitotic division, both cell and receptor numbers vary in the different populations. Thus, given the different absolute numbers, the data must be expressed as percentages. In general, the cpm bound ranged from 1,000 to 4,000 cpm, with a standard deviation of 5–10%. The nonspecific EGF binding was 100 cpm. The specifics for each experiment presented here are given in the figure legends.

## Preparation of Cell Extracts for Assay of Protein Kinase C

150 cc plates of confluent, quiescent Swiss 3T3 cells, which had been treated with 0, 10, 100, or 1,000 ng PDBu/ml for 48 hr at 37°C, 5.5% $CO_2$, were placed on ice and washed three times with phosphate-buffered saline. Cells were then lysed in 1 ml of 20 mM Tris, 2 mM EDTA, 0.5 mM EGTA, 5 mM dithiothreonine (DTT), 1 mM PMSF, 10 µg approtinin/ml, and 1% Triton-X 100, pH 7.4. Plates were then washed with 0.5 ml lysis buffer, and the lysate and wash were pooled. After centrifugation for 60 min at 35,000 rpm, the supernatants were loaded onto 1 ml DEAE 52 cellulose columns. These columns were washed extensively with 20 mM Tris, 2 mM EDTA, 0.5 mM EGTA, and 1 mM DTT, pH 7.4, before the samples were applied and with 50 ml of the same buffer after samples were loaded. Columns were eluted with a 40 ml continual salt gradient ranging from 0 to .15 M NaCl, and 1 ml fractions were collected. Samples were at all times kept on ice or at 4°C.

## Protein Kinase C Assays of Column Fractions

Fifty lambda of each fraction with 50 lambda reaction mix was assayed for protein kinase C activity at 30°C for 10 min. The reaction mixture contained 20 mM Tris, 5 mM $MgCl_2$, 0.5 mM $CaCl_2$, 0.8 mg histones/ml, 0.1 mM ATP, 2 µCi $\gamma$-$^{32}$P-ATP, 0.32 mg phosphatidyl serine/ml, and 400 ng PDBu/ml, pH 7.4. In control samples, the phosphatidylserine and PDBu were omitted. The reaction was started by addition of the reaction mixture and terminated by spotting 90 lambda of the reaction on phosphocellulose papers. The papers were washed extensively with 30 mM phosphoric acid, dried, and counted for histone-associated radioactivity. The activity of the pooled peak fractions was determined as described above except that, for these determinations, aliquots of 10, 20, 30, 40, and 50 lambda were assayed in triplicate.

## Protein Assays

Protein concentration of cell lysates and pooled peak fractions were determined by absorbance at 595 µm using Bio-Rad protein reagent and BSA as a standard.

## RESULTS

To determine whether the medium from v-*sis*–transformed NRK cells could modify EGF receptors in murine Swiss 3T3 cells, changes in EGF receptor binding were monitored. Cells were initially treated at 37°C for 60 min with either PDBu (100 ng/ml), which activates protein kinase C, or v-*sis*–medium (100 units/ml). These cells were then incubated at 4°C with $^{125}$I-EGF and cell-associated radioactivity determined. Scatchard analysis revealed two populations of EGF receptors in untreated cells with $K_d$ values of approximately $4 \times 10^{-11}$ M and $1 \times 10^{-9}$ M corresponding to approximmately $3 \times 10^2$ and $2 \times 10^4$ receptor molecules per cell, respectively (Fig. 1). Treatment of cells with either medium from v-*sis*–transformed

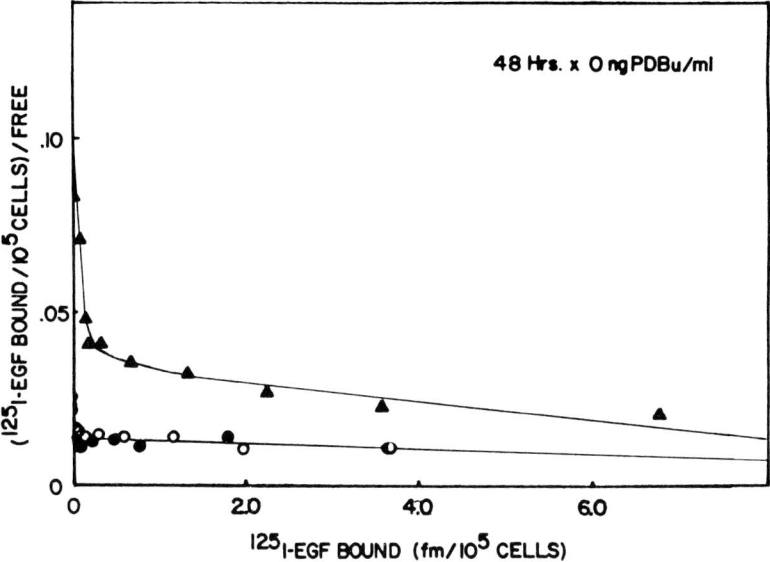

Fig. 1. Scatchard analysis of 4°C $^{125}$I-EGF binding to Swiss 3T3 cells. Confluent quiescent Swiss 3T3 cells were treated at 37°C for 15 min with control medium (▲), PDBu (100 ng/ml; ○), or medium from v-*sis*-transformed cells (100 units/ml; ●). These cells were then assayed for 4°C $^{125}$I-EGF binding as described in Materials and Methods. Analysis of two independent experiments gives high affinity $K_d$ values of $3.7 \times 10^{-11}$ and $2.1 \times 10^{-11}$, and low affinity $K_d$ values of $1.3 \times 10^{-8}$ and $0.91 \times 10^{-8}$, respectively.

cells or PDBu inhibited EGF binding to the high affinity receptor without significantly affecting low affinity EGF binding.

To determine whether the action of this medium on the EGF receptor is dependent on protein kinase C, we investigated the effect of the medium from v-*sis*-transformed NRK cells on EGF binding in cells depleted of protein kinase C. Confluent quiescent Swiss 3T3 cells were depleted of protein kinase C to various extents by exposure to 0, 10, 100, or 1,000 ng/ml PDBu for 48 hr. We assayed the protein kinase C activity in these cells by determining the amount of $Ca^{2+}$ phospholipid-activated kinase actvity in DEAE cellulose fractions of cell extracts. The results indicate that this treatment depletes cells of protein kinase C in a dose-dependent manner (Fig. 2).

Changing cellular levels of protein kinase C did not significantly affect the ability of the media from v-*sis*-transformed cells to inhibit high affinity EGF binding. This lack of dependence of the medium on protein kinase C was observed over a 2 hr time course (Fig. 3) and a 100-fold range in concentration independent of dose (Fig. 4A). In contrast, depletion of protein kinase C in these cells dramatically reduced the ability of PDBu to inhibit high affinity EGF binding over a 100-fold range in concentration (Fig. 4B). These experiments demonstrate that the medium from v-*sis*-transformed cells modulates EGF receptor properties in a completely protein kinase C-independent manner.

To determine whether the PDGF-like factors in the medium from v-*sis*-transformed cells could be responsible for this effect, the action of PDGF was examined. PDGF (Fig. 4C) caused a rapid, dose-dependent loss of high affinity EGF binding

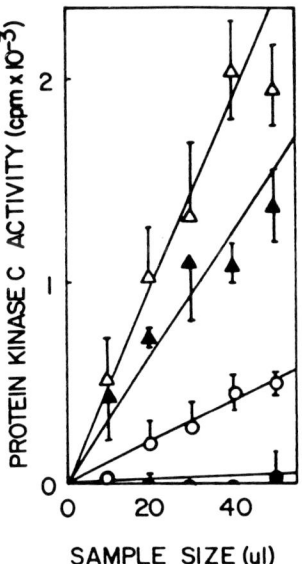

Fig. 2. Protein kinase C activity in pooled peak DEAE column fractions. Peak fractions of protein kinase C activity were pooled and assayed to determine the specific activity of protein kinase C following 48 hr treatment of confluent quiescent cells with 0 (△), 10 (▲), 100 (○), or 1,000 (●) ng/ml PDBu. Each point represents the mean of triplicate determinations of protein kinase C activity from which $Ca^{2+}$ phospholipid-independent kinase activity was subtracted. Standard deviations are indicated by bars.

(80%) in untreated Swiss 3T3 cells. This loss of high affinity EGF binding appeared to be dependent on protein kinase C, although some activity in the protein kinase C-depleted cells was noted. In cells with depleted levels of protein kinase C, PDGF treatment resulted in less than 50% inhibition of EGF receptor binding. These results indicate that PDGF cannot account for the protein kinase C-independent inhibition of high affinity EGF binding by the media from v-*sis*–transformed cells.

The action of PDGF is strikingly similar to that in the medium from v-*sis*–transformed cells following freeze-thaw treatment. After several cycles of freeze-thawing (Fig. 5A) or heat treatment at 100°C for 5 min (data not shown), the medium still caused a rapid, dose-dependent loss of high affinity EGF binding in normal Swiss 3T3 cells that was maintained for 2 hr. However, neither the freeze-thawed medium (Fig. 5A), PDGF (Fig. 5B), nor PDBu (Fig. 5C) was able to inhibit completely the EGF binding in the C kinase-deficient cells over the same time course. These results suggest that at least one component of the media from v-*sis*–transformed cells required for total protein kinase C-independent inhibition of EGF binding may not be related to the PDGF-like peptides.

To identify other potential components in the medium from v-*sis*–transformed cells that might mediate inhibition of EGF binding, we examined the action of transforming growth factor-β (TGF-β) and EGF. TGF-β is secreted into the medium of transformed cells, is labile to heat treatment under certain conditions, and blocks high affinity EGF binding in NRK cells [12,13]. However, TGF-β had little effect on high affinity EGF binding in Swiss 3T3 cells (Fig. 6A) and did not significantly increase the inhibition of EGF binding by PDGF in protein kinase C-depleted cells (Fig. 6B,C). TGF-α, an EGF-like factor, is also secreted into the medium of a

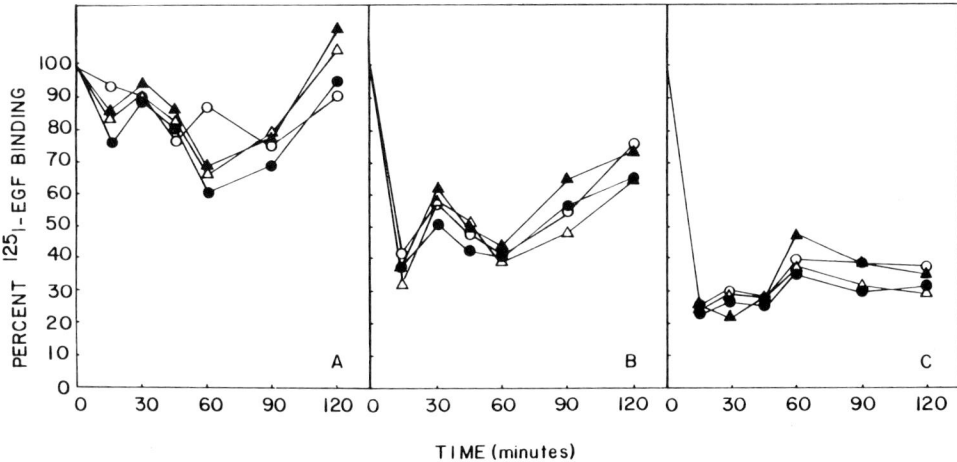

Fig. 3. Effect of protein kinase C depletion on the time course of action of medium from v-*sis*-transformed cells in blocking high affinity $^{125}$I-EGF binding to Swiss 3T3 cells. Confluent quiescent Swiss 3T3 cells were treated with 0 (●), 10 (○), 100 (▲), 1,000 (△) ng/ml PDBu for 48 hr. They were then exposed at 37°C to **A)** 1 unit/ml, **B)** 10 units/ml, or **C)** 100 units/ml of medium from v-*sis*-transformed cells for times ranging from 15 min to 2 hr. The data are expressed as the percentages of $^{125}$I-EGF binding in cells treated with either PDBu or medium from v-*sis*-transformed cells relative to untreated control cells. The specific binding of $^{125}$I-EGF constituting 100% for control and PDBu-pretreated cells ranged from 700 to 1,100 cpm. Nonspecific binding was approximately 100 cpm. Each point represents the mean of triplicate samples. Standard deviations were less than 10%. The concentration of $^{125}$I-EGF used in these studies was 0.05 nM.

number of cells and competes with EGF for binding to the receptor [14,15]. Although no TGF-α was detected in the medium from v-*sis*-transformed cells by competitive binding experiments (data not shown), we examined the effect of EGF pretreatment at 37°C on subsequent EGF binding at 4°C to high affinity receptors in both control and protein kinase C-depleted cells. The results indicate that EGF did not cause complete inhibition of EGF binding in protein kinase C-depleted cells either alone or in conjunction with PDGF (Fig. 7). Thus, neither PDGF, nor TGF-α type factors mimic the protein kinase C-independent action of the media from v-*sis*-transformed cells.

## DISCUSSION

The results presented here demonstrate that growth and transforming factors can alter the properties of the EGF receptor by a number of mechanisms. A scheme depicting some of these pathways is illustrated in Figure 8. PDGF, which stimulates phosphatidylinositol turnover and generation of the second messengers diacylglycerol and calcium, acts primarily through a protein kinase C-dependent pathway. It is possible that the released calcium slightly potentiates the residual protein kinase C in PDBu-pretreated cells or has other effects. EGF, through effects on EGF binding to the high affinity receptor or internalization, acts primarily in a protein kinase C-independent manner. Finally, there are at least two components to the action of media from v-*sis*-transformed cells on EGF receptor binding. The first factor, which is stable and presumably related to p28$^{sis}$, is mimicked by PDGF and appears to be

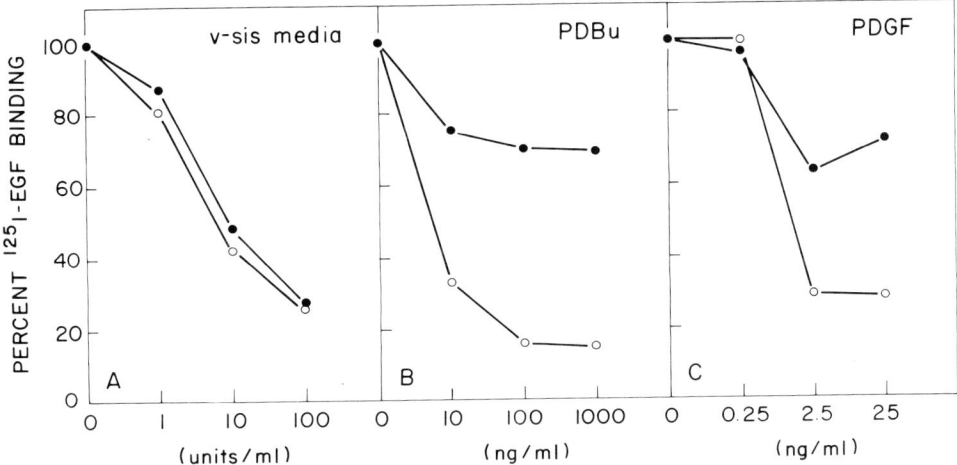

Fig. 4. Dose response of inhibition of $^{125}$I-EGF binding to high affinity receptors in Swiss 3T3 cells treated with PDBu, PDGF, or medium from v-*sis*-transformed cells. Confluent quiescent Swiss 3T3 cells that had been treated for 48 hrs with 0 (○) or 100 (●) ng/ml PDBu were treated for 15 min at 37°C with various doses of **A)** medium from v-*sis*-transformed cells, **B)** PDBu or **C)** PDGF and assayed for $^{125}$I-EGF binding at 4°C. The data are expressed as the percentages of $^{125}$I-EGF binding in cells treated with either PDBu or medium from v-*sis*-transformed cells relative to untreated control cells. In A, the specific binding of $^{125}$I-EGF constituting 100% ranged from 700 to 1,100 cpm for control and PDBu-pretreated cells; in B, representative values for the specific binding of EGF constituting 100% were 3,894 ± 184 cpm for control cells and 4,260 ± 321 cpm for PDBu-pretreated cells. Nonspecific binding was approximately 100 cpm. Each point represents the mean of triplicate samples. Standard deviations were less than 10%. The concentration of $^{125}$I-EGF used in these studies was 0.05–0.1 nM.

dependent on levels of protein kinase C in the cells. The labile, second factor in the media from v-*sis*-transformed cells is required for total protein kinase C-independent changes in EGF receptor properties. The evidence suggests this is a novel factor distinct from PDGF-like peptides (p28$^{sis}$), EGF-like peptides (TGF-α) or TGF-β.

In interpreting these results we have considered the limitations imposed by the technique used to deplete cells of protein kinase C. The properties of the pathways involved in modifying EGF receptor properties, which are depicted in Figure 8, may be altered as a consequence of prolonged exposure to PDBu. Clearly enzymatic and immunoprecipitation assays show that levels of protein kinase C are dramatically reduced by such treatment [16,17]. However, it is likely that there are other changes that directly affect the ability of the growth factors, transforming factors, and tumor promoters to modulate EGF receptor properties. We have observed that the cellular concentration of protein kinase C required for maximum inhibition of EGF receptor binding is four to five times greater in control cells than in cells treated with tumor promoters for prolonged periods of time [25]. Similar discrepancies with respect to $^3$H-PDBu binding and EGF receptor modification in control and C kinase-depleted cells have been reported [18]. Together, these findings indicate that secondary changes that sensitize cells to activators of protein kinase C can be induced by prolonged exposure to tumor promoters.

Despite the possibility of secondary changes, it does appear that protein kinase C is involved in the action of conditioned medium from v-*sis*-transformed cells and PDGF. In the present studies, the extent of inhibition of EGF binding in cells depleted

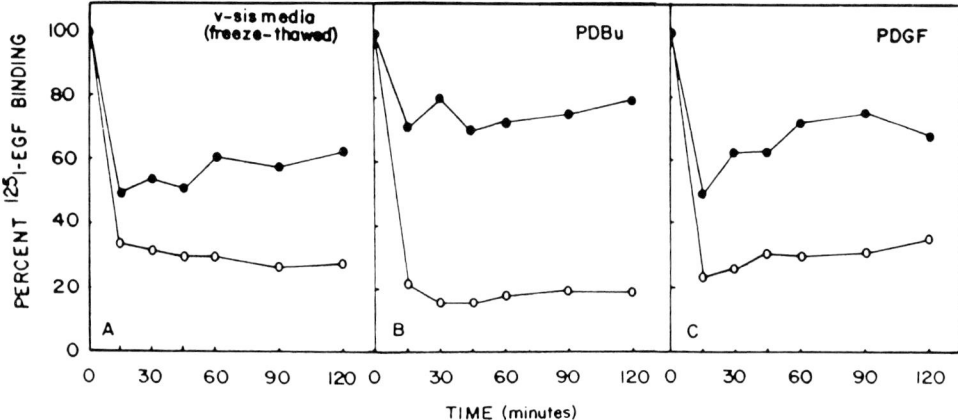

Fig. 5. Comparison between the time course of action of freeze-thawed medium from v-*sis*-transformed cells, PDBu, and PDGF in blocking high affinity $^{125}$I-EGF binding to Swiss 3T3 cells. Confluent quiescent Swiss 3T3 cells that had been incubated for 48 hr with 0 (○) or 100 (●) ng/ml PDBu were treated with **A)** medium from v-*sis*-transformed cells (50 units/ml), **B)** PDBu (100 ng/ml), or **C)** PDGF (25 ng/ml) for up to 2 hr at 37°C and then assayed at 4°C for $^{125}$I-EGF binding to high affinity EGF receptors. The data are expressed as the percentages of $^{125}$I-EGF binding in cells treated with freeze-thawed medium from v-*sis*-transformed cells relative to untreated control cells. Representative values for the specific binding of EGF constituting 100% were 3,894 ± 184 cpm for control cells and 4,260 ± 321 cpm for PDBu-pretreated cells. Nonspecific binding was approximately 100 cpm. Each point represents the mean of triplicate samples. Standard deviations ranged from 5 to 10%. The concentration of $^{125}$I-EGF used in these studies was approximately 0.1 nM.

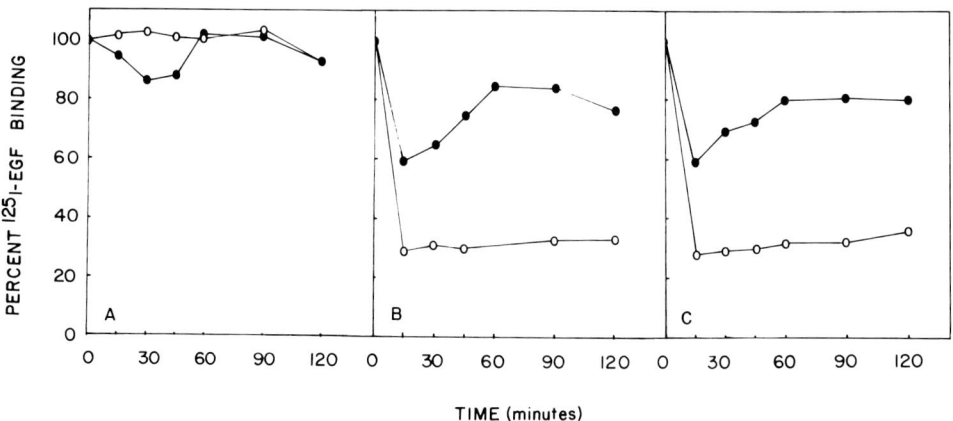

Fig. 6. Time course of specific binding of $^{125}$I-EGF to high affinity receptors following treatment with TGF-β and/or PDGF. Confluent quiescent Swiss 3T3 cells that had been incubated for 48 hr with 0 (○) or 100 (●) ng/ml PDBu were treated with **A)** TGF-β (8 ng/ml), **B)** TGF-β (8 ng/ml plus PDGF) [2.5 ng/ml], or **C)** PDGF (2.5 ng/ml) alone for up to 2 hr at 37°C. Cells were then assayed at 4°C for $^{125}$I-EGF binding to high affinity EGF receptors as described in Materials and Methods. Other conditions were as described in Figure 5.

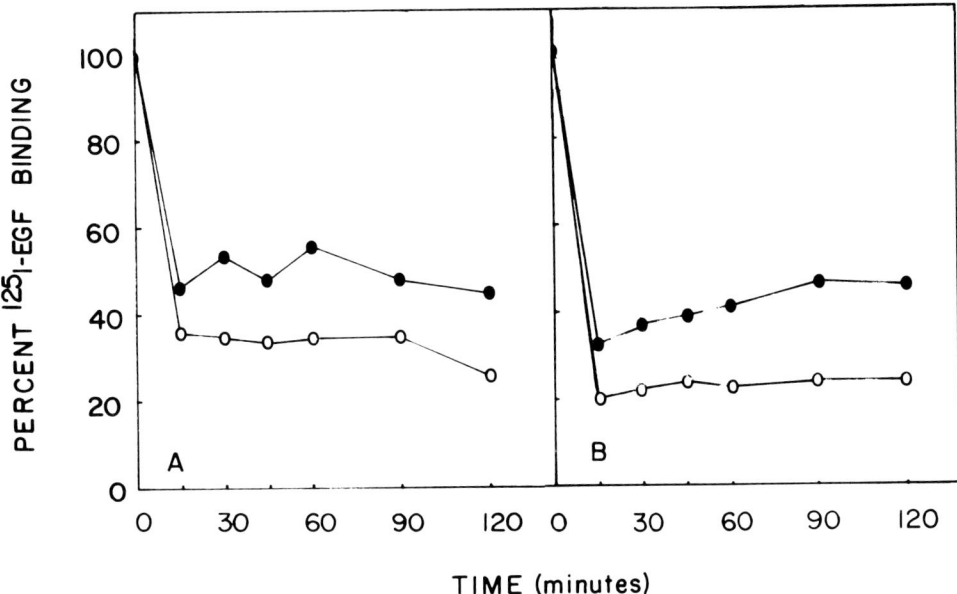

Fig. 7. Time course of specific binding of $^{125}$I-EGF to high affinity receptors following treatment of EGF alone or in combination with PDGF. Confluent quiescent Swiss 3T3 cells that had been incubated for 48 hr with 0 (○) or 100 (●) ng/ml PDBu were treated with **A)** EGF (1 ng/ml) or **B)** EGF (1 ng/ml) plus PDGF (2.5 ng/ml) for up to 2 hr at 37°C. Cells were then assayed at 4°C for $^{125}$I-EGF binding to high affinity EGF receptors. Other conditions were as described in Figure 5.

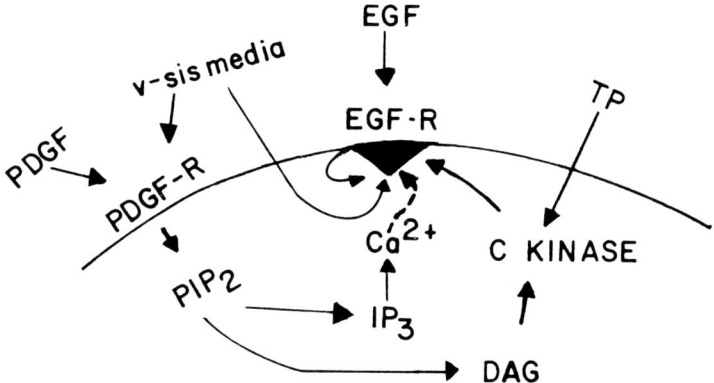

Fig. 8. Scheme depicting multiple pathways by which growth factors may modify the EGF receptor. Tumor promoters (TP) modulate EGF receptor properties through direct activation of protein kinase C. Presumably, the stable, PDGF-like factors in media from v-sis–transformed cells (v-sis media) bind to the PDGF receptor (PDGF-R), and by so doing stimulate phosphatidylinositol biphosphate (PIP$_2$) breakdown to diacylglycerol (DAG) and inositol triphosphate (IP$_3$). In addition to PDGF-like peptides, untreated media from v-sis–transformed cells contains a labile factor required for modification of EGF receptor properties in a protein kinase C-independent manner. The role of calcium in this pathway has not been elucidated.

of protein kinase C is less than that observed in control cells. Thus, these cells appear to be desensitized to the action of PDGF and freeze-thawed medium from v-*sis*-transformed cells on EGF receptor binding. PDGF and very likely the PDGF-like peptides in the medium from v-*sis*-transformed cells generate the endogenous activator of protein kinase C, diacylglycerol [19]. Therefore, the loss of potency of these agents in protein kinase C-depleted cells is consistent with the involvement of protein kinase C in their mechanism of action.

The role of protein kinase C in the action of EGF is not understood. In A431 cells, limited phosphatidylinositol turnover has been reported [20], leading us to postulate that protein kinase C may act in a limited manner as a feedback inhibitor of the EGF receptor [1]. Whether significant phosphatidylinositol turnover occurs in Swiss 3T3 cells is less certain. Habenicht et al [19] found a small but discrete level of diacylglycerol in this cell type following EGF treatment. However, recent evidence from a number of investigators suggests there is no significant phosphatidylinositol turnover in Swiss 3T3 cells treated with EGF [21,22]. The data we have obtained point to a limited role for protein kinase C, or some enzyme sensitive to PDBu down-modulation, in the action of EGF on its own receptor. Since, in Swiss 3T3 cells, EGF can trigger an increase in cellular calcium levels derived from extracellular sources [22], it is possible that protein kinase C is being activated by EGF in this cell type to a limited extent through a calcium-related mechanism. However, it is likely that there are other pathways that constitute the major route for desensitization or feedback inhibition of the EGF receptor by EGF.

The particular signal-transducing pathways that are activated by growth factors such as PDGF and TGF-$\beta$ may be cell-type specific. For example, in Balb/C 3T3 cells PDGF inhibits EGF binding in the complete absence of protein kinase C [26], whereas in Swiss 3T3 cells we note a dependence on this kinase. Similarly, in other cell types, TGF-$\beta$ causes a rapid inhibition of high affinity EGF binding [12,13] that is not observed in Swiss 3T3 cells (see Results). This lack of responsiveness occurs with concentrations of TGF-$\beta$ known to bind specifically to Swiss 3T3 cells [23]. Finally, we do not observe the synergism between TGF-$\beta$ and PDGF that has been reported for other cell types [24]. Thus, there are clearly multiple mechanisms for achieving similar endpoints, and not all these pathways are operational in every cell type.

## ACKNOWLEDGMENTS

We thank J. Porter and C. Stiles for providing the media from NRK-transformed cells, R. Assoian for the gift of purified PDGF, S. Decker for the antiEGF receptor antisera, P. McCaffrey for helpful comments, and E. Fahci for assistance in preparing the manuscript. This work was supported by National Cancer Institute Awards CA35541 and CA40407 to M.R.R.

## REFERENCES

1. Friedman B, Frackelton AR, Ross AH, Connors JM, Fujiki H, Sugimura T, Rosner MR: Proc Natl Acad Sci USA 81:3034–3038, 1984.
2. McCaffrey PG, Friedman B, Rosner MR: J Biol Chem 259:12502–12507, 1984.
3. Cochet C, Gill GN, Meisenhelder J, Cooper JA, Hunter T: J Biol Chem 259:2553–2558, 1984.

4. Iwashita S, Fox CF: J Biol Chem 259:2559–2567, 1984.
5. Davis RJ, Czech MP: J Biol Chem 259:8545–8549, 1984.
6. Decker S: Mol Cell Biol 4:1718–1723, 1984.
7. Nishizuka Y: Nature 308:693–698, 1984.
8. Bowen-Pope DF, Dicorleto PE, Ross R: J Cell Biol 96:679, 1983.
9. Collins MKL, Sinnett-Smith JW, Rozengurt E: J Biol Chem 258:11689–11693, 1983.
10. Davis RJ, Czech MP: Proc Natl Acad Sci USA 82:4080–4084, 1985.
11. Rosner MR, McCaffrey PG, Friedman B, Foulkes JG: In Feramisco J, Ozanne B, Stiles C (eds): "Cancer Cells 3: Growth Factors and Transformation." Cold Spring Harbor, NY: Cold Spring Harbor Laboratory, pp 347–351.
12. Assoian RK, Frolik CA, Roberts AB, Miller DM, Sporn MB: Cell 36:35–41, 1984.
13. Massague J: J Cell Biol 100:1508–1514, 1985.
14. DeLarco JE, Todaro GJ: Proc Natl Acad Sci USA 75:4001–4005, 1978.
15. Massague J: J Biol Chem 258:13614–13620, 1983.
16. Rodriguez-Pena A, Rozengurt E: Biochem Biophys Res Commun 120:1053–1059, 1984.
17. Blackshear PJ, Witters LA, Girard PR, Kuo JF, Quamo S: J Biol Chem 260:13304–13315, 1985.
18. Jaken S, Tashjian AH, Blumberg PM: Cancer Res 41:4956–4960, 1981.
19. Habenicht AJR, Glomset JA, King WC, Nist C, Mitchell CD, Ross R: J Biol Chem 256:12329–12335, 1981.
20. Sawyer ST, Cohen S: Biochemistry 20:6280–6286, 1981.
21. Besterman JM, Watson SP, Cuatrecases P: J Biol Chem 261:723–727, 1986.
22. Moolenaar WH, Aaerts RJ, Tertoolen LGJ, de Laat SW: J Biol Chem 261: 279–284, 1986.
23. Massague J, Like B: J Biol Chem 260:2636–2645, 1985.
24. Roberts AB, Anzano MA, Wakefield LM, Roche NS, Stern DF, Sporn MB: Proc Natl Acad Sci USA 82:119–123, 1985.
25. McCaffrey PG, Rosner MR: Cancer Res 47:1081–1086, 1986.
26. Olashaw NE, O'Keefe EJ, Pledger WJ: Proc Natl Acad Sci USA 83:3834–3838, 1986.

# Molecular Cloning of a Tumor Promoter-Inducible mRNA Found in JB6 Mouse Epidermal Cells: Induction Is Stable at High, but not at Low, Cell Densities

### James H. Smith and David T. Denhardt

*Cancer Research Laboratory, University of Western Ontario, London, Ontario N6A 5B7*

From the mouse JB6 epidermal cell line C122 we have isolated a cDNA clone representing a 1.6-kilobase mRNA, called 2ar, that exhibits a biphasic induction in response to 12-O-tetradecanoyl-phorbol-13-acetate (TPA). The first phase of induction in subconfluent cells is transient, peaking at 6 h after the addition of TPA and returning to noninduced levels by 24 h. When the cells reach plateau density, in the continued presence of TPA, this mRNA is reinduced and remains so upon continued exposure to the tumor promoter. Serum and certain growth factors also induce 2ar mRNA in serum-deprived quiescent fibroblasts. In vitro nuclear "run-on" transcription experiments indicate that the induction of 2ar mRNA is controlled at the transcriptional level.

**Key words:** TPA, tumor promotion, JB6 cells, transcriptional induction, inducible mRNA, 2ar mRNA, mRNA induction, serum-inducible, JB6, TPA, tumor promoter, nuclear "run-off" transcription, gene expression, colony screening, mRNA, growth factors, competence, 3T3 fibroblasts, protein kinase C

The biochemical events that contribute to the development of a tumor are of considerable interest. Although the evidence is good that most human cancers arise in the first instance as the result of an "initiating" genetic change in the DNA of a normal cell, it is nevertheless also evident that the development of a full fledged malignancy is inevitably accompanied by additional genetic and epigenetic changes in the cell; loosely speaking, the terms tumor promotion and tumor progression refer to

---

Abbreviations used: TPA, 12-O-tetradecanoyl-phorbol-13-acetate; DMSO, dimethylsulfoxide; p[+], sensitive to promotion; p[−], not sensitive to promotion; PDGF, platelet-derived growth factor; EGF, epidermal growth factor; SomC, somatomedin C; SDS, sodium dodecyl sulfate; FBS, fetal bovine serum; ssDNA, single-stranded DNA.

Received June 16, 1986; revised and accepted October 23, 1986.

© 1987 Alan R. Liss, Inc.

this process. Tumor progression occurs as the result of the selection of cell variants in the developing tumor that have a growth advantage [1]. Tumor promotion usually refers to a process by which tumorigenesis in initiated cells is facilitated by substances that are not by themselves tumorigenic for cells that have not been initiated [2].

The pre-eminent tumor promoter is 12-O-tetradecanoyl-phorbol-13-acetate (TPA), a potent activator of protein kinase C [3]. When mouse skin, provided that it has been previously initiated, is treated repeatedly with TPA, papillomas and ultimately carcinomas result [4]. We have chosen to study the changes in gene expression that occur in an in vitro paradigm of this system with the belief that from these studies we will gain a better understanding of important events occurring during tumor promotion. The in vitro model is based on the promotable ($p^+$) and non-promotable ($p^-$) clones of the nonclonal JB6 cell line derived from BALB/c mouse epidermal cultures developed by Colburn et al [5,6]. The $p^+$ derivatives, compared to $p^-$ derivatives, more readily acquire the ability to grow in soft agar and to form tumors in animals, when treated with TPA. In contrast to the $p^+$ lines, the $p^-$ lines show a decrease in trisialoganglioside synthesis [7], do not release fibronectin when treated with TPA [8], and do not contain active *pro* genes [9].

The strategy we adopted was based on the hypothesis (which is no longer a hypothesis, given the many examples) that TPA causes changes in gene expression, and that it is this altered gene expression that contributes, in part at least, to tumor promotion. There are numerous and diverse examples of effects of TPA on gene expression. Depending upon the specifics of the cell type, differentiation or maturation can be induced or blocked [reviewed in 10]. Gene expression can be affected at various levels. Transcription of c-*fos* [11], JD15 [12], 16C8 [13], and MEP [14], to name a few genes, is enhanced by TPA. Numerous reports have documented increases in the cytoplasmic abundance of a large number of mRNAs, possibly owing to new transcription, to altered processing/transport of the RNA, or to increased cytoplasmic stability [15–17]. Modification by phosphorylation [18,19] and poly(ADP)ribosylation [20] and enhanced secretion [21] of certain proteins have been detected. The research reported here is the initial report of a project designed to yield molecular clones of mRNA species that in response to TPA treatment are expressed exclusively (or at least to greater extents) in promotable JB6 (eg, clone 22) clones as compared to nonpromotable (eg, clone 30) controls. The particular mRNA identified here, although inducible by TPA, is, however, induced to a similar extent in nonpromotable cells.

## MATERIALS AND METHODS
### Cell Culture

The JB6 cell lines C121 and C122 (promotable) and C130 (nonpromotable), generously provided by Dr. N. Colburn, were grown in Eagles medium containing 8% fetal bovine serum (FBS) (Gibco Labs). Stock cell cultures were always maintained below plateau density. For experiments, cells were seeded at $1 \times 10^5$ cells/150-mm plate and grown for 5 days to obtain confluent cultures; details of specific experiments are given in the legends. TPA (LC Services Corp., Woburn, MA) dissolved in DMSO (Alfa products) was added to 10 ng/ml. PDGF was from Bioprocessing Ltd.; EGF was from Collaborative Research Inc. Swiss 3T3 cells were cultured and treated as described previously [22].

## RNA Isolation and Analysis

RNA was isolated as previously described [23]. For Northern blots the RNA was electrophoresed through 1.1% agarose gels in a 40 mM morpholinopropanesulphonic acid (MOPS)-acetate buffer (pH 7.0) containing 2.2 M formaldehyde [24]. RNA was transferred to nitrocellulose, and hybridization and autoradiography were performed as described [23]. Nick-translated DNA ($10^8$ cpm/$\mu$g) generated from either purified inserts or cDNA-containing plasmids was used to probe blots. When blots were reused, the previous probe was removed by incubation in hybridization buffer at 75°C for 5 min. RNA slot-blotting was performed as described [13]. Quantification of hybridization signals was accomplished by determining the amount of annealed radioactive DNA in a scintillation spectrometer.

## Preparation of the cDNA Library and Isolation of 2ar

The cDNA library was prepared using a modified Okayama-Berg procedure as described by Kowalski et al [25] using poly(A)RNA from C122 cells that had been grown for 5 days in the presence of 10 ng/ml TPA, reseeded again at $1 \times 10^5$ cells/150-mm plate, and grown for a further 5 days under the same conditions. The vector pSS24 [25] contains the origin of replication of the filamentous ssDNA bacteriophages, which enables the generation of ssDNA copies of the library when the latter is superinfected with mutant phage IR1. pSS24 serves as a vector primer for the cloning of the mRNA in a known orientation. This strategy permits the generation of clones carrying inserts with known polarity. Tailing of the first cDNA strand synthesized and linker-dependent recircularization of these clones optimizes the probability of obtaining full- to near-full-length cDNA clones. The "linker" used to recircularize the vector after cDNA synthesis was a synthetic oligonucleotide, 5'AGCTTGGGGGGG3', synthesized for us by F. Graham, McMaster University. (A more detailed presentation of this methodology is in reference [25].) The cDNA clones were obtained by transforming *Escherichia coli* strain R4 with the DNA preparation [26].

The library, consisting of $10^5$ clones in a single-stranded DNA form, was enriched for sequences that should be relatively more abundant in the TPA-treated C122 ($p^+$) cells by employing three rounds of "cascade" hybridization to a Rot value of 500–1,000 against a 200-fold mass excess of poly(A) RNA from C130 cells ($p^-$) that also had been treated with TPA for $2 \times 5$ days. The inserts in the single-stranded DNA were complementary to the mRNA. The hybridization was performed in 0.24 M phosphate buffer, pH 6.8, 0.4% SDS, at 65°C in a 40-$\mu$l volume. RNA/DNA hybrids were separated from nonhybridized nucleic acids on a hydroxyapatite column. After each round of hybridization the nonhybridizing single-stranded DNA was rendered double-stranded by annealing to a 1.9-kb XmnI/PvuII restriction fragment from pBR322, "filling in" with the *E coli* DNA polymerase large fragment, covalently closing with T4 DNA ligase, and amplifying by transfection into *E coli* RR1 [25]. The final nonhybridizing fraction in a double-stranded form was then used to make an enriched sublibrary in *E coli* RR1. To eliminate clones with no or small inserts, a plasmid preparation of the sublibrary was size-selected by agarose gel electroelution and transformed into *E coli* R4 [25].

The size-selected and enriched sublibrary was screened by probing duplicate colony-blots [23] on nitrocellulose with $^{32}$P-labeled cDNA probes made from poly(A)RNA derived from C122 and C130 cells, both treated with TPA. Colonies that gave differential signals after three independent screenings were used to make plasmid preparations. The plasmids from these clones were nick-translated and used to probe Northern blots or slot-blots of RNA preparations from TPA-treated and control cells.

### Nuclear "Run-On" Transcription

Nuclear "run-on" transcription assays were performed with isolated nuclei, from TPA-treated or untreated cells, according to Greenberg and Ziff [11] with a number of modifications [13]. Purified plasmid preparations representing specific mRNAs were blotted onto nitrocellulose (1 µg/slot) [27] and used to analyze the $^{32}$P-labeled nuclear "run-on" preparations. The results were analyzed by autoradiography.

## RESULTS
### Isolation of 2ar

Differential colony screening of the size-selected sublibrary (enriched for sequences present in TPA-treated C122 cells compared to TPA-treated C130 cells, see "Materials and Methods") with $^{32}$P-labeled cDNAs complementary to mRNA from C122 and C130 cells, both treated with 10 ng/ml TPA, yielded the strongly hybridizing cDNA clone 2ar. Seven colonies were picked out in the initial screen that gave a stronger signal when hybridized to the C122 probe versus the C130 probe. Nick-translated plasmid preparations from these seven isolates hybridized to a mRNA species in the same region of the gel and gave similar intensity patterns with various RNA preparations tested; restriction analysis and cross-hybridization of these clones confirmed that they represented at least five different individual cDNA clones of this mRNA (results not shown). 2ar was chosen because it was the longest cDNA, 1.4 kb.

To our surprise, the corresponding mRNA was found to be induced to similar, but variable, levels in the promotable and nonpromotable lines tested (Fig. 1). It was abundant in the sublibrary, because, for reasons we do not understand, the particular preparation of poly(A)RNA that was used to prepare the original library and that was employed in making one of the cDNA probes for colony screening was substantially enriched in this mRNA species (Fig. 1, lane g).

The 2ar mRNA migrates as a unique 1.6-kb species on a formaldehyde-agarose gel (see Fig. 2). The partial DNA sequence information we have obtained so far has not revealed strong homology to sequences in 1985 versions of GenBank or the NBRF protein data bank (see Note Added in Proof). Quantitative analyses of Southern blots indicates that 2ar corresponds to a unique gene in the mouse genome (A.M. Craig, personal communication).

### Induction of 2ar mRNA

Examination of the kinetics of induction of 2ar mRNA in proliferating C122 cells, treated for various times with TPA, revealed a relatively rapid induction of the mRNA that peaked around 6 h (Fig. 2A) and that returned to basal levels by 24 h. As it had already been established that 2ar mRNA was present at an increased abundance in confluent cells treated for 10 days with TPA, this observation led us to examine TPA treatments of longer duration. RNA was prepared from cells harvested after

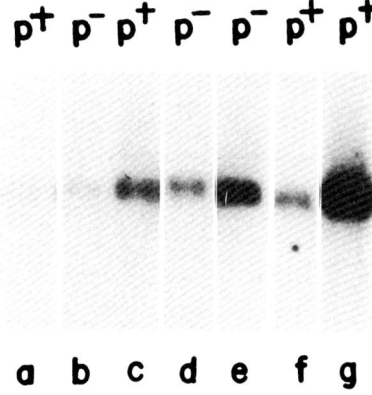

Fig. 1. Northern blot analysis of independent preparations of poly(A)RNA from different JB6 clones with or without TPA (10 ng/ml). Exposure to TPA was for a total of 10 days; after 5 days the cells were replated at $10^5$ cells/150-mm plate. Equal amounts of poly(A)RNA (1 μg per lane) were electrophoresed and transferred to nitrocellulose as described in "Materials and Methods." RNA preparations from different harvests show some variability in the relative amount of 2ar mRNA. **Lanes a** and **b** were from control cells exposed only to 0.01% DMSO; **lanes c–g** were from TPA-treated cells. The clones used were (a) C122; (b) C125; (c) C121; (d and e) C130; (f and g) C122. **Lane g** shows the preparation of poly(A)RNA used to prepare the library. The probe used was nick-translated 2ar in pSS24. $p^+$ and $p^-$ indicate whether the cells in that clone can form colonies in soft agar in the presence of TPA.

extended periods of exposure to TPA. A Northern blot analysis of these RNA preparations (Fig. 2B) disclosed that 2ar mRNA was reinduced after a period of 3 days and remained at levels of increased abundance, maintaining high levels for up to at least 8 days in the presence of TPA. TPA, therefore, induces 2ar mRNA in a biphasic manner in growing cultures. The early induction, when it takes place in cells at subconfluence, is transient in nature, whereas the later increase in confluent cells is sustained.

To ascertain whether the second phase of 2ar induction was dependent upon the duration of TPA exposure or the age or density of the cells, two sets of plates were seeded with C122 cells at low density. TPA was added to one set 1 day after seeding and to the other set 3 days after seeding. Cells from both sets of cultures were harvested at intervals, and 10 μg of total cytoplasmic RNA that was purified from each sample was immobilized on nitrocellulose using a slot-blot apparatus. The resulting blot was probed with the purified 2ar cDNA insert labeled by nick-translation. After autoradiography of the blot, the "slots" were excised, and the bound radioactivity was quantified. A graph of the level of 2ar mRNA versus time after seeding, for both sets of cultures, is shown in Figure 3A. From these data it is evident that the induction of 2ar mRNA is a function of cell concentration rather than of the duration of exposure of TPA.

A plot of the level of 2ar mRNA versus cell density (Fig. 3B) shows that the second wave of induction occurs at the point at which the control cells have reached a plateau density. In the presence of TPA, however, the cells divide one more time, and it is at this point that there is a substantial (10×) increase in the abundance of 2ar mRNA. The appearance of 2ar mRNA cannot be attributed to its having a function in

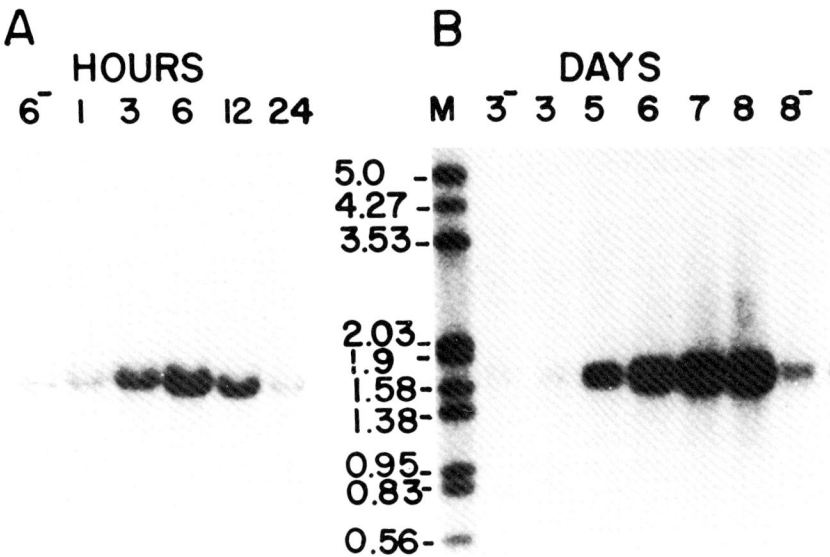

Fig. 2. Northern blot analysis of total cytoplasmic RNA (10 μg) from cells treated for various times with TPA (10 ng/ml) and probed with nick-translated 2ar. **A:** Cells were seeded at $1 \times 10^6$ cells/150-mm plate, incubated 18 h, exposed to TPA for the hours indicated, and then harvested; RNA from cells exposed to DMSO alone for 6 h is shown in the lane labeled 6−. **B:** Cells were seeded at $3.5 \times 10^4$ cells/150-mm plate, incubated for 24 h, and then treated with TPA. Cells from individual plates were harvested after the indicated number of days of TPA exposure; 3− and 8− are DMSO controls. The markers (denatured) are from a *Hind* III/*Eco*RI digest of λ DNA.

this round of replication only, as the level remains high even after the cells have reached a new plateau density and are no longer replicating. Induction at plateau density was also observed with C130 cells, which become confluent at a lower cell density (data not shown).

In order to ascertain the level of control of 2ar expression, nuclear "run-on" transcription analysis was carried out on isolated nuclei from untreated and TPA-treated C122 cells. Figure 4 shows that the transcription of the 2ar gene is enhanced after treatment with TPA for 4 h at a subconfluent cell density. Unfortunately the pSS24 vector alone, perhaps because of the presence of SV40 sequences, gives a positive, though clearly weaker, response. Induction is blocked by the inclusion of α-amanitin in the in vitro assay. This is evidence that 2ar mRNA is an RNA polymerase II transcript. The MGAP (mouse glyceraldehyde phosphate dehydrogenase) and 18S rRNA clones are controls for a noninducible RNA polymerase II transcript and an α-amanitin-resistant RNA polymerase I transcript, respectively.

## Induction of 2ar by Growth Factors in Quiescent Fibroblasts

The striking induction of 2ar mRNA in confluent cells and the growth factor-like properties of TPA led us to examine the possibility that this mRNA might be induced during the $G_0/G_1$ transition when quiescent cells are stimulated with either serum or growth factors that induce competence.

Swiss 3T3 fibroblasts were made quiescent in 0.5% serum for several days and then stimulated with fresh medium containing 10% fetal bovine serum. Subsequent to

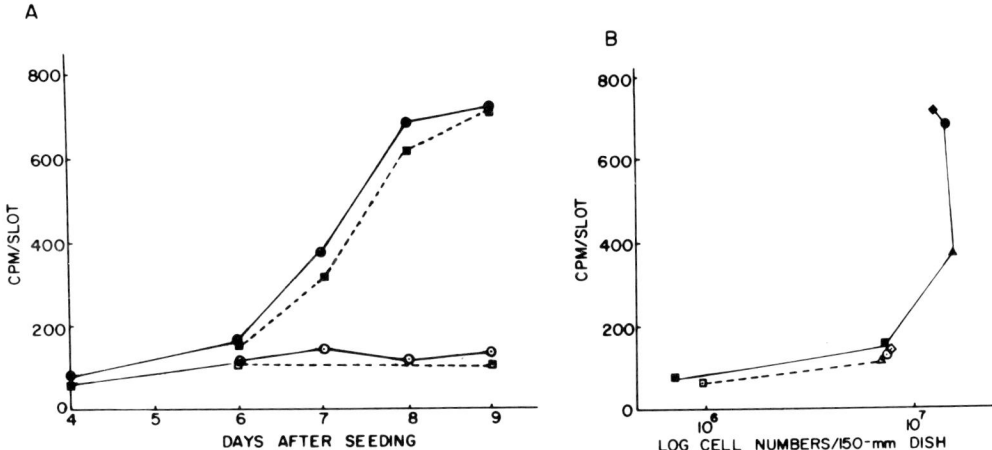

Fig. 3. Induction of 2ar as a function of TPA exposure and cell density. Cells were seeded at $3.5 \times 10^4$ cells/150-mm plate. One set of plates (circles) was left for 24 h and then exposed to TPA (10 ng/ml, solid symbols) or the DMSO solvent (0.01%, open symbols) for various times. The second set (squares) was left for 72 h before being treated. Total cytoplasmic RNA was prepared from plates from each set at the indicated times. The RNA was applied to the nitrocellulose (10 μg/slot), and the resulting blot was probed with 2ar. The hybridized probe was then quantified by scintillation counting as described in "Materials and Methods." **A** shows the amount of $^{32}$P-labeled 2ar DNA bound to each "slot blot" as a function of the number of days after seeding for each set of cells. **B** is a plot of $^{32}$P-labeled 2ar DNA bound to each "slot blot" versus the $\log_{10}$ of the cell number/150-mm plate for that set of cells treated 24 h after seeding. The solid triangle, circle, and square in this plot represent the cultures at 7, 8, and 9 days after seeding.

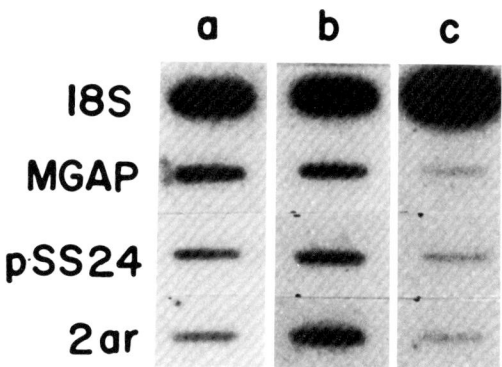

Fig. 4. Transcriptional analysis of the 2ar gene(s) in isolated JB6 C122 nuclei. Nuclei were prepared and the "run-on" transcription done as described in the "Materials and Methods." The filters with the indicated immobilized plasmid DNAs were hybridized with approximately $2 \times 10^6$ cpm/filter. MGAP is a cDNA clone in pBR322 corresponding to mouse glyceraldehyde 3-phosphate dehydrogenase mRNA, whose level is not influenced by growth factors [23]. pSS24 is the vector carrying 2ar. The columns are as follows: **a,** nuclei from control cells exposed only to DMSO; **b,** nuclei from cells exposed to TPA for 4 h; **c,** nuclei from cells exposed to TPA for 4 h and then allowed to generate run-on transcripts in the presence of 2 μg/ml α-amanitin.

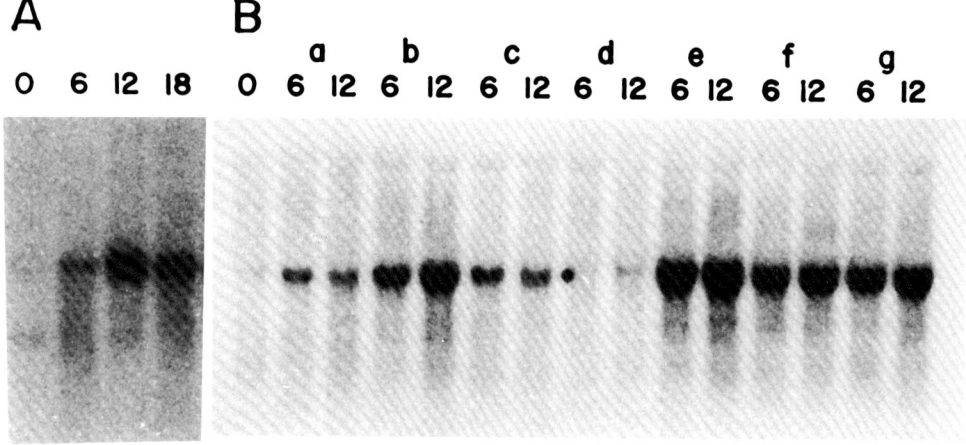

Fig. 5. A Northern blot analysis of total cytoplasmic RNA from Swiss 3T3 cells rendered quiescent and then stimulated with either 10% serum or various combinations of polypeptide growth factors for the duration indicated. The blot was probed with nick-translated 2ar. **A** shows the relative level of 2ar mRNA in quiescent cells and in cells at 6, 12, and 18 h after stimulation with 10% serum. **B** shows the induction of 2ar in quiescent 3T3 cells by treating for either 6 or 12 h with the indicated inducer(s). Lane **0**, no induction; **a**, 10% serum; **b**, PDGF; **c**, EGF; **d**, SomC; **e**, PDGF + EGF; **f**, PDGF + SomC; **g**, PDGF + EGF + SomC. The concentrations of growth factors were as follows: PDGF, 10 ng/ml; EGF, 20 ng/ml; SomC, 26 ng/ml.

stimulation a substantial induction of 2ar mRNA levels occurred (Fig. 5A). The enhancement of 2ar mRNA levels in quiescent Swiss 3T3 cells with individual growth factors, or combinations of growth factors, is shown in Figure 5B. PDGF is a more potent inducer than serum. EGF, a poor inducer alone, has an additive effect with PDGF, whereas somatomedin C has, if anything, a negative effect on induction. It appears that 2ar belongs in the class of mRNAs that are induced when cells are made competent. Protein synthesis is required for its induction (data not shown).

## DISCUSSION

We have isolated and partly characterized a cDNA clone of a mouse 1.6-kb mRNA that does not correspond to a known protein on the basis of the sequence information we have been able to analyze so far. Also, it does not appear identical with other serum- and growth factor-inducible mRNAs that have been described [reviewed in 28]. It is inducible by TPA and other compounds (teleocidin and aplysiatoxin) with promoting activity but not by compounds lacking promoting activity (phorbol-monoacetate and ethylphenyl propiolate) (Smith and Denhardt, unpublished); these two properties correspond roughly with the ability, or lack thereof, to activate protein kinase C. However, since 2ar RNA is induced in both promotable and nonpromotable cells it clearly does not distinguish the two phenotypes.

A particularly interesting feature of this mRNA is the contrast between the transient nature of its induction in proliferating cultures and the permanent expression seen in confluent, nonproliferating cultures. We are not aware of another cellular message with this property, and it will be interesting to elucidate its molecular basis. In the subconfluent cultures, peak mRNA levels were reached about 6 h after the addition of TPA. Protein synthesis was required, in contrast to many of the PDGF-inducible "competence mRNAs" whose induction was not inhibited, but was rather enhanced, by cycloheximide [28]. The cytoplasmic abundance of the 2ar mRNA declined slowly with an apparent half-life of roughly 6 h even though functional TPA was present in the medium. When the cells approached confluence, the cytoplasmic abundance again increased. We do not know whether this regulation is transcriptional, post-transcriptional, or at the level of mRNA turnover. In the confluent cultures in the presence of TPA the level of 2ar mRNA remained high for at least a week.

Other proteins that have been found to be expressed in a density-dependent fashion include bovine aortic smooth muscle collagenase [29] and the cytokeratins. Ben Ze'ev [30] observed that sparse monolayer cultures of Madin Darby bovine epithelial cells synthesized low levels of cytokeratins, whereas in dense cultures they were made at high levels. There was a good correlation between cytokeratin synthesis and the amount of translatable RNA, as assessed by in vitro translation in a reticulocyte lysate. We infer from hybridization analyses at moderately low stringency (single band on "Northern" blots of electrophoretically fractional mRNA) and from partial DNA sequence information (no evidence of similarities to keratins) that 2ar does not correspond to a characterized cytokeratin. Neither its characteristics nor its behaviour during induction resemble the keratin proteins studied by Toftgard et al [31] in TPA-treated mouse skin. We note, however, that keratin expression in vivo differs from that seen in cultured cells [32].

The relevance of this mRNA, if any, to tumor promotion remains to be determined. The fact that its synthesis is induced by TPA in dense, contact-inhibited cells of epithelial origin (JB6) and in serum-limited fibroblasts (3T3) is compatible with the possibility that it is relevant. To approach question more directly it will be necessary to ascertain whether the gene is expressed in mouse skin in vivo in response to TPA treatment, and, if so, whether suppression of its expression, for example with antisense RNA, affects tumor promotion.

## ACKNOWLEDGMENTS

This research was supported by grants from the National Cancer Institute of Canada and the Medical Research Council of Canada. J.H.S. was the recipient of a Terry Fox Training Centre Award. We thank N. Colburn for gifts of cells and information; Dylan Edwards and Jacek Kowalski for useful discussions; Martha Holman, Marilyn McLeod, and Nancy Ng for technical assistance; and Linda Bonis and Beth Orphan for preparation of the manuscript.

## REFERENCES

1. Nowell PC: Cancer Res 46:2203–2207, 1986.
2. Hecker E, Fusenig NE, Kunz W, Marks F, Thielmann HW (eds): "Carcinogenesis and Biological Effects of Tumor Promoters." New York: Raven Press, 1982.
3. Nishizuka Y: Nature 308:693–698, 1984.
4. Berenblum I, Armuth V: Biochim Biophys Acta 651:51–63, 1981.
5. Colburn NH, Bruegge WFV, Bates JR, Gray RH, Rossen JD, Kelsey WH, Shimada T: Cancer Res 38:624–634, 1978.
6. Colburn NH, Former BF, Nelson KA, Yuspa SH: Nature 282:589–591, 1979.
7. Srinivas L, Gindhart TD, Colburn NH: Proc Natl Acad Sci USA 79:4988–4991, 1982.
8. Zerlauth G, Wolf G: Carcinogenesis 6:73–78, 1985.
9. Lerman MI, Hegamyer GA, Colburn NH: Int J Cancer 37:293–302, 1986.
10. Vandenbark GR, Niedel JE: J Natl Cancer Inst 73:1013–1019, 1984.
11. Greenberg ME, Ziff EB: Nature 311:433–438, 1984.
12. Arya SK, Wong-Staal F, Gallo RC: Mol Cell Biol 4:2540–2542, 1984.
13. Edwards DR, Parfett CLJ, Denhardt DT: Mol Cell Biol 5:3280–3288, 1985.

14. Rabin MS, Doherty PJ, Gottesman MM: Proc Natl Acad Sci USA 83:357–360, 1986.
15. Angel P, Rahmsdorf HJ, Pöting A, Lücke-Huhle C, Herrlich P: J Cell Biochem 29:351–360, 1985.
16. Melber K, Krieg P, Fürstenberger G, Marks F: Carcinogenesis 7:317–322, 1986.
17. Lau LF, Nathans D: EMBO J 4:3145–3151, 1985.
18. Rozengurt E, Rodriguez-Pena M, Smith KA: Proc Natl Acad Sci USA 80:7244–7248, 1983.
19. Cochet C, Gill GN, Meisenhelder J, Cooper JA, Hunter T: J Biol Chem 259:2553–2558, 1984.
20. Singh N, Poirier G, Cerutti P: EMBO J 4:1491–1494, 1985.
21. Schorpp M, Mallick U, Rahmsdorf HJ, Herrlich P: Cell 37:861–868, 1984.
22. Parfett CLJ, Hamilton RT, Howell BW, Edwards DR, Nilsen-Hamilton M, Denhardt DT: Mol Cell Biol 5:3289–3292, 1985.
23. Edwards DR, Denhardt DT: Exp Cell Res 157:127–143, 1985.
24. Maniatis T, Fritsch EF, Sambrook J: "Molecular Cloning. A Laboratory Manual." Cold Spring Harbor, NY: Cold Spring Harbor Lab., 1982.
25. Kowalski J, Smith JH, Ng N, Denhardt DT: Gene 35:45–54, 1985.
26. Hanahan D: J Mol Biol 166:557–580, 1983.
27. Kafatos FC, Jones CW, Efstratiadis A: Nucleic Acids Res 7:1541–1552, 1979.
28. Denhardt DT, Edwards DR, Parfett CLJ: Biochim Biophys Acta 865:83–125, 1986.
29. Stepp MA, Kindy MS, Franzblau C, Sonenshein GE: J Biol Chem 261:6542–6547, 1986.
30. Ben-Ze'ev A: J Cell Biol 99:1424–1433, 1984.
31. Toftgard R, Yuspa SH, Roop DR: Cancer Res 45:5845–5850, 1985.
32. Roop DR, Hawley-Nelson P, Cheng CK, Yuspa SH: Proc Natl Acad Sci USA 80:716–720, 1983.

## NOTE ADDED IN PROOF

Comparison of the DNA sequence of 2ar with the sequence published for rat osteopontin (Oldberg A, Franzen A, Hinegard D: Proc Natl Acad Sci USA 83:8819–8823, 1986) reveals that 2ar is the murine equivalent.

# The Ha-ras-Induced Transformed Phenotype of Rat-1 Cells can be Suppressed in Hybrids With Rat Embryonic Fibroblasts

Klaus Willecke, Sabine Griegel, Wolfgang Martin, Otto Traub, and Reinhold Schäfer

*Institut für Zellbiologie, Universität Essen, Hufelandstrasse 55, 4300 Essen 1, Federal Republic of Germany*

Somatic cell hybrids were isolated from fusions of diploid embryonic rat fibroblasts with transformed Rat-1 cells which contained 4 to 5 copies of the transforming human Ha-ras 1 gene. In contrast to their transformed parental cells four hybrid clones showed normal morphology, long latency periods of tumorigenicity in newborn rats, anchorage requirement of proliferation, and an eightfold-reduced amount of secreted transforming growth factor activity. Thus these hybrids are called suppressed with regard to expression of the Ha-ras-induced transformed phenotype. Tumorigenic derivatives of the suppressed hybrids that had segregated chromosomes were isolated. Since two of the tumorigenic hybrid clones showed the similar low level of secreted transforming growth factors as the suppressed hybrids, decreased production of transforming growth factor activity is unlikely to be a sufficient criterion for suppression of malignancy. Whereas one of the suppressed hybrids expressed the transforming gene product p21 at a level similar to that of the transformed parental cells, other suppressed hybrids expressed less p21. This suggests that the suppressed phenotype can be regulated at the posttranslational level of p21 but that additional controls of expression of p21 are likely to exist. DNA of the suppressed hybrids transformed Rat-1 cells to proliferation in the presence of semisolid agar. Thus the activated human Ha-ras gene in the suppressed hybrids retained its biological activity even though it did not transform these cells to tumorigenicity.

**Key words:** cellular hybrids, tumor suppression, Harvey-ras oncogene

It is now well documented that the ability of transformed cells to form tumors can be suppressed in cell hybrids [for review, see 1]. This suppressed phenotype of hybrid cells is counterselected by the appearance of tumorigenic segregrants which

---

Dr. Schäfer's present address is Ludwig Institute for Cancer Research, Bern Branch, Inselspital, 3010 Bern, Switzerland.

Received June 6, revised and accepted November 10, 1986.

© 1987 Alan R. Liss, Inc.

have lost chromosomes of the normal parent. Several groups have reported that the presence of single chromosomes [2–4] or of two and more chromosomes of the normal parental cells is required for suppression of malignancy in somatic cell hybrids [5–8]. Among the tumorigenic cell lines for which suppression by somatic cell hybridization has been demonstrated are several virally transformed cells [9–12] and a human fibrosarcoma cell line containing an activated N-ras gene [8]. In none of these tumorigenic cell lines, however, is it known whether or not the tumorigenicity is due to a single gene mutation. If this is the case the mechanism of suppression may be less complicated than in a cell line that harbors different lesions which contribute to the expression of malignancy. Immortalized, nontumorigenic cells can be transformed to tumorigenicity by transfection with activated human ras genes [13–15]. In this case tumorigenicity appears to be dependent on the presence and expression of an activated ras gene which carries a single nucleotide mutation. We wished to study suppression of tumorigenicity in hybrids of established Rat-1 cells which were transformed with the activated human Ha-ras 1 gene from EJ bladder carcinoma cells. While our work was in progress Craig and Sager [16] published results of a similar approach, ie, the fusion of a Ha-ras transformed Chinese hamster cell line with nontransformed ("normal") established Chinese hamster cells. Similarly to these authors, we found stable suppression of the Ha-ras-induced transformed phenotype. If this effect is indeed due to the presence of a single suppressor gene ("antioncogene") it may eventually be characterized by DNA-mediated transfer rather than somatic cell hybridization.

## MATERIALS AND METHODS
### Cells and Cell Hybridization

Rat 208F cells are derived from Rat-1 cells and defective for hypoxanthine phosphoribosyl transferase activity [17]. The transformed parental cells FE6 and FE8 cells were isolated after transfection of 208F cells with the plasmid pEJ containing the 6.6-kb DNA fragment of the mutated human Ha-ras 1 gene [18]. Furthermore, FE6 and FE8 cells are resistant to the antibiotic G418 due to cotransfection with the pSV2 neoplasmid [19]. Rat embryonic fibroblasts (REFs) were prepared from BDIX rats and used for cell hybridization at the sixth passage. For cell fusions with polyethylene glycol 1500 a 20-fold excess of normal rat fibroblasts over transformed FE6 or FE8 cells was used. Hybrid cells were grown in Dulbecco modified Eagle medium containing 10% fetal bovine serum and selected in the presence of G418 (400 $\mu$g/ml) and HAT (hypoxanthine, aminopterine, thymidine) [20,21]. Hybrid clones FER5 and FER9 were derived from fusion of FE6 and REF cells; the other clones were derived from fusion of FE8 and REF cells. The original number of hybrid clones is not exactly known due to cross-feeding effects under these conditions of selection. Out of 18 isolated proliferating hybrids only the four hybrid clones described in this paper showed flat morphology and were analyzed for their DNA content by using an ICP11 pulse cytophotometer. All hybrids were found to have a tetraploid DNA content except hybrid FER5, which contained a near-triploid amount of DNA. These results were confirmed by karyotype analyses of hybrids FER1 and FER5.

### Southern Blot Hybridization

BamHI-digested DNA of parental and hybrid cells was electrophoresed in agarose gels, transferred to nylon filters, and hybridized following standard conditions [22] to the 2.9-kb Sac I fragment of pEJ, which had been labelled by nick translation [23].

### Immunoprecipitation

Cells were metabolically labelled with $^{35}S$ methionine. Cell lysates were subjected to a standard protocol of immunoprecipitation [24] by using the rat monoclonal antibody Y13-238 (Oncogene Science), which recognizes the p21 product of the Ha-ras gene. The immunocomplexes were electrophoresed and autoradiographed. The amounts of p21 in different cells were compared by densitometric analysis of the autoradiographs. For determination of p21 half-lifetime the cells were pulse labelled for 16 hr with $^{35}S$ methionine followed by different chase periods in the presence of unlabelled methionine [25].

### Growth Factor Analysis

Binding assays with $^{125}I$-epidermal growth factor (EGF, Amersham) were performed on subconfluent monolayers of parental and hybrid cells following the published procedure [26]. For detection of secreted growth factor activity conditioned media of the different cell lines were centrifuged at 100,000g [27] and tested with normal rat kidney (NRK) cells for induction of anchorage-independent growth [28].

### Tumorigenicity

Cells ($1 \times 10^6$) of each cell line were injected subcutaneously into the back of newborn BDIX rats. Tumors were scored positive when they had reached a size of about 0.5 cm$^3$.

### DNA Transfection

Calcium-phosphate-mediated transfections with DNA from cultured hybrid cells together with pSV2 neo-DNA were performed by using 208F or NIH 3T3 recipient cells [29]. G418-resistant colonies were selected and transferred to standard medium containing 0.15% agar (Difco). Colonies in semisolid agar were counted after 3 wk.

## RESULTS

Four somatic cell hybrids of FE6 or FE8 cells with rat embryonic fibroblasts were isolated that exhibited flat morphology and saturation densities of proliferation on plastic surfaces similar to their normal parental cells. Three of these hybrids had a tetraploid karyotype (average number of chromosomes, 82) and one (FER5) had a near-triploid DNA content. Several transformation parameters of the parental and hybrid cells are summarized in Table I. The hybrid cells show saturation densities similar to rat embryonic fibroblasts, ie, one order of magnitude lower than the transformed parental cells. Furthermore, all hybrid cells formed colonies in semisolid agar at frequencies three orders of magnitude lower than the transformed parental cells. When tumorigenicity of parental cells and hybrid cells was compared (Table I) it turned out that all hybrids eventually formed tumors after injection into newborn BDIX rats. The latency period of tumor formation, however, was widely different between the transformed parental cells and their hybrid derivatives. Whereas the transformed parental cells formed progressively growing tumors, palpable already after 4 days, the hybrid clones formed slowly growing tumors only after five- to

**TABLE I. Transformation Parameters**

| Cells | Saturation density[a] | Colony formation in semisolid agar (%)[b] | No. of tumors/ no. of injections[c] | Latency period (days) |
|---|---|---|---|---|
| Parental cells | | | | |
| FE6 | $1 \times 10^7$ | 66 | 7/7 | 4 |
| FE8 | $1 \times 10^7$ | 31 | 8/8 | 4 |
| REF | $1.1 \times 10^6$ | — | — | — |
| Suppressed hybrids | | | | |
| FER1 | $0.5 \times 10^6$ | 0.01 | 1/6 | 33 |
| FER5 | $1 \times 10^6$ | 0.01 | 6/6 | 26 |
| FER8 | $0.5 \times 10^6$ | 0.01 | 5/5 | 26 |
| FER9 | $0.5 \times 10^6$ | 0.03 | 5/5 | 19 |

[a]Number of attached growing cells per 25 cm$^2$.
[b]Cells ($10^2$–$10^4$) were plated into medium containing 0.15% (w/v) Difco Noble Agar.
[c]Per site $10^6$ cells were injected. Tumors were scored positive when they reached a size of about 0.5 cm$^3$.

**TABLE II. Expression of p21 in Hybrids and Parental Cell Lines**

| Cells | Copy no. of human Ha-ras 1[a] | Expression of p21[b] | Doubling time (hr) | Half-life of p21[d] (hr) |
|---|---|---|---|---|
| Parental cells | | | | |
| FE6, FE8 | 4–5 | + | 14 | 18 |
| REF | — | — | 20 | ND |
| Suppressed hybrids | | | | |
| FER1 | 4–5 | + | 31 | 37.5 |
| FER8 | 4–5 | + | 31 | ND |
| FER9 | 4–5 | + | 31 | ND |
| FER5 | 4–5 | + | 24 | 23 |
| Tumorigenic hybrid | | | | |
| FER5T[c] | ND[e] | + | 16 | 16 |

[a]Determined by comparison of Southern blot signals of DNA from cells with known Ha-ras 1 copy numbers.
[b]Ha-ras 1 gene product analyzed by immunoprecipitation.
[c]Cells, isolated from a tumor explant, isolated 5 wk after an injection of $1 \times 10^6$ cells in newborn BDIX rat.
[d]Determined by immunoprecipitation of pulse/chase-labelled p21.
[e]ND, not determined.

eightfold-longer latency periods. In comparison with the original hybrid cells, loss of chromosomes was noticed in tumorigenic hybrid derivatives which had been reestablished in culture for karyotype and biochemical analysis: Modal numbers of 54 chromosomes were found in FER5T and FER8T cells.

Southern blot analysis of DNA from hybrid clones showed that the human Ha-ras 1 gene (6.6-kb BamHI fragment) was present as 4 to 5 copies per genome (Table II). Immunoprecipitations with the rat monoclonal antibody revealed expression of p21 in all hybrid cell lines and in the transformed parental cells. The detailed quantitative comparison of p21 levels showed that only the suppressed FER5 hybrid contained about as much p21 as the transformed parental cells. The other hybrid cells, FER1, FER8, and FER9, contained 27, 11, and 24%, respectively, of the amount of p21 in the transformed parental cells. The reduction of the amount of p21 in these hybrid cells does not appear to be crucial for suppression of malignancy since the transformed phenotype is suppressed in these cells to a similar extent as in FER5 hybrid cells (Table I).

The doubling times of the suppressed hybrids FER1, FER8, and FER9 were significantly longer (ie, 31 hr) than that of the diploid rat embryonic fibroblasts (20 hr) or that of the hybrid FER5 (23 hr). The half life of p21 was compared in several hybrid and parental cells (Table II). In the parental FE6 cells as well as in FER5T tumor-forming hybrid cells half-lifetimes of 18 hr and 16 hr were determined. This is in line with previous results to the effect that the half-lifetime of p21 was 18 hr in cells transformed with Harvey sarcoma virus [25]. In the FER1 hybrid cells we measured a half-lifetime of 37.5 hr. This is about twice as long as found in the transformed parental cells FE6. Table II indicates that the half-lifetimes of p21 in the different hybrid and parental cells appear to be proportional to the doubling times of these cells. Thus we think it unlikely that an extension of the half-lifetime of the transforming gene product p21 is significant for suppression of malignancy in the hybrid cells.

Suppressed hybrids and tumorigenic derivatives were also compared with regard to their binding of $^{125}$I EGF (Table III). TGF$\alpha$ competitively inhibits binding of EGF to the EGF receptor [30]. The suppressed hybrids showed a fourfold increase of free EGF receptors compared to the tumor-forming hybrids. Relatively few free EGF receptors were found on tumorigenic hybrid cells as well as on cells of the transformed parental cell line FE6. This agrees with results previously reported [31,32]— that ras-transformed cell lines showed reduced EGF binding. This can be due to occupation of EGF receptors by TGF$\alpha$ secreted by the transformed cells or due to disappearance of EGF receptors from the cell surface. In order to decide between these two possibilities we measured the transforming growth factor activity secreted

TABLE III. Binding of EGF to Free EGF Receptor and Transforming Growth Factor Activity Secreted by Hybrids and Parental Cells

|  | Binding of $^{125}$I EGF[a] (%) | Cloning efficiency of NRK cells in semisolid agar[b] (%) |
|---|---|---|
| Parental cells |  |  |
| REF | 100 | 1 |
| Rat 1-208F | 100 | 1 |
| FE6 | 22 | 17 |
| Suppressed hybrid cells |  |  |
| FER1 | 50 | 2 |
| FER5 | 42 | 2 |
| FER8 | 72 | 3 |
| FER9 | 52 | 2 |
| Tumorigenic hybrid cells |  |  |
| FER5T | 10 | 2 |
| FER8T | 0 | 20 |
| FER9T | 14 | 1 |

[a]Analyzed with $10^6$ cells. The standard deviations of binding data were within 5% of the mean values for three independent experiments.
[b]NRK cells ($2 \times 10^3$) were cultivated in the presence of 100 $\mu$g protein from conditioned medium per ml of culture medium containing 3% fetal bovine serum in semisolid agar. Note that both TGF$\alpha$ and EGF or TGF$\beta$ must be present in order to induce the formation of large colonies of NRK cells in semisolid agar [33]. The standard deviations of cloning deficiency data were within 2% of the mean values for three independent experiments.

into the culture medium. Transforming growth factors can induce anchorage-independent proliferation of mammalian cells in culture [33]. Table III shows that the suppressed hybrids, like the normal parental REF cells, secreted virtually no transforming growth factor activity as measured by proliferation of normal rat kidney (NRK) cells in the presence of semisolid agar. In contrast, the tumorigenic hybrid derivative FER8T secreted much more transforming growth factor activity, similar to the amount secreted by the transforming parental cell line FE6. However, the data of Table III also indicate that two of the tumorigenic hybrid derivatives, FER5T and FER9T, produced about as little transforming growth factor activity as the corresponding suppressed hybrid cells. Thus we conclude that the lack of secretion of transforming growth factor activity is not a sufficient criterion for suppression of malignancy in hybrid cells.

## DISCUSSION

At the present state of tumor research two different observations appear to contradict each other. On the one hand, DNA transfections with transforming oncogenes isolated from tumor cells or retroviruses suggest that tumorigenesis appears to be due to dominantly acting genes. On the other hand, analysis of large numbers of somatic cell hybrids led to the conclusion that the tumorigenicity of transformed cells can be suppressed in somatic hybrids with normal cells. We have shown in this paper that even the malignant phenotype of transformed Rat-1 208F cells which is caused by the transfected Ha-ras gene from human bladder carcinoma cells can be suppressed in somatic cell hybrids with normal rat embryonic fibroblasts. The same activated Ha-ras gene appears to transform Rat-1 208F cells in a dominant fashion. In order to solve this dilemma one has to recall that neoplastic transformation of diploid rat cells by transfection with isolated oncogenes requires that the Ha-ras oncogene be under control of a strong promotor or that it cooperate with another cotransfected oncogene [34,35]. Furthermore, transformed colonies were selected via expression of a cotransfected resistance gene. Apparently in addition to the uptake and expression of the oncogene one or more further steps are required before a diploid rat fibroblast becomes tumorigenic.

It has been suggested [1,36] that certain gene loci ("suppressor genes" or "antioncogenes") need to be inhibited, inactivated, or deleted before a transforming gene can convert a normal recipient cell to a malignant one. In this context it is interesting that tumorigenic derivatives from Syrian hamster embryonic cells transformed by v-Ha-ras and v-myc oncogenes had consistently lost one copy of chromosome 15 [37]. The extent of suppression of the transformed phenotype differs between human-human and rodent-rodent somatic cell hybrids. Hybrids of human tumorigenic HeLa cells with normal human fibroblasts appear to be suppressed only in their tumorigenicity as measured after injection into nude mice [5,6]. In contrast, hybrids of tumorigenic and normal rodent cells [1,7] are also suppressed with regard to several additional transformation parameters (for example: morphology, requirement for growth factors, colony formation in semisolid agar, etc). These differences between human-human and rodent-rodent hybrids may be caused by different specificities of the products of the putative suppressor genes in the two experimental systems.

The mechanism of suppression of tumorigenicity is not completely understood at present. In several of the suppressed Chinese hamster hybrids [16] the Ha-ras product p21 was expressed to about the same level as in the corresponding transformed Chinese hamster parental cell lines, which suggests posttranslational control of suppression of tumorigenicity. Only one of the suppressed hybrids characterized in this paper showed the same level of p21 expression as the transformed parental cells and a tumorigenic hybrid derivative. This confirms that the phenotype of Ha-ras-transformed cells can be regulated at the posttranslational level of p21, but additional controls of expression of p21 are likely to exist. Apparently oncogene-induced tumorigenicity can be suppressed at several levels of oncogene expression. For example, flat, nontumorigenic revertants of Kirsten sarcoma virus–transformed cells contain elevated amounts of the p21 gene product and can be retransformed by Moloney murine sarcoma virus [38]. Furthermore, it has been shown in hybrids of Rous sarcoma virus–transformed cells and nontransformed cells that suppression of neoplastic transformation occurred at the level of transcription of the oncogene product pp60src [10,39]. In order to dissect the different molecular mechanisms of the suppression of tumorigenicity, cloning and reexpression of the putative suppressor genes in tumorigenic cells are required.

## ACKNOWLEDGMENTS

This work was supported by the Deutsche Forschungsgemeinschaft (SFB 102/A6 and Wi 270/13) and the Fonds der Chemischen Industrie.

## REFERENCES

1. Sager R: Adv Cancer Res 44:43, 1985.
2. Evans EP, Burtenshaw MD, Brown BB, Mennion R, Harris H: J Cell Sci 56:113, 1982.
3. Marshall CJ, Kitchin R, Sager R: Somatic Cell Genet 8:709, 1982.
4. Stoler A, Bouck N: Natl Acad Sci USA 82:570, 1985.
5. Stanbridge EJ, Der CJ, Doersen CJ, Nishimi RY, Peehl DM, Weissman BE, Wilkonson JE: Science 215:252, 1982.
6. Klinger HP, Shows TB: J Natl Cancer Inst 71:554, 1983.
7. Schäfer R, Hoffmann H, Willecke K: Cancer Res 43:2240, 1983.
8. Benedict WF, Weissman BE, Mark C, Stanbridge EJ: Cancer Res 44:3471, 1984.
9. Sager R, Anisowicz A, Howell N: Cell 23:41, 1981.
10. Dyson PJ, Quade K, Wyke JA: Cell 30:491, 1982.
11. Howell N: Cytogenet Cell Genet 34:215, 1982.
12. Noda M, Selinger Z, Scolnick EM, Bassin RM: Proc Natl Acad Sci USA 80:5602, 1983.
13. Reddy EP, Reynolds RK, Santos E, Barbacid M: Nature 300:144, 1982.
14. Tabin CJ, Bradley SM, Bargmann CJ, Weinberg RA, Papageorg AG, Scolnick EM, Dhar R, Lowy DR, Chang EH: Nature 300:143, 1982.
15. Taparowsky E, Suard Y, Fasano O, Shimizu K, Goldfarb M, Wigler M: Nature 300: 762, 1982.
16. Craig RW, Sager R: Proc Natl Acad Sci USA 82:2062, 1985.
17. Quade K: Virology 98:461, 1979.
18. Shih CS, Padhy LC, Murray M, Weinberg RA: Nature 290:61, 1981.
19. Southern PJ, Berg P: J Mol Appl Genet 1:327, 1982.
20. Littlefield JW: Science 145:709, 1964.
21. Calvo Riera F, Blam SB, Teng NNH, Kaplan HS: Somatic Cell Mol Genet 10:123, 1983.
22. Wahl GM, Stern M, Stark GR: Proc Natl Acad Sci USA 76:3683, 1979.
23. Rigby PW, Dieckmann M, Rhodes C, Berg P: J Mol Biol 113:237, 1977.
24. Furth ME, Davis LJ, Fleurdelys B, Scolnick EM: J Virol 43:294, 1982.

25. Ulsh LS, Shih TY: Mol Cell Biol 4:1647, 1984.
26. Todaro GJ, De Larco JE, Cohen S: Nature 264:5258, 1976.
27. Todaro GJ, Tryling C, De Larco JE: Proc Natl Acad Sci USA 77:5258, 1980.
28. De Larco JE, Todaro GJ: Proc Natl Acad Sci USA 75:4001, 1978.
29. Wigler M, Pelliger A, Silverstein S, Axel R: Cell 14:725, 1978.
30. Derynck R, Roberts AB, Eaton DM, Winkler ME, Goeddel DV: In Feramisco J, Ozanne B, Stiles C (eds): "Cancer Cells 3." Cold Spring Harbor, NY: Cold Spring Harbor Laboratory, 1985, pp 79–86.
31. Kamata T, Feramisco JR: In Levine AJ, Van de Woude GF, Topp WC, Watson JD (ed): "Cancer Cells 1." Cold Spring Harbor, NY: Cold Spring Harbor Laboratory, 1984, pp 11–16.
32. Stern DF, Roberts AB, Roche NS, Sporn MB, Weinberg RA: Mol Cell Biol 6:870, 1986.
33. Anzano MA, Roberts AB, Smith JM, Sporn MB, De Larco JE: Proc Natl Acad Sci USA 80:6264, 1983.
34. Land H, Parada LF, Weinberg RA: Nature 304:596, 1983.
35. Spandidos DA, Wilkie NM: Nature 310:469, 1984.
36. Knudson AG: Cancer Res 45:1437, 1985.
37. Osmimura M, Gilmer TM, Barrett JC: Nature 316:636, 1985.
38. Noda M, Selinger Z, Scolnick EM, Bassin RH: Proc Natl Acad Sci USA 80:5602, 1983.
39. Chiswell DJ, Enrietto PJ, Evans S, Quade K, Wyke JA: Virology 116:428, 1982.

# Suppression of Tumorigenicity in Somatic Cell Hybrids Does not Involve Quantitative Changes in Transcription of Cellular Ha-ras, Ki-ras, myc, and fos Oncogenes

R. Schäfer, S. Geisse, and K. Willecke

Institut fur Zellbiologie (Tumorforschung), Universität Essen (GH), D-4300 Essen 1, Federal Republic of Germany

The transcriptional activity of ten cellular oncogenes was analyzed in somatic cell hybrids that had been obtained after fusion of tumorigenic Chinese hamster cells and normal mouse fibroblasts. The hybrids showed either the tumorigenic or the nontumorigenic phenotype (suppression of tumorigenicity). Out of ten c-onc genes analyzed, four (c-Ha-ras, c-Ki-ras, c-myc, and c-fos) were found to be transcriptionally active at similar levels in tumorigenic as well as in nontumorigenic (suppressed) hybrids. Thus we conclude that suppression of tumorigenicity in Chinese hamster × mouse somatic cell hybrids does not correlate with quantitative changes in expression of these cellular oncogenes. The remaining six cellular oncogenes (c-abl, c-erb A and B, c-fes, c-myb, and c-sis) were not transcriptionally active in these hybrids.

Key words: protooncogenes, tumor suppression, somatic cell hybrids, transcription

Tumorigenicity is reduced or even abolished in many somatic cell hybrids generated in vitro after fusion of tumorigenic cells with normal cells (suppression of tumorigenicity; for review, see [1]). The molecular mechanisms of suppression of tumorigenicity in somatic cell hybrids are as yet unknown. Suppression of transformed and tumorigenic phenotypes is speculated to be caused by expression of "suppressor genes" or "antioncogenes." These genes could be structural genes, whose products directly affect proliferation of cells, or regulatory genes, which function by suppression of transforming gene products. Reexpression of transformed and tumorigenic phenotypes in derivatives of the originally nontransformed (suppressed) hybrids has been correlated with the loss of specific chromosomes or

R. Schäfer's present address is The Ludwig Institute for Cancer Research, Bern Branch, Inselspital, CH-3010 Bern, Switzerland.

Received June 5, 1986; revised and accepted December 18, 1986.

© 1987 Alan R. Liss, Inc.

combinations of two or more chromosomes of the normal parent (for review, see [1]). One may hypothesize that phenotypic tumor suppression is due to quantitative changes in expression of cellular oncogenes. If this hypothesis is correct, one would expect a high expression in tumorigenic parental cells and somatic cell hybrids and a reduced or largely impaired expression in suppressed hybrids. Candidate oncogenes for such alterations are the ras, myc, and fos protooncogenes, the transcriptional activity of which has been reported to be higher in malignant than in normal tissue [2,3]. Amplified oncogenes found in many tumor cells are also abundantly expressed at the RNA level (for review, see [4]). Furthermore, the Ha-ras protooncogene is able to induce oncogenic transformation in established NIH/3T3 cells when expressed at a high level [5,6].

Previously we described the suppression of the transformed and tumorigenic phenotype in somatic cell hybrids of spontaneously transformed, malignant Chinese hamster cells and normal mouse fibroblasts [7,8]. In this article we compare the level of transcription of ten different cellular oncogenes in these nontumorigenic somatic cell hybrids, in their tumorigenic derivatives, and in the parental cells. The transcript abundance of c-Ha-ras, c-Ki-ras, c-myc, and c-fos oncogenes was unaltered in transformed and suppressed somatic cell hybrids, whereas c-abl, c-fes, c-erb A and B, c-myb, and c-sis transcriptional activities were not detected.

## MATERIALS AND METHODS
### Cell Culture and Isolation of Somatic Cell Hybrids

Interspecific hybrids were generated by fusion of tumorigenic Chinese hamster cell lines (Wg3-h-o, CI-4, TK17-0, and E 36-o) with early passage BALB/c mouse embryonic fibroblasts as described previously [7,8]. The phenotypes and chromosomal constitutions of these hybrids have been described in detail [8]. Mouse chromosomes were identified by sequential staining with Giemsa (G-banding) and Hoechst 33258. For each hybrid clone, 20 or more metaphase spreads were analyzed. Cells were scored positive when at least 10% of the metaphases contained the particular chromosome. The presence in the hybrids of mouse chromosomes 6, 7, 12, and 15, to which cellular ras, fos, and myc oncogenes had been assigned, was confirmed by detection in cell extracts of the mouse isozymes triosephosphate isomerase (EC 5.3.1.1), glucosephosphate isomerase (EC 5.3.1.9), acid phosphatase 1 (EC 3.1.3.2), and superoxide dismutase 1 (EC 1.15.1.1), respectively [8,9].

### Northern Blot Analysis

Total cellular RNA was prepared from hybrid clones and parental cells by using the guanidine isothiocyanate method [10,11]. RNA was fractionated by electrophoresis through 1% agarose 2.2 M formaldehyde gels for 18 hr at 50 V and subsequently transferred to nitrocellulose filters [12]. Filters were hybridized with $^{32}$P-labeled oncogene-specific DNA fragments. Prehybridization was performed at 42°C for 16 hr in 50% (v/v) formamide, 5× SSC (1× SSC: 0.3 M NaCl, 0.03 M sodium citrate), 5× Denhardt's reagent, 50 mM $Na_2 HPO_4/NaH_2PO_4$, pH 6.5, 0.1% SDS, and 250 µg/ml salmon sperm DNA. Hybridization (36 hr at 42°C) was carried out in 50% (v/v) formamide, 5× SSC, 4× Denhardt's reagent, 20 mM $Na_2HPO_4/NaH_2PO_4$, pH 6.5, 0.1% SDS, 100 µg/ml salmon sperm DNA, and 10% (w/v) dextran sulfate containing the labeled probe. Filters were washed with 2× SSC, 0.1% SDS at room temperature for 30 min, followed by 0.1× SSC, 0.1% SDS at 60°C for 60 min.

## Cytoplasmic Dot Blot Hybridization

Logarithmically growing cells ($1–10 \times 10^6$) were trypsinized, harvested by centrifugation, counted, and lysed in 0.5% (v/v) Nonidet P-40 (Roth, Karlsruhe, FRG) as described [13]. Nuclei were removed by centrifugation (15,000g, 2.5 min). Nuclei-free extracts were denatured by incubation in the presence of 7.5% (v/v) formaldehyde in 15× SSC at 60°C for 15 min, and afterwards frozen at −70°C. Prior to hybridization analysis, extracts were thawed and serially diluted with 15× SSC in a 96-well microtiter plate to yield the indicated cell numbers in a final volume of 50 $\mu$l. Diluted extracts were dotted onto Biodyne A nylon filters (Pall, Glen Cove, NY) using a Manifold SRC-96 apparatus (Schleicher and Schuell, Keene, NH). Nylon filters were hybridized with a $\beta$ actin probe to confirm the appropriate dilution of cytoplasmic extracts (data not shown) and with oncogene-specific DNA fragments. Washed filters were exposed to Kodak XAR-5 films in the presence of intensifier screens for 48 hr at −70°C.

## Oncogene Probes

The following DNA fragments were prepared from plasmids, $^{32}$P-labeled by nick translation (specific activity $\geqslant 1 \times 10^8$ cpm/$\mu$g), and used as probes for the detection of transcripts related to cellular oncogenes: v-Ha-ras (BS9, 0.46 kb BglI/SalI fragment), v-Ki-ras (HiHi3, 0.8 kb HincII fragment), v-myc (1.5 kb PstI fragment), v-fos (1.3 kb BglII/PvuII fragment), v-sis (0.9 kb PstI/XbaI fragment), v-erb A and B (2.5 kb PvuII fragment), v-fes (0.5 kb PstI fragment), v-myb (HAX 4, 1.0 kb HaeI/XbaI fragment), and v-abl (pAB3 Sub3, 1.2 kb SmaI/BglII and 0.8 kb BglII fragments).

## RESULTS

Following somatic hybridization of malignant Chinese hamster cell lines with early passage mouse embryo fibroblasts, hybrids were isolated that exhibited either expression or nonexpression (suppression) of tumorigenicity in nude mice as well as of proliferation in semisolid agar medium, respectively. The phenotype and chromosomal constitution of these hybrids has been described in detail [7,8]. Tumorigenic parental cells and somatic cell hybrids formed tumors in nude mice after subcutaneous injection of less than 50 cells. Their cloning efficiency in semisolid agar medium ranged between 6 and 40%. Compared to these highly malignant cells, suppressed hybrids required a 100- to 50,000-fold inoculum of cells to initiate tumor growth in nude mice, and the latency periods were three- to sixfold longer. Similarly, the ability to proliferate without anchorage was found to be decreased (cloning efficiency in semisolid agar medium ranged from 0.2% to 0.01%). To determine the transcriptional activity of Ha-ras, Ki-ras, myc, and fos protooncogenes in hamster × mouse hybrids, total cellular RNA was prepared from cells in logarithmic growth phase and subjected to Northern blot analysis using v-Ha-ras-, v-Ki-ras-, v-myc-, and v-fos-specific DNA fragments as radioactive probes. Transcripts of expected size were detected (Fig. 1): 1.4 kb (c-Ha-ras); 5.2, 2.0, and 1.2 kb (c-Ki-ras); 2.7 kb (c-myc); and 3.5 and 2.0 kb (c-fos). The predominance of the 2.0 kb c-Ki-ras transcript relative to the other

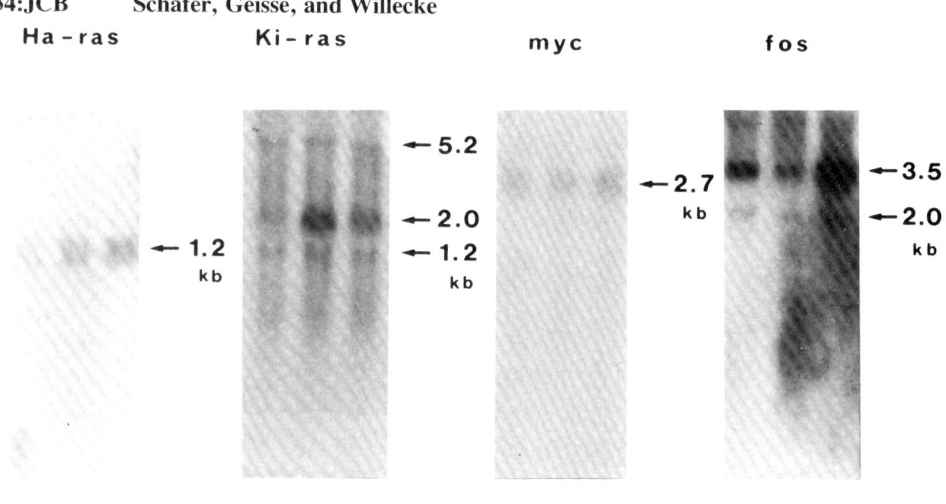

Fig. 1. Size of c-onc transcripts in Chinese hamster × mouse somatic cell hybrids. Northern blot analysis of total RNA from clones 50 BW-6 (**a**), 50 BW-6T (**b**), and 50 BW-6-1a (**c**) as described in Materials and Methods.

two transcripts was not constantly observed in all hybrids. The 3.5 kb transcript related to c-fos represents an unspliced mRNA precursor [14].

We then compared relative mRNA levels of these protooncogenes by dot blot hybridization of cytoplasmic extracts [13] from hybrids and parental cells using the same probes. All cytoplasmic cell extracts were prepared from logarithmically growing hybrid and parental cells. To compare the transcript abundance of c-onc genes in different clones, aliquots of cellular extracts corresponding to $2.5 \times 10^5$, $1.25 \times 10^5$, $6 \times 10^4$, and $3 \times 10^4$ cells from each clone were dotted onto nylon filters as described in Materials and Methods. Transcripts of c-Ha-ras, c-Ki-ras, c-myc, and c-fos were detectable in extracts corresponding to as few as $3 \times 10^4$ cells (Table I). As an example, hybridization of dotted extracts from parental and hybrid cells with a v-Ki-ras probe is shown in Figure 2. By cytoplasmic dot hybridization, transcripts of c-abl, c-erb A and B, c-fes, c-myb, and c-sis were not detectable in extracts corresponding to $2.5 \times 10^5$ or less cells.

Equal levels of mRNA related to c-Ha-ras, c-Ki-ras, c-myc, and c-fos were found in suppressed hybrid clones isolated early after fusion and in tumor outgrowths derived from them (Table I). The transcript abundance of these cellular oncogenes was also unchanged in another set of somatic cell hybrids (clones 2W3, 2W6, 2W14, and 6W4), which were as tumorigenic as the parental hamster cells. The suppressed hybrid clone 50 BW-6 was exceptional in that it had lower mRNA levels related to c-Ha-ras, c-Ki-ras, and c-myc than the other suppressed hybrids. Cells from a tumor outgrowth (50 BW-6T), however, showed again the same transcript abundance as the other hybrids. In general, mRNA levels related to c-Ha-ras, c-Ki-ras, c-myc, and c-fos frequently appeared to be at least twofold elevated in hybrids compared to the parental cells (Table I; exception: 50 BW-6). This result is probably explained by the fact that the hybrids contain a near-tetraploid or near-hexaploid hamster genome as well as a different number of mouse chromosomes [7,8].

The technique used did not allow us to distinguish directly between protooncogene transcripts of hamster or mouse origin. To find out whether the transcriptional

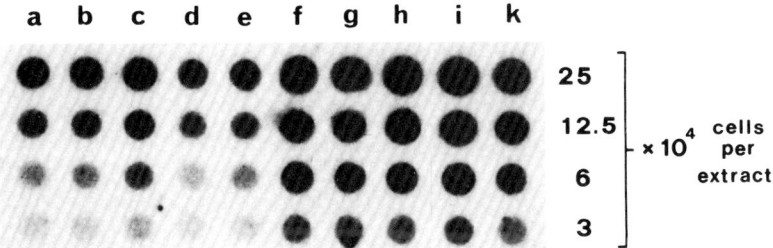

Fig. 2. Cytoplasmic dot hybridization of cellular extracts with a Ki-ras probe. Preparation of cellular extracts and conditions of dot hybridization are as described in Materials and Methods. Probe: 0.8 kb HincII fragment of plasmid HiHi3 (v-Ki-ras), $^{32}$P-labeled by nick translation (specific activity $\geq 1 \times 10^8$ cpm/$\mu$g). **a–d**) Tumorigenic Chinese hamster cell lines Wg3-h-o, CI-4, TK 17-o, and E 36-o, respectively; **e**) normal embryonic fibroblasts; **f–h**) tumorigenic Chinese hamster × mouse hybrids 2W3, 2W6, and 2W14, respectively; **i**) nontumorigenic (suppressed) Chinese hamster × mouse hybrid 20 BW-4; **k**) 20 BW-4T, a cell population isolated from a tumor derived after subcutaneous injection of $1 \times 10^7$ cells of clone 20 BW-4 into a nude mouse (latency period, 30 days).

activity of mouse protooncogenes contributed significantly to the overall expression found in the hamster × mouse hybrids, we compared transcript levels related to c-Ha-ras, c-Ki-ras, c-fos, and c-myc in those hybrids that had either lost or retained the mouse chromosomes, to which the corresponding protooncogene loci had been assigned. The c-Ha-ras, c-Ki-ras, c-fos, and c-myc loci have been assigned to mouse chromosomes 7, 6, 12, and 15, respectively [15–18]. As shown by karyotypic analysis, these mouse chromosomes were retained in hybrid clones 20 BW-4, 50 BW-12, and 2W23, whereas the other hybrid clones had either lost one (eg, clone 2W3), more than one (eg, clone 20 BW-4T), or all of them (eg, 2W14) (Table II). In cellular extracts prepared from clones that had lost a particular mouse chromosome we detected only the hamster isozyme activities of triosephosphate isomerase (gene locus assigned to mouse chromosome 6), glucosephosphate isomerase (mouse chromosome 7), acid phosphatase 1 (mouse chromosome 12), and superoxide dismutase 1 (mouse chromosome 15). Furthermore, there was no evidence of consistent chromosomal translocations involving any of these mouse chromosomes in the hybrids [7–9]. Despite the loss of any of these mouse c-onc loci, the hybrid clones showed the same overall transcriptional activity of protooncogenes as did hybrid clones that had retained the corresponding mouse chromosomes (cf. Table I).

## DISCUSSION

The comparison of oncogene transcription in tumor tissue with normal tissue is hampered not only by the heterogeneity of the tumor biopsy material but also by the presence of different cell types in the corresponding normal tissue. Nontumorigenic (suppressed) hybrids and their tumorigenic segregants are advantageous for such a comparative analysis in that they resemble clonal cell populations. In addition, their tumor-forming capacity can be exactly quantitated by injecting decreasing numbers of cells into immunosuppressed animals [cf. 8]. In human tumors transcriptional activity of c-Ha-ras, c-Ki-ras, c-myc, and c-fos oncogenes was frequently elevated compared to normal tissue [2,3]. We present evidence that the transcriptional activity of these cellular oncogenes was equal in tumorigenic and suppressed hybrids between

**TABLE I. Relative mRNA Levels Related to Cellular Oncogenes in Tumorigenic and Nontumorigenic Chinese Hamster × Mouse Hybrids and Their Parental Cells**

| Cells | Phenotype | c-Ha-ras | c-Ki-ras | c-myc | c-fos | c-abl, c-erb A/B, c-fes, c-myb, c-sis |
|---|---|---|---|---|---|---|
| Parental cells | | | | | | |
| Wg3-h-o | T | + | + | + | + | − |
| CI-4 | T | + | + | + | + | − |
| TK-17-o | T | + | + | + | + | − |
| E 36-o | T | + | + | + | + | − |
| Embryonic fibroblasts | N | + | + | + | + | − |
| Somatic cell hybrids | | | | | | |
| 50 BW-6 | N | + | + | + | ++ | − |
| 50 BW-6T | T | ++ | ++ | ++ | ++ | − |
| 50 BW-12 | N | ++ | ++ | ++ | ++ | − |
| 50 BW-12T | T | ++ | ++ | ++ | ++ | − |
| 20 BW-4 | N | ++ | ++ | ++ | ++ | − |
| 20 BW-4T | T | ++ | ++ | ++ | ++ | − |
| 2W23 | N | ++ | ++ | ++ | ++ | − |
| 2W23T | T | ++ | ++ | ++ | ++ | − |
| 2W3 | T | ++ | ++ | ++ | ++ | − |
| 2W6 | T | ++ | ++ | ++ | ++ | − |
| 2W14 | T | ++ | ++ | ++ | ++ | − |
| 6W4 | T | ++ | ++ | ++ | ++ | − |

Preparation of cellular extracts, conditions of dot hybridization, and oncogene probes as described in Materials and Methods. Intensities of dot hybridization were compared by visual inspection of X-ray films and with the help of a laser densitometer (2202 Ultroscan, LKB, Bromma, Sweden): +, transcripts detected in cell extracts corresponding to $3 \times 10^4$ cells; ++, at least twofold intensity of hybridization; −, no hybridization found. T, tumorigenic phenotype; N, nontumorigenic or partially suppressed phenotype expressed in the indicated hybrid clones or parental cells as described in the text. Hybrids 20 BW-4T, 50 BW-6T, 50 BW-12T, and 2W23 T were isolated from tumorigenic cell populations obtained after subcutaneous injection into nude mice of clones 20 BW-4, 50 BW-12, and 2W23 ($1 \times 10^7$ cells), respectively.

tumorigenic Chinese hamster cells and normal mouse cells. Furthermore, decreased rates of cell proliferation frequently found in suppressed hybrids [cf. 8] were not reflected in an appropriate decrease in transcript abundance of protooncogenes.

The overall mRNA levels related to c-Ki-ras, c-Ha-ras, c-fos, and c-myc were unaltered in those hybrids that had apparently lost the corresponding mouse structural genes as compared to those that still had retained them. Possibly, in hamster × mouse hybrids that retained mouse chromosomes 6, 7, 12, and 15, expression of the corresponding mouse c-oncogenes is suppressed. By analysis of polypeptides expressed in hamster × mouse hybrids (clones 20 BW-4 and 2W23) and separated by two-dimensonal gel electrophoresis, it has been demonstrated previously that the genome of the tumorigenic hamster parent can extinguish expression of a substantial number of mouse genes [9]. In the exceptional hybrid clone 50 BW-6, transcript levels related to protooncogenes were as low as in the parental cells. This clone has a modal number of 32 Chinese hamster and 22 mouse chromosomes [cf. 8]. A possible explanation is that there are trans-acting genes present in the mouse genome that down regulate the level of protooncogene expression in this hybrid clone. The loss of such regulatory genes together with the duplication of the Chinese hamster chromo-

**TABLE II. Loss and Retention of Mouse Chromosomes With c-onc Gene Loci in Chinese Hamster × Mouse Hybrids**

| Hybrid cells | Retention of mouse chromosomes with assigned c-oncogene loci | | | |
|---|---|---|---|---|
| | 6[a] | 7[b] | 12[c] | 15[d] |
| 20 BW-4 | + | + | + | + |
| 20 BW-4T | + | + | − | − |
| 50 BW-6[e] | + | + | + | ND |
| 50 BW-12 | + | + | + | + |
| 50 BW-12T | + | − | − | ND |
| 2W23 | + | + | + | + |
| 2W23T | + | + | − | + |
| 2W3 | + | + | + | − |
| 2W6 | + | + | − | + |
| 2W14 | − | − | − | − |
| 6W4 | − | − | + | − |

All hybrids exhibited equal mRNA levels related to c-Ha-ras, c-Ki-ras, c-fos, and c-myc (cf. Table I). Chromosome analysis and isozyme determinations were as described in Materials and Methods. +, Mouse chromosome found in at least 10% of metaphases analyzed, mouse isozyme activity detected in cell extracts; −, mouse chromosome not detected, isozyme activity not found. ND, karyotype not analyzed.

[a–d]Mouse chromosomes to which c-Ki-ras, c-Ha-ras, c-fos, and c-myc genes, respectively, have been assigned.

[e]Individual mouse chromosomes were not identified in hybrid 50 BW-6T (tumorigenic derivative of 50 BW-6).

some complement may result in an at least twofold increase in the level of protooncogene expression found in a tumorigenic segregant (50 BW-6T).

DNA from the tumorigenic hamster parental cells was used to transfect preneoplastic mouse NIH/3T3 cells. However, no activated oncogene capable of transforming these recipient cells with high efficiency was detectable (R. Schäfer, M. Dubbert, and K. Willecke, unpublished results). Other cellular oncogenes or yet unidentified transforming genes may directly contribute to expression of the transformed phenotype in the parental Chinese hamster cells and their corresponding somatic cell hybrids. The structure and function of putative suppressor genes contributed by the genome of normal cells in a hybrid genome are still unknown [cf. 1]. We favor the hypothesis that these suppressor genes may function via mechanisms not altering expression of oncogenes, eg, by inhibition of the function of oncogene products. This interpretation can be reconciled with the findings that tumorigenicity is suppressed in somatic cell hybrids even in the presence of activated cellular oncogenes [19–22] or viral oncogenes [23,24].

## ACKNOWLEDGMENTS

This work was supported by the Deutsche Forschungsgemeinschaft (SFB 102/A6). We thank Drs. R. Müller (Heidelberg) and R.A. Weinberg (Cambridge) for providing us with plasmids containing viral oncogenes.

## REFERENCES

1. Sager R: Adv Cancer Res 44:43, 1985.
2. Slamon DJ, DeKernion JB, Verma IM, Cline MJ: Science 224:256, 1984.
3. Yaswen P, Goyette M, Shank PR, Fausto N: Mol Cell Biol 5:780, 1985.
4. Alitalo K: TIBS, May 1985, p 194.
5. Chang EH, Furth ME, Scolnick EM, Lowy DR: Nature 297:479, 1982.
6. Pulciani S, Santos E, Long LK, Sorrentino V, Barbacid M: Mol Cell Biol 5:2836, 1985.
7. Schäfer R, Doehmer J, Drüge PM, Rademacher I, Willecke K: Cancer Res 41:1214, 1981.
8. Schäfer R, Hoffmann H, Willecke K: Cancer Res 44:2240, 1983.
9. Bravo R, Schäfer R, Willecke K, MacDonald-Bravo H, Fey SJ, Celis JE: Proc Nat Acad Sci USA 79:2281, 1982.
10. Glisin V, Crkvenjakov R, Buys C: Biochemistry 3:2633, 1974.
11. Chirgwin JM, Przybyla AE, MacDonald RJ, Ritter WJ: Biochemistry 18:5294, 1979.
12. Wahl GM, Stern M, Stark GR: Proc Natl Acad Sci USA 76:3683, 1979.
13. White BA, Bancroft FC: J Biol Chem 257:8569, 1982.
14. Müller R, Verma IM, Adamson ED: EMBO J 2:679, 1983.
15. Crews S, Barth L, Hood L. Prehn J, Calame K: Science 218:1319, 1982.
16. Pratcheva DD, Ruddle FH, Ellis RW, Scolnick EM: Somat Cell Genet 9:681, 1983.
17. Sakaguchi A, Lalley PA, Zabel BU, Ellis RW, Scolnick EM, Naylor SL: Proc Natl Acad Sci USA 81:525, 1984.
18. D'Eustachio P: J Exp Med 160:827, 1984.
19. Benedict WF, Weissman BE, Mark CE, Stanbridge J: Cancer Res 44:3471, 1984.
20. Craig RW, Sager R: Proc Natl Acad Sci USA 82:2062, 1985.
21. Griegel S, Traub O, Willecke K, Schäfer R: Int J Cancer 38:697, 1986.
22. Geiser AG, Der CJ, Marshall CJ, Stanbridge EJ: Proc Natl Acad Sci USA 83:5209, 1986.
23. Noda M, Selinger Z, Scolnick EM, Bassin RH: Proc Natl Acad Sci USA 80:5602, 1983.
24. Dyson PJ, Cook PR, Searle S, Wyke JA: EMBO J 4:413, 1985.

# Genes That Cooperate With Tumor Promoters in Transformation

## Nancy H. Colburn and Bonita M. Smith

*National Cancer Institute, Laboratory of Viral Carcinogenesis, Cell Biology Section, Frederick, Maryland 21701*

> Tumor-promoting phorbol esters, like growth factors, elicit pleiotropic responses involving biochemical pathways that lead to different biological responses. Genetic variant cell lines that are resistant to mitogenic, differentiation, or transformation responses to tumor promoters have been valuable tools for understanding the molecular bases of these responses. Studies using the mouse epidermal JB6 cell lines that are sensitive or resistant to tumor promoter-induced transformation have yielded new understanding of genetic and signal transduction events involved in neoplastic transformation. The isolation and characterization of cloned mouse promotion sensitivity genes *pro*-1 and *pro*-2 is reviewed. A new activity of *pro*-1 has been identified: when transfected into human cancer prone basal cell nevus syndrome fibroblasts but not normal fibroblasts mouse *pro*-1 confers lifespan extension on these cells. Recently, we have found that a *pro*-1 homolog from a library of nasopharyngeal carcinoma, but not the homolog from a normal human library, is activated for transferring promotion sensitivity. The many genetic variants for responses to tumor promoters have also proved valuable for signal transduction studies. JB6 P− cells fail to show the 12-O-tetradecanoyl-phorbol-13-acetate (TPA)-induced synthesis of two proteins of 15 and 16 kD seen in P+ cells. P−, P+, and TPA transformed cells show a progressive decrease in both basal and TPA-inducible levels of a protein kinase C substrate of 80 kD. P− cells are relatively resistant both to anchorage-independent transformation and to a protein band shift induced by the calcium analog lanthanum. It appears that one or more calcium-binding proteins and one or more *pro* genes may be critical determinants of tumor promoter-induced neoplastic transformation.

**Key words: genes specifying sensitivity, tumor promotion, neoplastic transformation**

Tumor-promoting phorbol esters, like growth factors, elicit pleiotropic responses. Some 40 biochemical or biological responses to tumor promoters have been described [1]. Emerging findings in several laboratories (see Table I) [refs. 2–38] suggest that subsets of tumor promoter-elicited responses may lead to mitogenesis, differentiation, or neoplastic transformation. The many genetic variants now available (Table I) can be valuable tools for assigning a response to one of these pathways. One

---

Received August 25, 1986; revised and accepted October 8, 1986.

© 1987 Alan R. Liss, Inc.

**TABLE I. Genetic Variants for Responses to Tumor Promoters**

| Cells or tissue | Variant parameter | References |
|---|---|---|
| HL-60, human promyelocytic leukemia | Phorbol diester receptor down modulation | [2] |
| El-4 mouse thymoma | Protein kinase C substrates | [3] |
| Balb/3T3 mouse | $K^+/Na^+/Cl^-$ transport | [4] |
| *Caenorhabditis elegans* | Shrinking movement responses | [5,6] |
| Initiation-promotion sensitive mouse strains: epidermis | Sustained hyperplasia | [7] |
| Nude mouse epidermis | Hyperplasia | [8] |
| Cultured mouse mammary glands: Mtv-2$^+$ | DNA synthesis stimulation | [9] |
| Human FP fibroblasts | Mitogenic response | [10] |
| Swiss 3T3 mouse | Mitogenic response | [11–14] |
| Swiss 3T3 mouse | Gene amplification | [15] |
| J6B mouse epidermal | Mitogenic response | [16,17] |
| EL-4 thymoma | Interleukin-2 induction | [3] |
| HL-60 | Differentiation responses | [18–20] |
| FELC | Differentiation responses | [21] |
| LC mouse kerantinocyte cell lines | Terminal differentiation | [22–24] |
| Human keratinocytes normal; transformed | Terminal differentiation | [25] |
| Human bronchial epithelial cells; normal; tumor | Terminal differentiation | [26] |
| Human ataxia telangectasia fibroblasts | Cytotoxicity response | [27] |
| Human lymphoma cells: EBV$^+$; EBV$^-$ | Cytotoxicity response | [28] |
| Ad-5 rat embryo fibroblasts | Transformation progression response | [29,30] |
| Rat fibroblasts + PyLT or *myc* | Transformation promotion response | [31] |
| 10 T 1/2 mouse embryo fibroblasts + T24 | Transformation promotion response | [32] |
| Rat embryo fibroblasts + activated *ras* | Transformation promotion response | [33] |
| JB6 mouse epidermal | Transformation promotion response | [34–37] |
| HSV infected NIH 3T3 cells | Promotion response | [38] |

can deduce, for example (Table I), that sensitivity to promotion of neoplastic transformation by phorbol esters can be specified by an activated *ras* [32,33] or by activated *myc* or polyoma large T [31] or by activated *pro* genes [37]. Just how it is that any of these genes cooperates with 12-O-tetradecanoyl-phorbol-13-acetate (TPA) to produce neoplastic transformation is not yet clear. The focus of this chapter is to describe what has been learned about promotion-relevant genes and signal transduction in studies using mouse JB6 cells.

As shown in Figure 1, the mouse epidermal JB6 model system shows several of the characteristics of second-stage tumor promotion in mouse skin [39–41]. Incomplete second-stage tumor promoters such as mezerein or retinoyl phorbol acetate (as

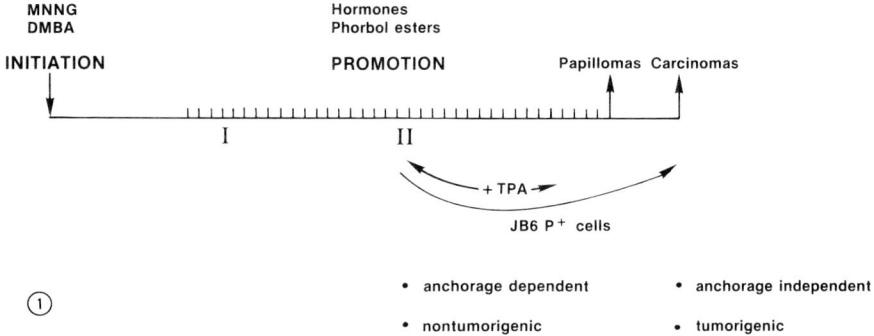

Fig. 1. The JB6 model system: an analog of second-stage mouse skin tumor promotion.

well as complete tumor promoters) induce the transition from nontumorigenic to tumorigenic phenotype. The process is blocked by inhibitors of second-stage tumor promotion such as retinoids but not by first-stage inhibitors such as antiproteases. The period during which the in vivo progression to carcinomas is dependent on continued exposure to tumor promoter ends with the benign papillomas [42], a subpopulation of which are precarcinomatous [43] and require only expression time to give rise to squamous carcinomas. Similarly, tumor-promoter exposure of JB6 promotion-sensitive ($P^+$) cells can be stopped prior to the appearance of soft agar colonies and coincident tumorigenicity without reduced yield of transformants. This indicates that the benign-to-malignant transition has been irreversibly set in motion by tumor-promoter exposure both in vivo and in JB6 cells at a stage that is premalignant.

## GENES THAT SPECIFY PROMOTION SENSITIVITY: *PRO*-1 AND *PRO*-2

Two new putative genes that specify sensitivity to promotion of neoplastic transformation have been cloned by sib selection from a genomic library of JB6 $P^+$ cells [37]. These sequences, designated *pro*-1 and *pro*-2, are different from each other and from known oncogenes.

Sib selection or successive subdivision of active pools first introduced by Cavalli-Sforza and Lederberg [44] offers an approach to isolating genes for which only a biological activity assay is available (ie, when specific molecular probes or antibodies are not available for screening a library). When the DNA from a genomic library is transfected and is shown to transfer the biological activity associated with genomic DNA from the parental cells, a sib selection applied to that library offers a straightforward unbiased way to retrieve and purify DNA sequences that can transfer the activity in question.

In the case of the *pro* genes, five cycles of sib selection identified 20–30% of the pools at each cycle as active [37]. Choosing the single most active pool at each cycle for further analysis finally yielded two active *pro* sequences. It is expected that other active pools not analyzed would yield additional *pro* sequences, as well as reisolates of *pro*-1 and *pro*-2. *Pro*-1 and *pro*-2 can each transfer promotion sensitivity to resistant JB6 $P^-$ cells with a similar specific activity [37]. The transfer of $P^+$ activity saturates at about $10^{-17}$ mol *pro* gene (about 200 pg plasmic DNA) per transfection dish ($6 \times 10^5$ cells), yielding a maximal number of about 100 TPA-induced transfected-DNA-dependent colonies per $10^5$ cells. Dose dependency of $P^+$ activity transferred occurs in the range of about $10^{-18}$ to $10^{-17}$ mol *pro* gene per transfection dish [37].

Table II shows a comparison of the characteristics of cloned mouse *pro*-1 and *pro*-2. *Pro*-1 is intronless with an open reading frame of a size that could specify a 7,000-dalton protein or an RNA of 1,000 nucleotides or less. *Pro*-1-hybridizable RNA levels are increased by exposure of JB6 cells to TPA. The stimulation is greater in $P^+$ than in $P^-$ cells [45]. Although mouse *pro*-1 is composed of sequences complementary to mouse repeats BAM 5 and B1, *pro*-1 appears to occur at a low copy number in the human genome [37,46]. *Pro*-2 appears to be a single-copy gene in both the mouse and the human genomes and to be expressed in mouse skin carcinomas and papillomas [37,46].

## ACTIVITIES OTHER THAN PROMOTION SENSITIVITY THAT ARE SPECIFIED BY *PRO* GENES: LIFESPAN EXTENSION IN HUMAN BCNS CELLS

A number of genes including transforming genes have been found to confer upon transfection different activities depending on the recipient cells used. *Myc*, for example, can confer immortalization or can collaborate with activated *ras* in transformation [47]; when overproduced it can induce lymphoid malignancy in tumorigenic mice [48]. What is apparently reflected is that a given gene can cooperate with other genes in alternative combinations, each leading to a different consequence. Since human genetically cancer-prone cells are presumably preneoplastic, the question arises as to whether a tumor-promoting stimulus would produce in these cells a progression to or toward a neoplastic state. As shown in Figure 2, transfection of basal cell nevus syndrome fibroblasts, but not age-, race-, and sex-matched normal fibroblasts with mouse *pro*-1 produced a substantial extension of lifespan of some 20 population doublings after the point at which control untransfected BCNS cells or transfected normal cells senesced [49]. The DNA transfer showed similar efficiency

**TABLE II. Characteristics of Cloned Mouse *Pro*-1 and *Pro*-2**

|  | *Pro*-1 | *Pro*-2 |
| --- | --- | --- |
| Length of minimum biologically active DNA sequences | 1.05 Kb | 3.7 Kb |
| Homology to other known sequences | Homologous to inverted complements of mouse Bam 5 repeat mouse Alu $B_1$ repeat | No known homologies |
| Estimated copy no. |  |  |
|   Mouse genome | $\sim 10^5$ | Single copy |
|   Human | $\sim 10$ | Single copy |
| Length of maximal ORF with pol II signals | 195 nucleotides | Not known |
| Introns present | No | Yes |
| Evidence for homologous RNAs (transcripts) | Yes, in both basal and TPA-induced $P^+$ cells | Yes, in mouse skin carcinomas |

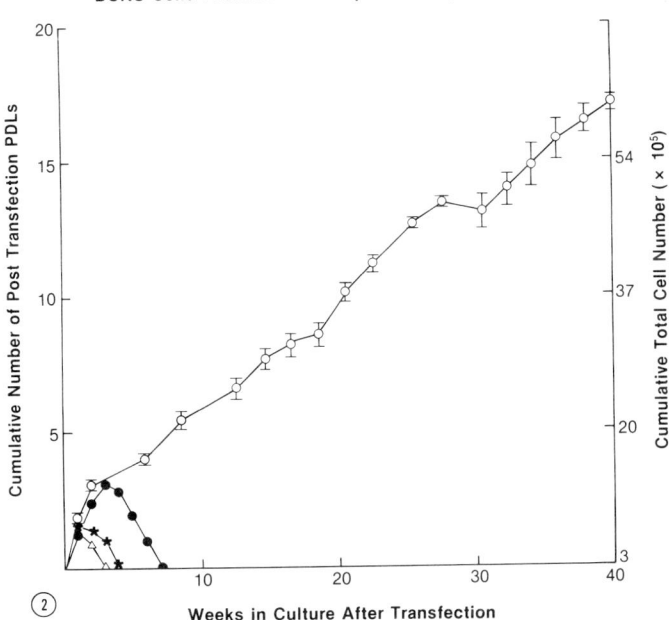

Fig. 2. Lifespan extension in human cancer-prone cells produced by transfection of *pro*-1. Transfection was performed as described [37,49]. BCNS GM 2098 cells transfected with *pro*-1 clone 26 [37] (○—○), BCNS cells transfected with inactive carrier DNA plasmic (★—★), nontransfected BCNS cells (●—●), paired normal GM 2912 cells transfected with *pro*-1 gene (△—△). For BCNS cells transfected with *pro*-1 each point represents the cumulative number of population doublings (PDLs) as well as the cumulative number of cells for the time in culture. Each point represents the mean for duplicate transfections with the range of the two values indicated by vertical bars through the points. In the three cases in which lifespan extension was not observed, the results are expressed only as the cumulative number of cells. Similar results were obtained with two additional sets of duplicate transfections.

in the BCNS and normal cells [49]. Thus *pro*-1 can apparently cooperate (in the absence of TPA) with BCNS gene(s) to produce partial immortalization. Addition of TPA or various growth factors such as EGF or PDGF to the *pro*-1 transfectants did not produce progression to a neoplastic endpoint [49]. Perhaps additional activated genes are required.

## HUMAN NASOPHARYNGEAL CARCINOMA AS AN ANALOG OF TPA-TRANSFORMED JB6 P$^+$ CELLS: ACTIVATED HUMAN *PRO*-1

We have recently reported that DNA from tumorigenic anchorage-independent transformants (T$^3$6274 and RT101) derived by exposure of JB6 P$^+$ cells to TPA can confer anchorage independence when transfected into JB6 P$^+$ cells [50]. This transforming activity appears to be determined by a gene(s) separate from *pro* genes. The DNA from these mouse transformed cells can not only transfer transforming activity into JB6 P$^+$ cells but can also transfer P$^+$ activity into JB6 P$^-$ cells [50], thus suggesting that both genes for induction (*pro* genes) and genes for maintenance (transforming genes) of the neoplastic phenotype must be present in activated form in the DNA of tumorigenic derivatives of JB6 P$^+$ cells. Like mouse transformants T$^3$6274 and RT101, these nasopharyngeal carcinoma cells produce carcinomas when injected into nude mice. In the case of $CNE_1$ and $CNE_2$ cells, the tumors are moderately to well-differentiated squamous carcinomas, whereas the carcinomas produced by RT101 or $T^36274$ are undifferentiated. Recent evidence has shown that the DNA from human nasopharyngeal carcinoma cell lines ($CNE_1$ and $CNE_2$) can transfer to mouse JB6 recipients either $P^+$ activity [46] or transforming activity (not shown).

To determine whether this CNE DNA might harbor activated homologs of mouse *pro*-1 or *pro*-2, a genomic library of $CNE_2$ was constructed and screened to retrieve homologs of both *pro*-1 and *pro*-2 [46]. $CNE_2$ *pro*-1, but not $CNE_2$ *pro*-2, turned out to be activated for $P^+$ activity. Figure 3 shows the specific activity of an activated $CNE_2$ cloned *pro* gene in comparison with cloned mouse *pro*-1 homolog from a normal human library. Human $CNE_2$ *pro*-1 clone i showed $P^+$ molar specific activity comparable to that of activated mouse *pro*-1, while the *pro*-1 homolog from a normal human library showed no $P^+$ activity (Fig. 3). Whether activated *pro*-1 plays a role in the etiology of nasopharyngeal carcinoma and how human *pro*-1 is activated are subjects of ongoing investigation. A working hypothesis is that activated *pro*-1 in concert with a tumor promoter acts to switch on constitutive expression of a transforming gene, which maintains the tumor cell phenotype in both mouse JB6 $P^+$ transformants and in human nasopharyngeal carcinoma cells.

## PROMOTION-RELEVANT SIGNAL TRANSDUCTION: C KINASE SUBSTRATES AND OTHER PROTEINS

Since JB6 $P^-$ cells are defective in a process that begins with the tumor promoter-receptor interaction and ends with the generation of a tumor cell endpoint,

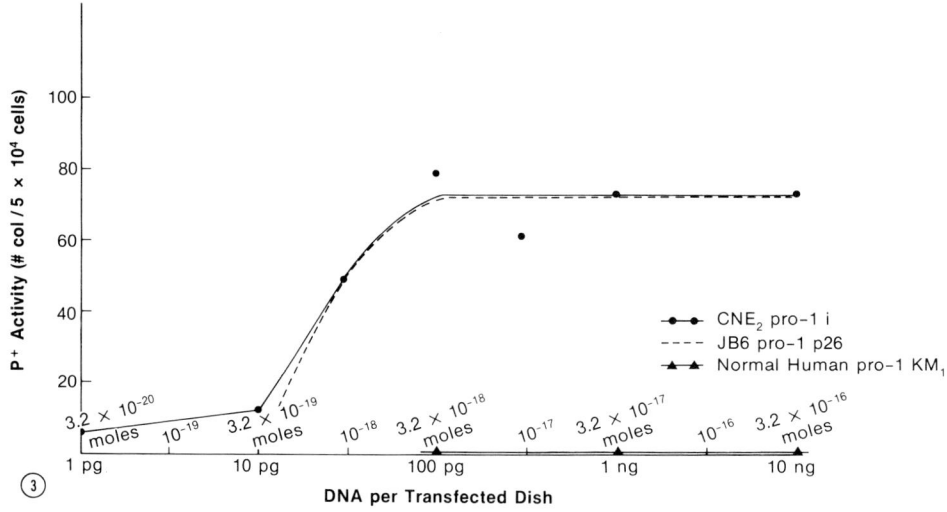

Fig. 3. *Pro*-1 homolog from nasopharyngeal carcinoma but not from normal human cells transfers $P^+$ activity. The $P^+$ assay was carried out as described [37]. One picogram to 10 ng $CNE_2$ *pro*-1 i DNA per dish together with 15 g carrier DNA from $P^-$ recipient cells was transfected into $P^-$ Cl 30 cells by calcium phosphate precipitation, followed by assay of TPA-inducible anchorage-independent colonies. Results are expressed as number of TPA-induced $P^+$ DNA-dependent agar colonies per $5 \times 10^4$ cells. The amount of $CNE_2$ DNA transfected is also expressed on a molar basis. Each point is the mean of duplicate agar dishes after transfection with $CNE_2$ *pro*-1 in a single experiment. Similar results were obtained in two additional experiments. For comparison the dashed line (points not shown) represents data for JB6 C1 22 *pro*-1 clone p26 [37]. Amount of mouse p26 DNA is plotted on a molar basis only.

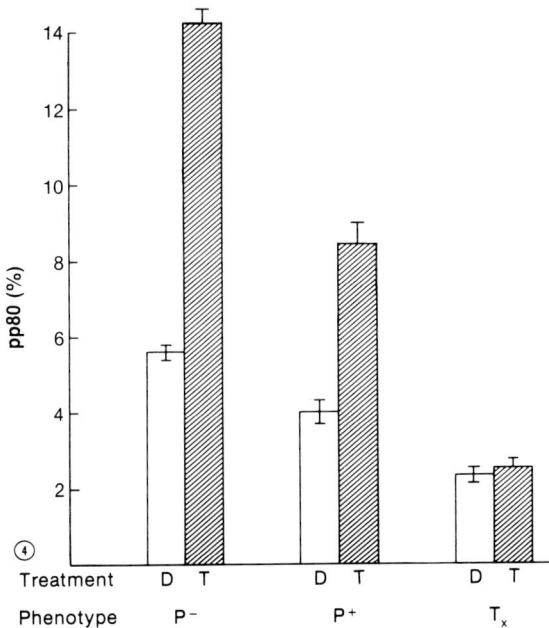

Fig. 4. Inverse relationship between levels of pp80 and stage of neoplastic progression. Cells were labelled for 1 hour in the absence or presence of TPA (10 ng/ml). Triton-X extracted >95% of pp80; results are expressed as a percentage of Triton-soluble cpm. Samples were analyzed by 12% SDS PAGE polyacrylamide gel electrophoresis. Each point represents the mean ± standard deviation of two or more lines of the same phenotype evaluated in two independent experiments (D, DMSO; T, TPA).

such $P^-$ cells might turn out to be defective at any one or more steps from the cell surface receptor binding to various cytoplasmic second messenger events (signal transduction) to altered gene expression in the nucleus. We have found that the $P^+$ and $P^-$ cells show no significant differences in the phorbol ester receptor number or binding affinity [51] or in C kinase activation or substrate availability [21,52]. Since transfer of an activated *pro* gene to $P^-$ cells is sufficient to confer promotion sensitivity [37], the $P^-$ cells are apparently not defective in any signals necessary for activated *pro* genes to function. Since there are undoubtedly other genes cooperating with *pro* genes to bring about transformation, the possibility arises that *pro* gene products might constitute signals for inducing other genes.

Differences in $P^+$ and $P^-$ cells can be distinguished at the level of C kinase substrate phosphorylation [52,53]. Both basal and TPA-inducible levels of an 80-kD phosphoprotein (pp80) were highest in $P^-$ cells, intermediate in $P^+$ cells, and nondetectable in transformed cells (Fig. 4). Thus, although $P^-$ and $P^+$ cells both showed TPA inducibility, there was a relationship between pp80 levels and progression to neoplastic transformation. This inverse relationship between pp80 levels suggests that this phosphoprotein may play a role in maintaining a preneoplastic phenotype, and thus may be transformation relevant. Perhaps the loss of pp80 allows induction to a neoplastic phenotype.

One should note that phosphoproteins that mediate TPA-induced signal transduction and that may account for the promotion sensitivity of the $P^+$ phenotype do not necessarily have to be substrates for PKC. The $P^+$ and $P^-$ phenotypes can also be distinguished by the $P^+$-specific TPA-induced synthesis of two proteins of 15 and 16 kD molecular weight [54]. These proteins are localized in the nucleus and show maximum induction at 20 hours.

## PROMOTION-RELEVANT SIGNAL TRANSDUCTION: LANTHANIDES AS TOOLS FOR UNDERSTANDING CALCIUM-REGULATED EVENTS

One interesting phosphoprotein has been identified that is a PKC substrate and is also sensitive to lanthanum, a transformation promoter that is a pharmacological analog of calcium but does not activate PKC [52]. Lanthanides readily induce neoplastic transformation in JB6 $P^+$ cells (with colony yields comparable to those with TPA of about 2,500 colonies per $10^4$ cells) (Fig. 5). Lanthanides also induce a response in the promotion-resistant cells that is about 20% of the response of $P^+$ cells (in contrast to TPA and other promoters that induce only about 1% of the $P^+$ response in $P^-$ cells). We have found that lanthanides must promote transformation by a mechanism other than C kinase activation. Although lanthanides will substitute for calcium in activating partially purified protein kinase C, these agents failed to activate PKC in intact JB6 cells, as measured by four independent experimental methods [52]. We have found, however, that there is a C kinase substrate (23 kD) found in both $P^+$

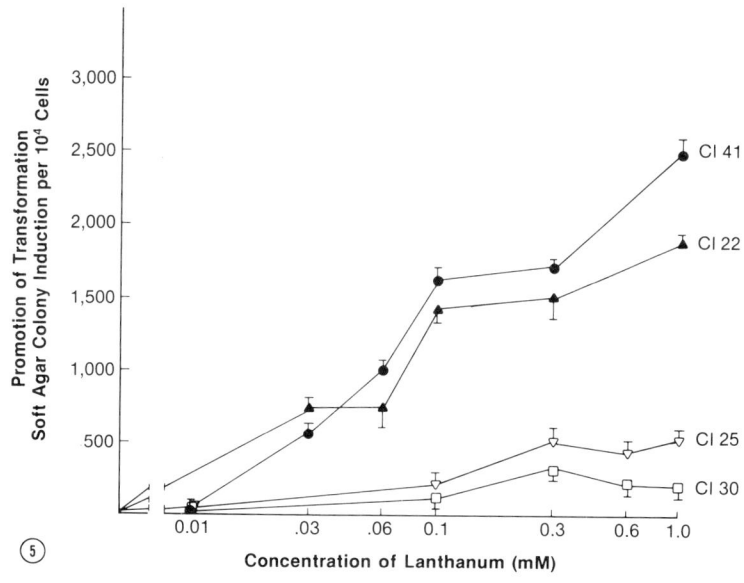

Fig. 5. Lanthanide induction of neoplastic transformation in JB6 $P^+$ and $P^-$ cells. TPA-resistant ($P^-$) or TPA-sensitive ($^+$) preneoplastic JB6 cell lines growing in logarithmic phase in monolayer culture were trypsinized (0.03% trypsin) and suspended in 0.33% agar medium containing 10% serum and DMSO or lanthanum. This agar suspension (1.5 ml containing 10,000 cells/sample) was layered over a bottom agar of 0.5% that contained the same concentration of DMSO or lanthanum. Anchorage-independent colony induction is expressed as number of colonies greater than eight cells in size induced per $10^4$ cells. Each point is the mean of at least three independent experiments each run in duplicate. Results are expressed as the mean ± standard error. Open symbols denote $P^-$ lines; closed symbols are $P^+$ lines.

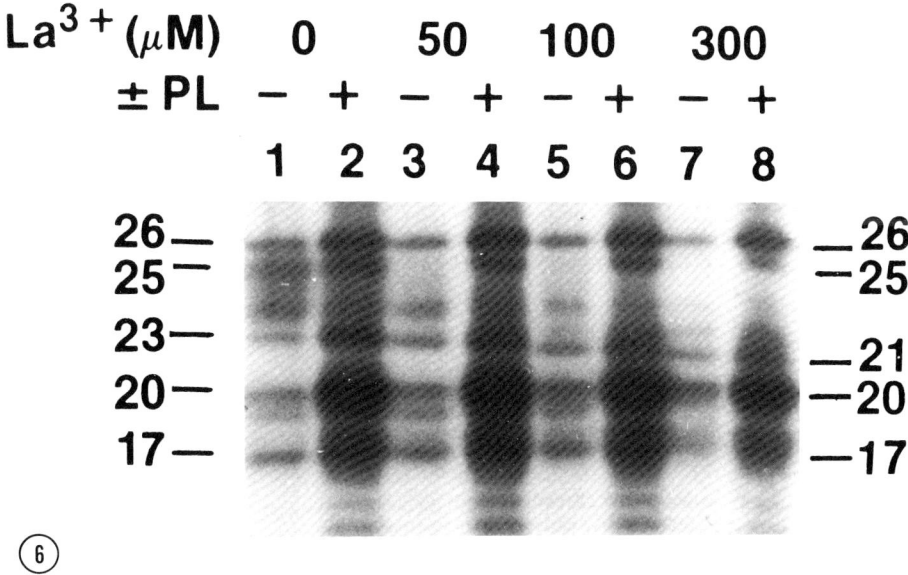

Fig. 6. Lanthanum sensitivity of a protein kinase C substrate. JB6 cell lysates were sonicated in Tris-HCl (20 mM) containing chelators (5.0 mM EGTA and 2.0 mM ethylenediamine tetraacetic acid) and centrifuged to remove nuclei and any remaining intact cells. The supernatant was centrifuged at 100,000g for 1 hour. The recovered supernatant was the cytosolic fraction and contained the calcium and phospholipid-dependent activity. The cytosolic fractions were incubated with required cations and $^{32}$P-ATP in the presence or absence of phospholipid, as indicated. Calcium and phospholipid-dependent phosphoproteins are indicated at molecular weight 26, 25, 23, 20, and 17 kD. Lanthanum, from 50 to 300 μM, was added prior to initiation of the reaction. Note that phosphorylation of the 23kD protein is calcium and phospholipid dependent, and that the 23 kD phosphoprotein exhibits increased migration in the presence of lanthanum. Reprinted with permission [52].

and P$^-$ cells that shows a band shift in response to lanthanum. Both P$^+$ and P$^-$ cells have up to 16 protein kinase C substrates (Smith et al, unpublished); only this one exhibits any change in the presence of lanthanum. Increasing concentrations of lanthanum produce an increased migration of a 23kD PKC substrate to 21 kD in sodium dodecyl sulfate (SDS) polyacrylamide gels (Fig. 6). P$^-$ cells showed a much smaller band shift.

The 23/21-kD lanthanum-sensitive protein kinase C substrate may represent a convergent, coincidental event on the promotion pathways of TPA and lanthanum. How might this potentially promotion-relevant PKC substrate act in the promotion of transformation process? The sensitivity of this substrate to lanthanum suggests a calcium link. Lanthanides substitute for calcium in numerous systems because of their high affinity for calcium-binding sites (10–1,000-fold higher than calcium) [55]. The 23/21-kD protein may in fact be a calcium-binding protein. Calcium-binding proteins in this molecular weight region are PKC substrates and are also known to exhibit altered electrophoretic mobilities in the presence of cations [56,57]. In fact, we know that calcium-regulated events are implicated in TPA-promoted neoplastic transformation. Employing chelators or calcium-deficient medium produces an almost complete inhibition of TPA-promoted transformation, an event that is reversible upon addition of calcium [58].

## MODELS FOR INTERACTION BETWEEN SIGNALS AND GENES

It appears that activated *pro* genes require calcium to function. Recent evidence has shown that in both parental $P^+$ cells and in *pro*-1 and *pro*-2 transfectants calcium depletion inhibits TPA-promoted transformation (Colburn et al, unpublished). Also, lanthanum induces transformation with a magnitude comparable to that observed in $P^+$ parental cells.

How might signals such as those described collaborate with *pro* genes? One possibility might involve a process by which TPA triggers a set of signals that induce *pro*-gene expression. These signals are expected to be identical in $P^+$ and $P^+$ cells (since $P^-$ cells are competent recipients for activated *pro* genes). The products of activated and nonactivated *pro* genes are different and determine whether a neoplastic endpoint is reached. These products may function as differential signals for inducing other genes such as separate transforming genes [50]. The *pro* gene products might be DNA-binding proteins that regulate a transcriptional promoter. A calcium-binding protein or the nuclear p15 and p16 discussed above might be candidates for such a DNA-binding protein.

Alternatively, TPA may trigger a set of signals that induce not *pro* gene expression but the expression of other genes that cooperate with constitutively expressed levels of *pro* genes. The possible status of *pro* gene products as DNA-binding proteins could be similar to that discussed above.

## GENES THAT DETERMINE SENSITIVITY TO TUMOR PROMOTION: WHAT FUNCTIONS DO THEY SPECIFY?

The above described results have made it clear that a dominantly acting single gene can confer sensitivity to promotion of neoplastic transformation by phorbol esters or various hormones. This gene can be an activated *onc* gene such as H-*ras* or one of several genes known to confer an "immortalizing" function such as c- or v-*myc*, polyoma large T, or adenovirus-5 E1a. Or this gene can be one of the recently described promotion sensitivity or *pro* genes that shows no homology to any known *onc* gene or other gene at the DNA level. It is of interest that v-*myc* transfers promotion sensitivity to JB6 promotion-insensitive cells with the same specific activity as *pro* genes (Shimada and Colburn, unpublished).

Can the promotion sensitivity function consist of immortalization? Probably not. Numerous spontaneously immortalized cell lines including mouse 3T3, 10T1/2, and prepromotable (prepassage 35) JB6 cells are not promotion sensitive [32,36]. Likewise, the promotion sensitivity function(s) appears not to simply consist of resistance to terminal differentiation. Differentiation-resistant, putatively initiated keratinocyte cell lines have not as yet been demonstrated to be promotion sensitive (Hennings, personal communication). Perhaps a clonal subpopulation of these cells will turn out to be promotable.

The promotion sensitivity found in mice sensitive to initiation-promotion carcinogenesis appears to be consistently associated with a sustained epidermal hyperplasia response to tumor promoters [7]. The promotion sensitivity found in JB6 $P^+$ cells is consistently associated with a decreased synthesis of ganglioside $G_T$ and a decreased activity of the superoxide anion-removing enzyme superoxide dismutase [1,34,36]. Both of these biochemical responses are dissociable from mitogenic response. Either of these might function as signal transducers for modulating gene expression. Synthesis of pp 80 and of 15 kD/16 kD proteins as well as a band shift in a 21-kD putative calcium-binding protein, discussed above, also distinguish $P^+$ from $P^-$ cells and may function as signal transducers for regulating gene expression. Such $P^+$-specific events could be consequences of *pro* gene expression. As for the function(s) related to promotion sensitivity that are specified by activated H-*ras*, immortalization is probably not involved, since with few exceptions [59] immortalization or establishment appears not to be achieved with this gene. Whether an adenylate cyclase modulation via "G" proteins is involved is not clear [60].

## ARE INITIATORS AND PROMOTERS ACTING ON THE SAME OR DIFFERENT GENETIC LOCI WHEN THEY PRODUCE CHANGES INVOLVED IN THE PROCESS OF PRENEOPLASTIC PROGRESSION?

The central dogma of tumor promotion has in the past held that promotion works only on initiated cells, not on normal or near normal cells that show only transient responses to tumor promoters. This suggested the possibility that tumor promoters might regulate the expression of genes mutated during the initiation event to produce preneoplastic progression. Several recent findings call for a re-examination of this assumption. The finding that mice bred for sensitivity to initiation-promotion skin carcinogenesis have apparently been bred specifically for promotion sensitivity [61–63] suggests that a gene for promotion sensitivity can be inherited independently of whether there exists an activated "initiation" gene. (Even if during the breeding the epidermis contained cells with activated "initiation" genes, such genes would not have been inherited in the germ line.) If promotion-sensitivity genes can be inherited independently of the presence of activated initiation genes, this suggests the possibility of two (or more) separate genetic loci. Another line of evidence that suggests separate genetic loci is that reported by Furstenberger et al, who found that first-stage tumor promotion can be achieved before—even 6 weeks before—initiation [64].

Evidence on gene cooperation [65,66] in transformation suggests 1) that two or more separate genes can cooperate or complement each other to produce a tumor cell and 2) that there is not an obligatory sequence for events that add up to neoplastic transformation. These experiments include the demonstration that *myc* and activated H-*ras* oncogenes can function together but not separately to transform embryo fibroblasts after transfection [65]. The *myc* function can alternatively be provided by other genes such as adenovirus E1a or polyoma large T. If these cooperating genes specify initiating and promoting events, respectively, in these cells, then separate loci are clearly involved. Balmain et al (this UCLA Symposia volume) have demonstrated that activated H-*ras* can function as an initiator of mouse skin carcinogenesis. The identity of the cooperating genes in this case is not clear, but they are presumably genes whose expression is elicited by tumor promoter exposure. In the case of *pro* genes, one can deduce (Table III) that they must be cooperating with gene(s) (in $P^-$ recipient cells) other than activated Ha-*ras* since V-Ha-*ras* will completely transform JB6 $P^-$ recipient cells without a requirement for activated *pro*-1 or *pro*-2 [58]. In this regard the JB6 $P^+$ cells may resemble "spontaneously" initiated human preneoplastic

**TABLE III. Possibilities for Gene Cooperation in Multistage Carcinogenesis***

| Initiation | Promotion |
|---|---|
| *ras*, mutated (activated) | *myc*, overexpressed |
| *myc* | *ras* |
| ? (Non-*ras*) | *pro* + other genes that are promoter-inducible |
| ? (Non-*ras*) | *myc* + other genes |

*Two or more genes in altered or overexpressed form cooperate to produce cancer. The sequence of expression may not be obligatory.

cells. Possibilities for TPA-inducible genes that may cooperate with activated *pro* genes (Table III) can be expected to emerge from the characterization of a hybrid selected cDNA library of TPA-induced JB6 $P^+$ cells reported by Smith and Denhardt (this volume). Other cooperating gene expression events may be elucidated by analysis of the "mal" genes isolated from a skin carcinoma cDNA library by Krieg and Bowden (this UCLA Symposia volume).

A final suggestion for a "separate gene-nonobligatory sequence" mechanism has been set forth by zur Hausen and coworkers [67,68] who suggest that one route to cervical carcinoma in women involves expression of Herpes virus sequences as initiator and of certain human papilloma virus genes as promoters. The Herpes virus expression produces DNA alterations that are characteristic of responses specifically elicited by chemical carcinogens, not tumor promoters, and can be complemented to produce carcinomas by various agents known to show tumor-promoting activity such as certain hormones [67]. Frequently the papilloma virus expression occurs prior to the Herpes virus expression [67,68], suggesting that a constitutively promoted state can be attained independently of initiation and there may not be an obligatory sequence of events.

## REFERENCES

1. Colburn NH: In Barrett JC (ed): "Mechanisms of Environmental Carcinogenesis." Boca Raton: CRC Press, 1986. In press.
2. Solanki TJ, Slaga M, Callahan R, Huberman E: Proc Natl Acad Sci USA 78:1722, 1981.
3. Kramer CM, Sando JJ: Cancer Res 46:3040–3045, 1986.
4. Sussman I, OBrien TJ: J Cell Physiol 121:153, 1985.
5. Lew KK, Chritton S, Blumberg PM: Teratogenesis Carcinog Mutagen 2:19, 1982.
6. Tubuse Y, Miwa, JA: Carcinogenesis 4:783, 1983.
7. Sisskin EE, Gray T, Barrett JC: Carcinogenesis 3:403, 1982.
8. Chambers DA, Cohen RL, Sando JJ, Krueger GG: Exp Cell Biol 52:125, 1984.
9. Rillema JA, Sluyser M: Horm Res 19:52, 1984.
10. Friedman E, Gillin S, Lipkin M: Cancer Res 44:4078, 1984.
11. Butler-Gralla E, Hershman HR: J Cell Physiol 107:59, 1981.
12. Herschman HR: Carcinogenesis 4:489, 1983.
13. Butler-Gralla E, Taplitz S, Herschman HR: Biochem Biophys Res Commun 111:194, 1983.
14. Butler-Gralla E, Herschman HR: J Cell Physiol 114:317, 1983.
15. Herschman HR: Mol Cell Biol 5:1130, 1985.
16. Colburn NH, Wendel EJ, Abruzzo G: Proc Natl Acad Sci USA 78:6912, 1981.
17. Copley M, Gindhart TD, Colburn NH: J Cell Physiol 114:173, 1983.
18. Lotem J, Sachs L: Proc Natl Acad Sci USA 76:5158, 1979.
19. Fisher PB, Schachter D, Abbott RE, Callaham MF, Huberman E: Cancer Res 44:5550, 1984.
20. Mascioli, DW, Estensen RD: Cancer Res 44:3280, 1984.
21. Yamasaki H, Enomoto T, Hamel E, Kanno Y: In Fujiki H (ed): "Cellular Interactions by Environmental Tumor Promoters." Tokyo: VNU Science Press, 1984, p 221.

22. Hennings H, Michael D, Cheng C, Steinert P, Holbrook K, Yuspa SH: Cell 19:245, 1980.
23. Yuspa SH, Morgan DL: Nature 293:72, 1981.
24. Yuspa SH, Ben T, Hennings H, Lichti U: Cancer Res 42:2344, 1982.
25. Parkinson EK, Grabham P, Emmerson A: Carcinogenesis 4:857, 1983.
26. Willey JC, Moser CE, Lechner JJ, Harris CC: Cancer Res 44:5124, 1984.
27. Shiloh Y, Tabor E, Becker Y: Mutat Res 149:283, 1985.
28. Bechet J-M, Guetard D: Int J Cancer 32:61, 1983.
29. Fisher PB, Dorsch-Hsaler K, Weinstein IB, Ginsberg HS: Nature 281:591, 1979.
30. Fisher PB, Bozzone JH, Weinstein IB: Cell 18:695, 1979.
31. Connan G, Rassoulzadegan M, Cuzin F: Nature 314:277, 1985.
32. Hsiao W-LW, Gattoni-Celli S, Weinstein IB: Science 226:552, 1984.
33. Dotto GP, Parada LF, Weinberg RA: Nature 318:473-475, 1985.
34. Gindhart TD, Nakamura Y, Stevens LA, Hegamyer GA, West MW, Smith BM, Colburn, NH: In Mass M (ed): "Tumor Promotion and Enhancement in the Etiology of Human and Experimental Respiratory Tract Cocarcinogenesis." New York: Raven Press, 1985, p 341.
35. Colburn NH, Former BF, Nelson KA, Yuspa SH: Nature 281:589, 1979.
36. Colburn NH, Lerman MI, Hegamyer GA, Wendel EJ, Gindhart TD: In Bishop M, Graves M, Rowley T (eds): "Genes and Cancer." New York: Alan R. Liss Inc., 1984, p 137.
37. Lerman MI, Hegamyer GA, Colburn NH: Int J Cancer 37:293-302, 1986.
38. Schlehofer JR, Matz B, Gissmann L, Heilbronn R, zur Hausen H: In Bishop M, Graves M, Rowley (eds): "Genes and Cancer." New York: Alan R. Liss, Inc., 1984, pp 185-190.
39. Slaga TJ, Fisher SM, Nelson K, Gleason GE: Proc Natl Acad Sci USA 77:3659-3663, 1980a.
40. Slaga TF, Klein-Szanto AJP, Fisher SM, Weeks CE, Nelson K, Major S: Proc Natl Acad Sci USA 77:2251-2254, 1980b.
41. Furstenberger GD, Berry DL, Sorg B, Marks F: Proc Natl Acad Sci USA 78:7722, 1981.
42. Hennings H, Shores R, Wenk ML, Spangler EF, Tarone R, Yuspa SH: Nature 304:67, 1983.
43. Burns FJ, Vanderlaan M, Snyder E, Albert RE: In Slaga TJ, Sivak A, Boutwell RK (eds): "Carcinogenesis, Vol 2, Mechanisms of Tumor Promotion and Cocarcinogenesis." New York: Raven Press, 1978, p 91.
44. Cavalli-Szorza LL, Lederberg J: Genetics 41:367-381, 1956.
45. Lerman MI, Colburn NH: In Cooper GM (ed): "Viral and Cellular Oncogenes." Boston: Martinus Nijhoff Publ., 1986, in press.
46. Lerman MI, Sakai A, Yao KT, Colburn NH: Carcinogenesis 8:121-127, 1987.
47. Land H, Parada LF, Weinberg RA: Nature 304:596, 1983.
48. Adams JM, Harris AW, Pinkert CA, Corcoran LM, Alexander WS, Cory S, Palmiter RD, Brinster RL: Nature 318:533-538, 1985.
49. Shimada T, Gindhart TD, Lerman M, Colburn NH: Int J Cancer 1986, in press.
50. Colburn NH, Lerman MI, Hegamyer GA, Gindhart TD: Mol Cell Biol 5:890-893, 1985.
51. Colburn NH, Gindhart TD, Hegamyer GA, Blumberg PM, Delclos KB, Magun BE, Lockyer J: Cancer Res 42:3093, 1982.
52. Smith BM, Gindhart TD, Colburn NH: Carcinogenesis 7:1949-1956, 1986
53. Gindhart TD, Stevens L, Copley MP: Carcinogenesis 5:1115-1121, 1984.
54. Hirano K, Smith BM, Colburn NH: Fed Proc 45:1581, 1986.
55. Martin RB, Richardson FS: Q Rev Biophys 12:181-209, 1979.
56. Mazzei GJ, Kuo JF: Biochem J 218:361-369, 1984.
57. Wall CM, Grand RJA, Perry SV: Biochem J 195:307-316, 1981.
58. Smith BM, Colburn NH, Gindhart TD: Cancer Res 46:701-706, 1986.
59. Yoakum GH, Lechner JF, Gabrielson EW, Korber BE, Malen-Shibley L, Willey JC, Valerio MG, Shamsuddin AM, Trump BF, Harris CC: Science 227:1174, 1985.
60. Gibbs JB, Sigal IS, Poe M, Skolnik EM: Proc Natl Acad Sci USA 81:5704, 1984.
61. Boutwell RF: Prog Exp Tumor Res 4:207, 1964.
62. Reiners J, Davidson K, Nelson K, Mamrack M, Slaga T: In Langenach R, Nesnow S, Rice JM (eds): "Organs and Species Specificity in Chemical Carcinogenesis." New York: Plenum Press, 1983, p 173.

63. Strickland JE, Strickland AG: Cancer Res 44:893, 1984.
64. Furstenberger G, Kinzel V, Schwarz M, Marks F: Science 230:76, 1985.
65. Land H, Parada LF, Weinberg RA: Nature 304:596, 1983.
66. Newbold RF, Overell RW: Nature 304:648, 1983.
67. Schlehofer JR, Gissmann L, Matz B, zur Hausen H: Int J Cancer 32:99, 1983.
68. Boshart M, Gissmann L, Ikenberg H, Kleinheinz A, Scheurlen W, zur Hausen: EMBO J 3:1151, 1984.

# Heterologous Regulation of EGF Receptors in Fibroblastic Cells

## Nancy E. Olashaw and W.J. Pledger

*Department of Cell Biology, Vanderbilt University School of Medicine, Nashville, Tennessee 37232*

Platelet-derived growth factor (PDGF) increases the mitogenic activity of epidermal growth factor (EGF) in several cells lines, including BALB/C-3T3. PDGF-treated BALB/C-3T3 cells manifest a reduced capacity to bind $^{125}$I-labeled EGF due to a loss of high affinity EGF receptors. Cholera toxin potentiates the ability of PDGF to both decrease EGF binding and initiate mitogenesis. Whether PDGF increases EGF sensitivity via its effects on EGF receptors is not known and requires a more complete understanding of the mechanism by which PDGF decreases EGF binding.

12-O-tetradecanoylphorbol 13-acetate (TPA) also reduces EGF binding in BALB/C-3T3 and other cells, presumably by activating protein kinase C and, consequently, inducing the phosphorylation of EGF receptors at threonine-654. PDGF indirectly activates protein kinase C, and EGF receptors in PDGF-treated WI-38 cells are phosphorylated at threonine-654. Thus, the effects of PDGF on EGF binding may also be mediated by protein kinase C. We investigated this hypothesis by comparing the actions of PDGF and TPA on EGF binding in density-arrested BALB/C-3T3 cells.

Both PDGF and TPA caused a rapid, transient, cycloheximide-independent loss of $^{125}$I-EGF binding capacity. The actions of both agents were potentiated by cholera toxin. However, whereas TPA allowed EGF binding to recover, PDGF induced a secondary and cycloheximide-dependent loss of binding capacity. Most importantly, PDGF effectively reduced binding in cells refractory to TPA and devoid of detectable protein kinase C activity. These findings indicate that PDGF decreases EGF binding by a mechanism that involves protein synthesis and is distinct from that of TPA.

**Key words: epidermal growth factor, platelet-derived growth factor, tumor promoters, growth stimulation, growth factor receptors, cyclic AMP**

The ability of nontransformed cells in culture to proliferate in response to a given mitogen is often dependent upon prior or concomitant sensitization by a second mitogen. Density-arrested BALB/C-3T3 cells, for example, do not respond proliferatively to epidermal growth factor (EGF) unlesss pretreated or co-incubated with

Received July 8, 1986; revised and accepted January 14, 1987.

© 1987 Alan R. Liss, Inc.

platelet-derived growth factor (PDGF). In this system, PDGF initiates proliferation by rendering quiescent cells "competent" to re-enter the cell cycle [1]. EGF in the presence of somatomedin C (SmC) mediates the subsequent $G_0/G_1$ traverse of competent cells [2]. EGF and SmC also promote the progression of quiescent NRK, AKR, 10T½ and Swiss 3T3 cells; PDGF, while not obligatory for the traverse of these cells, reduces the concentration of EGF required for their stimulation [3-5]. The mechanism by which PDGF sensitizes cells to EGF is not known.

Competence formation requires only a brief exposure (2-4 hr) of cells to PDGF. Once rendered competent, cells remain responsive to EGF and SmC for several hours following the removal of PDGF from the culture medium [1,6]. This observation implies the existence of stable PDGF-induced changes in the intracellular environment that denote the competent state. The addition of PDGF to target cells elicits a variety of responses including altered distribution of cytoskeletal components such as vinculin [7,8], preferential gene transcription and protein synthesis [9-13], and changes in ion transport [14-15], and phospholipid metabolism [16-19]. Although each of these events represents a potential competence signal, or may be involved in the generation of that signal, a relationship between these processes and the subsequent actions of EGF has yet to be defined.

## HETEROLOGOUS RECEPTOR MODULATION

PDGF and EGF communicate with target cells via specific receptors located in the plasma membrane; hormone-bound receptors are rapidly internalized and ultimately degraded [20,21]. Cellular capacity to bind $^{125}$I-EGF is reduced following treatment with PDGF, the result of an indirect action of PDGF on either the number or affinity of EGF receptors [22-27]. This finding is paradoxical, given the synergistic effects of PDGF and EGF on mitogenesis and contrasts with data showing increased SmC binding in PDGF-treated BALB/C-3T3 cells [28]. The ability of PDGF to modulate EGF binding demonstrates, however, that EGF receptors are targets of both PDGF and EGF and, thus, potential mediators of the effects of PDGF on EGF sensitivity.

In support of this hypothesis, Wharton et al [23] and Leof et al [29] have shown that cholera toxin and other agents that increase cellular levels of cyclic AMP enhance both PDGF-induced competence formation and the loss of EGF binding capacity produced by PDGF. Treatment of density-arrested BALB/C-3T3 cells with optimal levels of PDGF produced, at most, a 60-70% loss of EGF binding capacity, whereas PDGF in combination with cholera toxin reduced $^{125}$I-EGF binding greater than 90%. Moreover, cells treated with both PDGF and cholera toxin no longer required EGF for $G_0/G_1$ traverse [23].

In the absence of PDGF, cyclic AMP elevating agents also act synergistically with EGF to promote the $G_0/G_1$ traverse of BALB/C-3T3 cells and other cell lines [4,30-33]. In addition, comparison of three nontransformed cell lines revealed a direct correlation between cyclic AMP content at quiescence and ability to respond mitogenically to EGF [4]. In contrast to PDGF, however, agents increasing cellular cyclic AMP levels did not affect the binding of EGF to its receptors; down regulation and degradation of EGF-receptor complexes were also unaffected by these agents [4,34]. Thus, the mechanism by which cyclic AMP increases EGF sensitivity may differ from that of PDGF.

## COMPARISON OF THE EFFECTS OF PDGF AND TPA ON EGF BINDING

To address the question of whether PDGF potentiates EGF sensitivity by modulating EGF receptors, experiments were performed to determine the mechanism by which PDGF decreases EGF binding. Like PDGF, the potent tumor promoter, 12-O-tetradecanoylphorbol 13-acetate (TPA) also reduces EGF binding in several cell lines [35–48] and initiates the proliferation of density-arrested BALB/C-3T3 cells [49]. When added to intact cells, both agents induce the phosphorylation of EGF receptors at threonine-654 [50–51]. Threonine-654 is also phosphorylated in vitro by protein kinase C [52], the binding site and cellular mediator of TPA [53]. PDGF indirectly stimulates this kinase by increasing the turnover of phosphatidylinositol and the formation of diacylglycerol, the endogenous activator of protein kinase C [16–19]. Phosphorylation of purified EGF receptors by protein kinase C has been shown to reduce the affinity of these receptors for EGF [54]. These observations suggest that PDGF and TPA decrease EGF binding by a common process involving the activation of protein kinase C and the consequent phosphorylation of EGF receptors at threonine-654. Direct confirmation that TPA reduces EGF binding by this mechanism is provided by experiments showing that cells transfected with EGF receptor cDNA containing Ala-654 did not exhibit a loss of EGF binding capacity in response to TPA [55].

To determine whether PDGF acts like TPA, we compared the actions of PDGF and TPA on EGF binding in BALB/C-3T3 cells. In the first set of experiments, density-arrested cells were incubated at 37°C in medium containing 25 ng/ml PDGF or 500 nM TPA. At various times, the monolayers were rinsed and $^{125}$I-EGF binding was measured at 4°C. In response to PDGF, $^{125}$I-EGF binding decreased to within 30% of control (quiescent, nonrefed cultures) within 15 min, increased to 50% by 45–60 min, and between 1 and 3 hr, declined from 50% to 25% (Fig. 1). Binding capacity remained at this reduced level for at least 4 hr (data not shown). TPA also rapidly and transiently reduced EGF binding; the response elicited by TPA during the first hour of treatment was greater than, but qualitatively similar to, that observed with PDGF. In contrast to PDGF, however, $^{125}$I-EGF binding in the TPA-treated cells did not decline after 60 min but continued to slowly increase. PDGF, therefore, differs from TPA in its ability to effect a secondary, sustained decrease in EGF binding activity.

Previous studies have shown that prolonged exposure of cells to high levels of TPA renders cells refractory to TPA [56–59]. Cellular desensitization to TPA is accompanied by a dramatic decline in the number of binding sites for TPA [56,60]. In accord with the concept that protein kinase C is the receptor for TPA, the activity of this enzyme in TPA-pretreated cells is reduced or absent [61]. The loss of protein kinase C activity results from degradation, as opposed to inactivation, of the kinase [62]. To determine whether PDGF decreases EGF binding in the absence of protein kinase C, quiescent BALB/C-3T3 cultures were exposed to 600 nM TPA for 20 hr at 37°C under conditions nonpermissive for cell cycle traverse. The duration of the exposure was sufficient to allow EGF binding to recover to essentially control levels (Fig. 2). Such treatment resulted in the loss of detectable levels of protein kinase C activity and the consequent inability of cells to respond to TPA in terms of 1) decreasing $^{125}$I-EGF binding and 2) initiating DNA synthesis (Fig. 2 and data not shown). PDGF, however, reduced EGF binding to a similar extent and in a compa-

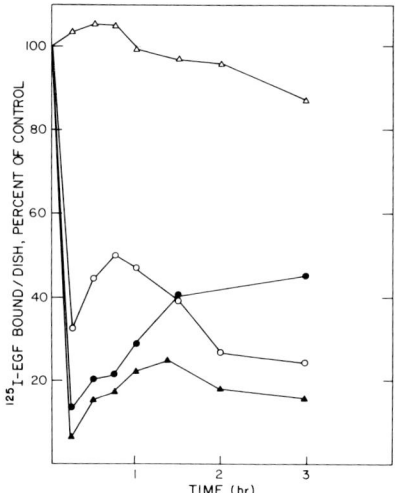

Fig. 1. Effect of PDGF or TPA on EGF binding. Density-arrested BALB/C-3T3 cells were incubated at 37°C in medium alone (△) or medium containing 25 ng/ml PDGF (○), 500 nM TPA (●), or both (▲). At various times after refeeding, the cells were rinsed with binding medium and $^{125}$I-EGF binding was determined at 4°C. The amount of $^{125}$I-EGF specifically bound by quiescent, nonrefed cultures represents 100%. Reproduced from Proc. Natl. Acad. Sci. USA 83:3834–3838, 1986, with permission.

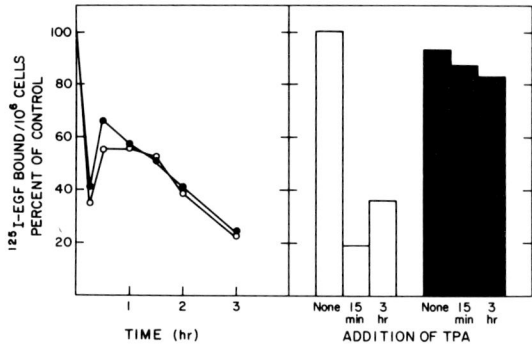

Fig. 2. Effect of PDGF or TPA on control or TPA-pretreated cells. Quiescent cultures were incubated for 20 hr in medium supplemented with 0.25% platelet-poor plasma and 600 nM TPA. TPA-pretreated (●) or nontreated control (○) cells were refed with medium containing 25 ng/ml PDGF (left panel) or 600 nM TPA (right panel). $^{125}$I-EGF binding was determined at the indicated times. Reproduced from Proc. Natl. Acad. Sci. USA 83:3834–3838, 1986, with permission.

rable manner in both control and TPA-pretreated cultures. In agreement with previous reports [59–63], cells desensitized to TPA also remained mitogenically responsive to PDGF (data not shown). The ability of PDGF to reduce EGF binding in cells refractory to TPA implies that the mechanism by which PDGF modulates EGF binding differs from that of TPA and thus, unlike that of TPA, is independent of protein kinase C. Whether PDGF activates other kinases that phosphorylate EGF receptors, and whether such phosphorylation is responsible for the loss of EGF binding capacity, are at present unknown.

## REQUIREMENT FOR PROTEIN SYNTHESIS AND EFFECT OF CHOLERA TOXIN

As described above, the secondary decrease in EGF binding induced by PDGF is not observed in cells receiving TPA. Unlike the initial reduction produced by TPA (or EGF) [64], this secondary loss of EGF binding capacity was strictly dependent on continued protein synthesis (Fig. 3, and data not shown). When added to cells in combination with PDGF, cycloheximide not only prevented this delayed decrease but allowed binding capacity to recover to essentially control levels. Cycloheximide had a similar effect on cells treated with both PDGF and cholera toxin, and 5,6-dichloro-$\beta$-ribofuranosylbenzimidazole, an inhibitor of mRNA synthesis, also reversed the loss of binding capacity produced by PDGF (data not shown). These observations suggest that PDGF may decrease EGF binding by two distinct processes that differ in their requirement for protein and RNA synthesis.

Although PDGF and TPA affect EGF binding dissimilarly and perhaps act via different mechanisms, the inhibitory actions of TPA on EGF binding, like those of PDGF, were enhanced by cholera toxin. When suboptimal amounts of TPA (5–50 nM) were added to cells in combination with cholera toxin, $^{125}$I-EGF binding decreased at a faster rate and to a greater extent than was observed in cells receiving TPA alone (Fig. 4). Higher concentrations of TPA (500 nM) elicited a rapid and maximum response and, thus, obscured these effects of the toxin. At this concentration of TPA, however, an additional effect of cholera toxin was evident; this agent antagonized the recovery of binding that began at 1 hr in the TPA-treated cells. Regardless of the dose of TPA or of the presence of cholera toxin, $^{125}$I-EGF binding increased after 3 hr; binding activity, however, remained lower in the toxin-treated cells (data not shown).

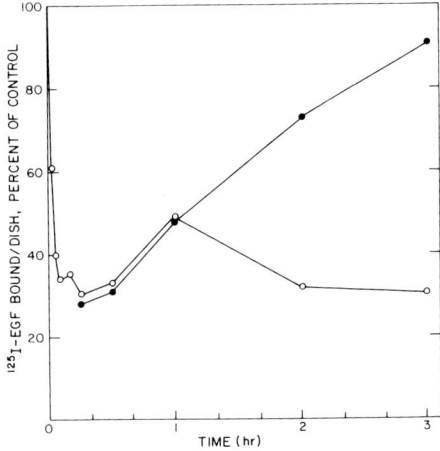

Fig. 3. Recovery of EGF binding in the presence of cycloheximide. Cycloheximide was added to cells in spent medium to a final concentration of 10 $\mu$g/ml. Fifteen minutes later, the cultures were refed with medium containing 25 ng/ml PDGF and 10 $\mu$g/ml cycloheximide (●). Cells not pre-exposed to cycloheximide received PDGF alone (○). $^{125}$I-EGF binding was determined at various times after refeeding. Cycloheximide at the concentration used inhibited protein synthesis greater than 95%. Reproduced from Proc. Natl. Acad. Sci. USA 83:3843–3838, 1986, with permission.

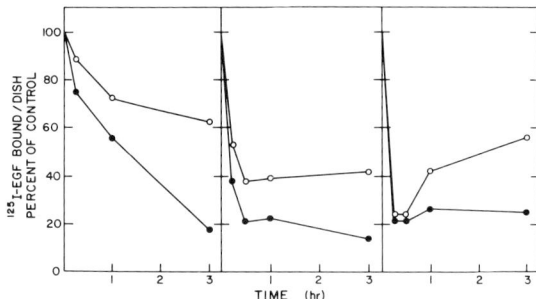

Fig. 4. Enhancement by cholera toxin of the TPA effect. Confluent cells were pretreated with medium containing 1 μg/ml cholera toxin for 2 hr. Nontreated (○) and toxin-treated cells (●) were then refed with medium supplemented with 5 nM (left panel), 50 nM (middle panel) or 500 nM (right panel) TPA; cells pre-exposed to cholera toxin also received the toxin at 1 μg/ml. $^{125}$I-EGF binding activity was measured at various intervals after refeeding. Reproduced from Proc. Natl. Acad. Sci. USA 83:3834–3838, 1986, with permission.

## CONCLUDING REMARKS

The data presented here and in previous reports [23,29] demonstrate that the effects of PDGF on EGF binding in density-arrested BALB/C-3T3 cells are 1) potentiated by agents that increase cellular levels of cyclic AMP, 2) are dependent, in part, on protein synthesis, and 3) presumably do not require protein kinase C. In addition, the data show that PDGF-induced competence formation and changes in EGF binding are correlated in two respects: 1) cholera toxin potentiated the actions of PDGF on both processes and 2) TPA pretreatment did not abrogate the effects of PDGF on either process. In contrast, as reported by Collins and Rozengurt [59], vasopressin decreased EGF binding and stimulated DNA synthesis in quiescent, but not in TPA-pretreated, Swiss 3T3 cells. A definitive relationship between heterologous receptor modulation and mitogenesis, however, has yet to be demonstrated and requires a more complete understanding of the mechanism by which PDGF decreases EGF binding and knowledge of the fate of EGF receptors in PDGF-treated cells.

## REFERENCES

1. Pledger WJ, Stiles CD, Antoniades, HN, Scher, CD: Proc Natl Acad Sci USA 74:4481–4485, 1977.
2. Leof EB, Van Wyk, JJ, O'Keefe EJ, Pledger WJ: Exp Cell Res 147:202–208, 1983.
3. Wharton W, Leof EB, Olashaw NE, O'Keefe EJ, Pledger WJ: Exp Cell Res 141:107–114, 1983.
4. Olashaw NE, Leof EB, O'Keefe EJ, Pledger WJ: J Cell Physiol 118:291–297, 1984.
5. Dicker P, Pohjanpelto P, Pettican P, Rozengurt E: Exp Cell Res 135:221–227, 1981.
6. Singh JP, Chaiken MA, Pledger WJ, Scher CD, Stiles CD: J Cell Biol 96:1497–1502, 1982.
7. Herman B, Pledger WJ: J Cell Biol 100:1031–1040, 1985.
8. Herman B, Harrington MA, Olashaw NE, Pledger WJ: J Cell Physiol 126:115–125, 1986.
9. Pledger WJ, Hart CA, Locatell KL, Scher CD: Proc Natl Acad Sci USA 78:4358–4362, 1981.
10. Kelly K, Cochran BH, Stiles CD, Leder P: Cell 35:603–610, 1983.
11. Hendrickson SL, Scher CD: Mol Cell Biol 3:1478–1487, 1984.
12. Greenberg ME, Ziff EB: Nature 311:433–437, 1984.
13. Kruijer W, Cooper JA, Hunter T, Verma IM: Nature 312:711–716, 1984.
14. Mendoza S, Wigglesworth NM, Pohjanpelto P, Rozengurt E: J Cell Physiol 103:17–27, 1980.
15. Cassel D, Rothenberg P, Zhuang Y-X, Deuel TF, Glaser L: Proc Natl Acad Sci USA 80:6224–6228, 1983.
16. Habenicht AJR, Glomset JA, King WC, Nist C, Mitchell CD, Ross R: J Biol Chem 256:12329–12335, 1981.

17. Brown KD, Blay J, Irvine RF, Heslop JP, Berridge MJ: Biochem Biophys Res Commun 123:377–384, 1984.
18. Berridge MJ, Heslop JP, Irvine RF, Brown KD: Biochem J 222:195–201, 1984.
19. Chu S-HW, Hoban CJ, Owen AJ, Geyer RP: J Cell Physiol 124:391–396, 1985.
20. Carpenter G: Mol Cell Endocrinol 31:1–9, 1983.
21. Bowen-Pope DF, Rosenfield ME, Seifert RA, Ross R: Int J Neurosci 26:141–153, 1985.
22. Wrann M, Fox CF, Ross R: Science 210:1363–1365, 1980.
23. Wharton W, Leof EB, O'Keefe EJ, Pledger WJ: Proc Natl Acad Sci USA 79:5567–5571, 1982.
24. Shupnik MA, Antoniades HN, Tashjian AH: Life Sci 30:347–353, 1982.
25. Collins MKL, Sinnett-Smith JW, Rozengurt E: J Biol Chem 258:11689–11693, 1983.
26. Bowen-Pope DF, DiCorleto PE, Ross R: J Cell Biol 96:679–683, 1983.
27. Brown KD, Blakely DM, McDonald M: Biosci Rep 3:659–666, 1983.
28. Clemmons DR, Van Wyk JJ, Pledger WJ: Proc Natl Acad Sci USA 75:2839–2843, 1980.
29. Leof EB, Olashaw NE, Pledger WJ, O'Keefe EJ: Biochem Biophys Res Commun 109:83–91, 1982.
30. Wharton W, Leof EB, Olashaw NE, Earp SH, Pledger WJ: J Cell Physiol 11:201–206, 1982.
31. Pruss RM, Herschman HR: J Cell Physiol 98:469–474, 1979.
32. Rozengurt E, Legg A, Strang G, Courtenay-Lack N: Proc Natl Acad Sci USA 78:4392–4396, 1981.
33. Rozengurt E: J Cell Physiol 112:243–250, 1982.
34. Hollenberg MD, Cuatrecasas P: Proc Natl Acad Sci USA 70:2964–2968, 1973.
35. Lee L, Weinstein IB: Science 202:313–315, 1978.
36. Murray AW, Fusenig NE: Cancer Lett 7:71–77, 1979.
37. Lee L, Weinstein IB: Proc Natl Acad Sci USA 76:5168–5172, 1979.
38. Shoyab M, DeLarco JE, Todaro GJ: Nature (London) 279:387–391, 1979.
39. Brown KD, Dicker P, Rozengurt E: Biochem Biophys Res Commun 86:1037–1043, 1979.
40. Chandler CE, Herschman HR: J Cell Physiol 105:275–285, 1980.
41. Magun BE, Matrisian LM, Bowen GT: J Biol Chem 255:6373–6381, 1980.
42. Saloman DS: J Biol Chem 256:7958–7966, 1981.
43. Hollenberg MD, Nexo E, Berhanu P, Hock R: In "Receptor-Mediated Binding and Internalization of Hormones." Middlebrook JL, Kohn LD (eds): New York: Academic Press, 1981, pp 181–195.
44. End D, Tolson N, Yu M, Guroff G: J Cell Physiol 111:140–148, 1982.
45. King CA, Cuatrecasas P: J Biol Chem 257:3053–3060, 1982.
46. Logsdon CD, Williams JA: Biochem J 223:893–900, 1984.
47. Boonstra J, Mummery CL, van der Saag PT, deLaat SW: J Cell Physiol 123:347–352, 1985.
48. Bequinot L, Hanover JA, Ito S, Richert ND, Willingham MC, Pastan I: Proc Natl Acad Sci USA 82:2772–2778, 1985.
49. Frantz CN, Stiles CD, Scher CD: J Cell Physiol 100:413–424, 1979.
50. Davis RJ, Czech MP: Proc Natl Acad Sci USA 82:1974–1978, 1985.
51. Davis RJ, Czech MP: Proc Natl Acad Sci USA 82:4080–4084, 1985.
52. Hunter T, Ling N, Cooper JA: Nature (London) 311:480–483, 1984.
53. Nishizuka Y: Nature (London) 308:693–697, 1984.
54. Downward J, Waterfield MD, Parker P: J Biol Chem 260:14538–14546, 1986.
55. Lin CR, Chen WS, Lazar CS, Carpenter CD, Gill G, Evans RM, Rosenfeld MG: Cell 44:839–848, 1986.
56. Collins MKL, Rozengurt E: J Cell Physiol 112:42–50, 1982.
57. Rozengurt E, Rodriguez-Pena M, Smith KA: Proc Natl Acad Sci USA 80:7244–7248, 1983.
58. Rozengurt E, Rodriguez-Pena A, Coombs M, Sinnett-Smith J: Proc Natl Acad Sci USA 81:5748–5752, 1984.
59. Collins MKL, Rozengurt E: J Cell Physiol 118:133–142, 1984.
60. Jaken S, Tashjian AH, Blumberg PM: Cancer Res 41:2175–2181, 1982.
61. Rodriguez-Pena A, Rozengurt E: Biochem Biophys Res Commun 120:1053–1059, 1984.
62. Ballester R, Rosen O: J Biol Chem 260:15194–15199, 1986.
63. Coughlin SR, Lee WMF, Williams PW, Giels GM, Williams LT: Cell 43:243–251, 1985.
64. Aharanov A, Pruss RM, Hershman HR: J Biol Chem 253:3970–3977, 1978.

# Vaccinia Virus and the EGF Receptor: A Portal for Infectivity?

### Y. Vivienne Marsh and Deborah A. Eppstein

*Institute of Bio-Organic Chemistry, Syntex Research, Palo Alto, California 94304*

We previously demonstrated that occupancy of the epidermal growth factor (EGF) receptor reduced the ability of vaccinia virus to infect L cells [Eppstein et al: Nature 318:663, 1985]. This result suggested that vaccinia virus was utilizing the EGF receptor as one pathway to infect cells. We have studied this system further, and now find that antibodies to the EGF receptor also reduce the ability of vaccinia virus to infect cells productively. Inclusion of both EGF and antibodies to the EGF receptor did not cause inhibition over that obtained by EGF alone, providing another line of evidence that the antiviral effects on vaccinia virus were at the level of the EGF receptor. The antiviral effects of EGF or synthetic peptides corresponding to the third disulfide loop of TGF-$\alpha$ or the vaccinia virus growth factor were specific to vaccinia virus and did not inhibit replication of herpes simplex virus type 2 or vesicular stomatitis virus. The inhibitory effects on replication of vaccinia virus were obtained when EGF (but not insulin or growth hormone) was present prior to, but not after, productive viral adsorption. These results provided further evidence that the antivaccinia viral effects of EGF were at the level of initial receptor occupancy. As interferon (IFN) treatment has been shown to interfere with the action of some growth factors, including EGF, we examined the effects of IFN treatment of cells on the antivaccinia viral activity of EGF. Our results show that the antivaccinia effect of IFN-$\beta$ either interfered with or partially coalesced with the inhibitory effects of EGF. The former interpretation is consistent with the report that IFN treatment results in a decrease both in the apparent number and affinity of cell-surface receptors for EGF [Zoon et al: Proc Natl Acad Sci USA 83:8226, 1986].

**Key words:** interferon, viral inhibition, vaccinia virus, EGF receptor

Cellular receptors for several viruses have been identified as being receptors for other specific physiological ligands. For example, Epstein-Barr virus infects B-lymphocytes via the complement receptor CR2 [1]; the AIDS virus (HIV-1) receptor on T-lymphocytes contains the T4 antigen [2–4]; rabies virus may utilize the acetylcholine receptor [5]; and the reovirus receptor has recently been identified as the cellular $\beta$-adrenergic receptor [6]. We asked the question of whether vaccinia virus (VV) utilizes the epidermal growth factor (EGF) receptor to bind to and infect cells,

---

Received August 28, 1986; revised and accepted February 4, 1987.

© 1987 Alan R. Liss, Inc.

in light of the observed sequence homology between an early protein encoded by VV (vaccinia growth factor, VGF) and EGF or transforming growth factor-$\alpha$ (TGF-$\alpha$) [7–9], all of which bind to the cellular EGF receptor [10–13].

Treatment of murine L-cells with EGF and synthetic decapeptides [14,15] corresponding to the third disulfide loop of TGF-$\alpha$ resulted in a dose-dependent reduction of vaccinia virus replication, as monitored by a plaque reduction assay [16]. The ability of EGF to inhibit VV replication correlated with its affinity for the EGF receptor on the L-cells. These results indicated that occupancy of the EGF receptor reduced infectivity by VV, and suggested that VV was utilizing the EGF receptor as one pathway to infect cells. We have now further characterized the interaction of VV with the EGF receptor, and our results provide additional evidence that interaction of VV with the EGF receptor aids viral infectivity.

## MATERIALS AND METHODS
### Chemicals

Eagle's minimal essential medium (EMEM) was from K.C. Biological (Lenexa, KS); fetal bovine serum was from Hyclone (Logan, UT); Sea Plaque agarose was from Marine Colloid Division, FMC Corp. (Rockland, ME); penicillin (10,000 U/ml)-streptomycin (10,000 U/ml), neutral red, and insulin were from GIBCO (Grand Island, NY); EGF was from Collaborative Research (Lexington, MA); TGF-$\alpha$ was a generous gift from Genentech, Inc. (S. San Francisco, CA); monoclonal antibody (MAb) 29.1.1 (an $IgG_1$ raised against A431 cell EGF receptor) was from International Diagnostic Laboratories, Inc. (Chesterfield, MO); MAb 528 (an $IgG_{2a}$ raised against EGF receptors of A431 cells [17]) was from Oncogene Science, Inc. (Mineola, NY); murine interferon-$\beta$ (MuIFN-$\beta$) was from Lee Biomolecular (San Diego, CA); and synthetic decapeptides were prepared by J. Nestor, Jr. (Syntex Research, Palo Alto, CA) [14,15].

### Cells and Viruses

Vero cells, vaccinia virus (strain Western Reserve), herpes simplex virus type 2, strain G [HSV-2(G)], and A431 cells were from American Type Culture Collection (Rockville, MD); L cells and vesicular stomatitis virus (VSV), Indiana strain, were obtained from Dr. C. Samuel, University of California (Santa Barbara, CA).

### Preparation of Vaccinia Virus Stock

Confluent monolayers of L cells seeded in T-75 flasks were washed and inoculated with vaccinia virus at a multiplicity of infection (MOI) of 0.05 plaque forming units (PFU)/cell. Virus inoculum was aspirated off and cells were washed after 1-hr incubation at 37°C. Twenty-five milliliters of EMEM + 2% FCS was then added to the flask, and cells were incubated for 2–3 days until cytopathic effect (CPE) reached 80–90%. Cells were scraped off the flask, homogenized with a Dounce homogenizer, and cell debris was removed by centrifugation at 2,000 rpm for 10 min. Virus stock was stored at −80°C.

### Plaque Reduction Assay

Plaque reduction assay was performed in 24-well dishes, employing Vero cells ($5 \times 10^4$ cells/well) for HSV-2(G), and L cells ($10^5$ cells/well) for vesicular stomatitis virus (VSV) and vaccinia virus. Antibodies as specified were added to cells 4 hr before as well as during viral adsorption; EGF and other compounds were added 3 hr before viral adsorption. Cells were then inoculated with 30–40 PFU of virus per well. Virus inocula were aspirated off after 1 hr adsorption at 37°C in 5% $CO_2$ atmosphere, and the cells were washed and overlaid with 1 ml/well of EMEM with 1.5% Sea Plaque agarose, 10% fetal bovine serum, 100 U/ml penicillin, and 100 $\mu$g/ml streptomycin. The cells were incubated at 37°C in 5% $CO_2$ until plaque formation became apparent under light microscope observation (approximately 3 days for HSV-2(G), 1 day for VSV and 4 days for vaccinia virus). Wells were overlaid with 0.5 ml/well EMEM with 1.5% Sea Plaque agarose and 0.01% neutral red. Plaques were counted after ~24 hr of incubation with stain [18]. Statistical analyses were performed using Student's t-test.

### Single-Cycle Virus Yield Assay

Confluent monolayers of murine L-cells plated in Costar 24-well dishes ($10^5$ cells/well, incubated 48 hr before use) were treated with EGF (1 $\mu$g/ml = $1.7 \times 10^{-7}$M) or control media 3 hr prior to as well as during the 1 hr vaccinia viral adsorption (MOI of 1 PFU/cell). After 1-hr adsorption at 37°C in 5% $CO_2$ atmosphere, cells were washed, and EGF or control media was added again as indicated to triplicate test wells. In one set of virus control wells, viral adsorption was prolonged to 3 hr. In two other sets of virus control wells, cells were washed (in addition to the wash after 1 hr of adsorption) a second time, at 2 or 3 hr after initial addition of virus. Virus yield was determined 7 hr post-adsorption, by plaque assay in L-cells.

## RESULTS

The specificity of the inhibition of vaccinia virus replication by EGF and homologous synthetic peptides is illustrated in Table I. First, two polypeptides unrelated to EGF, ie, insulin and bovine growth hormone (BGH), which bind to their respective receptors on L-cells, were unable to inhibit vaccinia viral plaque formation. Second, replication of two unrelated viruses—VSV, an RNA virus; and HSV-2(G), a

**TABLE I. Effects of Peptides on Replication of Vaccinia Virus, VSV, and HSV-2**

| Virus (no. control plaques) | No. plaques, ± SD (% of control) | | | | |
|---|---|---|---|---|---|
| | Ac-TGF-$\alpha$[34–43] NHEt ($3 \times 10^{-5}$ M) | Ac-VGF[71–80] $NH_2$ ($3 \times 10^{-5}$M) | EGF (0.17 $\mu$M[a]) | Insulin (0.2 $\mu$M[a]) | BGH (0.05 $\mu$M[a]) |
| Vaccinia ($39 \pm 4$) | $26 \pm 1$* (68) | $28 \pm 3$* (71) | $22 \pm 3$* (43) | $35 \pm 2$ (90) | $35 \pm 3$ (91) |
| VSV ($31 \pm 2$) | $ND^b$ | $29 \pm 2$ (92) | $30 \pm 3$ (96) | ND | ND |
| HSV-2 ($19 \pm 1$) | ND | $20 \pm 2$ (105) | $20 \pm 2$ (105) | ND | ND |

[a]EGF, insulin, and BGH were tested at 1 $\mu$g/ml.
[b]ND = not determined.
*$P < 0.001$, vs virus control, Student's t-test. All other values were not significantly different from the corresponding virus control.

DNA virus—was not inhibited by EGF or the related synthetic peptide. In addition, TGF-α, which also binds to the EGF receptor, inhibited VV replication to the same degree as did EGF (data not shown).

Analyses of single-cycle growth studies (Table II) showed that 60–70% of vaccinia virions productively adhered to cells during the first hour of adsorption; additional washing of cells 2 or 3 hr after initial addition of virus reduced viral yield by 30–40% (Table II, B, C, vs A, $P \leqslant 0.01$). When $1.7 \times 10^{-7}$ M EGF (= 1 μg/ml) was added 3 hr prior to as well as being present during the initial 1-hr viral adsorption period, virus yield was reduced by 50% (Table II, E, $P < 0.001$). However, when addition of EGF was delayed until immediately after the initial 1-hr viral adsorption period, virus yield was reduced by 45% (Table II, F, $P = 0.04$). This latter reduction of virus yield was eliminated when EGF addition was delayed by 2 hr (Table II, G vs control C, $P = 0.5$), by which time viral adsorption was complete.

Incubation of L-cells with MAb 29.1.1 resulted in a dose-dependent inhibition of VV plaque formation ($P < 0.001$ vs control for [Ab] $\geqslant$ 1 μg/ml); unrelated IgG$_1$ MAb's had no inhibitory effect ($P > 0.1$) (Fig. 1). Similar results were obtained in A431 cells with MAb 29.1.1 as well as MAb 528 against vaccinia replication (data not shown). Pretreatment with EGF (1 μg/ml) resulted in approximately 50% inhibition of viral replication ($P < 0.001$); no further inhibition was obtained by simultaneous incubation with EGF plus the antibody to the EGF receptor ($P \geqslant 0.3$), suggesting that both EGF and the antireceptor antibody likely act through the same inhibitory pathway. These results are consistent with the hypothesis that VV can productively bind to cells via the EGF receptor.

Figure 2 summarizes the effects of MuIFN-β and EGF on VV plaque formation. Antiviral activity of IFN was obtained by a 26-hr ($P = 0.04$ for 0.1 U/ml; $P \leqslant 0.001$

**TABLE II. Single-Cycle Yield of Vaccinia Virus**

| Sample | Treatment schedule | Virus yield (PFU/ml) | % of std control | % of own control |
|---|---|---|---|---|
| Hr | −3 −2 −1 0 1 2 3 4 5 6 7 (Harvest) | | | |
| A | Virus ↑             ↓ <br> W | $(7.6 \pm 0.6) \times 10^3$ | 100 | |
| B | Virus ↑  ↑           ↓ <br> W  W | $(5.3 \pm 0.6) \times 10^{3*}$ | 70 | |
| C | Virus ↑   ↑         ↓ <br> W   W | $(5.0 \pm 0.5) \times 10^{3*}$ | 66 | 100 |
| D | Virus    ↑        ↓ <br> W | $(8.7 \pm 0.6) \times 10^3$ | 115 | |
| E | EGF     Virus ↑           ↓ <br> $(1.7 \times 10^{-7}$M)   W | $(3.8 \pm 0.6) \times 10^{3*}$ | 49 | |
| F | Virus ↑ EGF        ↓ <br> W   W | $(4.2 \pm 1.5) \times 10^{3*}$ | 55 | |
| G | Virus ↑   ↑ EGF      ↓ | $(4.5 \pm 0.9) \times 10^{3*}$ | 59 | 90** |

\*$P < 0.05$ vs A.
\*\*$P = 0.5$ vs C.

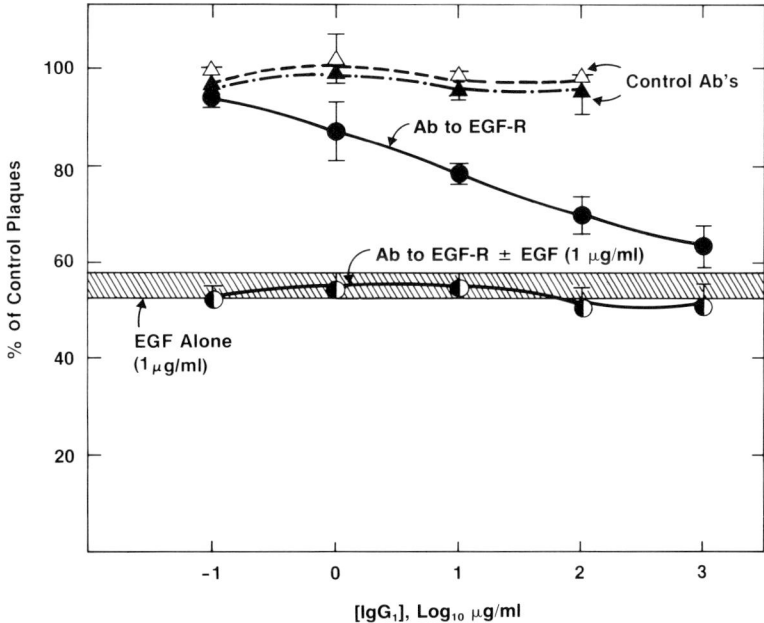

Fig. 1. Antivaccinia viral activity of EGF and antibody to the EGF receptor. L-cells were incubated with antibody to the EGF receptor (MAb 29.1.1) 4 hr prior to as well as during viral adsorption in a standard plaque reduction assay. EGF incubations were 3 hr before and during adsorption with vaccinia virus. The hatched area shows the level of viral plaques obtained by treatment with 1 µg/ml ($1.7 \times 10^{-7}$M) EGF. (●), MAb 29.1.1; (◐), MAb 29.1.1 and EGF (1 µg/ml); (△) and (▲), control MAb's of $IgG_1$ subclass made against unrelated antigens.

for $\geq$ 1 U/ml) but not a 3-hr pretreatment (P > 0.1), and its inhibitory effect did not increase with addition of EGF (P $\geq$ 0.4). These results suggest either that IFN and EGF may utilize at least in part a common mechanism in the inhibition of VV growth, or that IFN may have interfered with the antivaccinia viral activity of EGF.

## DISCUSSION

We have studied the vaccinia viral/EGF receptor system further and have obtained additional results showing that vaccinia virus can productively bind to cells via the EGF receptor. Two monoclonal antibodies made against the EGF receptor reduced the ability of VV to replicate. Inclusion of both EGF and antibody to the EGF receptor did not further increase the inhibition over that obtained by either EGF alone, or by the antibody alone, consistent with the hypothesis that the antivaccinia effects of EGF involved interaction at the EGF receptor. The activity of MAb's to the EGF receptor was specific as shown by inability of two control MAb's of the same $IgG_1$ subclass, made against unrelated antigens, to block vaccinia replication (Fig. 1). We have also shown that the antiviral effects of EGF and synthetic decapeptide analogs of the third disulfide loop of TGF-$\alpha$ or VGF were specific for vaccinia virus but not VSV or HSV-2 (Table I). In addition to inhibition of VV replication by pretreatment with EGF, we have found that TGF-$\alpha$ similarly inhibited vaccinia virus replication. Single-cycle virus growth studies indicated that the inhibitory effect of

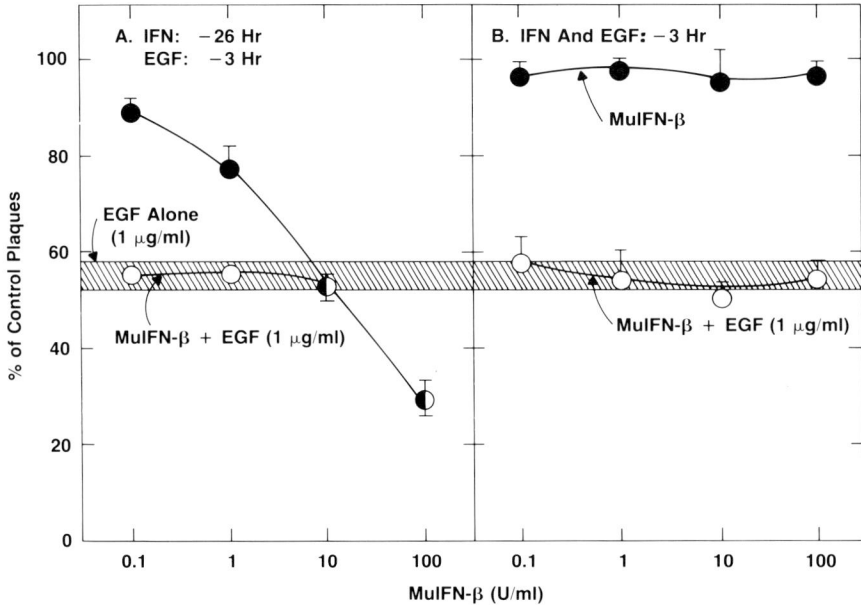

Fig. 2. Antivaccinia viral activity of MuIFN-$\beta$ and EGF. L-cells were incubated with MuIFN-$\beta$ either (A) 26 hr, or (B) 3 hr before viral adsorption in a standard plaque reduction assay. EGF (1 $\mu$g/ml) was added 3 hr before and during vaccinia virus adsorption. The hatched area shows the level of viral plaques obtained by treatment with 1 $\mu$g/ml (1.7 × $10^{-7}$M) EGF. (●), MuIFN-$\beta$; (○), MuIFN-$\beta$ + EGF (1 $\mu$g/ml); (◐), points at which MuIFN-$\beta$ and MuIFN-$\beta$ + EGF coincide.

EGF was at the early stages of the vaccinia viral replication cycle, acting mainly by inhibiting productive viral adsorption (Table II).

We have demonstrated that a relationship exists between vaccinia virus and the cellular receptor for EGF. EGF pretreatment of cells has an antiviral effect on VV replication. Such a pretreatment resulting in antiviral activity is reminiscent of the antiviral activity of IFN. In addition to its antiviral properties, IFN has inhibitory effects on growth of cells. IFN-$\beta$ has been shown specifically to inhibit the action of some growth factors, including platelet-derived growth factor (PDGF) [19,20] and EGF [21]. In light of the interaction observed between VV and the EGF receptor, we were thus interested in determining if any relationship existed between the antivaccinia viral activity of IFN and EGF, and the inhibitory effect of IFN on EGF activity. Dose-dependent antivaccinia activity for IFN was obtained with a 26- but not a 3-hr pretreatment of L-cells, consistent with the known need for IFN to establish an antiviral state in the cells to exert its effects. EGF alone (1 $\mu$g/ml) resulted in approximately 50% inhibition of VV replication. Additional antiviral activity was not observed with combined IFN-EGF treatment, suggesting that either these two inhibitory pathways interfered with each other, or that the two pathways at least partially coalesced. Recently, Zoon et al [23] have shown that IFN treatment of cells resulted in a decrease both in the number and affinity of cell surface receptors for EGF. Our results using combinations of IFN and EGF treatments are consistent with these findings, ie, that IFN treatment interfered with the inhibitory activity of EGF.

Studies with monoclonal antibodies to the EGF receptor have provided further support for the hypothesis that vaccinia virus can have an initial productive interaction with cells through the EGF receptor. Binding of the MAb to the EGF receptor can result in competition for binding of VV to the same receptor, thus reducing the viral infectivity. The fact that complete inhibition of vaccinia virus infection was not obtained with saturating concentrations (in terms of EGF receptor binding) of EGF suggests that the EGF receptor is not the only pathway by which VV can bind to and infect L-cells. This is consistent with the report of Stroobant et al [22] that vaccinia virus was able to infect NR-6 cells which lack detectable EGF receptors. However, our results suggests that in cells bearing the EGF receptor, an interaction of VV with the EGF receptor enhances the ability of VV to productively infect the cells.

## ACKNOWLEDGMENTS

We thank Carole G. Kurahara and Nicholas A. Bruno for expert tissue culture support, and Dr. John J. Nestor, Jr., for synthesis of the decapeptides. This paper is contribution number 248 from the Institute of Bio-Organic Chemistry.

## REFERENCES

1. Fingeroth JD, Weis JJ, Tedder TF, Strominger JL, Biro PA, Fearon DT: Proc Natl Acad Sci USA 81:4510, 1984.
2. Dalgleish AG, Beverly PCL, Clapham PR, Crawford DH, Greaves MF, Weiss RA: Nature 312:763, 1984.
3. Klatzmann D, Champagne E, Chamaret S, Gruest J, Guetard D, Hercend T, Gluckman JC, Montagnier L: Nature 312:767, 1984.
4. Maddon PJ, Dalgleish AG, McDougal JS, Clapham PR, Weiss RA, Axel R: Cell 47:333, 1986.
5. Lentz TL, Burrage TG, Smith AL, Crick J, Tignor GH: Science 215:182, 1982.
6. Co MS, Gaulton GN, Fields BN, Green MI: Proc Natl Acad Sci USA 82:1494, 1985.
7. Brown JP, Twardzik DR, Marquardt H, Todaro GJ: Nature 313:491, 1985.
8. Blomquist MC, Hunt LT, Barker WC: Proc Natl Acad Sci USA 81:7363, 1984.
9. Reisner AH: Nature 313:801, 1985.
10. Todaro GJ, Fryling C, DeLarco JE: Proc Natl Acad Sci USA 77:5258, 1980.
11. Reynolds FH, Jr, Todaro GJ, Fryling C, Stephenson JR: Nature 292:259, 1981.
12. Massagué J: J Biol Chem 258:13614, 1983.
13. Twardzik DR, Brown JP, Ranchalis JE, Todaro GJ, Moss B: Proc Natl Acad Sci USA 82:5300, 1985.
14. Nestor JJ, Jr, Newman SR, DeLustro B, Todaro GJ, Schreiber AB: Biochem Biophys Res Commun 129:226, 1985.
15. Nestor JJ, Jr, Newman SR, DeLustro BM, Schreiber AB: In Deber CM, Ruby VJ, Koppel K (eds): "Peptides: Structure and Function." Rockford, IL: Pierce Chem. Co., 1985, p. 39.
16. Eppstein DA, Marsh YV, Schreiber AB, Newman SR, Todaro GJ, Nestor JJ, Jr: Nature 318:663, 1985.
17. Kawamoto T, Sato JD, Le A, Polikoff J, Sato GH, Mendelsohn J: Proc Natl Acad Sci USA 80:1337, 1983.
18. Eppstein DA, Marsh YV, Schryver BB: Virology 131:341, 1983.
19. Einat M, Resnitsky D, Kimchi A: Proc Natl Acad Sci USA 82:7608, 1985.
20. Lin SL, Ts'o PO, Hollenberg MD: Biochem Biophys Res Commun 96:168, 1980.
21. Zullo JN, Cochran BH, Huang AS, Stiles CD: Cell 43:793, 1985.
22. Stroobant P, Rice AP, Gullick WJ, Cheng DJ, Kerr IM, Waterfield MD: Cell 42:383, 1985.
23. Zoon KC, Karasaki Y, zur Nedden DL, Hu R, Arnheiter H: Proc Natl Acad Sci USA 83:8226, 1986.

# Proliferative Control in Normal and Neoplastic Cells

## Arthur B. Pardee

*Dana-Farber Cancer Institute, Boston, Massachusetts 02115*

## OVERVIEW

Cell proliferation is a closely regulated process. The appearance of pairs of sister cells at division culminates a biochemical sequence extending over many hours. The cell cycle is partitioned by four readily detected events: initiation and termination of DNA synthesis, mitosis, and cell division. Some recent information on the biology and biochemistry by which proliferation is controlled will be summarized here (see [1] for a longer discussion).

Exogenous growth factors regulate proliferation of normal cells. Some of these factors stimulate cells to emerge from quiescence ($G_0$) and become competent to enter the cycle. Other factors are then required for progression through the $G_1$ phase, during which preparations are made for DNA synthesis.

Subsequently, growth factors are not required: cells deprived of all factors continue through S, $G_2$, and M, until they reach $G_1$, when a supply of factors is again required. Growth factors probably function sequentially, each being needed for a successive commitment step in the transit from quiescence to the onset of DNA synthesis.

Growth factors stimulate cells' biochemistry, activating general and specific transcriptional and post-transcriptional syntheses. Phosphorylations of key molecules such as growth factor receptors and ribosomal subunits increase as the stimulus is carried to the cell's machinery. A critical commitment event in G1, the restriction point, is reached under the influences of epidermal growth factor, somatomedin C, and nutritional conditions that permit rapid protein synthesis. An essential restriction point protein must be accumulated by untransformed cells. This is difficult owing to this protein's instability; therefore, rapid protein synthesis is required for its net accumulation and for cells to pass the restriction point and then initiate DNA synthesis.

Received May 28, 1986.

Presented at the Triton Biosciences-UCLA Symposium on Growth Factors, Tumor Promoters, and Cancer Genes, Steamboat Springs, Colorado, April 6–13, 1986.

© 1988 Alan R. Liss, Inc.

DNA synthesis starts about 2 hr beyond the restriction point. Enzymes essential for DNA replication increase dramatically along with the onset of DNA synthesis. The "induction" of these important enzymes is controlled by the same restriction point mechanism that turns on DNA synthesis. These enzymes assemble into a macromolecular replitase complex at the start of S. They cooperate within it to perform DNA synthesis.

The $G_0$ and $G_1$ processes that commit cells to DNA synthesis are the ones that regulate normal cell proliferation. Deranged control of these biochemical processes results in relaxed growth factor requirements in transformed and turmorigenic cells. This relaxed requirement for growth factors may result from misregulation of genes in pathways normally activated only when growth factor receptors are occupied.

## ONCOGENES

Several connections between oncogenes, growth factors, growth factor receptors, and cell proliferation have been discovered recently. Cells contain genes (proto-oncogenes) related to the oncogenes carried by retroviruses. These proto-oncogenes are being found to relate to growth control. They, like growth factors, function at different times in $G_1$. The myc and fos proto-oncogenes are activated by platelet-derived growth factor when fibroblastic cells emerge from quiescence into competence. The *ras* proto-oncogene activity increases at a later time in $G_1$ than does myc or fos. Constitutive expression or mutation of some proto-oncogenes has been found in transformed cells. This implies there are multiple genes in $G_1$ whose inappropriate expression can lead to deranged growth. Consequences of misregulation of a specific proto-oncogene may depend upon cell type.

## NORMAL CELLS NEED GROWTH FACTORS UNTIL A RESTRICTION (CONTROL) POINT THAT IS LOCATED 2 hr BEFORE S PHASE

The main control of progression to cell division of exponentially growing cells is located prior to DNA synthesis, in the $G_1$ part of the cell cycle. This is shown by the arrest of virtually all the cells in a population with a $G_1$ DNA content after growth factors are removed. In contrast, tumorigenic cells are not as readily arrested under the same conditions, so that many of them are still cycling.

Locations of the point in $G_1$ at which growth factors are not required for DNA synthesis to start has been an object of considerable study. For example, Zetterberg and Larsson [2] have shown it occurs approximately in the middle of the $G_1$ phase, approximately 3½ hr before the beginning of DNA synthesis in exponentially growing 3T3 fibroblasts. Similar results were obtained by Campisi et al [3] by using totally different techniques; they showed that exponentially growing cells located within about 3 hr of the beginning of S phase continued to enter DNA synthesis after serum was removed. Both of these data then suggest that serum factors are dispensable several hours prior to the actual onset of DNA synthesis. Transformed cells were defective in their response to removal of serum continuing to enter S phase for many hours.

## *RAS* GENE FUNCTION IS REQUIRED IN MID-LATE $G_1$

Relatively few events have been identified as occurring specifically in the latter part of the interval between quiescence or mitosis and the onset of DNA synthesis.

One of these is an increase in message for the *ras* oncogene [4]. This message appears to increase by about 8 hr after quiescent cells are stimulated to proliferate by addition of serum, which corresponds to about 4 or 5 hr before DNA synthesis starts. This result was found even with chemically transformed 3T3 cells.

Another sort of evidence indicates that *ras* becomes dispensable a few hours before the onset of DNA synthesis [5]. In this work cells were microinjected with anti-*ras* antibody at intervals after they were released from quiescence. Initially the antibody largely prevented entry of cells into DNA synthesis, but as the cells progressed more and more of them were able to initiate DNA synthesis so that about half became independent of the antibody at about 6 hr. Since the average cell entered DNA synthesis at about 10 hr, *ras* became dispensable approximately 4 hr before the onset of DNA synthesis. Undoubtedly other events are involved in the progression of cells to $G_1$ and largely remain to be discovered.

## NORMAL CELLS (BUT NOT TUMOR CELLS) REQUIRE RAPID PROTEIN SYNTHESIS TO PASS THE RESTRICTION POINT

We observed that when protein synthesis is inhibited partly (50%–70%) by cycloheximide or other agents, cells gradually become arrested with a $G_1$ DNA content [6]. This is very reminiscent of the effect seen after serum deprivation. Kinetic analysis of the cell cycle in the presence of cycloheximide revealed specific delay of cells during the early portion of $G_1$. Rates of transit through late $G_1$, S, G2, and M were unaffected. Other kinetic data indicated that rapid synthesis of an unstable protein is required during the cycloheximide-sensitive portion of $G_1$. Furthermore, the delaying effect of cycloheximide was much less in transformed cells [7]. This result led us to postulate that transformed cells have an abundance of this protein because it is more stable and/or produced in excess amounts.

## CHARACTERISTICS OF PROTEIN DETERMINING THE ABILITY OF BALB/c-3T3 CELLS TO PASS THE RESTRICTION POINT

Three characteristics of a protein essential for passage of cells into the S phase were proposed from our cell kinetic studies.
  1. Synthesis is dependent upon serum (growth factors) during $G_1$.
  2. It is a labile protein with a half-life of approximately 2.5 hr.
  3. The protein has an increased half-life and/or rate of synthesis in transformed cells.

Studies using two-dimensional (2D) gels revealed one spot out of at least 1,000 that had all of these characteristics [8]. Namely, this protein of 68,000 MW (p68) and pI 6.3 was synthesized during $G_1$ phase. Furthermore it was made more rapidly in transformed cells. It thus provides one of the few differences seen in cells transformed by a number of agents, both RNA viruses and chemicals, as compared to the untransformed cells. Furthermore in the untransformed cells it was unstable with a relatively short half-life, whereas in variously transformed cells it seemed to be quite stable. Thus, this protein is a good candidate for a substance that regulates cell growth and is altered upon transformation.

## ENZYMES AND DNA SYNTHESIS ARE COORDINATELY CONTROLLED AT ONSET OF S PHASE

As an approach to searching for the role of regulatory proteins such as p68 in the onset of DNA synthesis, we have turned to seeking a simpler assay for its action. For a long time it has been known that a variety of enzymes including thymidine kinase are induced at the time when DNA synthesis starts. It thus seems possible that the onset of DNA synthesis is only one of several events that occur coordinately at the beginning of S phase. Histone synthesis comes to mind as another of these processes.

Our investigation of the appearance of thymidine kinase has confirmed that its activity appears concurrently with the onset of DNA synthesis [9]. Furthermore, it is not dependent upon DNA synthesis, since inhibition of the latter with drugs such as hydroxyurea had no effect upon the induction of thymidine kinase. Conversely, thymidine kinase is not essential for the onset of DNA synthesis, since mutants defective in this enzyme still initiate DNA synthesis at the usual time.

This leads to the idea that one might measure an enzyme such as thymidine kinase rather than measuring DNA synthesis in order to investigate events that trigger the onset of both processes in parallel. We indeed showed that the timing of the onset of both processes was equally delayed by pulses of cycloheximide, and that these pulses had no effect on either process when transformed cells were used instead of untransformed cells [9]. Furthermore, studies with inhibitors and with growth factors revealed a highly similar response of thymidine kinase, thymidylate synthase, and the onset of DNA synthesis [10]. We therefore propose that the production of thymidine kinase, and to some extent its message, is triggered by the same $G_1$ regulatory mechanism as controls the overall process of DNA synthesis. Since the production of thymidine kinase, and particularly its message, is a much simpler process for molecular investigation, we believe that its study in depth should reveal events that are triggers of the S phase growth control processes and that are modulated in cancer cells.

## ACKNOWLEDGMENTS

This work was supported by grant GM24571 from the National Institutes of Health. I wish to thank Marjorie Rider for the preparation of this manuscript.

## REFERENCES

1. Pardee AB: DeVita VT, Jr, Hellman S, Rosenberg SA (eds): In "Cancer Principles and Practice of Oncology." Philadelphia, J.B. Lippincott, 1985, pp 3–22.
2. Zetterberg A, Larsson O: Proc Natl Acad Sci 82:5365–5369, 1985.
3. Campisi J, Pardee AB: Mol Cell Biol 4:1807–1814, 1984.
4. Campisi J, Gray HE, Pardee AB, Dean M, Sonenshein GE: Cell 36:241–247.
5. Mulcahy LS, Smith MR, Stacey DW: Nature 313:241–243, 1985.
6. Rossow PW, Riddle VGH, Pardee AB: Proc Natl Acad Sci 76:4446–4450, 1979.
7. Campisi J, Medrano EE, Morreo G, Pardee AB: Proc Natl Acad Sci 79:436–440, 1982.
8. Croy RG, Pardee AB: Proc Natl Acad Sci 80:4699–4703, 1983.
9. Coppock DL, Pardee AB: J Cell Physiol 124:269–274, 1985.
10. Yang HC, Pardee AB: J Cell Physiol 127:410–416, 1986.

# Growth-Regulated Genes and Human Leukemias

### Bruno Calabretta, Leszek Kaczmarek, and Renato Baserga

*Department of Pathology and Fels Research Institute, Temple University Medical School, Philadelphia, Pennsylvania 19140*

We have studied the expression of several growth-regulated genes in human leukemias. By comparing the mRNA level of growth-regulated genes to the mRNA level of the histone H3, the expression of which is restricted to the S phase of the cell cycle, we have developed a method to evaluate the expression of any growth-regulated gene according to the growth fraction of a given cell population.

We have found that the expression of most growth-regulated genes is dependent on the growth fraction of the leukemic RNAs we have analyzed. At variance with this pattern, the expression of the proto-oncogene c-myc and the growth-regulated sequence 2A9 appears truly deregulated in a significant number of patients with acute leukemias.

**Key words:** cell-cycle sequences, oncogenes

The phenotype manifestation of human neoplasia is usually reflected by the degree of maturation arrest of proliferative advantage of the neoplastic clone(s) [1]. Since our understanding of the impairment in the control of cell proliferation in human malignancies is limited, the identification and the study of genes involved in the control of cell-cycle progression in normal cells is a necessary first step toward that understanding.

In the past few years, a considerable number of animals genes have already been identified whose product is important for cell proliferation or whose expression is cell-cycle dependent, that is, genes whose RNA levels are markedly increased in a specific phase of the cell cycle or when quiescent cells are stimulated to proliferate.

The most fashionable and scrutinized members of a growing list of growth-regulated genes include the cellular homologues of at least six oncogenes (c-myc, c-fos, c-myb, c-H-ras, c-Ki-ras, and p53) [2–7]. Other well-known growth regulated genes are calmodulin [8], β-actin [6], thymidine kinase [9], ornithine decarboxylase [10], and histones [11,12]. Furthermore, other cell-cycle-dependent genes have been

Received April 21, 1986.

© 1988 Alan R. Liss, Inc.

identified as cDNA clones of yet unknown function by differential screening of cDNA libraries [13–16].

The fact that certain genes are expressed in a cell-cycle-dependent manner does not, of course, mean that they regulate cell-cycle progression.

However, the fact that some oncogenes are expressed in a cell-cycle-dependent manner [2–7] suggests that some cell-cycle-dependent genes, including those of which little is known, may play a regulatory role in the control of cell proliferation as growth-experimental evidence is suggesting for c-myc, c-ras, and p53 [17–21]. The example of P53 is particularly striking since this growth-regulated gene [7,22,23] was first found to play an important role in the control of cellular proliferation [24,25] and subsequently to cooperate with an activated human ras to transform primary embryonic cells [26–28]. Therefore it is possible that other growth-regulated genes will be found to behave in a similar way. A feature common to most oncogenes is that they were found to be overexpressed in a variety of human malignancies, including acute leukemias [29–32]. The altered expression of the oncogenes, however it might be achieved, is currently thought to be a major molecular event associated with the origin and progression of human malignancies [33]. Therefore, it is legitimate to ask whether cell-cycle-dependent genes different from known oncogenes are also overexpressed in human malignancies. This approach could widen our understanding of the proliferative advantage of neoplastic cells.

## MATERIALS AND METHODS
### Cell Culture of WI38

WI38, a strain of human diploid fibroblasts, at the 32nd population doubling level, was cultured as previously described [34]. The methods for serum stimulation, for making the fibroblasts quiescent, or for synchronizing them are described in the legend to Figure 1. Labeling with [$^3$H]thymidine and autoradiography were carried out according to standard procedures.

### Leukemic Cells

Leukemic cells were obtained from the M.D. Anderson Hospital in Houston, Texas, where the diagnosis was established. The patients were all untreated when the cells were collected.

### RNA Isolation

Total cellular RNA of leukemic cells was isolated according to Frazier [35]. Briefly, the cell pellet was homogenized in a Waring blender in the extraction buffer (75 mM NaCl, 20 mM EDTA/10 mM Tris-HCl, pH 8.0/0.2% NaDodSO4) mixed 1:1 with buffer-saturated phenol. The aqeous phase was recovered by centrifugation and reextracted with an equal volume of phenol/chloroform/isoamyl alcohol (25:24:1) and once again with chloroform/isoamyl alcohol (24:1). The nucleic acids were precipitated with ethanol, and DNA was removed by treatment with DNase I and precipitation with 3 M sodium acetate (pH 5.6).

### RNA Blot Analysis

Total RNA was denatured with 6.3% formaldehyde and 50% formamide at 65°C and then size fractionated on a 1.2% agarose gel containing 6.6% formalde-

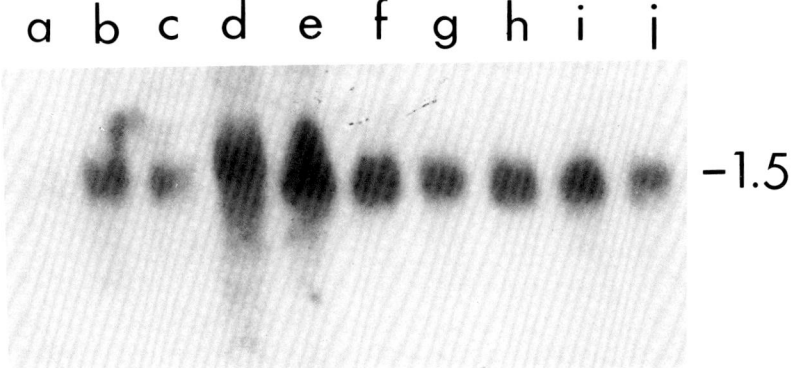

Fig. 1. Cell-cycle-dependent expression of pF1 cDNA clone in human diploid fibriblast. Human diploid fibroblasts (WI38) were plated at a density of $1 \times 10^4$ per $cm^2$ of plastic petri dish in Earle's minimal essential medium supplemented with 10% fetal calf serum, amino acids, vitamins, and antibiotics as described [70]. After 14 days in culture, cells were rendered quiescent and used for the experiments. Fibroblasts were either stimulated with fresh medium for 1, 6, 14, and 24 hr or replated at dilution 1:20 and after 3 days synchronized in S phase with 2.5 mM hydroxyurea. After overnight exposure to hydroxyurea, the drug was removed by extensive washing and cells were then harvested 4, 9, 13, 18, and 23 hr later. Total cytoplasmic RNA was extracted [15] from quiescent as well as from stimulated cells, and the cells were released from an hydroxyurea block. RNA (15 µg per lane) was separated on formaldehyde-containing agarose gels and transferred to a nitrocellulose filter as described by Thomas [36]. The filter was hybridized to p2F1 labeled by nick-translation [37]. **Lane a** represents hybridization to RNA from quiescent WI38 fibroblasts; **lanes b-e** represent hybridization to RNA obtained from cells stimulated with fresh medium for, respectively, 1, 6, 16, and 24 hr; **lanes f-j** represent hybridization to RNAs extracted from hydroxyurea-synchronized cells, 4, 9, 13, 18, and 23 hr, respectively, after release from early S phase block.

hyde. Blotting of RNA to nitrocellulose was done as described by Thomas [36]. Nick-translation [37] of the plasmid DNA at high specific activity was performed as previously described [38]. Prehybridization, hybridization, and posthybridization, washes for RNA blots were essentially as described by Wahl et al [39]. For autoradiography, the blots were exposed to Kodak XAR-5 film at 70°C in the presence of an intensifying screen.

## RESULTS
### Expression of Growth-Regulated cDNAs in Normal Cells

Our laboratory has been actively involved in the last few years in the identification and characterization of cDNA clones whose level of expression is increased when quiescent cells are induced to proliferate by the addition of serum. We have selected three such clones from a cDNA library prepared from poly $(A)^+$ mRNA derived from Syrian hamster ts13 cells 6 hr after serum stimulation [15]. These cDNA clones were called p2F1, p4F1, and P2A9. By Southern blot analysis it was shown that these sequences have distinct genomic organization, and Northern blot analysis has shown that each cDNA clone recognizes a distinct mRNA species [15].

The growth-dependent expression of these cDNA clones is maintained in closely related species such as mouse and rat [40,41]. Having this background information, we wanted to ascertain if we could detect hybridization of the hamster cDNA clones to human mRNA species and if the corresponding genes were also growth dependent in their expression in human cells. We have investigated the expression of the genes represented by our cDNA clones in two cellular systems: peripheral blood mononuclear cells (PBMC) stimulated by phytohemagglutinin (PHA) and human diploid fibroblasts (WI38) stimulated by serum. Our findings can be summarized as follows: the expression of P4F1, P2F1, and p2A9 is growth dependent in human diploid fibroblasts (WI38), as the mRNA specific to each cDNA clone is induced early during the addition of serum and reaches a peak in mid $G_1$ in WI38 cells [34,38]. The expression of p4F1 and p2F1 is also growth dependent in peripheral blood mononuclear cells stimulated by PHA, whereas p2A9 expression is barely detectable and does not change at different times of PHA stimulation [38]; p2F1 and p4F1 differ in their kinetic of induction as the expression of p4F1 is more transient than p2F1 expression [42]. These data have already been published and are summarized in Table I. In order to determine whether the expression of our cDNA clones changes during the different phases of the cell cycle of proliferating WI38 cells, we designed an experiment in which proliferating WI38 cells were synchronized in S phase with 2.5 mM hydroxyurea and, after a 16-hr exposure to the drug, released from the hydroxyurea block. This experimental design allowed us to investigate the expression of our cDNA clones (p2F1, p4F1, and p2A9) during the transition $S$-$G_2$-$M$-$G_1$-$S$ in comparison to the transition $G_0$-$G_1$-$S$ that occurs following the addition of serum to $G_0$ cells.

Histone H3, whose expression is restricted to the S phase of the cell cycle, was used as control. We have found that in growing WI38 released from the hydroxyurea block the level of p4F1, p2F1, and P2A9 is invariant throughout the cell cycle. Figure 1 shows that the mRNA sequence corresponding to p2F1 increases only after the serum stimulation of WI38 cells (lane b) and reaches the highest level after 16–24 hr (lanes d–e). In growing WI38 cells released from hydroxyurea block the level of p2F1 is constant throughout the cell cycle (lanes f–j) and higher than in quiescent cells (lane a). Similar results were obtained with p4F1 and p2A9. Our findings are in agreement with recently published reports showing that the expression of c-myc is induced during the transition $G_0$-$G_1$-$S$ but is constant in the cell cycle of proliferating cells [43,44].

Our results indicate that the expression of p4F1, p2F1, and p2A9 is growth dependent, although it does not appear to be regulated in cycling cells.

**TABLE I. Summary of the Results Obtained With the Gene Sequences Used in This Paper***

| Sequences | Levels of mRNA in human cells stimulated to proliferate | |
|---|---|---|
| | PBMC | WI38 |
| 2F1 | ↑ | ↑ |
| 4F1 | ↑↑ | ↑ |
| 2A9 | — | ↑↑↑ |
| c-myc | ↑↑ | ↑ |
| c-myb | ↑↑ | — |
| p53 | ↑↑ | ↑↑ |
| ODC | ↑↑ | ↑↑ |
| histone H3 | ↑↑ | ↑↑ |

*↑ = increased above $G_0$ levels (the number of arrows should not be interpreted as a quantitative measurement but simply as an indication of more or less increase); — = not detected.

## Expression Growth-Regulated Genes in Human Leukemias

We then asked if our growth-regulated genes are expressed in hematological malignancies. We have selected RNAs from patients with chronic lymphocytic leukemia (CLL), acute lymphocytic leukemia (ALL), chronic myelogenous leukemia (CML), and acute myelogenous leukemia (AML).

The results of our experiments are summarized in Table II. The levels of expression of p4F1 and P2F1 are similar: the highest expression is found in patients with CML and AML; the lowest expression is found in patients with CLL; intermediate levels are found in ALL patients.

The pattern of expression of p2A9 differs from that of p2F1 and p4F1; we have found the highest expression in CML patients and in some patients with AML (especially those with a mixed population of myelomonocytic cells). We did not detect expression of p2A9 in CLL and ALL patients. Of particular interest is the fact there is a very low level of expression of our growth-regulated cDNA clones in patients with chronic lymphocytic leukemia. This finding applies to all growth-dependent genes (including c-myb and p53) in the large majority of CLL RNAs we examined. These results are in agreement with the fact that CLL cells have a low mitotic index [45–47] and are poorly stimulated by mitogens [48]. The intriguing, although predictable, correlation of the lack of expression between growth-dependent genes and cell populations with a negligible growth fraction prompted us to investigate in a more detailed and systematic way how the expression of a growth-regulated gene is dependent on the growth fraction of a given cell population.

Since several reports have appeared indicating that certain oncogenes are overexpressed in human neoplasia and since it has been shown that the expression of at least six oncogenes (c-myc, c-myb, c-fos, c-H-ras, C-Ki-ras, and p53) is growth regulated, it is legitimate to ask whether the overexpression previously reported is truly overexpression or does not simply reflect the growth fraction of the tissue examined. In normal tissues, only a small percentage of cells are cycling, while the majority are in a quiescent state, which is called $G_o$ (for a review, see [49]. The fraction of cycling cells is called the growth fraction [50]. This growth fraction is often increased in tumors [51] and for reviews [49,52]). If, in a given cell population,

TABLE II. Expression of Growth-Dependent Genes in Human Hematological Malignancies*

| Sequences | CML | AML | CLL | ALL |
|---|---|---|---|---|
| 2F1 | +++ | ++ | − | +± |
| 4F1 | +++ | ++ | − | +± |
| 2A9 | +++ | ++± | − | − |
| c-myc | ++ | ++ | − | + |
| c-myb | +± | + | − | + |
| p53 | ND* | + | − | + |
| ODC | ND* | + | ND* | + |
| histone H3 | +++ | +± | − | + |

*The levels of expression were measured by densitometric readings of several Northern blots that were carried out under standardized conditions of RNA amounts, hybridization, and autoradiographic exposure. A minus sign indicates undetectable expression. ±–+++, average levels of expression measured by densitometric readings of the RNA level of each gene sequence in each patient within CML, AML, and ALL groups. ND = not done.

more cells are cycling, the expression of growth-dependent genes (like our cDNA clones and the oncogenes above) would be increased with respect to cell populations in which few cells are cycling. The overexpression of the growth-dependent genes mentioned above would then be simply a measure of the increased proliferating activity of tumors rather than a demonstration that their expression is deregulated. By comparing the levels of expression of growth-regulated genes to the level of expression of the S-phase-specific gene histone H3, we can distinguish the increased expression of a growth-regulated gene resulting from a true altered activation from the overexpression that simply refelcts an increase in the fraction of cycling cells. Using this approach, we have studied the expression of seven growth-regulated genes in human leukemias. We have purposely chosen three growth-regulated proto-oncogenes (c-myc, c-myb, and p53) that share similar biological features, notably nuclear localization of their protein [22,53] and cooperation with activated ras gene in transformation assays of primary embryo fibroblasts [26–28,54], and four gene sequences (ODC, p4F1, p2F1, and p2A9) that have in common the growth dependence of their expression in cells of different tissues and in different species [10,38,40,41].

Figure 2 provides an example of such analysis, showing a composite picture of the same filter hybridized first with c-myc and next with H3. It is obvious that the absolute amounts of H3 and c-myc mRNAs varied from one individual sample to another, as one would expect from the heterogenous populations of cells. Indeed, we have found a good correlation between the expression of histone H3 and the number of leukemic cells in S phase as determined by cytofluorimetric analysis (not shown). In particular, the highest expression of H3 is found in patients 1, 5, and 6, who had the highest number of leukemic cells in S phase (30%, 25%, and 27%, respectively) when the leukemic cells were collected. Similarly the lowest expression of histone H3 was found in patient 12, in whom only 2% of peripheral blood leukocytes were in the S phase of the cell cycle. It is also obvious that in patients with elevated expression of H3 there is a correspondingly high expression of c-myc. However, it should also be noted that an abnormally high expression of c-myc is detectable in several patients with a low H3 level (particulary patients 2, 7, 13, 14). The findings of these experiments are summarized in Table III and can be generalized as follows: a) The

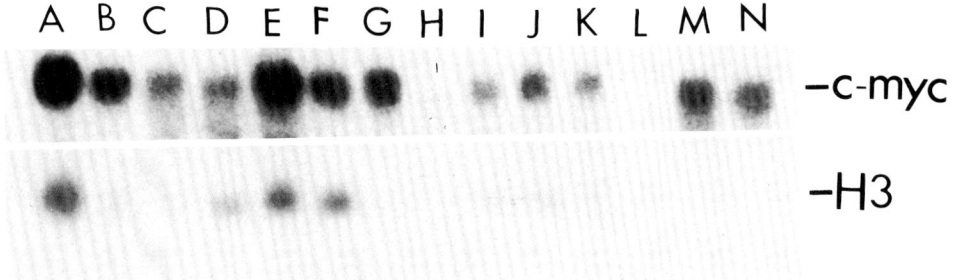

Fig. 2. Composite picture of autoradiographs from Northern blots of total RNA isolated from leukemic cells. Total RNA was isolated essentially as described by Frazier et al [35], and 15 μg/lane were electrophoresed on a 1.2% agarose-formaldehyde gel and subsequently transferred to nitrocellulose as described by Thomas [36]. The first hybrydization was carried out using a mixture of 5′ and 3′ ends of c-myc gene probe [71], in the presence of dextrane sulfate as described by Wahl et al [39]. After removal of residual radioactivity, the filter was hybridized to pF0422 carrying a histone H3 gene probe.

TABLE III. Expression Ratio of Growth-Dependent Genes (c-myc, c-myb, p53, 4F1, 2F1, 2A9) ODC to S Phase Gene (histone H3)*

| Genes | NBM | CML | AML | ALL |
|---|---|---|---|---|
| c-myc/H3 | 0.9 | 0.5 | >3.0 | >2.5 |
| c-myb/H3 | ND | ND | 1.9 | 1.1 |
| p53/H3 | ND | ND | 1.0 | 1.5 |
| pF1/H3 | 1.0 | 0.6 | 1.1 | 0.9 |
| 2F1/H3 | 1.1 | 0.5 | 1.0 | 0.9 |
| 2A9/H3 | 2.0 | 0.5 | 3.0 / 0.5 | ND |
| ODC/H3 | ND | ND | 1.0 | 0.8 |

*NBM = normal bone marrow cells; CML = chronic myelogenous leukemia; AML = acute myelogenous leukemia; ALL = acute lymphocytic leukemia. The expression of c-myc is >3.0 in AML and >2.5 in ALL in more than 50% of the patients. The expression ratio between 2A9 and H3 is >3.0 in a group of five patients with AML and 0.5 in a second group of four patients. The ratios listed above do not reflect absolute amounts of the concentration of each mRNA in the bulk of the cell population studied. They represent only a relative change of expression of each growth-dependent gene with respect to histone H3.

expression of most growth-regulated genes (including p53 and c-myb) both in normal and neoplastic tissues depends on the growth fraction of each population. b) Two reproducible exceptions are provided by c-myc and p2A9, whose expression is truly deregulated in a significant number of patients with acute leukemias.

## DISCUSSION

We have described studies on the expression of a number of growth-regulated gene sequences in normal and tumorous human cells. By measuring the levels of expression of the above sequences in association with the measure of the growth fraction of which histone H3 expression is an indicator, we can see that the extent of the overexpression of growth-regulated genes (including some oncogenes) has been greatly overestimated. However, we have found, particularly in human leukemias, that two growth-regulated genes (c-myc and p2A9) might be truly overexpressed.

What is the interpretation of these findings?

It is known that in patients with acute leukemias few blast cells are cycling, while the large majority lie in a dormant state that has been considered a $G_0$ state [55]. The altered expression of c-myc and p2A9 could be confined within the fraction of cycling cells, or alternatively, these genes might be abnormally expressed by out-of-cycle leukemic cells. In this regard, the observation by Campisi et al [6] that cell-cycle control of c-myc expression is lost in a chemically transformed cell line is consistent with our second interpretation. Morever, we have detected c-myc overexpressed in a patient with atypical chronic lymphocytic leukemia in the absence of detectable hybridization to histone H3 and without rearrangement of c-myc locus [56]. If we are correct in the prediction that some growth-regulated genes are inappropriately expressed by out-of-cycle neoplastic cells, it could be possible to identify at a molecular level two groups of leukemic patients.

A first group with a balanced ratio of $G_1$/S phase genes should be characterized by the presence of a population of proliferating cells randomly distributed in the different phases of the cell cycle and by a variable number of blast cells in a $G_0$ state.

The failure to detect an altered c-myc expression in these patients could be due to its occurrence in a very limited number of leukemic cells.

In the second group of patients the altered expression of growth-regulated genes suggests that these leukemic populations might be abnormally enriched for growth-arrested leukemic cells. The involvement of one or more growth-regulated genes in this event could reflect the molecular heterogeneity of the proliferative block, of which the most consistent feature seems to be the altered ratio of c-myc/H3 and p2A9/H3.

We would like to interpret the altered c-myc expression in the light of the proposed role of c-myc as a competence factor in normal cellular proliferation [17,18]. In the competence-progression model proposed for the cell cycle of 3T3 mouse fibroblasts [57], induction of DNA synthesis depends on the sequential activity of several growth factors. First, platelet-derived growth factor (PDGF) makes the cells "competent," and then PDGF-deprived serum (platelet-poor plasma, PPP) provides these cells with "progression" factors leading the fibroblasts to undergo DNA synthesis. Neither PDGF nor PPP alone can stimulate DNA synthesis in quiescent 3T3 cells. It has been shown that c-myc expression is induced by PDGF in these cells [12] and that the introduction of c-myc protein into quiescent 3T3 fibroblasts stimulates a portion of them to enter in the S phase of the cell-cycle provided that PPP is present in the medium [18]. These results suggest that cells with a high level of c-myc might not be necessarily proliferating, but simply advanced into the cell cycle to a stage of "competence." From this stage they could be more susceptible than their quiescent counterparts to proliferative stimuli.

If we extend this hypothesis to the altered expression of c-myc in leukemic cells, it is attractive to speculate that the abnormal activation of c-myc in the pool of noncycling blast cells could make these cells competent to reenter the cycle and therefore to maintain the pool of cycling cells.

In this regard, it is worth mentioning that the pool of proliferating blast cells is not self-maintaining during steady-state condition(s) [58] and the reentry of noncycling leukemic cells in the pool of cycling cells have been demonstrated [59]. As far as the altered expression of p2A9 is concerned, our interpretation remains even more tentative than it was for c-myc, owing to much less knowledge about the identity and the biological function of this sequence.

However, sequence-analysis data have provided evidence of an extensive homology between p2A9 and the $\alpha$- and $\beta$- subunits of the S100 protein [60].

S100 was originally identified as a brain-specific protein [61] of glial origin [62,63] and since then has been found in tissues outside the nervous system [64], including cells of the lymphoreticular system [65]. Furthermore S100 proteins are members of a class of structurally and functionally related calcium-modulated proteins [66] that bind calcium ions under physiological conditions. As the homology between p2A9 and S100 protein is particularly high in the two calcium binding domains [67,68], it is likely that p2A9 is also a calcium binding protein. As optimal levels of calcium might regulate the cell-cycle progression is normal cultured fibroblasts [69] both at the transition $G_o$-$G_1$ and in late $G_1$ [70], it is not too hazardous to predict that the constitutive expression of 2A9 in some leukemias might decrease the threshold of signals required for cell proliferation.

## ACKNOWLEDGMENTS

This work was supported by grants CA 25898 and GM 33694 from the National Institute of Health and by a grant from the Leukemia Research Foundation, Chicago.

## REFERENCES

1. Nowell PC: Science 194:23, 1976.
2. Kelly K, Cochran HB, Stiles CD, Leder P: Cell 35:241, 1983.
3. Greenberg ME, Ziff EB: Nature 311:433, 1984.
4. Cochran BH, Zullo J, Verma IM, Stiles CD: Science 226:1080, 1984.
5. Torelli G, Selleri L, Donelli A, Ferrari S, Emilia G, Venturelli D, Moretti L. Torreli U: Mol Cell Biol 5:2874, 1985.
6. Campisi J, Gray HE, Pardee AB, Dean M, Sonenshein GE: Cell 36:241, 1984.
7. Reich NC, Levine AJ: Nature (Lond) 308:199, 1984.
8. Chafouleas JG, Lagace L, Bolton WE, Boyd III, Means AR: Cell 36:73, 1984.
9. Liu H-T, Gibson CW, Hirschhorn RR, Rittling S, Baserga R, Mercer MW: J Biol Chem 260:3269,1985.
10. Kahana C, Nathans D: Proc Natl Acad Sci USA 80:3645, 1983.
11. Plumb M, Stein J, Stein G: Nucleic Acids Res 11:2391, 1983.
12. Hirschhorn RR, Maraschi F, Baserga R, Stein J, Stein G: Biochemistry 23:3731, 1984.
13. Linzer DIH, Nathans D: Proc Natl Acad Sci USA 80:4271, 1983.
14. Cochran BH, Reffel AC, Stiles CD: Cell 33:939, 1983.
15. Hirschhorn RR, Aller P, Yuan ZA, Gibson CW, Baserga R: Proc Natl Acad Sci USA 81:6004, 1984.
16. Arya SK, Wong-Staal F, Gallo RC: Mol Cell Biol 4:2540, 1984.
17. Armelin HA, Armelin MCS, Kelly K, Stewart T, Leder P, Cochran BK, Stiles CD: Nature 310:655, 1984.
18. Kaczmarek L, Hyland JK, Watt R, Rosenberg M, Baserga R: Science 228:1313, 1985.
19. Fermisco JR, Gross M, Kamata T, Rosenberg M, Sweet RW: Cell 38:109, 1984.
20. Hyland JK, Rogers CM, Scolnick EM, Stein RB, Ellis R, Baserga R: Virology 141:333, 1985.
21. Kaczmarek L, Oren M, Baserga R: Exp Cell Res 162:268, 1985.
22. Mercer WE, Baserga R: Exp Cell Res 100:31, 1985.
23. Milner J, Milner S: Virology 112:785, 1981.
24. Mercer WE, Nelson D, DeLeo AB, Old LJ, Baserga R: Proc Natl Acad Sci USA 79:6309, 1982.
25. Mercer WE, Avignolo C, Baserga R: Mol Cell Biol 4:276, 1984.
26. Eliyahu D, Raz A, Gruss P, Givol P, Oren M: Nature 312:643, 1984.
27. Parada LF, Hand H, Weinberg RA, Wolf D, Rotter V: Nature (Lond) 312:646, 1984.
28. Jenkins JR, Rudge K, Currie GA: Nature (Lond) 312:651, 1984.
29. Eva R, Robbins KC, Anderson PR, Srinivasan A, Tronick SR, Reddy EP, Ellmore NW, Galen AT, Lautenberg JA, Papas TS, Westin EH, Wong-Staal F, Gallo RC, Aaronson SA: Nature 245:166, 1982.
30. Westin EH, Wong-Staal F, Gelmann EP, Dalla Favera R, Papas TS, Lautenberg JA, Eva A, Reddy EP, Tronick SR, Aaronson SA, Gallo RC: Proc Natl Acad Sci USA 79:2490, 1984.
31. Rothberg GP, Erisman MD, Diehl RE, Rovigatti UG, Astrin SM: Mol Cell Biol 4:1096, 1984.
32. Slamon DJ, deKernion JB, Verma IM, Cline MS: Science 224:256, 1984.
33. Weinberg RA: Science 230:770, 1985.
34. Rittling SR, Brooks KM, Cristofalo VJ, Baserga R:Proc Natl Acad Sci USA (in press).
35. Frazier ML, Mars W, Florine LD, Montagna RA, Saunders GF: Mol Cell Biochem 56:113, 1983.
36. Thomas PS: Proc Natl Acad Sci USA 77:5201, 1980.
37. Rigby PWS, Dieckman M, Rhodes C, Berg P: J Mol Biol 113:237, 1977.
38. Calabretta B, Kaczmarek L, Mars W, Ochoa D, Gibson CW, Hirschhorn PR, Baserga R: Proc Natl Acad Sci USA 82:4463, 1985.
39. Wahl GM, Stein M, Stark GR: Proc Natl Acad Sci USA 76:3683, 1979.
40. Gibson CW, Rittling SR, Hirschhorn RR, Kaczmarek L, Calabretta B, Stiles CD, Baserga R: Mol Cell Biochem 71:61, 1986.
41. Ide T, Ninomiya-Tsuji J, Ferrari S, Philiponis V, Baserga R: Biochemistry 25:7041, 1986.

42. Kaczmarek L, Calabretta B, Baserga R: Proc Natl Acad Sci USA 82:5375, 1985.
43. Thompson CB, Challoner PB, Neiman PE, Groudine M: Nature (Lond) 314:363, 1985.
44. Persson H, Gray HE, Godeau F: Mol Cell Biol 5:2903, 1983.
45. Zimmerman TS, Godwin HA, Perry S: Blood 31:227, 1968.
46. Theml H, Trepel F, Schick P, Kaboth W, Begemann H: Blood 42:623, 1973.
47. Dormer P, Theml H, Lau B: Leuk Res 7:1983.
48. Smith LJ, Cowling CD, Barker RC: Lancet 1:229, 1972.
49. Baserga R: "The Biology of Cell Reproduction." Cambridge: Harvard University Press, 1985.
50. Mendelsonn ML: J Natl Cancer Inst 28:1015, 1962.
51. Bresciani F, Paoluzzi R, Benassi M, Nervi C, Casale C, Zipano E: Cancer Res 34:2405, 1974.
52. Steel GG: "Growth Kinetics of Tumors," Oxford: Clarendon Press, 1977.
53. Bishop JM: Annu Rev Biochem 52:301, 1983.
54. Land H, Parada LF, Weinberg R: Nature 304:536, 1983
55. Vincent PC: In Gunz FW, Henderson ES (eds): "in Leukemia." Grune and Stratton, 1983, pp 77–101.
56. Kaczmarek L, Elfenbein BI, Narni F, Vonderheid EC, Barry EW, Calabretta B: Cancer Genetics and Cytogenetics (in press).
57. Stiles CD, Capone GT, Scher CD, Antoniades HN, Wa Wyk JJ, Pledger WJ: Proc Natl Acad Sci USA 76:1279, 1979.
58. Gavosto F, Pileri A, Gabetti V, Masera P: Nature 216:188, 1967.
59. Saunders EF, Mauer AM: J Clin Invest 48:1299, 1969.
60. Calabretta B, Battini R, Kaczmarek L, de Riel JK, Baserga R: J Biol Chem 261:12628, 1986.
61. Moore BW: Biochem Biophys Res Commun 19:739, 1965.
62. Cicero TJ, Cowan WM, Moore BW, Suntreff V: Brain Res 18:25, 1970.
63. Bock E: J Neurochem 30:7, 1978.
64. Hidaka H, Endo T, Kowamoto S, Yamada E, Umekawa H, Tanabe K, Hara K:J Biol Chem 258:2705, 1985.
65. Takahaski K, Isobe T, Ohtsuki Y, Sonobe H, Takeda L, Akagi T: Am J Pathol 100:497, 1984.
66. Van Eldik IJ, Zendegui JG, Marshak DR, Waterson DM: Int Rev Cytol 77:1, 1982.
67. Isobe T, Okuyama T: Eur J Biochem 116:79, 1981.
68. Isobe T, Ishioka N, Okuyama T: Eur J Biochem 115:469, 1981.
69. Hazelon B, Mitchell B, Tupper J: J Cell Biol 83:487, 1979.
70. Critofalo VJ, Sharf B: Exp Cell Res. 76:419, 1973.
71. Dalla Favera R, Gelman RE, Martinotti VS, Franchini G, Papas TS, Gallo R, Wong-Staal F: Proc Natl Acad Sci USA 79:6497, 1982.

# Differentiation and Its Role in Carcinogenesis and Anticarcinogenesis

**Rodney L. Sparks, David N. Estervig, and Robert E. Scott**

Section of Experimental Pathology, Department of Biochemistry and Molecular Biology, Mayo Clinic/Foundation, Rochester, MN 55905

Carcinogenesis is a multistep process that we have proposed to be associated with the expression of defects in the integrated control of cellular differentiation and proliferation. To test this proposal, we have studied how the integrated control of differentiation and proliferation is regulated in normal cells and whether the induction of carcinogenesis with chemical or physical agents abrogates this regulatory process. We established that the integrated control of differentiation and proliferation is mediated at distinct arrest states and that a specific sequence of biological events must occur as part of this regulatory mechanism in both murine mesenchymal stem cells and normal human keratinocyte progenitor cells. In addition, we have identified and partially purified regulatory molecules that affect these processes. Physical and chemical initiators of carcinogenesis have also been shown to induce stable defects in the integrated control of cellular proliferation and differentiation in mesenchymal stem cells. Furthermore, we have demonstrated that mesenchymal stem cells that are completely transformed and human squamous carcinoma cells express defects in the integrated control of differentiation and proliferation. Our most recent studies suggest that for complete transformation cells must also demonstrate defects in the expression of anticancer/tumor suppressor activity. In this regard, we have obtained evidence that the expression of anticancer/tumor suppressor activity is regulated by differentiation-dependent mechanisms.

Key words: anticancer genes, cancer suppression, cell cycle-dependent differentiation, differentiation control, integration of differentiation and proliferation, neoplastic transformation, stem cells, tumor suppression, tumor suppressor gene

Clinical observations and experimental studies provide abundant evidence that the development of cancer is a multistep process. Although it has been proposed by many investigators that cancer is primarily a disease of uncontrolled cell proliferation, numerous studies argue against this conclusion. For example, in humans it typically takes 15 or more years for a single transformed cell to develop into a clinically detectable tumor [1] and it may take 5–10 years for a precancerous lesion to develop into an invasive and metastatic cancer [2]. Some benign cell types can in fact

Received June 26, 1986.

© 1988 Alan R. Liss, Inc.

proliferate at rates that are as fast or faster than those of their malignant counterparts. In the disease psoriasis, for example, epidermal cells proliferate at extremely rapid rates [3], and yet these patients show no increased incidence of squamous cell carcinoma [4]. Conversely, many malignant tumors exhibit very slow rates of cellular proliferation [5]. It thus is evident that cancer is not merely a disease of uncontrolled cell proliferation.

It has also been proposed that cancer is a disease primarily associated with aberrant differentiation [6,7]. However, the cells of many types of cancer actually differentiate quite efficiently. For example, it is difficult to distinguish many benign renal cell adenomas from renal small cell carcinomas based on their differentiation characteristics [8]. This is also true of other cell types. In addition, many studies have established that certain cancer cell types can actually terminally differentiate. These include leukemic cells [9–11], embryonal carcinoma cells [12–14], and squamous carcinoma cells [15].

Our concept of carcinogenesis suggests that for stem cells to be completely transformed a combination of several basic types of defects needs to occur. These include 1) the expression of defects in the integrated control of differentiation and proliferation, 2) the expression of partial or complete defects in the actual process of differentiation or proliferation, and 3) the expression of defects in the ability of cells to make decisions whether to initiate the above processes. In this regard, multiple types of "decision"-making processes may exist. These include 1) the types of decisions involved in so called "commitment" of stem cells to a specific lineage, 2) the types of decisions involved in the processes by which a progenitor cell initiates the differentiation process, and 3) the types of decisions involved in loss of proliferative potential, such as when senescence or terminal differentiation occurs.

## BIOLOGICAL MECHANISMS THAT REGULATE DIFFERENTIATION AND PROLIFERATION

A major focus of our studies on carcinogenesis has been to establish the biochemical and molecular mechanisms that regulate the integrated control of differentiation and proliferation of stem cells. Because previous reports using a variety of in vivo and in vitro systems suggested that these regulatory processes are mediated by cell-cycle dependent mechanisms [16–19], we have focused many of our studies in that direction.

We have employed two in vitro model cell systems to examine the mechanisms that regulate the integrated control of cellular differentiation and proliferation. Both murine BALB/c 3T3 T mesenchymal stem cells that are immortalized but nontransformed [20] and normal diploid human keratinocyte progenitor cells that are nonimmortalized and nontransformed [21] have been studied. In both cell systems we have established that there are specific regulatory mechanisms that control differentiation and proliferation and that these mechanisms are mediated at a distinct biological state(s).

Our most detailed studies have utilized the 3T3 T mesenchymal stem cells. These cells were originally subcloned from BALB/c 3T3 (clone A31) cells because of their propensity to differentiate into adipocytes [22]. BALB/c 3T3 T cells have been called proadipocytes, but a more appropriate name is mesenchymal stem cells because they can be used induced to differentiate into other cell types, including macrophages [23,24]. With these cells we have established that the integrated control of proliferation and differentiation is mediated at a complex of two specific arrest states in $G_1$ that we have designated $G_D/G_{D'}$. These states are distinct from other cell-cycle restriction points [16] such as those induced by growth factor deprivation or contact inhibition [18,19,25–27]. Figure 1 presents a cell-cycle model illustrating these points. This model is based on the results of a series of studies and demonstrates that there are two general classes of $G_1$ arrest states in 3T3 T stem cells. The first type of arrest state mediates the integrated control of differentiation and proliferation, whereas the second type of arrest state serves only to mediate the control of cellular proliferation, such as, by processes that are growth factor-dependent ($G_S$) or cell density-dependent/contact inhibition-dependent ($G_C$) [18,19,27–30]. It should be noted that in the past the term "$G_0$" has been employed to describe virtually all $G_1$ arrest states. With the advent of methodology to characterize multiple subtypes of cell cycle arrest states [16,18,19,25,27,28], we suggest the term $G_0$ should now be used only as a general term and not as a term to describe specific arrest states.

With respect to the biological mechanisms that integrally regulate the control of cellular differentiation or proliferation, Figure 1 shows that at least five distinct steps are involved. First, rapidly growing cells arrest their growth at the predifferentiation arrest state designated $G_D$ (arrow 1) when exposed to medium containing platelet-poor human plasma. Cells at the $G_D$ state are undifferentiated and retain their fibroblastlike morphology. Cells at the $G_D$ state possess the capability to make the decision either to remain quiescent, to reinitiate proliferation (arrow 5a), or to proceed through the differentiation process. If differentiation is induced, it involves at least two steps. The cells must first undergo nonterminal differentiation (arrow 2) and transition to the $G_{D'}$ state. Cells at the $G_{D'}$ state are fully morphologically differentiated and contain large, fat droplets and high levels of lipogenic enzymes [18,19]. Once cells are at the $G_{D'}$ state they can then make decisions to either remain quiescent at $G_{D'}$, to undergo terminal differentiation (arrow 3) to the TD state, to lose the

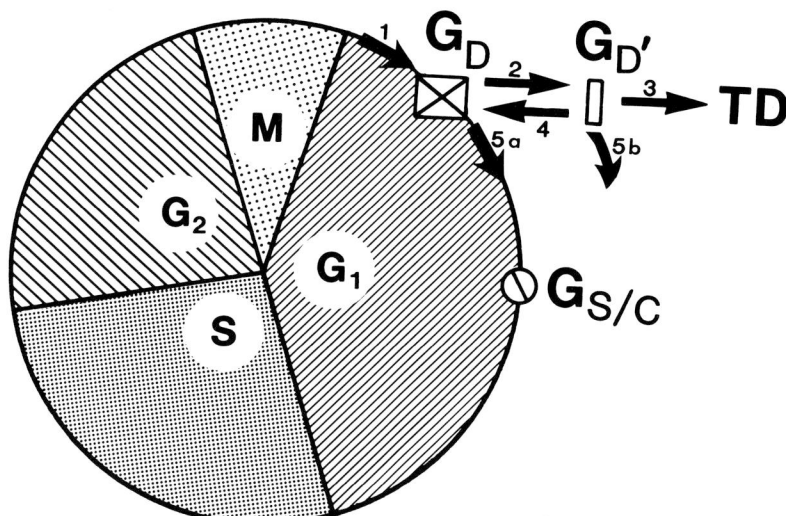

Fig. 1. Model for the cell-cycle dependent control of differentiation and proliferation in 3T3 T murine mesenchymal stem cells.

adipocyte phenotype and dedifferentiate (arrow 4), or to reinitiate proliferation with or without losing the differentiated phenotype (arrow 5b). Therefore the integrated control of cell proliferation and differentiation can be mediated at $G_{D'}$ as well as at $G_D$.

The integrated control of mesenchymal stem cell differentiation and proliferation (Fig. 1) has been extensively characterized in our studies, and a significant amount of information concerning each of the five steps reviewed above has been obtained [18,19,27–36]. Table I briefly itemizes some of the physiological and pharmacological factors that can influence each step present in the model (Fig. 1). For example, we have purified (20,000-fold) the human plasma protein that induces the terminal event in differentiation, ie, the $G_{D'} \rightarrow TD$ transition, and we have named this protein aproliferin. Significant progress has also been made in purifying the human plasma protein that induces predifferentiation arrest at the $G_D$ state, and we are currently using the term "$G_D$ arrestor" as a working designation for this protein. It has also been established that cells at each of these states have their own unique biological and biochemical characteristics, as summarized in Tables II–IV

In order to determine whether the mechanisms for the integrated control of cellular proliferation and differentiation in mesenchymal stem cells are representative of normal human cells, we have also performed experiments on cultured normal

**TABLE I. Factors That Mediate Specific Steps in the Integrated Control of Mesenchymal Stem Cell Differentiation and Proliferation***

| Step 1 | a) $G_D$ arrestor: Human plasma protein<br>Isoelectric point < 7.0<br>Trypsin sensitive<br>Acid labile, heat labile, DTT resistant<br>Does not adsorb to Affi-Gel Blue matrix nor to barium<br>Binds to heparin agarose<br>Molecular weight > 14,000<br>b) TPA (inhibitory) |
|---|---|
| Step 2 | a) Human plasma proteins that remain to be identified and various hormones, including insulin<br>b) Transforming growth factor type $\beta$ (inhibitory) |
| Step 3 | a) Aproliferin:<br>Human plasma protein<br>20,000 fold purified<br>Isoelectric point 7.6<br>Molecular weight ~40,000<br>Trypsin sensitive, acid labile, DTT resistant<br>b) Cachectin/tumor necrosis factor (inhibitory) |
| Step 4 | a) Isobutyl methyl xanthine<br>b) 8-bromo-cyclic AMP<br>c) Retinoic acid<br>d) TPA |
| Step 5a | a) Isobutyl methyl xanthine<br>b) EGF<br>c) PDGF<br>d) Serum |
| 5b | a) High serum concentrations plus insulin |

*DTT, dithiothreitol; EGF, epidermal growth factor; PDGF, platelet-derived growth factor; TPA, tetradecanoyl-phorbol-acetate.

## TABLE II. Characteristics of Cells at the Predifferentiation $G_D$ Arrest State

1. Possess the potential to make decisions to integrate the control of differentiation and proliferation
2. Is a state in $G_1$ that is distinct from those states induced by serum deprivation or contact inhibition which do not facilitate differentiation
3. Represent predifferentiated cells even though they do not express evidence of morphological or enzymatic differentiation
4. Characterized by the expression of unique mitogenic responsiveness to isobutyl methyl xanthine
5. Characterized by the expression of >100 distinct polypeptides detected in two-dimensional gel electrophoretic analysis
6. Characterized by the expression of distinct cell surface microvilli
7. Consists of at least two substates
8. Show predilection for extensive cell-cell clustering
9. Process of $G_D$ arrest can be inhibited by the tumor promoter designated TPA

## TABLE III. Characteristics of Cells at the Nonterminal $G_{D'}$ State of Differentiation

1. Possess the potential to make decisions to integrate the control of differentiation and proliferation
2. Represent differentiated cells that retain their proliferative potential
3. Express decreased responsiveness to growth factors
4. Consist of at least two substates
5. Can be induced to dedifferentiate and subsequently redifferentiate into another cell type (ie, transdifferentiation)
6. Show distinct polypeptide composition relative to $G_D$ and TD cells and more specifically contain a 35,000 MW polypeptide not present in TD cells

## TABLE IV. Characteristics of the Terminal Step in Differentiation and Terminally Differentiated Cells

Characteristics of the terminal step in differentiation
1. Requires 24–48 hrs to be completed but can be induced by brief pulse exposure to human plasma
2. Is induced by the human plasma protein designated aproliferin
3. Is blocked by protein synthesis inhibitors but not by inhibitors of RNA or DNA synthesis
4. Is not induced by any known physiological or pharmacological factors other than aproliferin
5. Can be inhibited by cachectin/tumor necrosis factor
6. Cell clones have been isolated and described that lack the potential to undergo the terminal event in differentiation efficiently

Characteristics of terminally differentiated cells
1. Cannot be distinguished from nonterminally differentiated cells except by proliferation and/or dedifferentiation assays; ie, TD cells cannot be induced to proliferate nor dedifferentiate
2. Contain six distinct polypeptides not present in nonterminally differentiated cells and lack the 35,000 MW proliferation-associated protein that is present in $G_{D'}$ cells
3. Do not mediate any aspect of the integrated control of differentiation and proliferation because they cannot modify their differentiation or proliferation characteristics even though they are viable cells

human keratinocyte progenitor cells isolated from neonatal foreskin. These cells are epithelial cells that can be induced to differentiate into mature keratinocytes. For initial studies on these cells, we utilized the basic protocol and medium developed by Ham and co-workers [37,38]. That is, the cells were grown in a serum-free/defined medium (ie, MCDB 153 medium), supplemented with ethanolamine 0.1 mM), phosphoethanolamine (0.1 mM), hydrocortisone (0.5 $\mu$M), EGF (10 ng/ml), insulin (5 $\mu$g/ml), and calcium (0.1 mM) $\pm$ bovine pituitary extract (140 $\mu$g/ml) [21]. This culture system does not require a feeder layer of cells or a specific extracellular matrix.

Preliminary studies established that normal human keratinocyte progenitor cells also possess mechanisms to integrate the control of cellular proliferation and differentiation. Briefly, it was found that proliferation and differentiation could be integrally regulated by modulating the concentrations of EGF and $Ca^{++}$ in the medium [21]. For example, in the presence of low EFG concentrations, but high $Ca^{++}$ concentrations, differentiation results. However, in the presence of high concentrations of EGF and low concentrations of $Ca^{++}$, proliferation is dominant.

Based on these observations, we sought to establish whether growth arrest at specific states are required for differentiation to occur. Briefly, the results of extensive studies show that there are two types of arrest states that can be identified in normal human keratinocyte progenitor cells [21, 39–41]. Table V summarizes these data and demonstrates that one type of arrest state is reversible, while the other type of arrest state is irreversible. Additional data show that during terminal differentiation, progression from a reversible to an irreversible state occurs. Therefore in keratinocyte progenitor cells two different types of predifferentiation growth arrest states exist and they can occur either in $G_1$ or $G_2$.

## NEOPLASTIC TRANSFORMATION AND THE CONTROL OF DIFFERENTIATION AND PROLIFERATION

The first part of this paper reviewed some of the clinical and experimental evidence suggesting that neoplastic transformation is associated with the expression of defects in the integrated control of cellular differentiation and proliferation, and not merely the development of defects in proliferation or differentiation. To test this

TABLE V. Evidence for Existence of a Distinct Predifferentiation Arrest State in Normal Human Keratinocyte Progenitor Cells

| | Growth characteristics (% LN)[a] | Predominant cell-cycle arrest state | Clonogenic potential after arrest (% CFE)[b] | Differentiation potential[c,d,e] |
|---|---|---|---|---|
| I. Control | 95 | — | >90 | — |
| II. Irreversible arrest state | | | | |
|   A. Razoxane | 5 | $G_2$ | <5 | + |
|   B. TPA | 8 | $G_1$ and $G_2$ | <5 | + |
|   C. Senescence | 1 | $G_1$ | <1 | + |
| III. Reversible arrest state | | | | |
|   A. Lymphokine | 2 | $G_1$ | >80 | + |
|   B. β-TGF | 3 | $G_1$ | >90 | + |
|   C. Isoleucine deficiency | 1 | $G_1$ | >80 | + |
|   D. Ethionine | 1 | $G_1$ | >90 | + |

[a]% Labeled nuclei: 24-hr [$^3$H]thymidine incorporation assayed and evaluated by autoradiography.
[b]% Colony forming efficiency in complete MCDB 153 containing EGF and insulin.
[c]Monoclonal antibodies 4F2 and BK and blue coloration by Ayoub and Shklar histochemical staining characterize undifferentiated/basal cells.
[d]Monoclonal antibodies HuK and A3.1-67, polyclonm antibody to involucrin and red coloration by Ayoub and Shklar histochemical staining characterize differentiated/suprabasal cells.
[e]To attempt to induce differentiation, arrested cells were typically refed growth factor depleted/high $Ca^{++}$ medium ± stated additives depending on the specific experimental protocol.

proposal more directly, two basic types of experiments have been performed. In one type of experiment we isolated clones of 3T3 T cells that expressed single or combined defects in their ability to integrate the control of differentiation and proliferation at $G_D/G_{D'}$ or their ability to regulate proliferation by growth factor/contact-dependent processes at $G_{S/C}$. We then performed tumorigenicity assays to determine which defect or defects are required for a cell to be completely transformed. In the second type of experiment, we sought to determine whether carcinogen treatment induces an acute effect in the control of differentiation or proliferation in these cells.

In the first type of study, nonmutagenic methods were used to select 23 clones of mesenchymal stem cells as described above [33,35,36]. Table VI briefly summarizes the results of these studies. The data show that clones that express combined defects in their ability to integrate the control of differentiation and proliferation and in their ability to regulate proliferation by growth factor-dependent processes are highly tumorigenic. In fact, every clone that expressed combined defects is tumorigenic; ie, 66% of animals injected with such cells develop tumors. By contrast, cell clones that expressed defects only in the integrated control process showed a tumor incidence of 22%, while clones that expressed defects only in the control of proliferation by growth factor-dependent processes showed a tumor incidence of only 10%. The parental 3T3 T stem cells and subclones that retain intact regulatory mechanisms were nontumorigenic [35,36].

In order to evaluate further whether the expression of defects in the integrated control of differentiation and proliferation represents an early step in carcinogenesis, the effects of a variety of carcinogenic agents on mesenchymal stem cells were evaluated. In this regard, mesenchymal stem cells were exposed to initiating dosages of ultraviolet (UV) irradiation [42] and they were then assayed for the effect of this treatment on their ability to $G_D$ arrest and differentiate and on their ability to arrest at the $G_{S/C}$ state [31,43]. The results established that low dosages of UV irradiation selectively inhibited differentiation control at the $G_D/G_{D'}$ complex in high percentages of mesenchymal stem cells (70%) without inducing proliferation control defects mediated at the $G_{S/C}$ state [43]. It is important to note that the UV-irradiation-induced defects are heritable and stable during continuous culture for >8 mo and at least 200

**TABLE VI. Suppression of Tumorigenicity by Cell-Cycle-Dependent Mechanisms**

| Cell-cycle phenotype | Tumor incidence[a,b] | No. of cell clones tested |
|---|---|---|
| Defective integrated control and defective proliferation control | 66% (21/32) | 6 |
| Defective integrated control and intact proliferation control | 22% (6/27) | 6 |
| Intact integrated control and defective proliferation control | 10% (2/19) | 3 |
| Intact integrated control and intact proliferation control | 0% (0/33) | 8 |

[a]Tumor incidence was calculated as the percentage of animals bearing tumors and was determined by including all of the cell clones expressing that particular phenotype. Syngeneic BALB/c mice were injected subcutaneously with $1 \times 10^7$ cells. Tumor formation was checked weekly and was declared to be negative if no tumor formed within 3 mo.
[b]The numbers in parentheses represent the number of animals that were positive for tumor formation relative to the number of animals injected.

population doublings. In more recent studies we have tested the effects of various dosages of chemical carcinogens and have obtained similar results. Table VII briefly summarizes the results of the effects of UV irradiation and chemical carcinogen treatment. In general, these data suggest that the expression of defects in the integrated control mechanism is correlated with an increased probability of transformation.

It has also been demonstrated that an initiating dose of chemical carcinogen can induce defects in the ability of cells to differentiate in vivo [44]. The fact that many metaplastic and dysplastic disease states that result from aberrant differentiation represent preneoplastic lesions also supports this conclusion. A variety of additional studies furthermore implicate the expression of aberrant differentiation to be of etiological significance in the transformation process [10,11,31,33,35,36,41,45–52].

Another critically important question that we have just begun to study is whether carcinogens can also inhibit the integrated control of differentiation and proliferation in human keratinocyte progenitor cells. Preliminary data already suggest that human squamous carcinoma cells show a variety of such defects.

## ANTICARCINOGENESIS AND THE SUPPRESSION OF TUMORIGENICITY

Our concept of carcinogenesis suggests that early in the process of transformation, stem cells must be induced to express defects in their ability to integrate the control of differentiation and proliferation, and thereafter they must express overt defects in the actual process of cellular differentiation and/or proliferation. The data reviewed above and published by other investigators support this conclusion [11,13,35,36,44,47,48,51,53]. Our concept of carcinogenesis, however, also suggests that normal cells contain another regulatory complex that functions to prevent/suppress carcinogenesis and that this process must also be abrogated early in the process of neoplastic transformation [24,31,54]. In the final section of this paper, we will briefly review this new area of research and present a summary of our recent data which support the conclusion that anticancer genes exist and function to prevent cell transformation and tumorigenicity.

TABLE VII. Inhibition of the Integrated Control of Cellular Differentiation and Proliferation Induced by Physical and Chemical Carcinogenic Agents in Mesenchymal Stem Cells and Its Association With an Increased Probability of Neoplastic Transformation*

| Carcinogen treatment and dosage | Inhibition of differentiation (%) | Neoplastic transformation assays (type III foci/$10^7$ cells) |
|---|---|---|
| Control | 0 | $\leq 1$ |
| UV irradiation (90 to 240 ergs/mm$^2$) | 70 | >50 |
| 4NQO (20 to 100 ng/ml) | 41 | 45 |
| DMBA (100 to 250 ng/ml) | 32 | 25 |
| MNNG (20 to 400 ng/ml) | 25 | 15 |

*DMBA, 7,12-dimethylbenz[a]anthracene; MNNG, N'-methyl-N'-nitro-N-nitrosoguanidine; 4NQO, 4, nitroquinoline.

The existence of anticancer/tumor suppressor genes is supported by a variety of previous studies. First, it has been demonstrated that the fusion of normal human cells and human carcinoma cells results in the suppression of tumorigenicity [51]. It has also been reported in similar types of studies that at least two complementation groups of anticancer/tumor suppressor genes exist and that they reside on different chromosomes [55]. Similar data have been obtained from studies on rodent cells [56–61]. In this regard, it has been proposed that the process of tumor suppression in vivo may result at least in part by mechanisms that involve the induction of differentiation [48,51,62].

The second type of study that supports the existence of anticancer/tumor suppressor genes involves analysis of the effect of implantation of malignant cells in specific embryonic microenvironments. These studies have demonstrated that the malignant phenotype of embryonal carcinoma cells [12,63,64], neuroblastoma cells [65], and acute myelogenous leukemia cells [66] can be suppressed. The mechanisms by which this process is mediated appear to involve the induction of nonterminal differentiation because such cancer cells have been shown to function in normal embryonic development following their implantation.

Finally, the existence of anticancer/tumor suppressor genes has been implicated by a variety of clinical and experimental studies which suggest that hereditary tumors including retinoblastoma, hepatoblastoma, osteosarcoma, and Wilms' tumor result from the deletion or inactivation of specific anticancer/tumor suppressor genes [67–74]. Retinoblastoma is the best understood of these models. Briefly, it has been found that in patients predisposed to develop retinoblastoma there appears to be a constitutional heterozygosity of the normal anticancer gene alleles such that one is mutated or lacking at a single locus. When the second normal allele of the "anticancer gene" is lost or mutated, a tumor then develops. In other words, it is the loss of function of the anticancer gene(s) that promotes or permits tumor formation. The chromosomal location for the putative anticancer gene in retinoblastoma has been determined to reside in chromosomal region 13q14. It should also be stressed that other steps in the process of retinoblastoma development may, however, also be important [62,73,75]. The possibility that stem cell differentiation or development plays a significant role in susceptibility or resistance to transformation in retinoblastoma is suggested by studies which show that once young patients exceed a certain age or develop to a certain state of maturity, their susceptibility to such a hereditary tumor markedly decreases [62,69,70].

Based on these data and our recent studies, we have recently begun to investigate the role of differentiation in regulating the susceptibility of stem cells to neoplastic transformation via activation of proported anticancer/tumor suppressor gene expression. For these studies, we have employed the 3T3 T murine mesenchymal stem cells described above because these cells are nontumorigenic yet can be transformed by UV irradiation or chemical carcinogens and because they express a well-defined differentiation pathway that facilitates isolation of clones of nonterminally differentiated cells (Fig. 1). Initial experiments show that the process of nonterminal differentiation of native 3T3 T cells induced expression of anticancer/tumor suppressor activity in approximately 25% of selected cell clones but that no significant anticancer/tumor suppressor activity was detected in cells at other growth/cell-cycle states such as during exponential growth or at the $G_D$ or $G_{S/C}$ states. For these studies, anticancer activity was defined as complete resistance to spontaneous transformation and transformation induced by UV irradiation and 4NQ0 treatment. We have also shown that differentiation-induced anticancer/cancer suppressor activity is stable for $\geqslant 50$ population doublings and it was also demonstrated that these cell clones were also nontumorigenic in vivo. We therefore have concluded that anticancer gene(s) must exist, and that their expression is regulated at least in part by differentiation-dependent processes. In this regard, studies are in progress to attempt to isolate and clone anticancer/tumor suppressor genes and to establish their precise biological mechanism of action.

## DISCUSSION AND FUTURE DIRECTIONS

In this paper, experimental and clinical evidence has been reviewed which supports two basic concepts concerning the process of neoplastic transformation. The first concept is that the development of defects in the ability of stem cells to integrate differentiation and proliferation is of etiological significance in carcinogenesis, as is the overt expression of specific differentiation and proliferation defects. The second concept is that anticancer/tumor suppressor genes exist in normal cells, that their expression is regulated by differentiation-dependent processes, and that their expression must be abrogated for a cell to be neoplastically transformed. Although our knowledge in these areas of cancer research has increased greatly in recent years, it has now become evident that the control of proliferation and differentiation in normal cells is exceedingly complex, as are the events that result in cell transformation.

We suggest that it is now most important to accept that cancer is not solely a disease of uncontrolled cell proliferation, nor is cancer a disease solely of uncontrolled cell differentiation. On the contrary, cancer must be a disease that results from the expression of defects in the control of both proliferation and differentiation and from the expression of defects in the coupling or integration of these two regulatory processes. We also suggest that it is most important to understand that cancer does not simply result from activation of genes that promote cell proliferation but also from modulation in the expression of genes that function to control cellular differentiation and regulate anticancer/tumor suppressor gene expression.

## ACKNOWLEDGMENTS

The authors would like to thank K. Connelly for typing the manuscript. We acknowledge the expert technical assistance of B.J. Hoerl, P.B. Maercklein, M.A. Zschunke, N.N. Swanson, M. Edens, and B.M. Hsu. This work was supported in part by NIH grants CA28240 and CA21722 to R.E.S.; a grant from the Council for Tobacco Research-USA, Inc., to R.E.S.; and by the Mayo Foundation. R.L.S. and D.N.E. also received partial support from NIH training grant CA09441.

## REFERENCES

1. Johnson LD, Nickerson RJ, Easterday CL, Stuart RS, Hertig AT: Cancer 22:901–914, 1968.
2. DeCosse JJ: Cancer Surveys 2:347–357, 1983.
3. Grove GL: Int J Dermatol 18:111–122, 1979.
4. Pittelkow MR, Perry HO, Muller SA, Maughan WZ, O'Brien PC: Arch Dermatol 117:465–468, 1981.

5. Ackerman LV, Rosai J: "Surgical Pathology", Fifth Edition. St. Louis: CV Mosby Co, 1974.
6. Anders A, Anders F: Biochim Biophys Acta 516:61–95, 1978.
7. Gateff E: Science 200:1448–1459, 1978.
8. Evans RW: "Histological Appearance of Tumors." Baltimore: Williams and Wilkins, 1968, pp 1–1223.
9. Lotem J, Sachs L: Proc Natl Acad Sci USA 79:4347–4351, 1982.
10. Sachs L: Nature 274:535–539, 1978.
11. Sachs L: Proc Natl Acad Sci USA 77:6152–6156, 1980.
12. Mintz B, Illmensee K: Proc Natl Acad Sci USA 72:3585–3589, 1975.
13. Pierce GB: Fed Proc 29:1248–1254, 1970.
14. Pierce GB: Shikes R, Fink LM: "Cancer: A Problem of Developmental Biology." Englewood Cliffs, New Jersey: Prentice-Hall, 1978, pp 68–87.
15. Pierce GB, Wallace C: Cancer Res 31:127–134, 1971.
16. Pardee AB, Dubrow R, Hamlin JL, Kletzien RF: Annu Rev Biochem 47:715–750, 1978.
17. Prescott DM: Adv Genet 18:99–177, 1976.
18. Scott RE, Florine DL, Wille JJ, Jr, Yun K: Proc Natl Acad Sci USA 79:845–849, 1982.
19. Scott RE, Hoerl BJ, Wille JJ, Jr, Florine DL, Krawisz BR, Yun K: J Cell Biol 94:400–405, 1982.
20. Scott RE, Boone CW: J Natl Cancer Inst 66:733–736, 1981.
21. Wille JJ, Jr, Pittelkow MR, Shipley GD, Scott RE: J Cell Physiol 121:31–44, 1984.
22. Diamond L, O'Brien TG, Rovera G: Nature 269:247–249, 1977.
23. Krawisz BR, Florine DL, Scott RE: Cancer Res 41:2891–2899, 1981.
24. Sparks RL, Seibel-Ross EI, Wier ML, Scott RE: Cancer Res 46:5312–5319, 1986.
25. Pledger WJ, Stiles CD, Antoniades HN, Scher CD: Proc Natl Acad Sci USA 75:2839–2843, 1978.
26. Leof EB, Van Wyk JJ, O'Keefe EJ, Pledger WJ: Exp Cell Res 147:202–208, 1983.
27. Wille JJ, Jr, Scott RE: J Cell Physiol 112:115–122, 1982.
28. Krawisz BR, Scott RE: J Cell Biol 94:394–399, 1982.
29. Hoerl BJ, Wier ML, Scott RE: Exp Cell Res 155:422–434, 1984.
30. Scott RE, Yun K, Florine DL: Exp Cell Res 143:405–414, 1983.
31. Scott RE, Wille JJ, Jr, Pittelkow MR, Sparks RL: In Barrett JC, Tennant RW (eds): "Carcinogenesis, Vol 9—Mammalian Cell Transformation." New York: Raven Press, 1985, pp 67–80.
32. Sparks RL, Scott RE: Exp Cell Res 165:345–352, 1986.
33. Wier ML, Scott RE: Cancer Res 45:3339–3346, 1985.
34. Wier ML, Scott RE: J Cell Biol 102:1955–1964, 1986.
35. Wille JJ, Jr, Scott RE: Int J Cancer 37:875–881, 1986.
36. Wille JJ, Jr, Maercklein PB, Scott RE: Cancer Res 42:5139–5146, 1982.
37. Boyce ST, Ham RG: J Invest Dermatol [Suppl] 81:33–40, 1983.
38. Tsao MC, Walthall BJ, Ham RG: J Cell Physiol 110:219–229, 1982.
39. Pittelkow MR, Wille JJ, Jr, Scott Re: J Invest Dermatol 86:410–417, 1986.
40. Shipley GD, Pittelkow MR, Wille JJ, Jr, Scott RE, Moses HL: Cancer Res 46:2068–2071, 1986.
41. Wille JJ, Jr, Pittelkow MR, Scott RE: Carcinogenesis 6:1181–1187, 1985.
42. Mondal S, Heidelberger C: Nature 260:710–711, 1976.
43. Scott RE, Maercklein PB: Proc Natl Acad Sci USA 82:2995–2999, 1985.
44. Yuspa SH, Morgan DL: Nature 293:72–74, 1981.
45. Fibach E, Landau T, Sachs L: Nature New Biol 237:276–278, 1972.
46. Nakano S, Ts'o POP: Proc Natl Acad Sci USA 78:4995–4999, 1981.
47. Nakano S, Ueo H, Bruce SA, Ts'o POP: Proc Natl Acad Sci USA 82:5005–5009, 1985.
48. Peehl DM, Stanbridge EJ: Int J Cancer 30:113–120, 1982.
49. Rheinwald JG, Beckett MA: Cell 22:629–632, 1980.
50. Scott RE, Florine DL: Am J Pathol 107:342–348, 1982.
51. Stanbridge EJ, Der CJ, Doersen C-J, Nishimi RY, Peehl DM, Weissman BE, Wilkinson SE: Science 215:252–259, 1982.
52. Yen A: Exp Cell Res 156:198–212, 1985.
53. Pierce GB: Am J Pathol 77:103–118, 1974.
54. Scott RE, Wille JJ, Jr, Wier ML: Mayo Clin Proc 59:107–117, 1984.
55. Weissman BE, Stanbridge EJ: J Natl Cancer Inst 70:667–672, 1983.
56. Craig RW, Sager R: Proc Natl Acad Sci USA 82:2062–2066, 1985.
57. Harris H, Miller OJ, Klein G, Worst P, Tachibana T: Nature 223:363–368, 1969.

58. Howell N, Sager R: Somatic Cell Genet 5:129–143, 1979.
59. Klinger HP, Shows TB: J Natl Cancer Inst 71:559–569, 1983.
60. Oshimura M, Gilmer TM, Barrett JC: Nature 316:636–639, 1985.
61. Stoler A, Bouck N: Proc Natl Acad Sci USA 82:570–574, 1985.
62. Sager R: Cancer Res 46:1573–1580, 1986.
63. Pierce GB: Am J Pathol 113:117–124, 1983.
64. Pierce GB, Aguilar D, Hood G, Wells RS: Cancer Res 44:3987–3996, 1984.
65. Podesta AH, Mullins J, Pierce GB, Wells RS: Proc Natl Acad Sci USA 81:7608–7611, 1984.
66. Gootwine E, Webb CG, Sachs L: Nature 299:63–65, 1982.
67. Cavenee WK, Dryja TP, Phillips RA, Benedict WF, Godbout R, Gallie BL, Murphree AL, Strong LC, White RL: Nature 305:779–784,1983.
68. Hansen MF, Koufos, A, Gallie BL, Phillips RA, Fodstad O, Brogger A, Gedde-Dahl T, Cavenee WK: Proc Natl Acad Sci USA 82:6216–6220, 1985.
69. Knudson AG, Jr: Prog Nucleic Acid Res Mol Biol 29:17–25, 1983.
70. Knudson AG, Jr: Cancer Res 45:1437–1443, 1985.
71. Koufos A, Hansen MF, Copeland NG, Jenkins NA, Lampkin BC, Cavenee WK: Nature 316:330–334, 1985.
72. Koufos A, Hansen MF, Lampkin BC, Workman ML, Copeland NG, Jenkins NA, Cavenee WK: Nature 309:170–172, 1984.
73. Murphree AL, Benedict WF: Science 223:1028–1033, 1984.
74. Orkin SH, Goldman DS, Sallan SE: Nature 309:172–174, 1984.
75. Chaum E, Ellsworth RM, Abramson DH, Haik BG, Kitchin FD, Chaganti RSK: Cytogenet Cell Genet 38:82–91, 1984.

# Aberrant Regulation of Differentiation in Epidermal Carcinogenesis

S.H. Yuspa, U. Lichti, J. Strickland, S. Jaken, D. Lowy, J. Harper, D. Roop, and H. Hennings

*Laboratories of Cellular Carcinogenesis and Tumor Promotion (S.H.Y., U.L., J.S., J.H., D.R., H.H.) and Cellular Oncology (D.L.), National Cancer Institute, and Center for Drugs and Biologics, FDA, Division of Virology, National Institutes of Health (S.J.), Bethesda, Maryland 20892*

The induction of mouse skin carcinomas by initiation-promotion protocols involves three operationally distinct stages. Initiation and promotion result in papillomas which can be converted to carcinomas by a promoter-independent event. Initiation appears to be a genetic change which results in an altered basal cell response to differentiation signals allowing the altered cells to proliferate away from the basement membrane. In cell culture, epidermal basal cells proliferate in medium with $Ca^{2+}$ concentrations <0.1 mM and cease proliferating and terminally differentiate in medium with higher concentrations of $Ca^{2+}$. $Ca^{2+}$-induced differentiation appears to be mediated via stimulation of phosphatidylinositol lipid metabolism and subsequent activation of protein kinase C. This signal transduction system may modulate the level of intracellular ions, particularly $K^+$, which provides a permissive environment for differentiation to proceed. Cells isolated from initiated skin, from carcinogen-treated basal cell cultures, and from mouse papillomas are able to proliferate in medium with >0.1 mM $Ca^{2+}$, and this may represent the initiated phenotype. Phorbol esters also induce terminal differentiation in normal keratinocytes, presumably by activating protein kinase C directly. Initiated cells are resistant to phorbol-ester-induced terminal differentiation and have a selective growth advantage during tumor promotion. Activated transforming genes, recognized by transfection into several test systems, have not been identified for all initiated cells. However, an activated *ras* gene may result in an initiated cell phenotype (resistance to terminal differentiation) which requires tumor-promoter treatment for mitogenic stimulation in vitro under high $Ca^{2+}$ conditions. In contrast, the introduction of an activated *ras*$^{Ha}$ gene into initiated cells which do not contain an active transforming gene reproducibly accomplishes the malignant-conversion step. This suggests that the *ras* locus may be a target for genetic damage during both initiation and malignant conversion in vivo.

Key words: mouse epidermal cells, tumor promotion, papilloma cells, aberrant differentiation and carcinogens, phosphatidylinositol metabolism, calcium, *ras* oncogene

Received June 19, 1986.

© 1988 Alan R. Liss, Inc.

Cancer is commonly considered a disease in which rapid, unregulated proliferation causes the expansion of tumor cells to a clinically relevant end point. Indeed, many of the current molecular approaches to understanding the genetics of carcinogenesis (oncogene studies) have focused on genes which are recognized because they have been transduced by retroviruses and cause rapid proliferation in their experimental target cells leading to tumorlike end point. However, the transformation-sensitive target cells for these transduced "proliferation" genes have generally been simple cell types (fibroblasts) or cells which are already abnormal by virtue of a previous treatment or extended cultivation in vitro. In vivo, the usual target sites for cancer development in humans or carcinogen-exposed experimental animals are epithelial cells, most commonly lining epithelia such as those of the skin, bronchus, gastrointestinal tract, pancreatic duct, mammary duct, or urogenital tract. These epithelia are complex in structure and may have a stringently regulated program of differentiation which counteracts aberrant proliferative activity. For example, psoriasis is a hyperproliferative disease in skin which is not neoplastic because the intact program of epidermal differentiation inevitably leads to the death of the hyperproliferative cells. Such considerations suggest that the carcinogenesis process requires an alteration in the program of terminal differentiation, and the development of this alteration must consititute a rate-limiting step in cancer development in lining epithelia.

## MOUSE EPIDERMAL CELL CULTURE AS A MODEL SYSTEM TO STUDY CARCINOGENESIS

Mouse skin is the prototype model for the study of carcinogenesis in lining epithelia and successful cultivation of mouse epidermal cells has provided a unique opportunity to explore the cellular and molecular changes which occur during neoplastic transformation.

The extracellular calcium concentration is a critical regulator of the growth and differentiation of cultured mouse epidermal cells. When cultured in medium with 0.02–0.1 mM calcium (low $Ca^{2+}$), primary epidermal cells proliferate as a monolayer of basal cells without desmosomal connections and with a high rate of DNA synthesis and a high growth fraction [1]. When medium calcium is increased to >0.1 mM (high $Ca^{2+}$), desmosomes rapidly form between cells, proliferation decreases to a low level, cells begin to stratify and cornify (produce a rigid cross-linked cell envelope), and are shed from the culture dish. The synthesis of several markers of epidermal differentiation is increased in the maturing cultures [2]. Inhibitor studies have suggested that $Ca^{2+}$-induced differentiation is independent of $Ca^{2+}$ or $Na^+$ fluxes [3]. However, the increase in medium $Ca^+$ produces an elevation of both intracellular $Na^+$ and $K^+$ within 12 hr. The $Ca^{2+}$-induced increase in intracellular $K^+$ appears to be the more relevant to these changes since the increase was blocked by ouabain and four other inhibitors of epidermal differentiation [3]. The five inhibitors had no consistent effect on intracellular $Na^+$. Thus, elevated intracellular $K^+$ may be necessary for the later stages of epidermal differentiation. However, none of the inhibitors affected the assembly of desmosomes, the earliest ultrastructural change noted after increasing medium $Ca^{2+}$ [4]. This rapid change in cell-cell contact, beginning within minutes after $Ca^{2+}$ elevation, appears to be independent of changes in $Na^+$ and $K^+$, but may instead be modulated by increased $Ca^{2+}$ at the cell surface.

Regardless of the mechanism of the $Ca^{2+}$ effect, the optimization of medium $Ca^{2+}$ for cell growth or differentiation has increased the usefulness of epidermal cultures as an in vitro model of epithelial carcinogenesis.

## MOLECULAR CONTROL OF EPIDERMAL DIFFERENTIATION

A role for protein kinase C as a regulator of epidermal differentiation was suggested when we discovered that phorbol esters induce differentiation of cultured keratinocytes [5]. Subsequently it was shown that exogenously added phospholipase C (which increases cellular diacylglycerol) mimics phorbol ester action [6]. This suggested that the external $Ca^{2+}$ signal may lead to C-kinase activation via stimulation of cellular phospholipase C activity. The effect of the external $Ca^{2+}$ signal on phospholipase C was studied in cultures prelabeled with [$^3$H]-inositol. Within 2 min after addition of 1 mM $Ca^{2+}$ to the culture medium of epidermal basal cells in low $Ca^{2+}$ medium, an increase in inositol phosphates (IPs) was measured [7]. This correlated with a decrease in radiolabeled phosphoinositides suggesting that these were the source of the increased IPs. After 3 hr in 1 mM $Ca^{2+}$ medium, each of the IPs remained elevated to 140% of control levels. This effect appears to be mediated by a rise in intracellular $Ca^{2+}$ because the ionophore A23187 causes a similar rise in IP levels. These results link the regulation of epidermal differentiation to the phosphatidylinositol-diacylglycerol-protein kinase C second messenger system.

## CARCINOGENS MODIFY THE RESPONSE OF NORMAL KERATINOCYTES TO THE $CA^{2+}$ SIGNAL FOR DIFFERENTIATION

Based on the biology of benign skin tumors, we have proposed that an early effect of carcinogens in mouse skin is to alter the differentiation program of some keratinocytes, and this would constitute the cellular basis for initiation [8]. To test this possibility, we studied the response of carcinogen-treated keratinocyte cultures to changes in extracellular $Ca^{2+}$. Basal epidermal cells were cultured in low $Ca^{2+}$ medium, exposed to carcinogens, and then changed stepwise at weekly intervals into higher $Ca^{2+}$ medium over 4 wk. Carcinogen exposure in vitro resulted in the focal persistence of keratinocytes which proliferated in high $Ca^{2+}$ medium and could be easily recognized because they stained dark red with rhodamine. A number of characteristics of this assay suggest it selects for initiated cells: 1) cultures of basal cells isolated from initiated mouse skin and selected in high $Ca^{2+}$ medium yield identical foci [9,10]; 2) the number of foci increases with higher doses of carcinogen in vivo or in vitro [10–13]; 3) stronger initiators yield more foci than weaker initiators for both in vivo and in vitro exposures [10,11]; 4) for benzo[a]pyrene, the number of foci directly correlates with the extent of DNA binding after in vitro exposure [14]; 5) the focus-forming potential of initiated skin persists for at least 1 wk after initiation [10]; 6) spontaneous foci develop in cultures from SENCAR mice, in which 12-O-tetradecanoylphorbol-13-acetate (TPA) promotion induces papillomas in uninitiated mice [9,10]; 7) foci are not tumorigenic when first formed, but cell lines derived from foci may progress to produce carcinomas upon in vivo testing [15].

Cell lines derived from these foci may be considered putative initiated cells. Three of these cell lines, designated 308, D, and F, have been characterized for their response to calcium. The formation of cornified cells and the activity of epidermal transglutaminase were utilized as markers of epidermal differentiation. Calcium did not increase transglutaminase activity or cornification of any of the three lines. Proliferation was estimated by the [$^3$H]thymidine-labeling index, by incorporation of [$^3$H]thymidine into DNA, and by a clonal growth assay. Unlike primary normal cultures, raising the calcium level of the medium did not markedly reduce the rate of proliferation of any of the three putative initiated cell lines.

To compare these cells to authentic initiated cells, methods to culture cells from papillomas induced by an initiation-promotion protocol in SENCAR mice were developed, and the resultant cell lines have been characterized [16]. Using Eagle's medium with 0.05 mM $Ca^{2+}$ conditioned by dermal fibroblasts and supplemented with 1 ng/ml epidermal growth factor (EGF) in culture dishes coated with collagen and fibronectin, six cell lines (PA, PB, PC, PD, PE, and PF) were established from separate pools of papillomas. When tested for tumorigenicity in nude mice by injection of a cell suspension or implantation of cells growing on a plastic liner, two of the lines (PC and PF) produced no tumors at any passage. In contrast, cells of the lines PA and PE produced highly differentiated squamous cell carcinomas from the earliest passage tested. The results with PB and PD were variable on tumorigenicity testing with some passages positive and others negative. When tested for responsiveness to $Ca^{2+}$ (>0.1 mM) as a differentiation stimulus, all lines responded. In the higher $Ca^{2+}$ medium there was a 50-95% decrease in colony-forming efficiency, a slight decrease in $^3$H-thymidine incorporation (except for PA), and an increase in the number of cornified cells (except for early-passage PF). Epidermal transglutaminase activity was increased in the presence of medium with $Ca^{2+}$ >0.1 mM. However, unlike normal cells, only a fraction of the cells from each of the papilloma-derived cell lines terminally differentiated in response to $Ca^{2+}$, while the remaining cells continued to proliferate, although at a slower rate. Thus papilloma cells and putative initiated cells have in common the ability to resist terminal differentiation induced by $Ca^{2+}$, but papilloma cells retain the capacity to respond partially to the $Ca^{2+}$ signal. When phospholipid metabolism was studied in response to the $Ca^{2+}$ signal, there was a substantial increase in phosphatidylinositol turnover in some papilloma cell lines, while others were less responsive.

In comparing the properties of papilloma-derived cell lines to the properties of putative initiated cells derived by carcinogen treatment in vitro, we have found that the differentiation-altered ("initiated") cell lines demonstrate a much higher growth potential in high $Ca^{2+}$ medium, produce very few cornified cells (<0.1%) in that medium, and transglutaminase is not induced. There are several explanations for these differences. It is possible that the differentiation-altered cells produced by carcinogen treatment represent a different change from that which produces papilloma cells. Further studies are required to analyze this possibility. Different selection methods are used to produce each type of cell line. In the case of papilloma cells, which are a highly selected population, cultivation was exclusively in low $Ca^{2+}$ culture conditions until cell lines evolved. For the production of differentiation-altered cell lines, cultivation was exclusively in high $Ca^{2+}$ medium once the selection process was instituted, and prolonged culture time was required to establish these lines including 6 weeks to identify foci and several months to isolate lines [15]. Thus, progressive changes may have taken place during the development of the differentiation-altered cell lines, though in most cases not leading to the tumorigenic state ([15] and unpublished). In support of this possibility, studies on late-passage papilloma cell line PD indicated a loss of both differentiated characteristics and sensitivity to the effects of $Ca^{2+}$ for inducing differentiation. Thus, the propensity for cells to change progressively in cell culture may explain the apparent differences in the cell types. Perhaps papilloma cells are most like the cellular foci which evolve early in the altered differentiation assay [9–12]. These foci slowly expand in size when maintained in high $Ca^{2+}$ medium but also display marked stratification and shed fully differentiated cells into the medium. Prolonged culture in high $Ca^{2+}$ medium may select for those cells in the initiated foci that no longer respond to $Ca^{2+}$.

## THE CELLULAR BASIS FOR PHORBOL-ESTER-MEDIATED TUMOR PROMOTION

The biology of phorbol-ester-mediated tumor promotion in mouse skin is consistent with a process of cell selection, leading to clonal expansion of initiated cells [8]. Mouse keratinocyte cultures were used to explore the cellular basis for selection. Cultured basal cells respond heterogeneously to tumor-promoting phorbol esters [17]. Distinct subpopulations are either induced to differentiate terminally or are stimulated to proliferate. The basis for heterogeneity in the population appears to be the maturation state of the basal cell at the time of exposure since 1) proliferative responders give rise to differentiative responders [17]; 2) advancing the maturation of basal cells by culture in high $Ca^{2+}$ medium leads to loss of proliferative responses and enhancement of differentiative responses [18]; 3) altered keratinocytes which fail to mature in response to high $Ca^{2+}$ medium are resistant to the differentiation-inducing effects of phorbol esters [17]. At the molecular level mediation of each response could be via distinct classes of phorbol ester receptors which also are modulated by maturation state [19]. In vivo, phorbol esters also appear to accelerate differentiation of maturing keratinocytes [20]. Cells are rapidly lost from the differentiating compartment after mouse skin is exposed to phorbol esters since mRNA transcripts for differentiation-associated keratins rapidly decrease when epidermal RNA is probed with cDNAs for these genes [21]. Taken together these results indicate that phorbol esters produce an imbalance in epidermal homeostatis owing to accelerated loss of one subpopulation and selective growth of another. Our results further predict that initiated cells form a compartment which is resistant to the differentiative influences of phorbol esters. This compartment would selectively expand in cell number with each promoter exposure ultimately to yield a benign tumor.

## INITIATED AND PAPILLOMA CELLS RESPOND DIFFERENTLY THAN NORMAL CELLS TO PHORBOL ESTERS

12-O-Tetradecanoylphorbol-13-acetate (TPA) and other promoters failed to increase transglutaminase activity or the number of cornified cells in any of the putative initiated cell lines. For some, but not all, of the initiated cell lines phorbol esters were mitogenic at both clonal and high cell density. Results with initiated cells were similar if TPA exposure was in low or high $Ca^{2+}$ medium. Responsiveness to phorbol ester tumor promoters was also examined in papilloma cell lines [16]. TPA treatment increased colony-forming efficiency, DNA synthesis, and colony size in all lines studied in medium with either high or low $Ca^{2+}$. TPA treatment also increased ornithine decarboxylase activity in all lines, even at the higher $Ca^{2+}$ concentration, although normal keratinocytes respond only when grown in medium with low $Ca^{2+}$. TPA treatment caused only a slight increase in the number of cornified cells and no increase in epidermal transglutaminase activity in papilloma cells, while it caused tenfold or greater increases in these differentiation markers in normal keratinocytes. Thus, papilloma cells and initiated cells appear to differ from normal keratinocytes in their ability to maintain a proliferating population under conditions favoring terminal differentiation, their proliferative response to phorbol esters under these same conditions, and their reduced sensitivity to phorbol-ester-induced terminal differentiation. All of these properties should provide a growth advantage to these cells during tumor promotion.

## THE ACTIVATION OF A *RAS* ONCOGENE AS AN EARLY OR LATE EVENT IN SKIN CARCINOGENSIS

The isolation of an activated form of the c-*ras*$^H$ gene from skin papillomas has provided presumptive evidence that this gene may be a target for a mutation which could constitute the initiating mutation in skin carcinogenesis [22]. Further support for this idea was provided by our studies indicating that the v-*ras*$^H$ gene could impart a conditional initiated phenotype on cultured keratinocytes by blocking their ability to differentiate terminally and arresting them in a late basal cell stage of maturation [23,24]. Such cells were reverted back to a proliferative basal cell phenotype by exposure to phorbol ester tumor promoters [24]. When the v-*ras*$^H$ gene of Harvey murine sarcoma virus (Ha-MuSV), a replication-defective transforming retrovirus, is introduced into cultured keratinocytes by a defective retroviral vector [25], skin grafts constructed with cells carrying the mutated *ras* oncogene produce papillomas on athymic nude mouse recipients [26]. Furthermore, the exogenous oncogene appears to be regulated at the transcriptional level and is not expressed in the differentiated portions of the benign tumor as determined by in situ hybridization studies using the v-*ras*$^H$ gene as a probe [26].

The regulation of *ras* gene expression in tumor material was surprising considering that the defective viral vector contained the v-*ras*$^H$ gene MoMuLV long terminal repeat (LTR) regions both 3' and 5' to the *ras* gene coding sequences. A previous study has demonstrated that endogenous MoMuLV LTR transcripts are not expressed in benign mouse skin tumors but are transcribed in malignant tumors [27]. In culture, the switch from low $Ca^{2+}$ to high $Ca^{2+}$ growth conditions is not sufficient to repress exogenous *ras* gene expression, although the proliferative effects of the p21 protein are completely suppressed [23]. However, high $Ca^{2+}$ culture conditions do not induce the complete program of terminal differentiation in epidermal cells expressing an activated *ras* gene [23,24]; and the cells remain blocked in a late basal cell stage and never express the high molecular weight keratins. In contrast, cells in benign tumors do progress to the later stages of differentiation and express the 67-kilodalton keratin [21]; and it is these cells in which v-*ras*$^H$ gene expression is diminished or message stability decreased. The regulation of a mutated c-*ras* oncogene by the differentiated portions of benign tumors could account for the benign phenotype in chemically induced tumors where *ras* gene activation is the initiating event. However, studies to test this idea require an analysis of the mutated gene under the influence of its own regulatory sequences.

The chemical induction of tumors on mouse skin usually proceeds through a series of stages resulting first in benign tumors (papilloma) which subsequently convert to malignancy (carcinoma). Based on the mutagenic action required for specific chemicals to produce each tumor type, at least two genetic events occur prior to carcinoma formation [28]. While *ras* activation may be the first event, it is clear that not all papillomas contain an activated *ras* oncogene, and thus the benign tumor phenotype can also be achieved via other routes.

When DNA from six papilloma cell lines was tested in the NIH-3T3 transfection assay, active transforming activity was not detected [29]. However, when the EJ *ras*$^H$ gene was introduced into papilloma cells (lines PB, PC, PD) by DNA transfection, transfectants showed an enhanced capacity to proliferate at clonal density under high $Ca^{2+}$ culture conditions and formed rapidly growing, anaplastic carcinomas in nude mice. Thus, in papilloma cells, a genetic change distinct from *ras*$^H$ activation may produce an altered differentiation program associated with the initiation step, and this genetic alteration may act in a cooperating fashion with an activated *ras* gene to result in malignant transformation.

## REFERENCES

1. Hennings H, Michael D, Cheng C, Steinbert P, Holbrook, K, Yuspa SH: Cell 19:245, 1980.
2. Yuspa SH: In Skerrow D, Skerrow C (eds): "Methods in Skin Research." Sussex: John Wiley and Sons, Limited, 1985, pp 213–249.
3. Hennings H, Holbrook K, Yuspa SH: J Cell Physiol 116:265, 1983.
4. Hennings H, Holbrook K: Exp Cell Res 143:127, 1983
5. Yuspa SH, Ben TB, Hennings H, Lichti U: Biochem Biophys Res Commun 97:70, 1980.
6. Jeng AY, Lichti U, Strickland JE, Blumberg PM: Cancer Res 45:5714, 1985.
7. Jaken S, Yuspa SH: (submitted for publication).
8. Yuspa SH, Hennings H, Lichti U: J Supramol Struct Cell Biochem 17:245, 1981.
9. Yuspa SH, Morgan DL: Nature 293:72, 1981.
10. Kawamura H, Strickland JE, Yuspa SH: Cancer Res 45:2748, 1985.
11. Kilkenny AE, Morgan D, Spangler EF, Yuspa SH: Cancer Res 45:2219, 1985.
12. Kulesz-Martin M, Koehler B, Hennings H, Yuspa SH: Carcinogenesis 1:995, 1980.
13. Kulesz-Martin M, Yoshida MA, Prestine L, Yuspa SH, Bertram JS: Carcinogenesis 6:1245, 1985.
14. Nakayama J, Yuspa SH, Poirier MC: Cancer Res 44:4087, 1984.
15. Kulesz-Martin M, Kilkenny AE, Holbrook KA, Digernes V, Yuspa SH: Carcinogenesis 4:1367, 1983.
16. Yuspa SH, Morgan DL, Lichti U, Spangler EF, Michael D, Kilkenny A, Hennings H: Carcinogenesis 7:949–958, 1986.
17. Yuspa SH, Ben T, Hennings H, Lichti U: Cancer Res 42:2344, 1982.
18. Yuspa SH, Ben T, Hennings H: Carcinogenesis 4:1413, 1983.
19. Dunn JA, Jeng AY, Yuspa SH, Blumberg PM: Cancer Res 45:5540, 1985.
20. Reiners JJ, Slaga TJ: Cell 32:247, 1983.
21. Toftgard R, Yuspa SH, Roop DR: Cancer Res 45:5845, 1985.
22. Balmain A, Ramsden M, Bowden GT, Smith J: Nature 307:658, 1984.
23. Yuspa SH, Vass W, Scolnick E: Cancer Res 43:6021, 1983.
24. Yuspa SH, Kilkenny AE, Stanley J, Litchti U: Nature 314:459, 1985.
25. Mann R, Mulligan RC, Baltimore D: Cell 33:153, 1983.
26. Roop DR, Lowy DR, Tambourin PE, Strickland J, Harper JR, Balaschak M, Spangler EF, Yuspa SH: Nature 323:822–824, 1986.
27. Housey GM, Kirschmeier P, Garte SJ, Burns F, Troll W, Bernstein IB: Biochem Biophys Res Commun 127:391, 1985.
28. Hennings H, Shores R, Wenk M, Spangler EG, Tarone R, Yuspa SH: Nature 304:67, 1983.
29. Harper JR, Roop DR, Yuspa SH: Mol Cell Biol 6:3144–3149.

# Growth and Differentiation Programs of Normal and Transformed Human Bronchial Epithelial Cells

Tohru Masui, John F. Lechner, George E. Mark III, Andrea M.A. Pfeifer, Masao Miyashita, George H. Yoakum, James C. Willey, Dean L. Mann, and Curtis C. Harris

*Laboratory of Human Carcinogenesis, Division of Cancer Etiology, National Cancer Institute, Bethesda, Maryland 20892*

A serum-free culture system was established for normal human bronchial epithelial (NHBE) cells, and a set of differentiation inducers (12-O-tetradecanoylphorbol-13-acetate [TPA] and type-$\beta$ transforming growth factor [TGF-$\beta$]) and their antagonists (epinephrine, cholera toxin, and pertussis toxin) have been identified. Based on the results with these antagonists, we suggest that GTP binding protein systems and/or the cAMP system may have important roles in controlling programs of growth and differentiation in NHBE cells. We have also summarized effects of TGF-$\beta$ as a negative-growth-regulatory molecule in normal epithelial cells. Further, oncogenes v-Ha-*ras* and v-*raf* were tested for transformation activity in NHBE cells and showed differential effects suggesting that *ras* may play a role in lung carcinogenesis.

**Key words:** normal human bronchial epithelial (NHBE) cells, terminal squamous differentiation, GTP binding protein systems, cyclic AMP, epinephrine, cholera toxin, pertussis toxin, 12-O-tetradecanoylphorbol-13-acetate, type-$\beta$ transforming growth factor, serum-free medium, oncogenes, *ras*, *raf*, nickel ion, cigarette smoke condensate

With the exception of stem cells, normal cells in vivo are usually in quiescent or differentiating stages [1]. In this background malignant cells may have a selective growth advantage by being relatively insensitive to exogenous signals that exert negative growth control and/or induce terminal differentiation of normal cells. Therefore, aberrant interactions of microenvironmental stimuli with the transformed cells may allow their persistent growth by at least two mechanisms: 1) lost or decreased sensitivity to negative regulatory signals and 2) lost requirement for and increased sensitivity to growth-promoting stimuli [2–6].

The overlapping concepts of oncogenes and autocrine growth have been of great interest to cancer researchers (see reviews [4–12]). Gastrin-releasing peptide is an

Received September 16, 1986.

© 1988 Alan R. Liss, Inc.

example of an autocrine growth factor that enhances the growth of small cell lung carcinoma cells [13]. Although the loss of growth-factor dependency of fibroblasts has been observed [14–16], the majority of human cancers are derived from epithelial cells that normally terminally differentiate [17], and a diminished response to terminal differentiation inducers is a hallmark in vitro phenotypic marker of neoplastic cells [2, 18–27]. Recently a number of differentiation inducers and/or growth inhibitors that may control the behavior of normal cells have been isolated from normal and neoplastic sources [28–33], and the way a cell responds to these regulatory factors during carcinogenesis may also play an important role in tumor progression [34–38]. Therefore, studies of the regulatory mechanisms involving differentiation inducers and growth inhibitors and the way by which cells become insensitive to these factors is indispensable to understanding the mechanism of carcinogenesis.

In this paper we will summarize our studies investigating the growth and differentiation programs of normal human bronchial epithelial (NHBE) cells and will discuss results from transformation studies of NHBE cells using by different types of agents.

## RESULTS AND DISCUSSION
### Serum-Free Culture of NHBE Cells *In Vitro*

We have developed and extensively described a serum-free culture system for NHBE cells [2,39–41]. The LHC culture media are based on modified MCDB 151 [42] and are typically supplemented with insulin, 5 $\mu$g/ml; epidermal growth factor (EGF), 5 ng/ml; transferrin, 10 $\mu$g/ml; hydrocortisone, 0.2 $\mu$M; gentamicin, 50 $\mu$g/ml; and bovine pituitary extract, 35 $\mu$g protein/ml.

Until recently serum has been conventionally considered as an essential component of culture medium [43, 44]. In contrast, we found that serum specifically inhibits growth of NHBE cells by inducing terminal squamous differentiation [2,39,45]; we will discuss in a later section the fact that this serum inhibitory effect is widely observed in culture systems of normal epithelial cells (Table I) [2,39, 46–55].

### Identification of Endogenous and Exogenous Differentiation Inducers for NHBE Cells and Their Differential Effects on Human Lung Carcinoma Cells *In Vitro*

Because serum induces NHBE cells but not their malignant counterparts to differentiate terminally [2], we initially focused our efforts on identifying the differ-

**TABLE I. Examples of Growth Inhibition of Serum on Normal Epithelial Cell Culture Systems**

| Species | Tissue | References |
|---|---|---|
| Human | Bronchus | [2,39] |
| Human | Epidermis | [46] |
| Human | Mammary gland | [47] |
| Human | Ureter, bladder | [48] |
| Bovine | Parathyroid | [49] |
| Rabbit | Trachea | [50] |
| Rat | Liver | [51] |
| Rat | Thyroid | [52] |
| Mouse | Vagina (collagen gel) | [53] |
| Mouse | Epidermis | [54,55] |

entiation-inducing serum factor(s). Our initial efforts culminated with the identification of type-$\beta$ transforming growth factor (TGF-$\beta$) as the primary factor [27]. TGF-$\beta$ induces the following markers of terminal squamous differentiation in NHBE cells: 1) irreversible inhibition of DNA synthesis (Fig. 1); 2) cessation of clonal growth; 3) increase in cell surface area; and 4) increase in Ca-ionophore-induced formation of cross-linked envelopes. These markers are also induced by serum [2,39]. In addition, the IgG fraction of rabbit anti-TGF-$\beta$ antiserum neutralizes inhibition of DNA synthesis caused by either TGF-$\beta$ or serum in a dose-dependent manner (Fig. 2) and also prevents squamous differentiation of NHBE cells in the presence of TGF-$\beta$ or serum. These data indicate that TGF-$\beta$ is playing a primary role in the differentiation-induction process.

In the mouse skin carcinogenesis model, two-stage carcinogenesis (initiation and promotion) has been intensively studied, and 12-$O$-tetradecanoylphorbol-13-acetate (TPA) is a potent tumor promoter [56–58]. Although proliferation of initiated cells

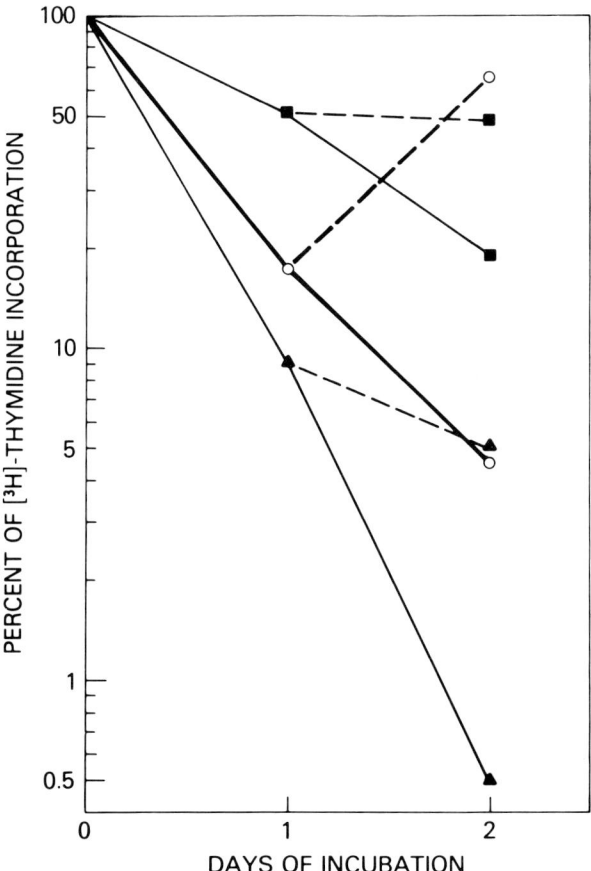

Fig. 1. Irreversible nature of DNA synthesis inhibition caused by TGF-$\beta$. NHBE cells were incubated in the presence of TGF-$\beta$ or serum for 24 hr. Then test media were changed to fresh TGF-$\beta$- or serum-free medium (broken line) or the same fresh test media (solid line). With serum, DNA synthesis was inhibited up to 20%, but inhibitory effect was reversible. In contrast, TGF-$\beta$ inhibited DNA synthesis irreversibly. DNA synthesis was measured by [$^3$H]-thymidine incorporation. TGF-$\beta$: 4 pM, ▲; 0.4 pM, ■. Serum: 8%, ○.

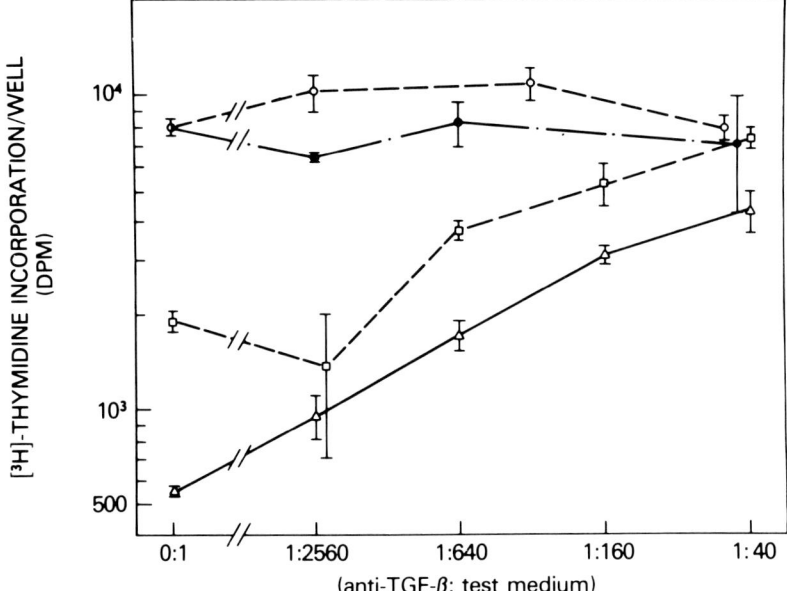

Fig. 2. Effect of IgG fraction of anti-TGF-$\beta$ antiserum on inhibition of DNA synthesis by TGF-$\beta$ and serum. DNA synthesis was measured by [$^3$H]-thymidine incorporation assay. The effects of the IgG fraction of anti-TGF-$\beta$ antiserum on TGF-$\beta$ 1.2 pM ($\square$) and serum 8% ($\triangle$). The inhibitory effects of TGF-$\beta$ or serum were neutralized by the IgG fraction of anti-TGF-$\beta$ antiserum. The IgG fractions of anti-TGF-$\beta$ antiserum ($\bullet$) or normal preimmune rabbit serum ($\bigcirc$) alone did not affect the DNA synthesis of NHBE cells. Abscissa reflects the dilution rate of IgG fraction with test media.

is essential for promotion, TPA induces terminal differentiation of normal mouse and human keratinocytes in vitro [59–61]. We have shown that TPA also induces terminal squamous differentiation of NHBE cells [62]. Thus, endogenous and exogenous inducers of differentiation of NHBE cells, ie, TGF-$\beta$ and TPA, respectively, have been identified.

Interestingly human lung carcinoma cell lines are relatively insensitive to these differentiation inducers [24,27]. We have determined that the loss of sensitivity was not due to any observable differences in the receptors for either TGF-$\beta$ or TPA on the surface of the lung carcinoma cells [27,63,64]. Therefore, lung carcinoma cells acquire differentiation-resistant phenotypes at a step of the differentiation-induction pathway other than at ligand binding of the surface receptors for the differentiation inducers.

Serum is growth inhibitory in primary cell cultures of various epithelia (Table I). In some of these systems TGF-$\beta$ is also growth inhibitory, ie, NHBE cells [27], human epidermal keratinocytes [65, 66], rat hepatocytes [51,67,68], and mouse epidermal keratinocytes [55]. Further, we have shown that anti-TGF-$\beta$ antibody neutralizes the differentiation-inducing effect of serum [27]. Therefore, it is plausible that TGF-$\beta$ is the serum factor inhibiting growth of these epithelial cells.

## Antagonists to Induction of Differentiation

In the presence of both EGF and bovine pituitary extract the growth rate of NHBE cells is enhanced by epinephrine and other cAMP enhancers (3-isobutyl-1-

methyl-xanthine, dibutyryl cyclic AMP, and cholera toxin) [69]. In some epithelial cell types increased levels of cAMP elevate growth of cells cultured in the presence of serum [70,71]. We have also found that cholera toxin antagonizes the growth-inhibitory effect of serum on NHBE cells [45]. Therefore, the effect of epinephrine on differentiation induction caused by TGF-$\beta$ was examined (see Fig. 3) [27]. Epinephrine and TGF-$\beta$ are mutually antagonistic; epinephrine neutralizes the inhibition on DNA synthesis and inhibits the induction of squamous differentiation caused by TGF-$\beta$, whereas TGF-$\beta$ inhibits the growth enhancement caused by epinephrine. Further, we have measured the cAMP levels in NHBE cells exposed to TGF-$\beta$ and/or epinephrine. As expected, epinephrine causes increased levels of cAMP. On the other hand, without altering the cAMP levels in the cells, TGF-$\beta$ cancels the growth enhancement caused by epinephrine but, by itself, induces differentiation of NHBE cells. Therefore, epinephrine and TGF-$\beta$ appear to affect different intracellular pathways that control growth and differentiation processes of NHBE cells. Further, we found that the differentiation-inducing effect of TPA is partially antagonized by epinephrine [72]. This observation supports the hypothesis that TGF-$\beta$ and TPA may share common steps in the pathways of differentiation induction.

In several cell types interactions of growth factors seem to be mediated by membrane receptors [73,74]. However, epinephrine does not alter the observable characteristics of TGF-$\beta$ receptors on the NHBE cells [27]. Interestingly, in addition to epinephrine, cholera toxin, and pertussis toxin show the similar pattern of antagonism against TGF-$\beta$ (Table II) [72]. Various types of biological effects are caused by these two toxins on GTP binding protein systems that control adenylate cyclase [75,76]. The commonality of these three reagents suggested that high cAMP levels in the cells are antagonistic to the differentiation-inducing effect of TGF-$\beta$.

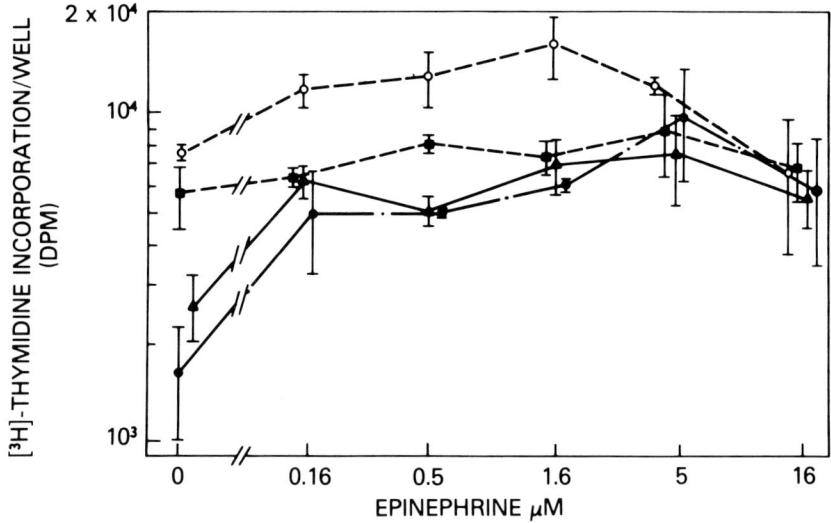

Fig. 3. Effect of epinephrine on DNA synthesis of NHBE cells in the presence or absence of TGF-$\beta$. DNA synthesis was examined by [$^3$H]-thymidine incorporation assay. Epinephrine completely neutralized the inhibitory effect of TGF-$\beta$ on DNA synthesis at high concentrations up to the control level. TGF-$\beta$ was a very potent inhibitor of DNA synthesis enhancement caused by epinephrine in NHBE cells. TGF-$\beta$: 0.12 pM, ■; 0.4 pM, ▲; 1.2 pM, ●. Control: ○.

**TABLE II. CAMP Enhancers Antagonize Differentiation-Inducing Effect of TGF-$\beta$ on NHBE Cells†**

|  | Control | + TGF-$\beta$ (1.2 pM) |
|---|---|---|
| Control | 100 | 13*** |
| + epinephrine (1.6 $\mu$M) | 250*** | 120 |
| + cholera toxin (10 ng/ml) | 140* | 80 |
| + pertussis toxin (5 ng/ml) | 190* | 140*** |

†Growth of NHBE cells was examined in triplicate with a single [$^3$H]-thymidine incorporation assay. Significant P-values indicate significant difference from control (Student's t-test).
*$P < 0.05$.
***$P < 0.005$.

Initially the *ras* oncogene product p21 has been suggested as a member of the GTP binding protein systems that may regulate adenylate cyclase [77–80]. Further, it has been shown that *ras* p21 does not regulate adenylate cyclase in a reconstituted system [81] and has other biological effects not directly relating to adenylate cyclase [82–84]. Our data show that the factors that control GTP binding protein systems at different levels may activate adenylate cyclase [75,76] and can introduce transient malignant phenotypic traits upon NHBE cells, ie, resistance to differentiation-inducing factors. Therefore, it is conceivable that growth and differentiation programs in NHBE cells may be indirectly mediated by GTP binding protein systems and/or cAMP systems, and dysregulation of the pathways that normally control growth and terminal differentiation may be involved in the transformation.

## Effects of Carcinogenic Chemical and Physical Agents on NHBE Cells In Vitro

Epidemiological studies have shown that cigarette smoke and to a lesser extent the inhalation of nickel compounds enhances the risk of respiratory tract cancers [85–87]. Therefore, the effects of nickel ions [88] and cigarette smoke condensate (CSC) [64,89] on NHBE cells were examined.

NHBE cells were continuously exposed to 10–20 $\mu$g/ml of NiSO$_4$, concentrations that reduced their colony-forming efficiency by 30–80%. After 40–75 days of exposure, mitotic cells were observed at a rate of one colony per 100,000 cells originally at risk. The cells were then maintained in medium without NiSO$_4$. No colonies appeared in either control cultures or in cultures exposed to <5 $\mu$g/ml of NiSO$_4$. Most of the NiSO$_4$-altered cell lines have an increased population-doubling potential and some showed preneoplastic phenotypes, ie, loss of sensitivity to differentiation-inducing effect of serum, loss of the requirement for EGF for clonal growth, aneuploidy, and marker chromosomes. However, none of them had completely malignant phenotypes such as tumorigenicity in athymic nude mice.

CSC and its fractions were tested for differential effects using NHBE cells and human lung carcinoma cell lines [64]. Since CSC and some of its fractions have been shown to have promoter effects in the mouse skin carcinogenesis system [90], the differentiation-inducing effect of CSC and its fractions were tested [64,89]. Further, inhibitory effects of CSC and its fractions on phorbol 12,13-dibutyrate (PDBu) binding [91] to NHBE cells has been examined to detect TPA-like activity in them [64,89]. Cytotoxicity tests revealed that NHBE cells are 2–3 times more sensitive to CSC and its fractions when compared to lung carcinoma cells. CSC and the neutral fraction strongly induce squamous morphology and inhibit PDBu binding to NHBE cells. TPA also induces squamous differentiation and inhibits PDBu binding to NHBE cells [62,64]. Therefore, CSC has TPA-like activity that mainly resides in the neutral fraction of CSC.

## Transformation of NHBE Cells by Oncogene Transfection

Certain oncogenes have been shown readily to transform fibroblastic cells, eg, NIH 3T3 cells [8], and oncogene expression at least partially plays a role in carcinogenesis [92,93]. To date three families of oncogenes—*ras, myc,* and *raf*— have been associated with human lung carcinomas [94–101].

Initially NHBE cells were transfected by v-Ha-*ras* oncogene by protoplast fusion [102] and selected for the phenotype that is resistant to serum-induced differentiation, since previous studies had shown that lung carcinoma cells are relatively resistant to differentiation induction and grow better in the presence of serum. Transformed foci resulted, and one cell line (TBE-1) has been studied intensively [103]. The transfected v-Ha-*ras* oncogene caused NHBE cells to be immortal and after a long latency period, they became malignant as judged by their continued growth, aneuploidy, and tumorigenicity in athymic nude mice. Though we do not know the mechanism of *ras* transformation of NHBE cells, one proposed hypothesis is that the v-Ha-*ras* oncogene initiated this multistage process by causing inherited genetic instability of the transfected NHBE cells [104].

In contrast to the above results following the introduction of the v-Ha-*ras* gene into NHBE cells, the transfection of the v-*raf* oncogene into the cells is without observable consequence [100]. Expression of the c-*raf*-1 proto-oncogene may, however, be related to the growth characteristics of small cell lung carcinoma (SCLC) cells. All SCLC cells examined to date, both cell lines (>12, including classic and variant phenotypes) and fresh metastatic tissues, express elevated levels of c-*raf*-1 RNA compared with normal bronchial epithelia and fibroblasts [100]. Analysis of the c-*raf*-1 locus in these cell lines revealed no evidence of DNA amplification or genomic rearrangements. We conclude, therefore, that activation is the result of the expression of a trans-active regulator.

C-*raf*-1 is expressed routinely in neural crest-derived cells and cells of hematopoietic origin [105–107]. As these cells differentiate, the level of *raf*-1 mRNA decreases. Recently, reports have described the expression of hematopoietic cell surface antigens on SCLC cells [108–111]. Therefore, we have compared surface antigens on NHBE and SCLC cells. Several B cell surface antigens and the macrophage monocyte antigens were detected. In addition, the SCLC cells possess markers found only on undifferentiated monomyelocytes [112]. We, therefore, concluded that SCLC cells have a pleiotropic phenotype. The possible role of these aberrant cell surface antigens in altering the malignant behavior of the highly metastatic SCLC cells is being investigated.

## The Possible Autocrine Mechanisms of Transformation

Autocrine growth regulation of transformed cells was first proposed in the positive sense, and after the identification of TGF-$\beta$ as a negative regulator, escape from negative signals has also become an important concept in carcinogenesis [3,4,113,114]. In support of the positive regulation concept, Minna and his colleagues have shown that gastrin-releasing peptide is an autocrine growth factor for SCLC [13]. In the negative regulation mechanism, the effect of epithelial inhibitor, ie, TGF-$\beta$, on growth of BSC-1 cells is an example [28,114,115].

In a search for factors that specifically regulate growth of NHBE cells, we tested whether or not ectopic hormones produced by lung carcinomas might be growth factors. For example, gastrin-releasing peptide (bombesin) is frequently found in SCLC [116,117] and human chorionic gonadotropin is detected in many non-SCLCs [118]. These hormones weakly enhance the growth of NHBE cells in vitro [119,120]. We also found that v-Ha-*ras*-transformed NHBE cells produce a growth factor(s) that stimulates their own growth [103]. Identifying and characterizing the growth factor activity found in the conditioned medium from these cells will be of interest.

Because TGF-$\beta$ has been identified as an endogenous differentiation inducer [27], NHBE cell-conditioned medium was tested for the amount and form of TGF-$\beta$. NHBE cells secrete considerable amounts of TGF-$\beta$ in a latent form [121]. The latent form of TGF-$\beta$ has been found in conditioned media of various cell types and also in platelets [122,123]. The latent form can be activated to the biologically active form by various conditions—eg, acidification, alkalinisation, exposure to urea, and heating to 100°C for 3 min [122]. Although the in vivo activation process of latent TGF-$\beta$ is not known and is considered an important controlling point, it is feasible that TGF-$\beta$ is an autocrine differentiation factor in NHBE cells.

## CONCLUSIONS

Though growth factors and positive regulations are suggested to be important in carcinogenesis, insensitivity to differentiation-inducing stimuli, and/or resistance to cytotoxic substances are common phenotypic traits of malignant cells. Using a serum-free culture system for NHBE cells, we have identified differentiation inducers—ie, TPA and TGF-$\beta$—and their extrinsic antagonists—ie, epinephrine, cholera toxin, and pertussis toxin. Because human carcinoma cells acquire resistance to differentiation inducers without detectable changes in the ligand binding characteristics of the receptors for these agents, we have been interested in mechanisms that dysregulate growth and differentiation pathways. Since the antagonists of these agents may influence GTP binding protein systems and cAMP systems, the importance of these signal transduction systems in the control of growth and differentiation pathways is suggested. The role of *ras* oncogene was also discussed in this context.

Exogenous carcinogenic chemicals cause both abnormal phenotypes in NHBE cells and may also allow selective growth of malignant cells by their differential cytotoxicity. Finally, oncogene transfection experiments have been done to assess the direct contribution of oncogenes in human lung carcinogenesis. The v-Ha-*ras* oncogene has been shown to transform NHBE cells to carcinoma cells in a multistep process. The possible roles and candidates of positive and negative autocrine regulation have been also discussed. Throughout this paper, escape from negative regulatory mechanisms has been considered as a primary process in carcinogenesis. In this sense, it is interesting that TGF-$\beta$ was found in the conditioned medium of NHBE cells in the latent form. However, since the biological activation step of the latent form is not known, the biological significance of latent TGF-$\beta$ in NHBE cell system remains to be established.

## ACKNOWLEDGMENTS

The authors wish to thank Dr. R.D. Sekura for pertussis toxin and Drs. L.M. Wakefield, A.B. Roberts, and M.B. Sporn for their helpful discussion.

## REFERENCES

1. Pierce GB: Fred Proc 29:1248, 1970.
2. Lechner JF, McClendon IA, LaVeck MA, Shamsuddin AM, Harris CC: Cancer Res 43:5915, 1983.
3. Sporn MB, Roberts AB: Nature 313:745, 1985.
4. Goustin AS, Leof EB, Shipley GD, Moses HL: Cancer Res 46:1015, 1986.
5. Harris CC: Cancer Res 47:1, 1987.
6. Roberts AB, Sporn MB: Cancer Surveys 4:683, 1985.
7. Sporn MB, Todaro GJ: N Engl J Med 303:878, 1980.
8. Cooper GM: Science 218:801, 1982.
9. Land H, Parada LF, Weinberg RA: Science 222:771, 1983.
10. Heldin C-H, Westermark B: Cell 37:9, 1984.
11. Bishop JM: Cell: 42:23, 1985.
12. Salomon DS, Perroteau I: Cancer Invest 4:43, 1986.
13. Cuttitta F, Carney DN, Mulshine J, Moody TW, Fedorko J, Fischler A, Minna JD: Nature 316:823, 1985.
14. Dulbecco R: Nature 227:802, 1970.
15. Holley RW: Nature 258:487, 1975.
16. Scher CD, Pledger WJ, Martin P, Antoniades H, Stiles CD: J Cell Physiol 97:371, 1978.
17. Green H: Cell 11:405, 1977.
18. Rheinwald JG, Beckett MA: Cell 22:629, 1980.
19. Kaighn ME, Narayan KS, Ohnuki Y, Jones LW, Lechner JF: Carcinogenesis 1:635, 1980.
20. Yuspa SH, Morgan DL: Nature 293:72, 1981.
21. Wille JJ, Jr, Maercklein PB, Scott RE: Cancer Res 42:5139, 1982.
22. Weissman BE, Aaronson SA: Cell 32:599, 1983.
23. Parkinson EK, Grabham P, Emmerson A: Carcinogenesis 4:857, 1983.
24. Willey JC, Moser CE, Jr, Lechner JF, Harris CC: Cancer Res 44:5124, 1984.
25. Kawamura H, Strickland JE, Yuspa SH: Cancer Res 45:2748, 1985.
26. Yuspa SH, Kilkenny AE, Stanley J, Lichti U: Nature 314:459, 1985.
27. Masui T, Wakefield LM, Lechner JF, LaVeck MA, Sporn MB, Harris CC: Proc Natl Acad Sci USA 83:2438, 1986.
28. Holley RW, Böhlen P, Fava R, Baldwin HJ, Kleeman G, Armour R: Proc Natl Acad Sci USA 77:5989, 1980.
29. McMahon JB, Farrelly JG, Iype PT: Proc Natl Acad Sci USA 79:456, 1982.
30. Assoian RK, Komoriya A, Meyers CA, Miller DM, Sporn MB: J Biol Chem 258:7155, 1983.
31. Iype PT, McMahon JB: Mol Cell Biochem 59:57, 1984.
32. Böhmer FD, Lehmann W, Schmidt HE, Langen P, Grosse R: Exp Cell Res 150:466, 1984.
33. Nomura S, Yamagoe S, Kamiya T, Oishi M: Cell 44:663, 1986.
34. Bell GI, Science 192:569, 1976.
35. Potter VR: Prog Nucleic Acid Res Mol Biol 29:161, 1983.
36. Vandenbark GR, Niedel JE: J Natl Cancer Inst 73:1013, 1984.
37. Gescher A: Biochem Pharmacol 34:2587, 1985.
38. Wang JL, Hsu Y-M: TIBS 11:24, 1986.
39. Lechner JF, Haugen A, McClendon IA, Pettis EW: In Vitro 18:633, 1982.

40. Lechner JF, LaVeck MA: J Tissue Culture Methods 9:43, 1985.
41. Lechner JF, Stoner GD, Yoakum GH, Willey JC, Grafstrom RC, Masui T, LaVeck MA, Harris CC: In Schiff LJ (ed): "In Vitro Models of Respiratory Epithelium." Boca Raton: CRC Press, 1986, pp 143–159.
42. Peehl DM, Ham RG: In Vitro 16:526, 1980.
43. Barnes D, Sato G: Anal Biochem 102:255, 1980.
44. Weinstein R: BioTechniques 1:61, 1983.
45. Lechner JF, Haugen A, McClendon IA, Shamsuddin AM: Differentiation 25:229, 1984.
46. Wille JJ, Jr, Pittelkow MR, Shipley GD, Scott RE: J Cell Physiol 121:31, 1984.
47. Hammond SL, Ham RG, Stampfer MR: Proc Natl Acad Sci USA 81:5435, 1984.
48. Kirk D, Kagawa S, Vener G, Narayan KS, Ohnuki Y, Jones LW: In Vitro 21:165, 1985.
49. Brandi ML, Fitzpatrick LA, Coon HG, Aurbach GD: Proc Natl Acad Sci USA 83:1709, 1986.
50. Wu B, Smith D: In Vitro 18:800, 1982.
51. Hayashi I, Carr BI: J Cell Physiol 125:82, 1985.
52. Ambesi-Impiombato FS, Parks LAM, Coon HG: Proc Natl Acad Sci USA 77:3455, 1980.
53. Iguchi T, Uchima F-DA, Ostrander PL, Bern HA: Proc Natl Acad Sci USA 80:3743, 1983.
54. Bertolero F, Kaighn ME, Gonda MA, Saffiotti U: Exp Cell Res 155:64, 1984.
55. Bertolero F, Kaighn ME, Camalier RF, Saffiotti U: In Vitro 22:423. 1987.
56. Hecker E: In Slaga TJ, Sivak A, Boutwell RK (eds): "Carcinogenesis—A Comprehensive Survey." New York: Raven Press, 1978, Vol 2, pp 11–48.
57. Weinstein IB, Lee LS, Mufson A, Yamasaki H: J Supramol Struct 12:195, 1979.
58. Blumberg PM: CRC Crit Rev Toxicol 8:153, 1980.
59. Yuspa SH, Ben T, Hennings H, Lichti U: Biochem Biophys Res Commun 97:700, 1980.
60. Yuspa SH, Ben T, Hennings H, Lichti U: Cancer Res 42:2344, 1982.
61. Parkinson EK, Emmerson A: Carcinogenesis 3:525, 1982.
62. Willey JC, Saladino AJ, Ozanne C, Lechner JF, Harris CC: Carcinogenesis 5:209, 1984.
63. Gescher A, Reed DJ: Cancer Res 45:4315, 1985.
64. Miyashita M, Willey JC, Sasajima K, Lechner JF, LaVoie EJ, Hoffmann D, Harris CC: (submitted).
65. Moses HL, Tucker RF, Leof EB, Coffey RJ, Jr, Halper J, Shipley GD: Cancer Cells 3:65, 1985.
66. Shipley GD, Pittelkow MR, Wille JJ, Jr, Scott RE, Moses HL: Cancer Res 46:2068, 1986.
67. Carr BI, Hayashi I, Branum EI, Moses HL: Cancer Res 46:2330, 1986.
68. McMahon JB, Richards WL, Del Campo AA, Song M-KH, Thorgeirsson SS: Cancer Res 46:4665, 1986.
69. Willey JC, LaVeck MA, McClendon IA, Lechner JF: J Cell Physiol 124:207, 1985.
70. Green H: Cell 15:801, 1978.
71. Marcelo CL: Exp Cell Res 120:201, 1979.
72. Masui T, Lechner JF, Harris CC: (unpublished observations).
73. Assoian RK, Frolik CA, Roberts AB, Miller DM, Sporn MB: Cell 36:35, 1984.
74. Bowen-Pope DF, Dicorleto PE, Ross R: J Cell Biol 96:679, 1983.
75. Hewlett EL, Cronin MJ, Moss J, Anderson H, Myers GA, Pearson RD: In Greengard P, Robison GA, Paoletti R, Nicosia S (eds): "Advances in Cyclic Nucleotide and Protein Phosphorylation Research." New York: Raven Press, 1984, Vol 17, pp 173–182.
76. Gilman AG: Cell 36:577, 1984.
77. Hurley JB, Simon MI, Teplow DB, Robishaw JD, Gilman AG: Science 226:860, 1984.
78. McGrath JP, Capon DJ, Goeddel DV, Levinson AD: Nature 310:644, 1984.
79. Toda T, Uno I, Ishikawa T, Powers S, Kataoka T, Broek D, Cameron S, Broach J, Matsumoto K, Wigler M: Cell 40:27, 1985.
80. Kataoka T, Powers S, Cameron S, Fasano O, Goldfarb M, Broach, J, Wigler M: Cell 40:19, 1985.
81. Beckner SK, Hattori S, Shih TY: Nature 317:71, 1985.
82. Birchmeier C, Broek D, Wigler M: Cell 43:615, 1985.
83. Levitzki A, Rudick J, Pastan I, Vass WC, Lowy DR: FEBS Lett 197:134, 1986.
84. Fleischman LF, Chahwala SB, Cantley L: Science 231:407, 1986.
85. Doll R: Environ Health Perspect 22:23, 1978.
86. Doll R: Cancer Res 38:3573, 1978.
87. Sunderman FW, Jr: Environ Health Perspect 40:131, 1981.
88. Lechner JF, Tokiwa T, McClendon IA, Haugen A: Carcinogenesis 5:1697, 1984.
89. Willey JC, Grafstrom RC, Moser CE, Jr, Ozanne C, Sundqvist K, Harris CC: Cancer Res 47:2045, 1987.

90. Hoffmann D, Hecht SS, Wynder EL: Environ Health Perspect 50:247, 1983.
91. Blumberg PM, Dunn JA, Jaken S, Jeng AY, Leach KL, Sharkey NA, Yeh E: In Slaga TJ (ed): "Mechanisms of Tumor Promotion." Boca Raton: CRC Press, 1984, Vol 3, pp 143–184.
92. Schwab M, Alitalo K, Varmus HE, Bishop JM, George D: Nature 303:497, 1983.
93. Sukumar S, Notario V, Martin-Zanca D, Barbacid M: Nature 306:658, 1983.
94. Yuasa Y, Srivastava SK, Dunn CY, Rhim JS, Reddy EP, Aaronson SA: Nature 303:775, 1983.
95. Nakano H, Yamamoto F, Neville C, Evans D, Mizuno T, Perucho M: Proc Natl Acad Sci USA 81:71, 1984.
96. Little CD, Nau MM, Carney DN, Gazdar AF, Minna JD: Nature 306:194, 1983.
97. Gazdar AF, Carney DN, Nau MM, Minna JD: Cancer Res 45:2924, 1985.
98. Nau MM, Brooks BJ, Battey J, Sausville E, Gazdar AF, Kirsch IR, McBride OW, Bertness V, Hollis GF, Minna JD: Nature 318:69, 1985.
99. Nau MM, Brooks BJ, Jr, Carney DN, Gazdar AF, Battey JF, Sausville EA, Minna JD: Proc Natl Acad Sci USA 83:1092, 1986.
100. Mark GE III, Pfeifer AMA, Harris CC: (unpublished observations).
101. Kurzrock R, Gallick GE, Gutterman JU: Cancer Res 46:1530, 1986.
102. Yoakum GH, Korba BE, Lechner JF, Tokiwa T, Gazdar AF, Seeley T, Siegel M, Leeman L, Autrup H, Harris CC: Science 222:385, 1983.
103. Yoakum GH, Lechner JF, Gabrielson EW, Korba BE, Malan-Shibley L, Willey JC, Valerio MG, Shamsuddin AM, Trump BF, Harris CC: Science 227:1174, 1985.
104. Harris CC, Lechner JF, Yoakum GH, Amstad P, Korba BE, Gablielson E, Grafstrom R, Shamsuddin A, Trump BF: In Barrett JC, Tennant RW (eds): "Carcinogenesis." New York: Raven Press, 1985, Vol 9, pp 257–269.
105. Rosenberg YJ, Malek TR, Schaeffer DE, Santoro TJ, Mark GE, Steinberg AD, Mountz JD: J Immunol 134:3120, 1985.
106. Mountz JD, Mushinshi JF, Mark GE, Steinberg AD: J Mol Cell Immunol 2:121, 1985.
107. Toskos M: (personal communication).
108. Ruff MR, Pert CB: Science 225:1034, 1984.
109. Cole SPC, Mirski S, McGarry RC, Cheng R, Campling BG, Roder JC: Cancer Res 45:4285, 1985.
110. Ball ED, Sorenson GD, Pettengill OS: Cancer Res 46:2335, 1986.
111. Bunn PA, Linnoila I, Minna JD, Carney D, Gazdar AF: Blood 65:764, 1985.
112. Mark GE III, Mann D: (unpublished observations).
113. Roberts AB, Anzano MA, Wakefield LM, Roche NS, Stern DF, Sporn MB: Proc Natl Acad Sci USA 82:119, 1985.
114. Tucker RF, Shipley GD, Moses HL, Holley RW: Science 226:705, 1984.
115. Holley RW, Armour R, Baldwin JH: Proc Natl Acad Sci USA 75:1864, 1978.
116. Erisman MD, Linnoila RI, Hernandez O, DiAugustine RP, Lazarus LH: Proc Natl Acad Sci USA 79:2379, 1982.
117. Moody TW, Pert CB, Gazdar AF, Carney DN, Minna JD: Science 214:1246, 1981.
118. Trump BF, Wilson T, Harris CC: In Ishikawa S, Hayata Y, Suematsu K (eds): "Lung Cancer 1982." Amsterdam: Excerpta Medica, 1982, pp 101–124.
119. Willey JC, Lechner JF, Harris CC: Exp Cell Res 153:245, 1984.
120. Harris CC, Yoakum GH, Lechner JF, Willey JC, Gerwin B, Banks-Schlegel S, Masui T, Mark G: In Harris CC (ed): "Biochemical and Molecular Epidemiology of Human Cancer." New York: Alan R. Liss, Inc., 1986, pp 213–226.
121. Wakefield LM, Smith DM, Masui T, Harris CC, Sporn MB: J Cell Biol (in press).
122. Lawrence DA, Pircher R, Jullien P: Biochem Biophys Res Commun 133:1026, 1985.
123. Pircher R, Jullien P, Lawrence DA: Biochem Biophys Res Commun 136:30, 1986.

# Analysis of Murine Homeo Box Genes and Their Expression During Development

**Stephan D. Voss, Anamaris M. Colberg-Poley, and Peter Gruss**

Zentrum für Molekulare Biologie der Universität Heidelberg, D-6900 Heidelberg, Federal Republic of Germany

> Highly conserved copies of homeo box genes, responsible for controlling a number of developmental processes in *Drosophila*, are found in a wide variety of animals, including mouse and man. A great deal of research in mammalian systems has recently been directed toward investigating the possibility that mammalian homeo-box-containing genes perform morphogenetic functions similar to those observed in the fruit fly. The focus of this attention has mainly been on homeo boxes isolated from the mouse genome. We present here an overview of our analysis of a cluster of three homeo boxes located on mouse chromosome 6, as well as a summary of the analyses of other laboratories. Consistent with the notion that mammalian homeo boxes may play an analogous role during development, evidence is presented showing temporal expression of a number of murine homeo boxes during embryogenesis. Moreover, the data also indicate a tissue-specific restriction in the expression of some of these genes during postnatal development.

**Key words:** cellular differentiation, chromosomal locations, expression during embryogenesis, interaction of gene products, tissue specific expression, transcription

Embryonic development is a reproducible process which appears to take place according to a very specifically programmed pattern of events. The fact that pluripotent embryonic cells undergo specific changes eventually leading to their determination and differentiation into tissue-specific cells is neither coincidental nor unpredictable. Perhaps the best evidence of this predictability and specificity during cellular developmental processes is found in (i) E.B. Lewis's "regulatory rules," which correlate genetic data compiled from a mutational analysis of the bithorax complex (BX-C) with the effects of these mutations on the developmental fate of

---

Received April 15, 1986.

Stephan D. Voss's present address is Department of Human Oncology, 600 Highland Ave., University of Wisconsin, Madison, WI 53792.

Anamaris M. Colberg-Poley's present address is Central Research and Development, Experimental Station Bldg. 328, Rm. B31c, E.I. DuPont de Nemours and Co., Inc., Wilmington, DE 19898.

© 1988 Alan R. Liss, Inc.

mutated fly embryos [1], and (ii) the work of Horvitz and colleagues with *Caenorhabditis elegans* (the soil nematode), in which they have showed by mutant analyses that disruption of the timing of developmental processes resulted in specific structural changes in this organism.

Despite overt physiological differences, it has been argued that *all* cells, regardless of their origin, undergo a similar series of developmental decisions [3] and that individual cells, upon reaching a certain irreversible point, follow an apparently preprogrammed set of decisions through determination and ultimately differentiation. This developmental program, according to Dienstman and Holtzer [4], proceeds in a linear fashion such that any one cell must choose either of two fates. Thus, an understanding of even a single event in the complex developmental program of well-characterized organisms (eg, *Drosophila* and *C. elegans*) may allow us access to similar events in higher mammals. Through this understanding of such single events, we might hope to gain insight into the overall process(es) of development.

In order to gain an understanding of the regulation of the events leading to the differentiation of precursor stem cells along specific pathways into particular cell types, one requires the identification of the genes involved in these processes. Owing to genome complexity and life-cycle length, however, thorough genetic analyses of mouse developmental genes have not been as feasible as those performed in *Drosophila*. It has been possible through genetic studies of *Drosophila* to identify developmental genes, and within these, a conserved DNA element—the homeo box [5,6]. The developmental genes containing this element are involved in the determination of segment number or segment identity [7]. The discovery that this small (180 nucleotides) genetic element was present in higher organisms, including mouse and man, led to its use as a "probe" to identify mammalian genes involved in potentially analogous developmental programs [7]. Such a probe perhaps provides one with a unique opportunity to isolate mammalian developmental genes and to begin to dissect the likely myriad of molecular interactions resulting in a particular developmental decision.

**Why Homeo Boxes?**

In *Drosophila*, homeo box sequences were initially found to be contained within the coding sequences of genes specifying segment number (segmentation genes) and segment identity (homeotic genes) in the developing fly embryo. These genes were shown to be present in two main complexes, the bithorax complex (BX-C)—specifying identity of thoracic and abdominal segments [8,9]—and the antennapedia complex (ANT-C), which determines identity of the head and anterior thoracic segments [10–15]. Since that time this conserved sequence has been detected in at least seven additional loci in the *Drosophila* genome [7, 16], while homeo box sequences have been located on at least four mouse chromosomes (see Table I), with the number of distinct and characterized boxes presently being 10. A similar analysis of the human genome has revealed homeo box sequences contained in genomic contexts structurally similar to corresponding mouse homeo boxes [17–20].

What physiological role these conserved domains play in mammalian development remains for the moment speculative and limited to descriptive rather than functional comparisons. Determining the role these genes play in development remains difficult owing to the relatively few defined loci known to affect morphogenesis [21]; and, for the present, genetic analyses will likely be restricted to demonstrating

**TABLE I. Chromosomal Locations of Murine Homeo Boxes**

| Chromosome | Homeo box | References |
|---|---|---|
| 1 | Mo-en.1 | Joyner et al [45] |
| 6 | Mo-10 (Hox-1)<br>m6, m5, m2 | McGinnis et al [46];<br>Colberg-Poley et al [43];<br>Bućan et al |
| 11 | Hox-2.1 (Mu1, H24.1)<br>Hox-2.2-2.4 | Hart et al [20]; Hauser et<br>al [18]; Jackson et al [48];<br>Hart et al [20] |
| 15 | Hox-3 (m31) | Awgulewitsch et al [50];<br>Breier et al [49] |

linkage of homeo box genes to known developmental mutations. Thus, we wish to present data from a number of studies on murine homeo boxes in the context of the considerable knowledge already compiled, and still being compiled, about the role genes containing these domains might play in Drosophila development.

In Drosophila, homeo boxes have continued to be found in genes which had previously been genetically described as affecting segment identity (antennapedia, Antp; ultrabithorax, Ubx; sex combs reduced, Scr; deformed, Dfd; infra-abdominal-2, iab-2; infra-abdominal-7, iab-7), segment number (fushi tarazu, ftz), or compartmentalization (engrailed, en; invected, inv) [7,22,23]. Indeed, the presence of the majority of these genes in large chromosomal clusters facilitated their identification, and in each case in situ analyses showed gene transcripts localized to the region(s) of the embryo affected in the respective mutants [24–30]. Furthermore, homeo box sequences were recently used to isolate caudal, a previously uncharacterized gene, which displays maternal as well as zygotic expression during Drosophila development [16]. Thus the plausibility of the use of homeo boxes to isolate new developmentally significant genes, even in Drosophila, and in the absence of confirmatory genetic data, is clear.

### The Possible Regulatory Role of Homeo Boxes

A conceptual translation of the conserved homeo box sequence reveals a characteristic array of highly basic amino acids. Significant similarities between this consensus homeo domain and the yeast mating-type proteins a-1 and $\alpha$-2 [31,32], proteins structurally similar to prokaryotic DNA binding proteins [33] and thought to play a regulatory role in yeast mating-type determination [34], have led to speculation that homeo box domains may bind DNA. Such properties, consistent with proteins thought to regulate groups of other genes, are supported by evidence of localization of Ubx, ftz, and en proteins in the nuclei of Drosophila embryonic cells [35–38]. In addition, more direct evidence provided recently by Desplan et al [39] showed that fusion peptides containing engrailed homeo box sequences bind to specific regions of DNA 5' to the site of transcriptional initiation of the engrailed gene, while lac Z fusions of sequences 5' of the ftz gene have shown the presence of a number of morphogenetic control elements upstream of ftz coding sequences [40].

## RESULTS AND DISCUSSION
### Isolation of Mouse Homeo Boxes

In an effort to identify genes involved in the control of mammalian development, we and a number of other laboratories sought to isolate and functionally characterize murine homeo box sequences. It is unlikely that mouse homeo-box-containing genes will function in a manner analogous to those of *Drosophila*, owing to fundamental differences between the development of *Drosophila* and mammals [41]. However, one can still expect putative regulatory genes to be spatially and temporally expressed during embryogenesis, as well as during in vitro differentiation of embryonal carcinoma (EC) cells. Moreover, a role in postnatal cellular and tissue development is also possible.

## m6 Cluster

In analogy to the organization of *Drosophila* homeotic genes, we have identified a cluster of three homeo-box-containing genes—m6, m5, and m2—lying within 15 kilobases (kb) of one another (Fig. 1), which are expressed during EC cell differentiation [42–44]. These genes were also shown to be temporally regulated during murine embryogenesis and to exhibit tissue-specific expression in the adult mouse [43].

A comparison of the three homeo boxes present in the m6 cluster is shown in Figure 1. The m5 and m6 homeo boxes show the greatest homology to each other and to the *Antp* sequences used to isolate them, whereas m2 appears more homolgous to Mu1. Noteworthy also is the significant decline in homology between sequences of the m6 cluster and the mouse *engrailed*-like homeo box [45], and see later). A similar drop in homology was also observed in *Drosophila*, between the two homeo boxes of the EN-C and those of the ANT-C and BX-C. As shown in the margin of Figure 1, the m6 cluster has been localized to chromosome 6 and lies between Tcrβ and IgK (Bućan et al, in preparation). Its position is, however, distinguishable from that of the Hox-1 (Mo-10) cluster [46], also located on chromosome 6 [47].

Differential expression of genes containing each of the homeo box sequences in the m6 cluster has been demonstrated, both in EC cells and during mouse embryogenesis (see Fig. 2 for summary of transcriptional pattern within m6 cluster). A 2.4-kb

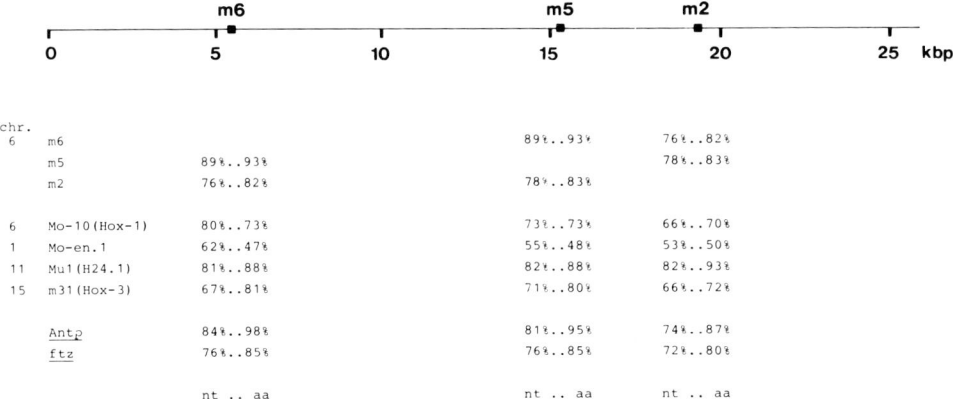

Fig. 1. Sequence homologies of murine homeo boxes. The nucleotide (nt) and amino acid (aa) sequences of the clustered homeo boxes m6, m5, and m2 ([42,43], and unpublished results) were compared with the sequences of other murine homeo box genes. The percentages of homologies between the clustered homeo boxes and Mo-en.1 [49], Mu1l (H24.1 [18,48, respectively]) Mo-10 [46] and m31 (Hox-3 [50,51 respectively]), as well as those from the *Drosophila* probes used to isolated this murine DNA (*Antp*, *ftz* [54]) are shown. The comparisons of m6 are based on corrections made in the original sequence at nt positions 91–94, making the sequence homologous to *Antp* [50].

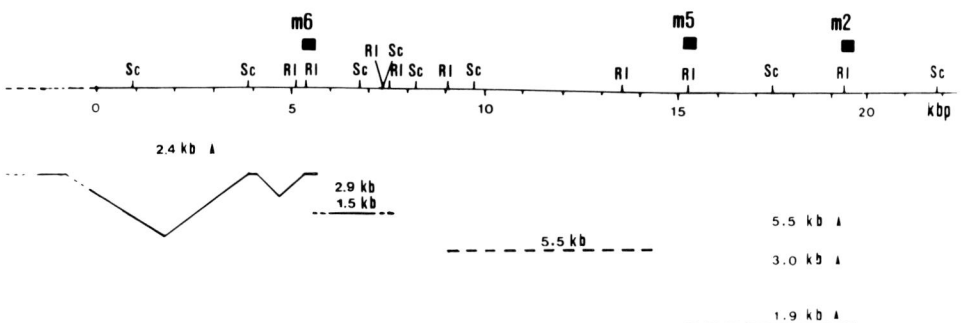

Fig. 2. Transcription of the m6 region in F9 stem and differentiating cells. The drawing is a graphic representation of the stable transcriptions present in F9 stem cells or in F9 cells induced to differentiate by treatment with $5 \times 10^{-7}$ M retinoic acid and $10^{-3}$ M dibutyryl cAMP for 24 h and detected by hybridization with genomic fragments from this region. The solid lines indicate the minimum span and the dashed lines the maximum span of transcription as deduced by hybridization. The sizes of the transcripts are indicated: m6, 2.4 kb; downstream of m6, the 2.9- and 1.5-kb RNAs; upstream of m5, a 5.5-kb and between m5 and m2 5.5-, 3.0-, and 1.9-kb transcripts were detected. The arrows indicate the RNAs induced upon differentiation of F9 cells. The left portion shows the structure of a cDNA which encodes the 2.4- kb m6 RNA. Some of the sequences included in the cDNA are encoded by sequences which lie outside of the region represented here.

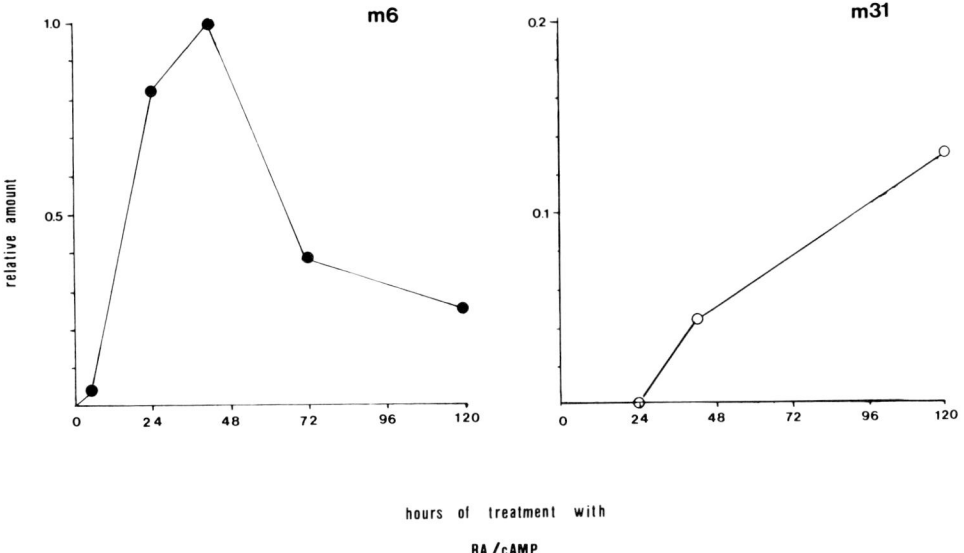

Fig. 3. Temporal expression of two murine homeo box genes, m6 and m31, during F9 cell differentiation. The relative amounts of stable m6 2.4-kb RNA (**left**) and m31 2.7-kb RNA (**right**) detected at various times of differentiation are shown. The maximal expression of m6 RNA at 42 h was designated a unit and the values of each RNA were compared to that value. Note that the scale of the relative amounts of m31 RNA has been amplified to show the details of its induction.

m6 homeo-box-containing transcript has been shown to accumulate as early as 6 h following differentiation of F9 EC cells into parietal endoderm. This transcript is absent in F9 stem cells and reaches a maximal abundance 24–42 h following induction of differentiation, after which its expression drops dramatically [42], and see Fig. 3). This differential expression of m6-containing sequences is accompanied by the constitutive (in stem cells as well as their differentiated derivatives) expression of flanking sequences immediately 3' to m6 as 2.9- and 1.5-kb mRNAs. The expression of m6 sequences does not, however, appear to be sufficient to direct a cell into a tissue-specific differentiation program and has, rather, a more likely general impact, since the 2.4-kb m6 message is also induced during differentiation of F9 and PC13 cells into visceral endoderm [44] and of P19 cells into myogenic tissue (C. Dony and P. Gruss, in preparation).

Consistent with the EC cell data, m6-containing transcripts were found to be temporally regulated during embryogenesis [43]. Examination of embryonic and extra-embryonic tissue at various prenatal time points showed a maximal accumulation of m6- and 3' flanking sequence-containing transcripts in embryonic tissue following 12 days of gestation. Furthermore, m6-containing sequences display tissue-specific patterns of expression in the adult mouse accumulating primarily in the kidney, testis, and ovary, suggesting an involvement of these sequences in postnatal cellular and tissue development, as well as during embryogenesis.

Sequences containing the m5 homeo box, in contrast to m6, do not detect transcripts in F9 stem cell populations (see, however, [43], whereas upon differentiation at least three transcripts were observed: a 5.5-kb doublet and a 1.9-kb transcript [43]. Unique sequences hybridizing to the 5.5-kb transcripts have been mapped to the region between the m5 and m6 homeo boxes (see Fig. 2), although the use of a m5 homeo-box-specific probe has confirmed the presence of conserved sequences in these transcripts (unpublished results). The 1.9-kb transcript, however, is detected by unique sequences residing between m5 and m2 (see Fig. 2). This location and the presence of several transcripts hybridizing to m5 homeo box sequences makes unequivocal identification of the 1.9-kb inducible transcript difficult. This is further complicated by data showing hybridization of a 350-base pair (bp) m2 homeo-box-specific probe to similarly sized 1.9-kb RNA induced in differentiated F9 cells (unpublished results).

The expression of m5 homeo-box-containing sequences during embryogenesis is considerably less complicated than that of m6, with m5-containing transcripts only being detected in 12-day embryonic tissue. Tissue-specific expression in the adult mouse has also been observed for m5 sequences, although limited to novel-sized transcripts in the testis and kidney.

## Hox-2 Cluster

The Hox-2 homeo box cluster on mouse chromosome 11 was originally characterized by virtue of its strong homology to the Hu1 and Hu2 homeo boxes on human chromosome 17 [47]. Further characterization of both clusters has revealed the presence of four murine homeo boxes (designated Hox-2.1, 2.2, 2.3, 2.4) contained within a 20-kb segment of the mouse genome [20], while Hu1 and Hu2 have been joined by Hu5 [18], although it is not obvious from comparisons of restriction data that Hu5 has a corresponding Hox-2 locus homologue. Hauser et al [18] also identified a mouse homologue of Hu1 (Hox-2.1 of [20]), which they named Mu1. Sequence comparisons between Hu1 and Mu1 showed that the conserved sequences between the mouse and human loci extended well beyond the 180-nucleotide boundary of the box. Thirteen differences in nucleotide sequences resulted in no amino acid changes, while sequences 3' of the respective homeo boxes share 15 identical amino acids and a stop codon, confirming the mapping of these loci to the same (respectively) conserved genetic linkage group [17,19]. Hogan and co-workers have also isolated what appears to be the equivalent of Hox-2.1/Mu1, termed H24.1, the sequence of which is identical with Mu1 [49]; see also Fig. 1).

The most complete analyses of the temporal expression of transcripts from the Hox-2.1 (H24.1, Mu1) locus were performed by Hogan and co-workers [49] and by Hauser et al [18], who examined expression of Hu1 in teratocarcinoma cells and Mu1 in EC cells and during mouse embryogenesis. H24.1-containing transcripts were shown to be expressed in HD-14 EC cells only following in vitro differentiation of these cells, with maximal accumulation of transcripts seen after 32 days of differentiation, by which time the cells have differentiated into a variety of different cell types. These data are consistent with our examination of m6 cluster expression during EC cell differentiation, as is the observation that the highest levels of H24.1 transcription in embryos, beginning 7.5 days postcoitum (pc), reaches its highest levels 11.5 to 12.5 days pc. This peak of transcription was shown to be confined primarily to 12-day spinal cord and 12-day brain tissue, while comparable levels of transcription were also seen in the adult kidney.

These data have been supported by Hauser et al [18], who have shown that Hu1-(H24.1/Mu1/Hox-2.1 homologue) containing transcripts are only detectable after the human teratocarcinoma line NT2/D1 is induced to differentiate into a variety of tissue types by addition of retinoic acid. The more abundant of these transcripts begins to accumulate 2 days after induction, approximately the same time morphological changes begin to take place, although interestingly, removal of the retinoic acid inducer results in a rapid decrease in the transcript level, despite the fact that stem-cell-specific cell surface markers are no longer expressed. The Hu1 homologue, Mu1, was also analyzed directly using PSA-1 EC cells and RNA isolated from embryonic tissue at various stages during development. Not entirely consistent with the H24.1 data was the observation that the 2.3-kb Mu1 transcript was expressed in the embryo from 10.5 days through 17.5 days, with no obvious peak at 12 days. Furthermore, Mu1 expression was not detectable in PSA-1 EC cell cultures, even followng differentiation into the stages believed to represent 7.5 days and 9.5–11.5 days of embryogenesis, although this may simply indicate differentiation into embryonic cell types not expressing Mu1 sequences (consistent with the localization of H24.1 transcripts to specific embryonic tissues).

Finally the data summarizing expression of the Hox-2 locus homeo-box-containing transcripts is completed by the study of Hart et al [20], who have shown expression of each of the Hox-2 locus homeo boxes in day 13 embryonic tissue. The analysis of Hox-2.1, however, indicated the presence of two transcripts—1.7 and 1.9 kb—which is inconsistent with the results of Hauser et al, who described a single, prominent 2.3-kb Mu1 transcript present in 13-day embryonic tissue. Despite these inconsistencies, the clustered organization and expression of these homeo boxes during embryogenesis (in analogy to the m6 and *Drosophila* clusters) strengthen the argument that these mammalian homeo box complexes perform important developmentally significant functions.

### Hox-1 "Cluster"

The Hox-1.1 homeo box (originally Mo-10 [46], located on mouse chromosome 6, also apparently lies within a cluster composed of at least three homeo boxes [20,45] and references therein); Hox-1.1 is the least conserved of the ANT-C/BX-C -like murine homeo boxes, and no expression data relating this homeo box domain to a developmental function have been described.

**Mo-en Locus**

In keeping with our model of using *Drosophila* homeo box genes to isolate possible functionally analogous murine genes, Joyner et al [45] have used the *engrailed* gene of *Drosophila* to isolate a related murine *engrailed*-like region. In *Drosophila*, engrailed is the namesake of the EN-C complex, which can be structurally differentiated from ANT-C and BX-C by virtue of chromosomal location (chromosome 2 rather than 3 for *en*) and the structure of the homeo boxes found within the cluster. EN-C homeo boxes share approximately 87% homology with one another, while BX-C/ANT-C homeo boxes are 75% to roughly 90% homologous. However, EN-C homeo box sequences are at best 50% homologous to corresponding sequences from ANT-C/BX-C genes.

Analogous to *Drosophila*, sequence similarity of Mo-en.1 was found to extend 63 nucleotides 3' of the defined 180-bp homeo box. Mo-en.1 sequences, located distal to band C2 on chromosome 1, were most homologous to other engrailed sequences (73%), with homology to ANT-C-like murine homeo box sequences falling considerably lower (see Fig. 1).

Analysis of the expression of Mo-en.1 during murine development was performed using PSA-1 EC cells and RNAs isolated at various stages during embryonic development. In undifferentiated PSA-1 cells, no Mo-en.1 expression was detectable, whereas differentiation of these cells by suspension culturing and embryoid body formation was accompanied by the induction of a 3.1-kb major transcript in all stages examined. This transcript was also observed to be expressed throughout embryogenesis, with a peak of expression between 10.5 and 12.5 days of gestation. A less abundant 2.0-kb transcript was also observed in all embryonic stages examined and did not appear to be temporally regulated. Finally the presence of a second *engrailed*-like murine gene is suggested based on genomic Southern data, although the genomic clones containing the putative Mo-en.2 will be needed to confirm this observation.

**Hox-3/m31**

The murine homeo box most recently described, isolated by virtue of its cross-hybridization to *Drosophila* probes in our laboratory [50] and by Awgulewitsch et al [51], has been mapped to mouse chromosome 15 (see Table 1). Sequence analysis of m31 (Hox-3 in [51] showed the expected high degree of homology to other murine homeo boxes (see Fig. 1, [50,51], while analysis of the transcriptional activity of this novel homeo box indicated temporal expression during differentiation of F9 cells into parietal endoderm [50], Fig. 3) as well as temporal restriction expression during embryogenesis [51]. Furthermore, based on preliminary Southern hybridization data, the m31 homeo domain may lie within a cluster containing at least another homeo box domain (Breier and Gruss, unpublished).

Examination of F9 cells revealed the absence of m31 transcripts in stem cell populations, while the differentiation of these cells resulted in the gradual appearance of a 2.7-kb m31- specific mRNA, with maximal expression observed 5 days following addition of inducer. This pattern of m31 expression was compared to that of m6 (see Fig. 3), in which maximal accumulation of m6 transcripts occurs 24–42 h following induction of differentiation. The level of m31 transcription remained, however, at least a factor of ten below the peak of m6 transcription, even at the latest time points examined. This sequential pattern of expression suggests a number of possibilities for the interaction and "self-regulation" of homeo-box-containing genes, a suggestion also supported by evidence from Hafen et al [52] that deletions in one *Drosophila* homeotic gene cluster (BX-C) affect the spatial distribution of another homeo box gene (*antp*), implying regulatory interactions between different homeo box genes (see also Conclusions and Outlook).

In recently published work, Awgulewitsch et al [51] have presented convincing data showing the spatial restriction of Hox-3 (m31) homeo box expression to the central nervous system of the newborn and adult mouse, while tissue-specific expression was also observed to occur during embryogenesis. Maximal expression of Hox-3 was observed between 11 days and 13 days of gestation, with a noticeable reduction in the abundance of the 2.7-kb Hox-3 transcript in later stages examined, consistent with data from our laboratory showing m31 expression in 12-day embryonic tissue [50].

Hox-3 transcription was also examined using RNA isolated from a number of different tissues dissected from newborn and adult animals. The strongest accumulation of the 2.7-kb message in newborn mice was found to be restricted to the spinal column, while faint, similarly sized transcripts were also observed in RNAs isolated from new born kidney. Other tissues examined, including brain, spleen, liver, and muscle, were negative for Hox-3 expression. A similar pattern of restricted distribution was observed in RNAs prepared from the same panel of tissues isolated from adult mice. In contrast to newborn mice, maximal expression of a 2.1-kb Hox-3 transcript was observed in spinal cord RNA, although the 2.7-kb transcript previously observed was also present in spinal cord RNA, as well as RNA prepared from adult kidney.

Subsequent dissection of the spinal cord of newborn and adult mice into cervical, thoracic, and lumbar/sacral/caudal sections showed strong hybridization in both cases to RNAs isolated from the cervical and thoracic regions, whereas detection of Hox-3 transcripts in RNA samples from the brain and lumbar region were negative and weak, respectively. These results suggested spatial restriction of Hox-3 expression to anterior regions of the spinal column and were further corroborated by in situ analyses of spinal column cross sections, which confined the region of Hox-3 expression to the posterior cervical and anterior thoracic region of the spinal cord. Expression in more anteriorly or posteriorly located sections was either not observed or very weak, respectively, indicating a localized expression of homeo box sequences within the mouse CNS, results analogous to the spatially restricted expression of homeo-box-containing homeotic genes in the *Drosophila* CNS [24,53]. These studies and other studies of the expression of murine homeo box sequences during mouse development are summarized in Table II.

## CONCLUSIONS AND OUTLOOK

Although no *direct* evidence for the regulatory function of murine homeo-box-containing genes is available so far, we have presented here an accumulation of evidence which is consistent with these genes' playing a crucial role in the development of the mouse. These genes share with the homeo-box-containing developmental genes of *Drosophila* a number of properties, including the conserved homeo box

**TABLE II. Expression of Murine Homeo Box Genes During Development**

| Homeo box | Differentiated EC cells | Embryonic tissue | Adult tissue | References |
|---|---|---|---|---|
| Mo-en.1 | PSA1 | 10.5–12.5[a] | | Joyner et al [45] |
| m6 | F9,PC13,P19 | 12[a] | Kidney, testes, ovary | Colberg-Poley et al [42–44] |
| m5 | F9, P19 | 12[a] | Kidney, testes | Colberg-Poley et al [43] |
| Hox-2.1 | | 13 | | Hart et al [20] |
| Mu1 | — | 11.5–13.5[a] | | Hauser et al [18] |
| H24.1 | | 11.5–12.5[a] | Kidney | Jackson et al [48] |
| Hox-2.2 | | 13 | | Hart et al [20] |
| 2.3 | | 13 | | |
| 2.4 | | 13 | | |
| Hox-3 | | 11–13[a] | Spinal cord, kidney | Awgulewitsch et al [50] |

[a]Time of maximal expression (in days).

protein coding domain, and structural similarity, ie, clustering of genes containing the homeo boxes and their presence on several chromosomes. Moreover, these genes are associated with critical events such as differentiation of EC cells and are temporally regulated during embryogenesis. Perhaps the most striking correlation with the properties of the *Drosophila* homeotic genes is the fact that some of these genes have been shown to be spatially restricted in their expression during late prenatal and postnatal development. It is, however, difficult to exclude the possibility that these genes encode housekeeping enzymes which are necessary for the cell to survive during differentiation and development. Therefore research in this field must address the functional role these genes play during the time of their expression, and the possibility that the interaction of several of the homeo box gene products may affect the determination of a cell to proceed along a particular developmental pathway.

## ACKNOWLEDGMENTS

The authors wish to thank Georg Breier, Barbara Zink, Michael Kessel, and Marija Bućan for helpful discussions and for making available results prior to publication. S.D.V. was supported by a Fulbright Fellowship and a grant from Fonds der Chemischen Industrie. The research was supported by the Deutsche Forschungsgemeinschaft (DFG Ba 384/18-4).

## REFERENCES

1. Lewis EB: Nature 276:565, 1978.
2. Greenwald IS, Sternberg PW, Horvitz HR: Cell 34:435, 1983.
3. Slack JMW: "From Egg to Embryo." Cambridge: Cambridge University Press, 1983.
4. Dienstman SR, Holtzer H: In Reinert S, Holtzer H (eds) "Cell Cycle and Cell Differentiation." Heidelberg: Springer Verlag, 1975, pp 1–25.
5. McGinnis W, Levine MS, Hafen E, Kuroiwa A, Gehring WJ: Nature 308:428, 1984.
6. Scott MP, Weiner AJ: Proc Natl Acad Sci USA 81:4115, 1984.
7. Gehring WJ: Cell 40:3, 1985.

8. Bender W, Akam M, Karch F, Beachy P, Pfeifer M, Spierer P, Lewis EB, Hogness DS: Science 221:23, 1983.
9. Sanchez-Herrero E, Vernos I, Marco R, Morata G: Nature 313:108, 1985.
10. Kaufman TC, Lewis R, Wakimoto B: Genetics 94:115, 1980.
11. Lewis RA, Kaufman TC, Denell RE, Tallerico P: Genetics 95:367, 1980.
12. Lewis R, Wakimoto B, Denell R, Kaufman T: Genetics 95:383, 1980.
13. Wakimoto BT, Turner FR, Kaufman TC: Dev Biol 102:147, 1984.
14. Struhl G: J Embryol Exp Morphol 76:297, 1983.
15. Struhl G: Nature 308:454, 1984.
16. Mlodzik M, Fjose A, Gehring WJ: EMBO J 4:2961, 1985.
17. Joyner AL, Lebo RV, Kan YW, Tjian R, Cox DR, Martin GR: Nature 314:173, 1985.
18. Hauser CA, Joyner AL, Klein RD, Learned TK, Martin GR, Tjian R: Cell 43:19, 1985.
19. Rabin M, Hart CP, Ferguson-Smith A, McGinnis W, Levine M, Ruddle FH: Nature 314:175, 1985.
20. Hart CP, Awgulewitsch A, Fainsod A, McGinnis W, Ruddle F: Cell 43:9, 1985.
21. Green MC: "Genetic Variants and Strains of the Laboratory Mouse." Stuttgart: Fischer, 1981.
22. Poole SJ, Kauvar LM, Frees B, Kornberg T: Cell 40:37, 1985.
23. Fjose A, McGinnis W, Gehring WJ: Nature 313:284, 1985.
24. Levine M, Hafen E, Garber RL, Gehring WJ: EMBO J 2:2037, 1983.
25. Kornberg T, Sidén I, O'Farrell P, Simon M: Cell 40:45, 1985.
26. Hafen E, Kuroiwa A, Gehring WJ: Cell 37:833, 1984.
27. Akam ME: EMBO J 2:2075, 1983.
28. Karr T, Ali Z, Drees B, Kornberg T: Cell 43:591, 1985.
29. Kuroiwa A, Kloter U, Baumgartner P, Gehring WJ: EMBO J 4:3757, 1985.
30. Regulski M, Harding K, Kostriken R, Karch F, Levine M, McGinnis W: Cell 43:71, 1985.
31. Sheperd JWC, McGinnis W, Carrasco AE, DeRobertis EM, GehringWJ: Nature 310:70, 1984.
32. Laughon A, Scott MP: Nature 310:25, 1984.
33. Pabo CD, Sauer RT: Annu Rev Biochem 53:293, 1984.
34. Nasmyth K: Annu Rev Genet 16:439, 1982.
35. White RAH, Wilcox M: EMBO J 4:2035, 1985.
36. Caroll SB, Scott MP: Cell 43:47, 1985.
37. Beachy PA, Helfand SL, Hogness DS: Nature 313:545, 1985.
38. DiNardo S, Kuner JM, Theis J, O'Farrell PH: Cell 43:59, 1985.
39. Desplan C, Theis J, O'Farrell P: Mature 318:630, 1985.
40. Hiromi Y, Kuroiwa A, Gehring WJ: Cell 43:603, 1985.
41. Hogan B, Holland P, Schofield P: Trends Genet 1:67, 1985.
42. Colberg-Poley AM, Voss SD, Chowdhury K, Gruss P: Nature 314:713, 1985.
43. Colberg-Poley AM, Voss SD, Chowdhury K, Stewart CL, Wagner EF, Gruss P: Cell 43:39, 1985.
44. Colberg-Poley AM, Voss SD, Gruss P: Cold Spring Harbor Symp Quant Biol 50: (in press).
45. Joyner AL, Kornberg T, Coleman KG, Cox DR, Martin GR: Cell 43:29, 1985.
46. McGinnis W, Hart CP, Gehring WJ, Ruddle F: Cell 38:675, 1984.
47. Bućan M, Yang-Feng T, Colberg-Foley A, Wolgemuth D, Guenet JL, Francke U, Lehrach H: EMBO J 5:2899–2905.
48. Levine M, Rubin GM, Tjian R: Cell 38:667, 1984.
49. Jackson IJ, Scholfield P, Hogan B: Nature 317:745, 1985.
50. Breier G, Bućan M, Francke U, Colberg-Poley AM, Gruss P: EMBOJ 5:2209–2215.
51. Awgulewitsch A, Utset MF, Hart CP, McGinnis W, Ruddle FH: Nature 320:328, 1986.
52. Hafen E, Levine M, Gehring W: Nature 307:287, 1984.
53. Harding K, Wedeen C, McGinnis W, Levine M: Science 229:1236, 1985.
54. McGinnis W, Garber RL, Wirz J, Kuroiwa A, Gehring WJ: Cell 37:403, 1984.

# Induction of c-*fos* Is Mediated by Diverse Biochemical Pathways

## Tom Curran and James I. Morgan

*Departments of Molecular Oncology (T.C.) and Neurosciences (J.I.M.), Roche Institute of Molecular Biology, Roche Research Center, Nutley, New Jersey 07110*

A variety of extracellular stimuli elicit a rapid but transient induction of the *fos* proto-oncogene (c-*fos*) and a series of antigenically related proteins. Polypeptide mitogens cause an early increase in expression prior to the onset of mitogenesis. Other polypeptide growth factors which induce cellular differentiation factors give a similar induction. In addition, agents such as phorbol esters and calcium ionophores, although linked to different biological responses in different cell types, also stimulate c-*fos* expression. In PC12 pheochromocytoma cells, activation of voltage-dependent calcium channels results in c-*fos* induction. In this case, the receptor-genome coupling mechanism involves a calcium-dependent stimulation of calmodulin. Thus, transcriptional activation of c-*fos* is a common early nuclear event which is effected by diverse messenger systems.

**Key words:** proto-oncogene, calmodulin, second messengers, calcium channels, neuronal cells

Proto-oncogenes are involved in many of the key regulatory steps in the control of cell growth. The c-*sis* gene is related to platelet-derived growth factor [1,2], the c-erb$^B$ gene to the epidermal growth factor (EGF) receptor [3], and the c-*fms* gene to the CSF-1 receptor [4]. The proto-oncogenes which encode nuclear proteins, c-*fos* and c-*myc,* have been shown to be induced dramatically following activation of many growth factor receptors [5–14]. There has been a great deal of interest and speculation in the mechanism which couples extracellular signals to alterations in gene expression. While many early events associated with receptor activation have been described, such as alterations in ion flux and protein phosphorylation [15,16], it has proven more difficult to establish biochemical links to effects on gene expression. The induction of c-*fos* in growth-factor-treated cells is the most rapid nuclear event, at the level of transcription, yet described. Thus, c-*fos* expression provides a very convenient marker with which to study signal transduction systems. We have followed c-*fos* expression in many different situations; in quiescent cells stimulated to resume growth [10], in cells stimulated to differentiate [11,13], and in neuronal cells treated with agents that cause depolarization [17]. In particular, we have chosen to concentrate on the PC12 pheochromocytoma cell line. In these cells, we have distinguished at least three

Received May 20, 1986.

© 1988 Alan R. Liss, Inc.

distinct biochemical pathways which lead to the induction of c-*fos*. These different messenger systems have been dissected by judicious choice of stimulatory agents and by employing a series of pharmacological antagonists. Taken together, the data suggest that c-*fos* may play a role as a nuclear switch, in many different response pathways, which couples short-term signals at the cell membrane to long-term alterations in gene expression.

## MATERIALS AND METHODS
### Cells and Materials

The origin and maintenance of 208F cells was as previously described [18]. PC12 cells [19] were maintained as described [13]. The induction of c-*fos* in fibroblasts [10] and PC12 cells [13] has been described. The origin and use of pharmacological agents was as described [17].

### Detection of c-*fos* Protein

[$^{35}$S]-methionine labelling, preparation of cell lysates, and immunoprecipitation with *fos*-specific antibodies was as described [20]. SDS-polyacrylamide gel electrophoresis and autoradiography were essentially as described [21].

## RESULTS AND DISCUSSION

A number of laboratories have demonstrated induction of the c-*fos* and c-*myc* proto-oncogenes following treatment of quiescent fibroblasts with polypeptide mitogens [5–12]. Although different cell lines have been used, the experimental approach employed in each instance was quite similar. Cell cultures were maintained for 2–3 days in 0.5% fetal calf serum (FCS), or in platelet-poor plasma, and then treated with polypeptide mitogens or with whole serum. This procedure led to a synchronous reentry of cells into the S phase of the cell cycle approximately 12 to 14 hr later. The presence of c-*fos* and c-*myc* mRNAs or proteins was then determined at various times poststimulation by standard procedures. An example of c-*fos* protein induction in serum-stimulated rat fibroblasts is shown in Figure 1. In this case, serum-deprived rat fibroblasts were treated for 1 hr with 10% FCS, labelled with [$^{35}$S]-methionine for 15 min, and then chased for 30 min in the presence of excess unlabelled methionine. A clarified cell lysate was treated with *fos*-specific peptide antibodies, and the products were analyzed on an SDS-polyacrylamide gel. In a 15-min pulse-labelling period the *fos* product appears as a series of bands in the 55–62-kilodalton (kDa) range as a consequence of varying degrees of post-translational modification [21]. By chasing for 30 min with unlabelled methionine the mature forms of c-*fos* are detected. This result is consistent with the time-course reported for the appearance of c-*fos* mRNA [10]. In stimulated fibroblasts, c-*fos* mRNA levels reach a maximum at 30–45 min post-stimulation and decline quite sharply thereafter. Peak levels of c-*myc* mRNA are usually reached after 60 min of stimulation and generally persist at elevated levels for several hours [10].

It has been suggested that c-*fos* and c-*myc* are members of a set of coordinately regulated genes which contribute to the mitogenic response [7]. Stiles and co-workers have identified a series of between 10 and 30 genes, termed competence genes, that are induced in platelet-derived growth factor (PDGF)-treated Balb/c-3T3 cells [22].

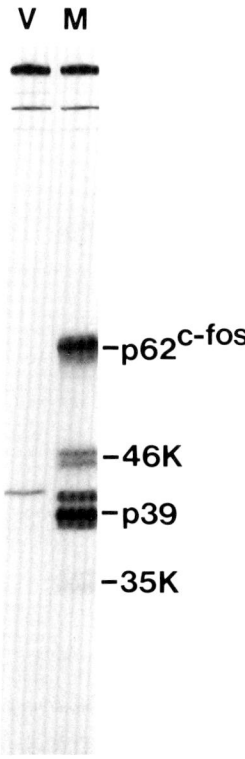

Fig. 1. Immunoprecipitation of fos-related proteins from serum-stimulated fibroblasts. Rat fibroblasts (208F) were maintained in DMEM plus 0.5% fetal calf serum (FCS) for two days. Growth medium was removed and replaced with 1 ml of DMEM lacking methionine supplemented with 20% dialyzed FCS, and the cells were incubated for 1 hr at 37°C. [$^{35}$S]-methionine (300 µCi) was added for a 15-min labelling period after which the cells were washed in DMEM plus 20% FCS, and incubation was continued in the same growth medium, lacking [$^{35}$S]-methionine, for 30 min. A cell lysate was prepared and treated with either v-fos specific (V) or fos-specific (M) antibodies as previously described [20]. The immunoprecipitation products were separated on an 8% SDS-polyacrylamide gel and visualized by autoradiography. The positions of the p62$^{c\text{-}fos}$, 46 K, p39, and 35 K proteins are indicated.

Although there are a number of good indications that c-*myc*, at least, can contribute to the mitogenic process [23,24], it is possible that c-*fos* and c-*myc* serve as nuclear signals in a more general sense. Induction has been observed in many different cell types not specifically associated with mitogenesis (Table I). Indeed, in some situations induction is associated with differentiation processes [11,13]. In addition, there is no specific association with the $G_0$-$G_1$ transition period, since induction may occur at any point in the cell cycle, even in S phase [25]. Thus, c-*fos* and c-*myc* are perhaps more akin to the "immediate early" genes of some DNA viruses [26]. Perhaps they play a role as intermediaries in the regulation of gene expression following cell surface stimulation.

As shown in Figure 1, a number of other proteins are detected in c-*fos* immunoprecipitates. One of these proteins, which appears as a series of bands because of post-translational modifications, is p39, a cellular protein which is complexed with *fos* product. The other proteins 46 K (Fig. 1), 35 K (Fig. 1), and a 30 K protein which is occasionally detected (data not shown), appear to be *fos*-related antigens.

**TABLE I. Situations in Which c-*fos* Induction Occurs**

| Agent | Cells | Comment | Reference |
|---|---|---|---|
| Serum | Fibroblasts | Mitogenic | [10, 25] |
| PDGF, FGF | Fibroblasts | Induces "Competence" | [10] |
| EGF | A431 Clones | Not associated with inhibition or stimulation of growth | [12] |
| A23187 | A431 | Toxic | [12] |
| CSF-1 | Macrophages | Required for survival | [11] |
| TPA | HL60 | Monocyte differentiation | [11] |
| NGF | PC12 Cells | Neurite outgrowth | [13] |
| 50 mM $K^+$ Veratridine BK8644 | PC12 Cells | Promote $Ca^{++}$ influx via voltage-dep $CA^{++}$ channel | [17] |

These proteins are not detected by tumor-bearing rat serum (TBRS); and, unlike p39, they are precipitated from denatured cell lysates, suggesting that they are not complexed with the c-*fos* product. They are not detected in unstimulated cells but are seen in the many and varied situations in which c-*fos* induction occurs (Table I). One or more of these proteins may be candidate products of the r-*fos* gene [7]. This gene is a member of the competence family and was isolated from a cDNA library made from RNA extracted from PDGF-treated Balb/c3T3 cells [22]. Nucleotide sequence analysis revealed a short region of homology to the c-*fos* gene, particularly within the third exon. Coincidently, this is the region of c-*fos* used to predict the peptide to which antisera were raised [20]. These data suggest that there may be aset of genes, some of which are related, which are induced in synchrony by a variety of extracellular stimuli.

As shown in Table I, many different stimuli elicit the induction of c-*fos* expression. Indeed, in some cell types a variety of agents seem to promote the same response. Figure 3 summarizes some of the intracellular signaling systems which may be responsible for transmitting signals from the cell surface to the nucleus. We have focused our studies on PC12 cells in an attempt to identify the pathways involved in c-*fos* activation.

We have previously shown that depolarization of PC12 cells leads to a calcium influx that stimulates a calmodulin-dependent induction of c-*fos* [17]. The role of a voltage-dependent calcium channel is implicated since induction is blocked by nisoldipine, a calcium channel antagonist [27]. In addition, the BAY K8644 compound, which is a calcium channel agonist, stimulates c-*fos* expression (Fig. 2) [17]. It is thought that the calcium influx activates calmodulin since agents, such as trifluoperazine and chlorpromazine, which are antagonists of calmodulin, block calcium-dependent c-*fos* induction [17]. Nerve growth factor (NGF) also stimulates c-*fos* expression in PC12 cells, however, in this case, induction does not require extracellular calcium and it is not inhibited by antagonists of calmodulin [17]. Other candidate cellular enzymes which might mediate the NGF response are phospholipase C (PLC, Fig. 2) [28], protein kinase C (PKC, Fig. 2) [29], and cyclic AMP-dependent kinase (cAMP-dependent kinase, Fig. 2) [30]. We have investigated the effects of each of these enzymes on NGF-stimulated c-*fos* expression in PC12 cells. Phospholipase C is a phosphodiesterase that can generate two intracellular messenger molecules, diacylglycerol (DAG) and inositol-trisphosphate ($IP_3$), from $PIP_2$ (Fig. 2). DAG is known to stimulate protein kinase C, the action of which will be discussed below. $IP_3$ has been

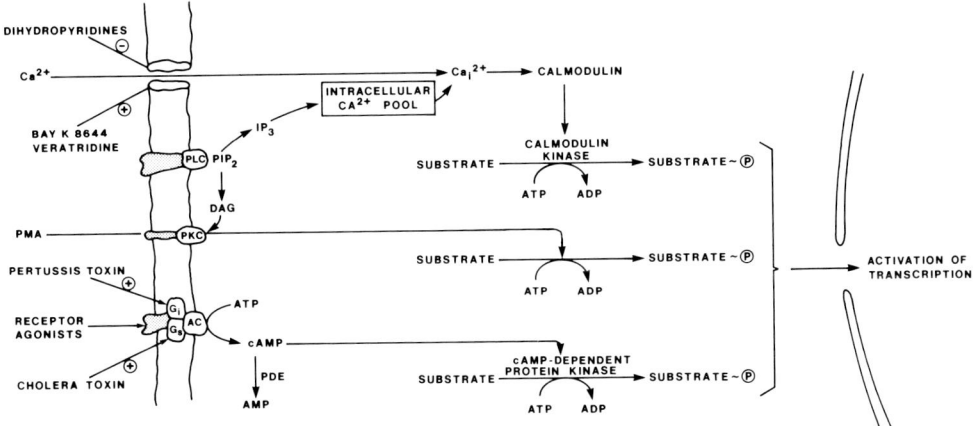

Fig. 2. Schematic representation of signal transduction pathways that can affect c-*fos* expression.

shown to cause the release of calcium from intracellular stores [31]. Conceivably, the increase in the intracellular free calcium pool could activate calmodulin as indicated in Figure 2. However, since NGF induction of c-*fos* is calmodulin-independent, this is not likely to be the case in PC12 cells [17]. These data do not preclude an alternative target for either calcium or $IP_3$ which could mediate c-*fos* activation.

The activation of protein kinase C by DAG can be mimicked by using phorbol esters such as phorbol-myristate-acetate (PMA) (Fig. 2) [32]. Phorbol esters do induce c-*fos* in PC 12 cells (data not shown); however, this is a very inefficient process compared to the NGF response. Prolonged treatment of cells with phorbol esters leads to a down-regulation of protein kinase C [33]. PC12 cells pretreated with TPA for 24 hr still exhibit an efficient induction of c-*fos* following addition of NGF. Thus, it is unlikely that protein kinase C alone mediates NGF-stimulated c-*fos* induction.

An alternative pathway would involve the cAMP-dependent protein kinase. This enzyme is activated by elevation of intracellular cAMP levels. One way in which this can occur is by activation of adenylate cyclase using cholera toxin (AC, Fig. 2) [34]. Alternatively, inactivation of the $G_i$ subunit of adenylate cyclase using pertussis toxin can also be utilized to identify agents that normally act to reduce cyclase activity. In our hands pertussis toxin does not affect NGF-induced c-*fos* expression. In addition, bypassing the adenylate cyclase system by adding dibutyryl cAMP to PC12 cells gives only a modest c-*fos* induction compared to NGF addition. Thus, it is unlikely that the cAMP-dependent kinase alone is the mediator of c-*fos* induction by NGF.

In each of the pathways illustrated in Figure 2, we have proposed that stimulation ultimately leads to the phosphorylation of a substrate which serves as a nuclear signal for the transcriptional activation of c-*fos*. At present, we do not know if the substrate is the same for each pathway, but is only inefficiently phosphorylated by the protein kinase C or cAMP-dependent kinase systems, or if each pathway activates specific substrates which act on different targets.

After transcriptional activation, c-*fos* mRNA is processed and transported to the cytoplasm, where it is translated into the c-*fos* protein (Fig. 3). The c-*fos* protein undergoes some phosphorylation in the cytoplasm before being conveyed to the

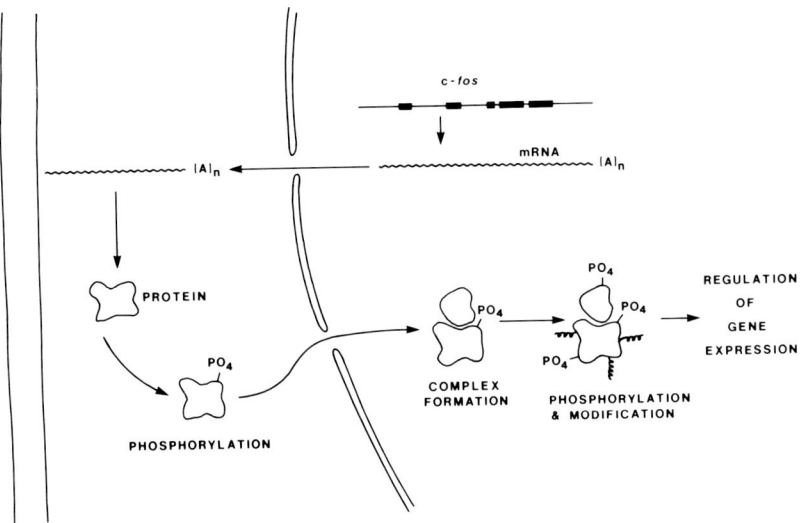

Fig. 3. Schematic representation of the synthesis and processing of the c-*fos* protein.

nucleus. Almost immediately upon entry into the nucleus, the *fos* protein forms a complex with p39 and rapidly undergoes extensive post-translational modification of an unknown nature [10,21]. We propose that the c-*fos* protein complex is then involved in the regulation of gene expression. Since c-*fos* induction is observed in many cell types, it is possible that other cellular components, post-translational modifications, or chromatin structure in the cells in which c-*fos* is induced, determine its specificity of action.

The picture we have presented is, of necessity, quite simplified. It is likely that a multitude of signal transduction pathways interact at many different points. In addition, there is no way of knowing yet how many steps occur at each of the stages indicated by arrows in Figure 2. However, this rather naive view of intracellular communication forms the basis of a working hypothesis. This hypothesis is testable and will be embellished in the light of future experimentation.

## REFERENCES

1. Doolittle RF, Hunkapiller MW, Hood LE, Devare SG, Robbins KC, Aaronson SA, Antoniades NH: Science 221:275, 1983.
2. Waterfield MD, Scrace GT, Whittle N, Stroobant P, Johnson A, Wasteson A, Westermark B, Heldin CH, Juang JS, Duel T: Nature 304:35, 1983.
3. Downward J, Yarden Y, Mayes E, Scrace G, Totty N, Stockwell P, Ullrich A, Schlessinger J, Waterfield MD: Nature 307:521, 1984.
4. Sherr CJ, Rettenmier CW, Sacca R, Roussel MF, Look AT, Stanley ER: Cell 41:665, 1985.
5. Kelly K, Cochran BH, Stiles CD, Leder P: Cell 35:603, 1983.
6. Campisi J, Gray HE, Pardee AB, Dean M, Sonenshein GE: Cell 36:242, 1984.
7. Cochran BH, Zullo J, Verma IM, Stiles CD: Science 226:1080, 1984.
8. Greenberg ME, Ziff EB: Nature 311:433, 1984.
9. Kruijer W, Cooper JS, Hunter T, Verma IM: Nature 312:711, 1984.
10. Muller T, Bravo R, Burkhardt J, Curran T: Nature 312:716, 1984.
11. Muller R, Curran T, Muller D, Guilbert L: Nature 314:546, 1985.

12. Bravo R, Burckhardt J, Curran T, Muller R: EMBO J 4:1193, 1985.
13. Curran T, Morgan JI: Science 229:1265, 1985.
14. Greenberg ME, Greene LA, Ziff EB: J Biol Chem 260:14101, 1985.
15. Halegoua E, Patrick J: Cell 22:571, 1980.
16. Rozengurt E, Heppel GA: Proc Natl Acad Sci USA 82:7330, 1985.
17. Morgan JI, Curran T: Nature 322:552, 1986.
18. Quade K: Virology 98:461, 1979.
19. Greene LA, Tischler AS: Proc Natl Acad Sci USA 73:2424, 1976.
20. Curran T, Van Beveren C, Ling H, Verma IM: Mol Cell Biol 5:167, 1985.
21. Curran T, Miller AD, Zokas L, Verma IM: Cell 36:259, 1984.
22. Cochran BH, Reffel AC, Stiles CS: Cell 33:939, 1983.
23. Armelin HA, Armelin MCS, Kelly K, Stewart T, Leder P, Cochran BH, Stiles CS: Nature 310:655, 1984.
24. Kaczmarek L, Hyland JK, Watt R, Rosenberg M, Baserga R: Science 228:1313, 1985.
25. Bravo R, Burckhardt J, Curran T, Muller R: EMBO J 5:695, 1986.
26. Rakusanova T, Ben-Porat T, Himeno M, Kaplan AS: Virology 46:877, 1971.
27. Kazda S, Garthoff B, Meyer H, Schlossmann K, Stoepel K, Towart R, Vater W, Wehinger E: Arzneimittel Forschung 30:2144, 1980.
28. Mitchell RH: Biochim Biophys Acta 415:81, 1975.
29. Takai Y, Kishimoto A, Takai Y, Nishizuka Y: J Biol Chem 252:7603, 1977.
30. Brostrom CO, Corbin JD, King CA, Krebs EG: Proc Natl Acad Sci USA 68:2444, 1971.
31. Streb H, Irvine RF, Berridge MJ, Schultz T: Nature 306:67, 1983.
32. Castangna M, Takai Y, Kaibuchi K, Sano K, Kikkawa Y, Nishizuka Y: J Biol Chem 257:7847, 1982.
33. Collins MKL, Rozengurt E: J Cell Physiol 112:42, 1982.
34. Gill DM, Meran R: Proc Natl Acad Sci USA 75:3050, 1978.

# Role of Protein Kinase C in Cell Surface Signal Transduction

## Ushio Kikkawa

*Department of Biochemistry, Kobe University School of Medicine, Kobe 650, Japan*

Information from a group of extracellular signals flows from the cell surface into the cell interior through two routes, $Ca^{2+}$ mobilization and protein kinase C activation. At an early phase of cellular responses, inositol-1,4,5-trisphosphate mobilizes $Ca^{2+}$, whereas diacylglycerol activates protein kinase C. These two signal mediators are generated from the receptor-mediated hydrolysis of phosphatidylinositol-4,5-bisphosphate. It has been shown that two routes are both essential and often act synergistically to induce many cellular responses. Like these short-term responses, both protein kinase C activation and $Ca^{2+}$ mobilization appear to be essential for long-term responses as well. However, additional receptor occupation by growth factor is necessary to induce full activation of cell proliferation, and the signal pathway through protein kinase C appears to be separate from and synergistic with that via growth factors. Immunohistochemical studies with monoclonal antibodies raised against protein kinase C suggest that this enzyme seems to be absent or present in very low concentrations in the nucleus; probably another step in signal translation is needed for the ultimate activation of nuclear events. Several functional proteins in many tissues have been reported to serve as substrates of protein kinase C. The phosphorylation of some of these proteins such as membrane receptors is apparently related to down-regulation or negative feedback control of activation of cellular functions. It is possible that protein kinase C has dual actions in the positive as well as the negative phase of regulation, depending on the function of each target substrate protein. Results are summarized of further studies on the mode of activation, intracellular localization, and possible role of this unique protein kinase system.

**Key words:** calcium, phorbol ester, tumor promoter

In the action of a group of hormones, some neurotransmitters, and many other biologically active substances, signal-induced degradation of inositol phospholipids may generate important intracellular second messengers. This species of phospholipids is a relatively minor component of mammalian cell membranes, and the involvement of inositol phospholipids in the receptor mechanism was suggested first by Hokin and Hokin [1]. A small portion of the phospholipid contains an additional one or two phosphates at position 4 or positions 4 and 5 of the inositol moiety. These minor phospholipids, phosphatidylinositol 4-phosphate and phosphatidylinositol 4, 5-

Received April 17, 1986.

© 1988 Alan R. Liss, Inc.

bisphosphate (PIP$_2$), are produced through sequential phosphorylation of phosphatidylinositol (PI). Although PI was initially regarded as the primary target for stimulus-transduction, PIP$_2$ has been more intensely studied in recent years because it generally disappears more rapidly than PI in stimulated cells, and its water-soluble product, inositol-1,4,5-trisphosphate, has been shown to serve as an intracellular mediator for the release of Ca$^{2++}$ from internal stores [2]. A series of studies in our laboratory has provided some evidence that 1,2-diacylglycerol, the other product of PIP$_2$ breakdown, remains in the membrane and initiates the activation of a specialized protein kinase, protein kinase C [3–5]. Thus, the diacylglycerol pathway is separate from, but may be synergistic with, the pathway that increases intracellular Ca$^{2+}$ as outlined in Figure 1. This paper will describe the physiological roles of inositol phospholipid turnover, primarily those of protein kinase C, in cell surface signal transduction.

## ACTIVATION OF PROTEIN KINASE C

Protein kinase C is distributed widely in tissues and organs. In most tissues, the enzyme is recovered mainly in soluble fraction by subcellular fractionation, in some other tissues a large quantity of the enzyme is associated with membranes [6]. Recent analysis in this laboratory using the immunocytochemical procedure with monoclonal antibodies against protein kinase C indicates that this enzyme is present in cytoplasm or associated with membranes, and is seemingly absent or poorly represented in the nucleus as shown in Figure 2.

The activation of protein kinase C normally depends on Ca$^{2+}$ as well as phospholipids in addition to diacylglycerol. However, diacylglycerol dramatically

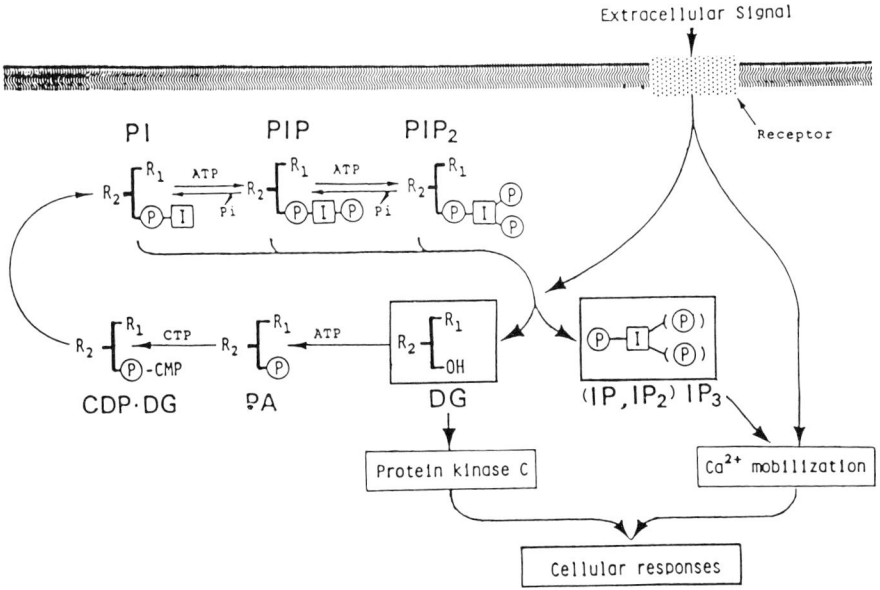

Fig. 1. Inositol phospholipid turnover and signal transduction. PI, phosphatidylinositol; PIP, phosphatidylinositol 4-phosphate; PIP$_2$, phosphatidylinositol 4,5-bisphosphate; DG, 1,2-diacylglycerol; IP$_3$, inositol-1,4,5-trisphosphate; PA, phosphatidic acid; CDP-DG, cytidine diphosphodiacylglycerol; I, inositol moiety; P, phosphoryl group. (Adapted from [4].)

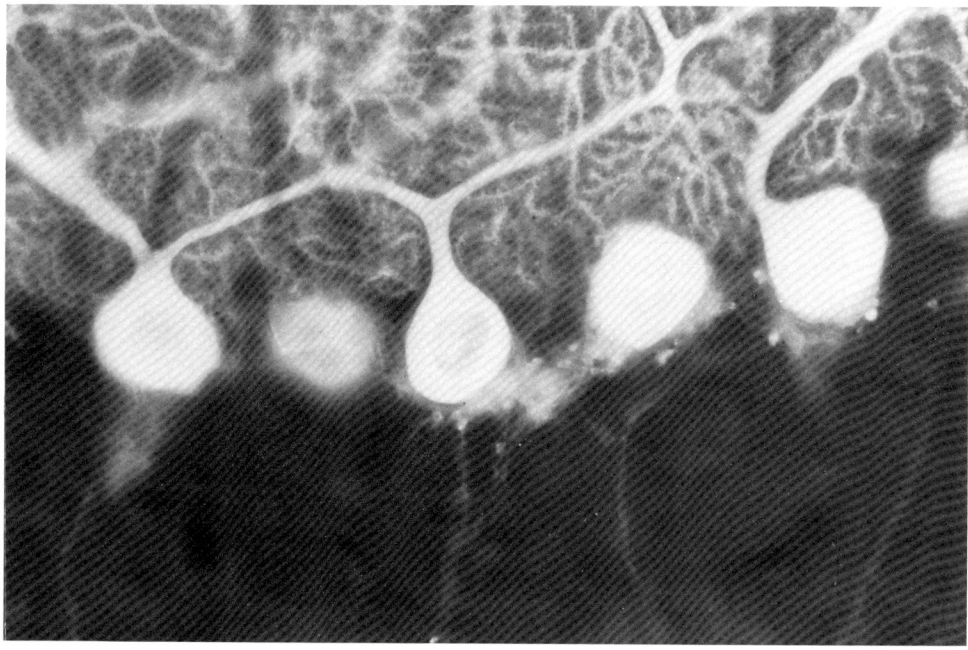

Fig. 2. Immunofluorescence staining of rat brain Purkinje cells with monoclonal antibodies against protein kinase C. (Adapted from [79].)

increases the affinity of this enzyme to $Ca^{2+}$, and thereby renders it fully active without a net increase in the $Ca^{2+}$ concentration [7]. Thus, the activation of this protein kinase is biochemically dependent on $Ca^{2+}$, but physiologically does not require a change in $Ca^{2+}$ concentration. Diacylglycerols having various fatty acyl moieties are capable of activating protein kinase C. In in vitro assay, diacylglycerols that contain at least one unsaturated fatty acyl moiety at either the first or second position are active [7]. When one fatty acyl moiety is replaced by a short chain, the resulting diacylglycerol, such as 1-oleoyl-2-acetyl-glycerol, are also active to support enzyme activation [8]. However, all active diacylglycerols contain 1,2-sn-configuration, and other stereoisomers are inactive, suggesting that a highly specific lipid-protein interaction is needed for this enzyme activation [9].

To explore a link between inositol phospholipid breakdown and protein kinase C activation as well as its role in stimulus-response coupling, a synthetic diacylglycerol such as 1-oleoyl-2-acetyl-glycerol has been often used, since it is readily intercalated into intact cell membranes and activates protein kinase C directly [10]. Also, 1,2-dioctanoylglycerol and 1,2-didecanoylglycerol have been shown to be effective permeable diacylglycerols [11]. Synthetic diacylglycerols having 2,3-sn-configuration are not active for intact cell systems [12].

## PROTEIN KINASE C AND TUMOR PROMOTERS

Tumor-promoting phorbol esters, such as 12-*O*-tetradecanoyl-phorbol-13-acetate (TPA), have a structure very similar to that of diacylglycerol in their molecules and activate protein kinase C directly both in vitro and in vivo [13,14]. Phorbol esters, like diacylglycerol, increase the affinity of this enzyme for $Ca^{2+}$, resulting in its full activation at physiological concentrations of this divalent cation. Several lines of evidence suggest that protein kinase C is probably a prime target of tumor promoters [13–19]. Kinetic analysis with purified protein kinase C suggests that roughly one molecule of tumor promoter can activate one molecule of this protein kinase [19]. Other classes of tumor promoters such as mezerein [20,21], and teleocidin and *Aplysia* toxin [22,23], also activate protein kinase C, suggesting that a diacylglycerol-like structure is not always essential, and that many tumor promoters induce a membrane perturbation analogous to that caused by diacylglycerol.

In experiments with phorbol esters, the resulting cellular responses sometimes appear to cast doubt as to their suitability for studies on cell biology. Diacylglycerol, the physiological activator of protein kinase C, is present only transiently in membranes as noted above, while TPA is hardly metabolizable and persists for longer periods of time. The possibility arises, therefore, that some limited phase of the cellular response might be extended, thereby distorting the normal sequence of events. In addition, the concentration of phorbol ester employed should be given special consideration when attempting to evaluate its exact contribution to physiological responses. At higher concentrations, tumor promoters per se can induce significant biological effects. There is no proof at present for protein kinase C's being the sole target of tumor promoters, and the possibility exists that these compounds act as membrane perturbers or fusigens, particularly at higher concentrations. To interpret any experimental result for phorbol ester correctly, the parallel demonstration of other parameters such as phosphorylation of some endogenous proteins is obviously desirable, as described in our earlier experiments for platelets [14] and lymphocytes [24].

## POSSIBLE ROLES OF PROTEIN KINASE C IN CELLULAR RESPONSES AND FEEDBACK CONTROL

Protein kinase C activation and $Ca^{2+}$ mobilization can be induced selectively and independently. The application of a permeable diacylglycerol or phorbol ester for the activation of protein kinase C and a $Ca^{2+}$-ionophore for $Ca^{2+}$ mobilization. By using this procedure it has been shown that the two signal pathways are both essential to elicit full ceilular responses. Such a role of protein kinase C in stimulus-response coupling was first demonstrated for the release of serotonin from platelets [10], and subsequently shown for release reactions, secretion, and exocytosis from a wide variety of cell types as well as for many other cellular processes [5]. The activation of cellular responses by this protein kinase C pathway appears to be separate from but often synergistic to those activated via an increase of intracellular $Ca^{2+}$.

These two signal pathways may play some roles in long-term responses such as gene expression and cell proliferation. It has been shown with macrophage-depleted human peripheral lymphocytes that the two pathways are both essential and act synergistically for promoting DNA synthesis. However, in order to stimulate rapid cell proliferation, some growth factor even in a low concentration is still needed, indicating that an additional signal pathway is involved in eliciting the full activation of cell proliferation [24]. Presumably interleukin 1 utilizes the two signal pathways to promote T-lymphocyte proliferation by initiating the induction of the receptor of interleukin 2 [25–27] as well as by facilitating the release of this growth factor [27,28]. Synergism between a permeable diacylglycerol and $Ca^{2+}$-ionophore also leads to proliferation of human B-lymphocytes as well, and a parallel situation may exist in the mechanism of growth promotion for these cells [29,30]. Tumor promoters and permeable diacylglycerol both induce expression of some genes such as those of ornithine decarboxylase [31–33], histidine decarboxylase [34], prolactin [35], and probably γ-interferon [36,37], and plasminogen activator [38]. some proto-oncogenes such as *c-fos* [33,39–43] and *c-myc* [33,41,43–47] are potential targets of protein kinase C action. By analogy to lymphocyte activation discussed above, it is attractive to imagine that protein kinase C may induce expression of those genes which are related to the action of growth factors, thereby promoting cell proliferation in the long-term. Figure 3 illustrates hypothetical pathways of signal transduction for short-term and long-term cellular responses. However, as noted above, protein kinase C seems to be absent or has a weak presence in the nucleus. Presumably, an additional step of signal translation is needed prior to the ultimate activation of nuclear events.

Another major function of protein kinase C appears to be intimately related to such feedback control or down-regulation. The stimulation of platelets by thrombin induces inositol phospholipid breakdown which initiates aggregation and release reactions of platelets. Recent evidence obtained strongly suggests that protein kinase C exerts negative feedback control on thrombin receptor and inhibits the $Ca^{2+}$ mobilization [48–51]. Similar feedback control by protein kinase C over its own receptor has been suggested for many other signalling systems in various cell types such as hepatocytes [52–54], neutrophils [55], pituitary cells [56], PC12 cells [57], and astrocytoma cells [58]. The biochemical basis of such a negative feedback control remain to be clarified, the feedback control or down-regulation by protein kinase C is not confined to its own receptors but appears to extend growth factor receptors. For instance, this enzyme is shown to phosphorylate the epidermal growth factor receptor with a concomitant decrease in both its tyrosine-specific protein kinase and growth-factor-binding activities [59–62]. In this receptor, threonine-654 is shown to be phosphorylated which is located at the N-terminal side between the membrane-spanning and tyrosine-specific protein kinase domains [63]. In an analogous fashion, the receptors of other growth factors, such as insulin [64], somatomedin C [64], transferrin [65,66], and interleukin 2 [25] are proposed to be possible targets of protein kinase C. However, the site of phosphorylation and logical consequence of these receptor phosphorylation reactions are not known.

## RECEPTOR INTERACTIONS

Although there are many variations in the signalling systems from tissue to tissue, most tissues seem to have at least two major receptor classes for transducing

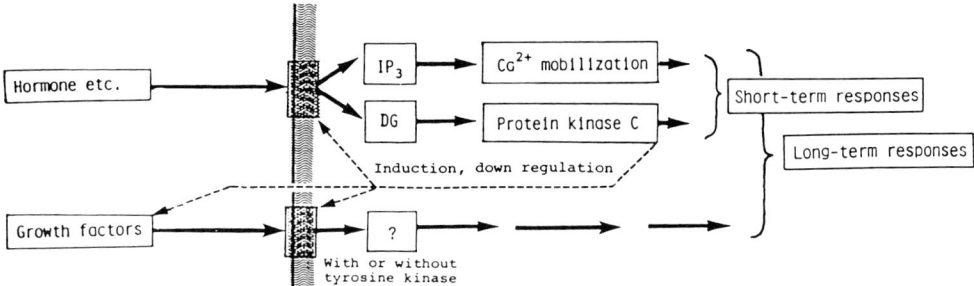

Fig. 3. Hypothesis for signal transduction of short-term and long-term cellular responses. (Adapted from [80].)

information across the membrane. One class is related to cyclic AMP, while the other induces rapid turnover of inositol phospholipids as well as mobilization of $Ca^{2+}$. Cellular responses may be divided tentatively into several modes as given schematically in Figure 4. In *bidirectional control systems* the two classes of receptors appear to counteract each other, whereas in *monodirectional control systems* one receptor class may potentiate the other one. For instance, in some tissues such as platelets, lymphocytes, mast cells, and many others, the signals that induce inositol phospholipid breakdown promote the activation of cellular functions, but the signals that produce cyclic AMP usually antagonize such activation. In such tissues the signal-induced breakdown of inositol phospholipids and the subsequent events eventually leading to cellular responses are all profoundly blocked by cyclic AMP [3,4]. Inversely, in another group of cell types such as Leydig cells, [67–69], hepatocytes [54,70], glioma C6 cells [71], protein kinase C inhibits, and desensitizes the adenylate cyclase system. In several other cell types including pinealocytes [72,73], pituitary cells [74,75], and lymphoma S49 cells [76–78], protein kinase C greatly potentiates cyclic AMP production. On the other hand, there is no example as yet of a tissue in which cyclic AMP potentiates signal-induced turnover of inositol phospholipids. However, there two signal transduction pathways frequently act in concert in many endocrine cells. The evidence presented thus far is still incomplete, but it is reasonable to assume that various combinations of the two receptor signalling systems may operate positively and sometimes be intensified in many physiological processes. Further exploration of such receptor interactions in individual cell types appears to be extremely important for understanding the molecular basis of signal-induced cell growth and differentiation.

## CONCLUSIONS

The evidence available at present seems to suggest the crucial role of protein kinase C in signal transduction for the activation of many cellular functions including

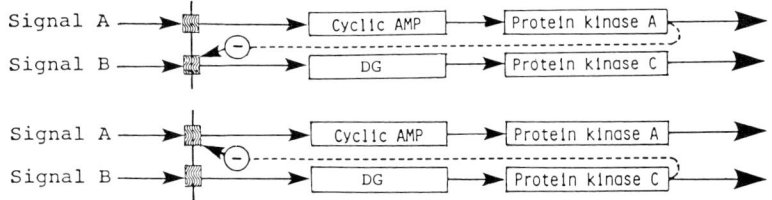

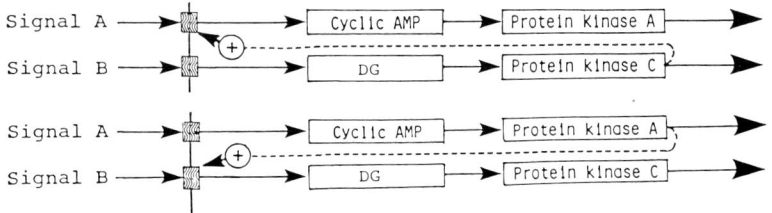

Fig. 4. Interaction of receptor signalling systems. DG, 1,2,-diacylglycerol; (Adapted from [5].)

proliferation, particularly at an early phase of responses. Perhaps, the signal-induced breakdown of inositol phospholipids initiates a cascade of events starting with $Ca^{2+}$ mobilization and protein kinase C activation. However, it seems still premature to discuss the detailed relationship between the role of $Ca^{2+}$ and that of protein kinase C. Each of these signal pathways may play diverse roles in controlling biochemical reactions. Most probably, the protein phosphorylation catalyzed by protein kinase C may exhort "profound modulation" of various $Ca^{2+}$-mediated processes. It is hoped that further exploration of the role of this protein kinase may provide clues to understand the mechanism of trans-membrane control of biological reactions.

## ACKNOWLEDGMENTS

I would like to express indebtedness to my collaborators Professor Y. Nishizuka, Drs. T. Kitano, H. Nomura, M. Go, J. Koumoto, T. Hashimoto, K. Ase, and K. Ogita; and I am also grateful to Professor C. Tanaka and Dr. N. Saito, Department of Pharmacology, Kobe University School of Medicine. Skillful secretarial assistance on the part of Mrs. S. Nishiyama, Miss S. Fukase, and Mrs. T. Nishikori is gratefully acknowledged.

This investigation has been supported in part by research grants from the Ministry of Education, Science, and Culture, Japan.

## REFERENCES

1. Hokin MR, Hokin LE: J Biol Chem 203:967, 1953.
2. Berridge MJ, Irvine RF: Nature 312:315, 1984.
3. Nishizuka Y: Nature 308:693, 1984.
4. Kikkawa U, Nishizuka Y: "The Enzyme," 3rd edition. Volume 17. Academic Press, 1986, pp 167–189.
5. Kikkawa U, Nishizuka Y: Annu Rev Cell Biol 2:149, 1986.
6. Kikkawa U, Takai Y, Minakuchi R, Inohara S, Nishizuka Y: J Biol Chem 257:13341, 1982.
7. Kishimoto A, Takai Y, Mori T, Kikkawa U, Nishizuka Y: J Biol Chem 255:2273, 1980.
8. Mori T, Takai Y, Yu B, Takahashi J, Nishizuka Y, Fujikura T: J Biochem 91:427, 1982.
9. Rando RR, Young N: Biochem Biophys Res Commun 122:818, 1984.
10. Kaibuchi K, Takai Y, Sawamura M, Hoshijima M, Fujikura T, Nishizuka Y: J Biol Chem 258:6701, 1983.
11. Lapetina EG, Reep B, Ganong BR, Bell RM: J Biol Chem 260:1358, 1985.
12. Nomura H, Nakanishi H, Ase K, Kikkawa U, Nishizuka Y: Prog Hemost Thromb 8:143, 1986.
13. Castagna M, Takai Y, Kaibuchi K, Sano K, Kikkawa U, Nishizuka Y: J Biol Chem 257:7847, 1982.
14. Yamanishi J, Takai Y, Kaibuchi K, Sano K, Castagna M, Nishizuka Y: Biochem Biophys Res Commun 112:778, 1983.
15. Niedel JE, Kuhn LJ, Vandenbark GR: Proc Natl Acad Sci USA 80:36, 1983.
16. Sando JJ, Young MC: Proc Natl Acad Sci USA 80:2642, 1983.
17. Leach KL, James ML, Blumberg PM: Proc Natl Acad Sci USA 80:4208, 1983.
18. Ashendel CL, Staller JM, Boutwell RK: Biochem Biophys Res Commun 111:340, 1983.
19. Kikkawa U, Takai Y, Tanaka Y, Miyake R, Nishizuka Y: J Biol Chem 258:11442, 1983.
20. Couturier A, Bazgar S, Castagna M: Biochem Biophys Res Commun 121:448, 1984.
21. Miyake R, Tanaka Y, Tsuda T, Kaibuchi K, Kikkawa U, Nishizuka Y: Biochem Biophys Res Commun 121:649, 1984.
22. Fujiki H, Tanaka Y, Miyake R, Kikkawa U, Nishizuka Y, Sugimura T: Biochem Biophys Res Commun 120:339, 1984.
23. Arcoleo JP, Weinstein IB: Carcinogenesis 6:213, 1985.
24. Kaibuchi K, Takai Y, Nishizuka Y: J Biol Chem 260:1366, 1985.
25. Shackelford DA, Trowbridge IS: J Biol Chem 259:11706, 1984.

26. Depper JM, Leonard WJ, Kronke M, Noguchi PD, Cunningham RE, Waldmann TA, Green WC: J Immunol 133:3054, 1984.
27. Truneh A, Albert F, Golstein P, Schmitt-Verhulst AM: Nature 313:318, 1985.
28. Isakov N, Bleackly RC, Shaw J, Altman A: Biochem Biophys Res Commun 130:724, 1985.
29. Guy GR, Gordon J, Michell RH, Brown G: Biochem Biophys Res Commun 131:484, 1985.
30. Nel AE, Wooten MW, Goldschmidt-Clermont PJ, Miller PJ, Stevenson HC, Galbraith RM: Biochem Biophys Res Commun 128:1364, 1985.
31. Otani S, Matsui I, Kuramoto A, Morisaw S: Eur J Biochem 147:27, 1985.
32. Jetten AM, Ganong BR, Vandenbark GR, Shirley JE, Bell RM: Proc Natl Acad Sci USA 82:1941, 1985.
33. Greenberg ME, Greene LA, Ziff EB: J Biol Chem 260:14101, 1985.
34. Watanabe T, Taguchi Y, Sasaki K, Tsuyama K, Kitamura Y: Biochem Biophys Res Commun 100:427, 1981.
35. Murdoch GH, Waterman M, Evans RM, Rosenfeld MG: J Biol Chem 260:11852, 1985.
36. Wilkinson M, Morris A: Biochem Biophys Res Commun 111:498, 1983.
37. Johnson HM, Vassallo T, Torres BA: J Immunol 134:967, 1985.
38. Degen JL, Estensen RD, Nagamine Y, Reich E: J Biol Chem 260:12426, 1985.
39. Greenberg ME, Ziff EB: Nature 311:433, 1984.
40. Kruijer W, Cooper JA, Hunter T, Verma IM: Nature 312:711, 1984.
41. Muller R, Bravo R, Burckhardt J, Curran T: Nature 312:716, 1984.
42. Mitchell RL, Zokas L, Schreiber RD, Verma IM: Cell 40:209, 1985.
43. Bravo R, Burckhardt J, Curran T, Muller R: EMBO J 4:1193, 1985.
44. Kelly K, Cochran BH, Stiles CD, Leder P: Cell 35:603, 1983.
45. Coughlin SR, Lee WMF, Williams PW, Giels GM, Williams LT: Cell 43:243, 1985.
46. de Bustros A, Baylin SB, Berger CL, Ross BA, Leong SS, Nelkin BD: J Biol Chem 260:98, 1985.
47. Faletto DL, Arrow AS, Macara IG: Cell 43:315, 1985.
48. Rittenhouse SE, Sasson JP: J Biol Chem 260:8657, 1985.
49. MacIntyre DE, McNicol A, Drummond AH: FEBS Lett 180:160, 1985.
50. Watson SP, Lapetina EG: Proc Natl Acad Sci USA 82:2623, 1985.
51. Zavoico GB, Halenda SP, Sha'afi RI, Feinstein MB: Proc Natl Acad Sci USA 82:3859, 1985.
52. Lynch CJ, Charest R, Bocckino SB, Exton JH, Blackmore PF: J Biol Chem 260:2844, 1985.
53. Cooper RH, Coll KE, Williamson JR: J Biol Chem 260:3281, 1985.
54. Garcia-Sainz AJ, Mendlovic F, Martinez-Olmedo A: Biochem J 228:277, 1985.
55. Naccache PH, Molski TFP, Borgeat P, White JR, Sha'afi RI: J Biol Chem 260:2125, 1985.
56. Drummond AH: Nature 315:752, 1985.
57. Vincentini LM, di Virgilio F, Ambrosini A, Pozzan T, Meldolesi J: Biochem Biophys Res Commun 127:310, 1985.
58. Orellana SA, Solski PA, Brown JH: J Biol Chem 260:5236, 1985.
59. Cochet C, Gill GN, Meisenhelder J, Cooper JA, Hunter T: J Biol Chem 259:2553, 1984.
60. Iwashita S, Fox CF: J Biol Chem 259:2559, 1984.
61. McCaffrey PG, Friedman B, Rosner MR: J Biol Chem 259:12502, 1984.
62. Davis RJ, Czech MP: J Biol Chem 259:8545, 19844.
63. Hunter T, Ling N, Cooper JA: Nature 311:480, 1984.
64. Jacobs S, Sahyoun NE, Saltiel AR, Cuatrocasas P: Proc Natl Acad Sci USA 80:6211, 1983.
65. May WS, Jacobs S, Cuatrecasas P: Proc Natl Acad Sci USA 81:2016, 1984.
66. Klausner RD, Harford J, van Renswoude J: Proc Natl Acad Sci USA 81:3005, 1984.
67. Mukhopadhyay AK, Schumacher M: FEBS Lett 187:56, 1985.
68. Papadojoulos V, Carreau S, Drodsdowsky MA: FEBS Lett 188:312, 1985.
69. Rebois RV, Patel J: J Biol Chem 260:8026, 1985.
70. Heyworth CM, Whetton AD, Kinsella AR, Houslay MD: FEBS Lett 170:38, 1984.
71. Kassis S, Zaremba T, Patel J, Fishman PH: J Biol Chem 260:8911, 1985.
72. Sugden D, Vanecek J, Klein DC, Thomas TP, Anderson WB: Nature 314:359, 1959.
73. Zatz M: J Neurochem 45:637, 1985.
74. Cronin MJ, Canonico PL: Biochem Biophys Res Commun 129:404, 1985.
75. Quilliam LA, Dobson PR, Brown BL: Biochem Biophys Res Commun 129:898, 1985.

76. Bell JD, Buxton ILO, Brunton LL: J Biol Chem 260:2625, 1985.
77. Katada T, Gilman AG, Watanabe Y, Bauer S, Jakobs KH: Eur J Biochem 151:431, 1985.
78. Kiss Z, Steinberg RA: J Cell Physiol 125:200, 1985.
79. Kikkawa U, Kitano T, Saito N, Kishimoto A, Taniyama K, Tanaka C, Nishizuka Y: "Calcium and the Cell (Ciba Foundation Symposium 122)." John Wiley & Sons Ltd., Chichester, UK, 1986, pp 197–211.
80. Nishizuka Y: JNCI 76:363, 1986.

# Characterization of the Epidermal Growth Factor- and Vanadate-Activated Calcium Influx in A431 Cells

## Ian G. Macara and George M. Gray

*Division of Toxicology, Department of Biophysics, University of Rochester Medical Center, Rochester, New York 14642*

Epidermal growth factor (EGF) and vanadate can activate the uptake of Ca in A431 epidermal carcinoma cells by 2–5-fold with no detectable lag period. Preincubation of the cells with the tumor-promoting phorbol ester tetradecanoyl phorbol acetate (TPA) or with 1-oleoyl-2-acetylglycerol inhibits the activation by both agents. TPA alone has no effect on the basal rate of Ca uptake. Pre-incubation with EGF to down-regulate the EGF receptor prevents the subsequent stimulation by EGF but not that by vanadate, nor does it block the inhibition of vanadate-stimulated uptake by TPA. Ca uptake is sodium-independent and is not activated by depolarization in high KCl. On the contrary, the EGF- or vanadate-stimulated uptake is completely inhibited by decreasing the plasma membrane potential from about −65 to −30 mV. These results demonstrate that the EGF receptor is not itself functioning as a Ca channel, that vanadate is not acting at the level of EGF receptor, and that Ca uptake can be inhibited via protein kinase C at a site distinct from the receptor. The results also indicate that the Ca transport system is potential-sensitive, and in this respect it is similar to the IgE-activated Ca channel in mast cells.

**Key words: tyrosine kinase, receptor, phosphatidylinositol**

Unlike many other mitogens, epidermal growth factor (EGF) appears to induce rapid increases in cytosolic $Ca^{2+}$ by raising the plasma membrane permeability to $Ca^{2+}$ [1,2] rather than by triggering the release of intracellular $Ca^{2+}$ stores [3]. Neither the coupling mechanism between the growth factor receptor and the change in $Ca^{2+}$ permeability nor the mechanism of $Ca^{2+}$ transport are yet understood.

In particular, it is not known whether the intrinsic tyrosyl kinase activity of the EGF receptor is required, nor whether the coupling mechanism involves the activation of protein kinase C via an increase in phosphatidylinositol (PI) turnover. Previous studies have reported changes in diacyglycerol levels and in the distribution of kinase C in A431 epidermal carcinoma cells in response to EGF [4], but the activation of PI

Received November 4, 1986.

© 1988 Alan R. Liss, Inc.

turnover appears to be an effect rather than a cause of increased $Ca^{2+}$ uptake by the cells [2]. Moreover, the activation of protein kinase C by tumor-promoting phorbol esters or synthetic diacylglycerols results in the phosphorylation and inactivation of the EGF receptor [5,6]. We have therefore begun to characterize the effects on $Ca^{2+}$ influx into A431 cells of EGF and activators of kinase C. The effects of vanadate on this system have also been studied since vanadate is known to be a potent phosphotyrosine phosphatase inhibitor [7] and to significantly increase the level of autophosphorylation of the EGF receptor in A431 cell membrane preparations [8]. Moreover, the EGF-activated $Na^+/H^+$ exchange system in A431 cells is also activated by vanadate, in a partially additive manner [9]. It is not clear, however, whether the effect of vanadate is mediated by the EGF receptor.

## MATERIALS AND METHODS
### Cells and Materials

A431 epidermal carcinoma cells were kindly provided by S. Cohen (Vanderbilt, TN) and cultured in Dulbecco's modified Eagle's medium (DMEM) with 10% calf serum (Hyclone, Logan, UT). EGF was from Sigma (St. Louis, MO) and Collaborative Research (Lexington, MA), tetradecanoyl phorbol acetate (TPA) was from L.C. Systems (Woburn, MA), and 1-oleoyl-2-acetyl glycerol (OAG) was from Avanti Polar Lipids (Birmingham, AL). $^{45}CaCl_2$ was from Amersham (Arlington Heights, IL).

### $^{45}$Ca Uptake Assays

Uptake of $^{45}Ca^{2+}$ was performed with confluent monolayers of A431 cells grown in 96-well plates, as described previously [10], using 100 $\mu$M $Ca^{2+}$. At this concentration Ca-activated phosphatidylinositol breakdown is largely eliminated so that secondary responses to activation of protein kinase C are minimized. Cell protein was determined by the method of Lowry et al [11] after extraction from the wells in 0.5 N NaOH, 1% SDS. The number of cells per well was determined by electronic counting (Coulter) after removal of the cells with trypsin-EDTA. The mean value was 26,000 $\pm$ 2000 ($\pm$ 1 SD, n = 6).

## RESULTS

Both EGF and vanadate were found to stimulate the initial rate of uptake of $Ca^{2+}$ into A431 cells by 2–5-fold (Fig. 1 and [10]). No detectable lag period preceded the stimulation by either agent, and the vanadate stimulation appeared to be transient. EGF stimulation is also apparently transient, since the control uptake was linear for >60 min (not shown) and the increase in cytosolic $Ca^{2+}$ observed by Moolenar et al [12] rapidly decreased to basal levels. Stimulation by vanadate and EGF was not additive, indicating that both agents act on the same $Ca^{2+}$ transport system. The apparent $K_m$ and $V_{max}$ for the basal influx of $Ca^{2+}$ by the cells were 83 $\mu$M and 0.26 fmol/cell/min, respectively (data not shown). Assuming that the unidirectional influx, measured using $^{45}Ca^{2+}$ as tracer, represents only a net flux of $Ca^{2+}$, this initial rate could carry a maximum inward current of about 0.8 pA (basal rate of influx).

Pre-incubation of A431 cells with the tumor-promoting phorbol ester TPA or the synthetic diacylglycerol OAG, both of which can activate protein kinase C,

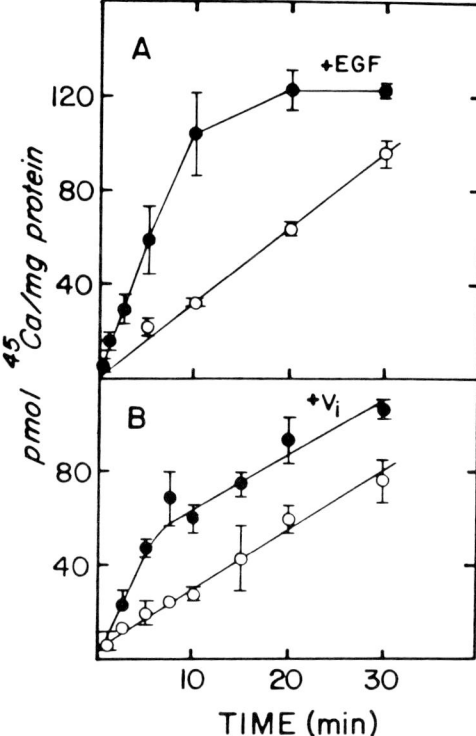

Fig. 1. Uptake of $Ca^{2+}$ by EGF- or vanadate-treated A431 cells. Cells were grown to confluence in 96-well plates in DMEM/10% calf-serum, then starved of serum overnight. The wells were rinsed with a buffered saline, pH 7.4, containing 100 $\mu$M $CaCl_2$ and incubated at 37°C. Uptake was initiated by addition of $^{45}Ca^{2+}$ in saline (○) with EGF or vanadate as required (●). Incubations were terminated by washing through 4 × 1 liter of 0.1 M $MgCl_2$, 10 mM Tris (pH 7.4), 1 mM EGTA (4°C). Wells were drained, extracted, and counted for $^{45}$Ca. Error bars are ± 1 SD (n = 3). Trapped counts (equivalent to 16.4 pmol/mg) were subtracted from the data in each experiment. A: 200 ng/ml EGF. B: 200 $\mu$M vanadate. (Reprinted from [10] with permission of the publisher.)

inhibited the $Ca^{2+}$ response to either EGF or to vanadate (Table I). Phorbol ester alone had no effects on basal $Ca^{2+}$ uptake.

To determine whether vanadate acts through the EGF receptor, the effect was studied of down-regulation of the receptor induced by pre-incubation with EGF (Table I). We found that pre-incubation with EGF blocked the subsequent stimulation of $Ca^{2+}$ influx by EGF but not that by vanadate. Nor did it prevent inhibition of the vanadate-stimulation by TPA. Pre-treatment with vanadate had no effect on the activation of $Ca^{2+}$ uptake by either agent. These results suggest that vanadate acts at a site distinct from or distal to the EGF receptor.

The best-characterized $Ca^{2+}$ transport systems that might increase $Ca^{2+}$ influx into a cell are the voltage-activated $Ca^{2+}$ channel [13] and the $3Na^+/Ca^{2+}$ exchange system [4]. To determine whether either of these systems is activated in A431 cells, we examined the effects of depolarization in high KCl and of altering extracellular or intracellular $Na^+$. Since $3Na^+/Ca^{2+}$ exchange is electrogenic, depolarization might be expected to increase $Ca^{2+}$ influx via this system. The channel is also activated by depolarization. However, as can be seen from Table II and [15], 40 mM KCl

**TABLE I. Activation and Inhibition of $Ca^{2+}$ Influx Into A431 Cells**

| Pre-incubation | Treatment | Initial rate of $^{45}Ca^{2+}$ influx (% of control) |
|---|---|---|
| — | EGF (200 ng/ml) | $279 \pm 21.4$[a] |
| — | Vanadate (200 µM) | $216 \pm 18.1$ |
| — | EGF + vanadate | $318 \pm 8.7$ |
| TPA (100 nM)[b] | EGF | 119 |
| OAG (50 µM)[b] | EGF | 119 |
| TPA | Vanadate | 107 |
| OAG | EGF | 129 |
| — | TPA | $97 \pm 12.6$ |
| EGF (500 ng/ml)[c] | EGF | $109 \pm 6.4$ |
| EGF (500 ng/ml)[c] | Vanadate | $306 \pm 17$ |
| EGF + TPA[c] | EGF | $92 \pm 3.6$ |
| Vanadate (200 µM)[c] | EGF | $278 \pm 7.5$ |
| Vanadate (200 µM)[c] | Vanadate | $160 \pm 19.0$ |

[a]Values shown are means of duplicates or of triplicates $\pm$ 1 SD. (Data taken from [10] with permission of the publisher.)
[b]Pre-incubation with TPA or OAG was for 30 min at 37°C.
[c]Pre-incubation with EGF or vanadate was for 90 min at 37°C. Monolayers were then rapidly washed twice and incubated for a further 5 min with fresh medium before initiation of influx by addition of $^{45}Ca$, with or without the agents indicated. Influx was measured as defined in legend to Figure 1.

**TABLE II. Effects of $Na^+$ and Depolarization on Vanadate-Stimulated $Ca^{2+}$ Influx in A431 Cells**

| Pre-incubation | Treatment | Initial rate of $^{45}Ca^{2+}$ influx (% of control) |
|---|---|---|
| — | Vanadate (200 µM) | $225 \pm 23.4$ |
| 40 mM KCl | Vanadate (200 µM) | $72 \pm 5.1$ |
| 50 mM choline Cl | Vanadate (200 µM) | $294 \pm 29.3$ |
| 5 µM valinomycin | Vanadate (200 µM) | $77 \pm 29.7$ |
| 100 µM ouabain[a] | Vanadate (200 µM) | $58 \pm 5.5$ |

[a]Cells with pretreated with ouabain for 30 min at 37°C prior to initiation of $^{45}Ca^{2+}$ influx. All other pre-incubations were for 3–5 min only. Influx was measured as described in legend to Figure 1. (Data taken from [15] with permission of the publisher.)

completely blocked the stimulatory effect of vanadate, even reducing uptake to below the basal rate. Moolenar et al [12] have demonstrated that the plasma membrane potential in A431 cells is dominated by the $K^+$ permeability, and we calculate that 40 mM $K^+$ would cause a fall in the potential from about $-68$ mV to $-33$ mV. The $Ca^{2+}$ influx was not dependent on extracellular $Na^+$, since partial replacement of the NaCl by choline chloride had no effect (Table II and [15]). Moreover, the influx was not being inhibited simply by competition with high extracellular $K^+$, since depolarization by pre-incubation with ouabain in low KCl medium also inhibited uptake. Again, the inhibition decreased uptake to well below the basal rate of influx. Since treatment with ouabain would also increase intracellular $Na^+$, which would increase $Ca^{2+}$ influx via a $3Na^+/Ca^{2+}$ exchange, it appears very unlikely that such an exchange system mediates a significant $Ca^{2+}$ flux in A431 cells. Valinomycin also inhibited the vanadate-stimulate uptake of $Ca^{2+}$ (Table II and [15]), an effect similar to that reported in leukemic T cells for the T cell receptor-associated $Ca^{2+}$ influx [16]. The mechanism of inhibition is not clear. It is unlikely to be due to hyperpolarization as suggested for T cells [16], however, since the potential in A431 cells is already controlled by $K^+$ [12]. Valinomycin might conceivably be causing a rapid depolarization, or inhibit the $Ca^{2+}$ transport system directly.

To characterize the effects of depolarization in more detail, vanadate-stimulate $Ca^{2+}$ uptake was measured over a range of external KCl concentrations, and expressed as a function of the membrane potential, calculated from the Goldman equation [12]. The results (Fig. 2 and [15]) indicate that the inhibitory effect of depolarization is half-maximal at about $-55$ mV (interior negative).

## DISCUSSION

We have found that both EGF and vanadate can stimulate $Ca^{2+}$ influx into A431 cells via the same transport system, and that this stimulation is blocked by the activation of protein kinase C. EGF and vanadate appear to operate through distinct mechanisms, since down-regulation of the EGF receptor is without effect on the stimulation by vanadate. This result also demonstrates that the EGF receptor is not itself acting as the $Ca^{2+}$ transport system.

The properties of the transport system are distinct from those expected of a $3Na^+/Ca^{2+}$ exchange system or a voltage-activated $Ca^{2+}$ channel. The mode of $Ca^{2+}$ influx remains obscure. It is, however, reminiscent of the IgE-activated $Ca^{2+}$ channel found in mast cells [16] and of the $Ca^{2+}$ transport system associated with the T3-T cell receptor in leukemic T cells [17]. Definitive evidence that the system

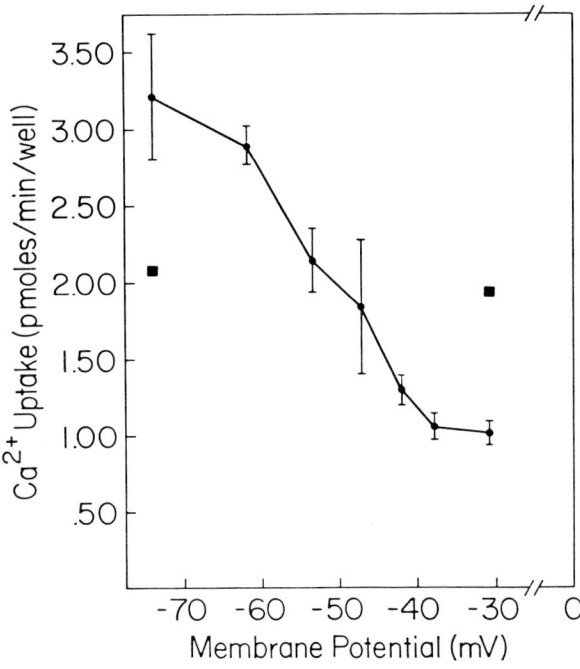

Fig. 2. Effect of membrane potential on vanadate-stimulated $Ca^{2+}$ influx (●) in A431 cells. The plasma membrane potential was altered by partial replacement of NaCl with KCl in the buffered saline used for the uptake assays. Absolute potentials were calculated from the Goldman equation given by Moolenar et al [12]. Initial rates of $Ca^{2+}$ influx were measured using $^{45}Ca^{2+}$ as tracer, as described in Figure 1. Error bars are ± 1 SD (n = 3). Basal influx was measured at two points (■). (Reprinted from [15] with permission of the publisher.)

represents a novel form of voltage-sensitive $Ca^{2+}$ channel might prove difficult to obtain, since the calculated conductance per cell will be less than 4 pA even under fully activated conditions. The approach used to demonstrate that the IgE-stimulated mast cells $Ca^{2+}$ transport system behaves as a channel required its isolation and insertion into planar bilayers, and depended upon the specific binding of the inhibitor cromolyn [18]. Cromolyn has no effect on the A431 cell $Ca^{2+}$ influx (unpublished observation) and no specific inhibitors have yet been identified.

The mechanism by which vanadate activates the $Ca^{2+}$ flux is unclear, particularly since no lag is detected before the onset of stimulation, and EGF receptors do not appear to be required. It is possible that the agent is acting on a distinct receptor system, and that entry into the cell is not required. Since vanadate has been widely used as an inhibitor of phosphotyrosol phosphatases to study the action of oncogenes and growth factors [eg, 7,8,19] it will be of some importance to determine with what other systems it can interact.

## ACKNOWLEDGMENTS

These studies were supported by PHS grant number CA-38888 awarded by the National Cancer Institute, DHHS.

## REFERENCES

1. Hesketh TR, Moore JP, Morris JDH, Taylor MV, Roger J, Smith GA, Metcalfe JC: Nature (Lond) 313:481, 1985.
2. Sawyer ST, Cohen S: Biochem 20:6280, 1981.
3. Berridge MJ, Irvine RF: Nature (Lond) 312:316, 1984.
4. Smith KB, Losonczy I, Sahai A, Pannerselram M, Fehrel P, Solomon DS: J Cell Physiol 17:91, 1983.
5. Hunter T, Ling N, Cooper JA: Nature (Lond) 311:480, 1984.
6. McGaffrey PG, Friedman BA, Rosner MR: J Biol Chem 259:12502, 1984.
7. Swarup G, Speeg KV, Cohen S, Garbers DL: J Biol Chem 257:7298, 1982.
8. Swarup G, Cohen S, Garbers DL: Biochem Biophys Res Commun 107:1104, 1982.
9. Cassell D, Zhuang Y-X, Glaser L: Biochem Biophys Res Commun 118:675, 1984.
10. Macara IG: J Biol Chem 261:9321, 1986.
11. Lowry OH, Rosenbrough NJ, Farr AL, Randall RS: J Biol Chem 193:265, 1951.
12. Moolenar WH, Aerts RJ, Tertoolen LGJ, deLaat SW: J Biol Chem 261:279, 1986.
13. Tsien RWA: Physiol Rev 45:341, 1983.
14. Mullins LJ: Am J Physiol 5:C103, 1979.
15. Macara IG, Gray G: J Cell Biochem 34:125, 1987.
16. Kanner BI, Metzger H: J Biol Chem 259:10188, 1984.
17. Oettgen HC, Terharst C, Cantley LC, Rosoff PM: Cell 40:583, 1985.
18. Mazurek N, Schindler H, Schurholz TH, Pecht I: Proc Natl Acad Sci USA 81:6841, 1984.
19. Collett MS, Belzer SK, Kamp LE: J Cell Biochem 26:95, 1984.

# Oxidant Tumor Promoters

### Peter A. Cerutti

*Department of Carcinogenesis, Swiss Institute for Experimental Cancer Research, 1066 Epalinges s/ Lausanne, Switzerland*

Tumor promotion is the sum total of the action of xenobiotics or endogenous factors on the *entire* target tissue—ie, initiated and noninitiated epithelial cells, surrounding fibroblasts, inflammatory leukocytes, etc. In many cases promoters interact with the initiated cell itself; in others, they mostly disturb short-and long-range cellular interactions. Long-range interactions are mediated by paracrine and clastogenic signals. Multiple cellular and molecular mechanisms contribute to the end result, and their relative importance depends on the promoter and the tissue; some mechanisms may be obligatory; others may merely facilitate the process. Major mechanisms are [1] modulation of the expression of growth- and differentiation-related genes; this may result from the direct interaction of the promoter with the target cell or from the action of paracrine or clastogenic mediators which originate from nontarget cells [2] selective toxicity to the noninitiated surrounding tissue by the promoter itself or by clastogenic mediators from nontarget cells, in particular inflammatory leukocytes.

We can attempt to distinguish classes of promoters for which a particular mechanism predominates. For polypeptide hormones and phorbolester-type promoters modulation of the expression of growth- and differentiation-related genes may be in the foreground, while selective toxicity may play a crucial role for bona-fide oxidant promoters, promotional inflammatory irritants, and mineral fibers. However, in most instances multiple overlapping mechanisms are expected to be at play. For example, phorbol-myristate-acetate (PMA), which in part uses pathways reminiscent of polypeptide hormones [1], also induces a cellular pro-oxidant state with its consequences [2], stimulates the formation of several chemotactic and growth-modulating arachidonic acid metabolites [3], and causes the infiltration into mouse skin of inflammatory leukocytes [4]. Upon stimulation these phagocytic cells release active oxygen and a multitude of clastogenic and paracrine mediator molecules. Bonafide oxidant promoters, such as active oxygen and organic peroxides, while undoubtedly toxic, also may affect signal transduction and modulate the expression of a family of "pro-oxidant genes" in target cells [2,5]. Oxidant promotion is expected to play a

Received April 17, 1986.

© 1988 Alan R. Liss, Inc.

role in skin carcinogenesis because near-ultraviolet light in solar radiation is known to produce active oxygen by photosensitization reactions. Similarly, traces of active oxygen and ozone in the air may exert a promotional effect in lung carcinogenesis. Irritants such as mineral fibers and particulates which cause a strong inflammatory response may resemble oxidant promoters because they stimulate macrophages to release active oxygen and arachidonic acid hydroperoxides.

In this article I will concentrate on the mechanism of action of bona fide oxidant promoters, the role of inflammation in carcinogenesis, and the consequences of the establishment of a cellular pro-oxidant state by phorbolester-type compounds.

## MECHANISMS OF ACTION OF OXIDANT PROMOTERS

There is a long list of oxidants with tumor-promoting properties. On mouse skin or cultured mouse epidermal cells the following agents are active: near UV-light, $H_2O_2$, peroxyacetic acid, chlorobenzoic acid, benzoylperoxide, decanoylperoxide, cumene-hydroperoxide, p-nitro-perbenzoic acid, $IO_4^-$ [7,8]. We found that superoxide ($O_2^-$) plus $H_2O_2$ when produced extracellularly was a weak complete carcinogen but a potent promoter for mouse embryo fibroblasts C3H 10T1/2 [9]. There is evidence from whole-animal experiments that hyperbaric oxygen possesses promotional or cocarcinogenic activity in the lung [10]. The mechanism of action of oxidants is unknown but is expected to differ substantially from that of phorbolester-type promoters and growth factors. The oxidation of cellular constituents may play a role. This notion is supported by the finding of Colburn, Gindhart and colleagues [8,11]. Benzoylperoxide treatment of promotable mouse epidermal cells JB6(Cl 41) resulted in the rapid oxidative destruction of the membrane ganglioside $GT_1$. The loss of cell-cell communication following treatment with oxidant promoters [12] could be a consequence of the oxidation of constituents on the cell surface.

In addition to these direct effects, oxidant promoters set in motion a multitude of complex secondary reactions. Among them, the stimulation of the metabolism of arachidonic acid (AA) is of particular interest. It is known that the oxidation of membrane lipids activates phospholipase $A_2$ (and C?) [13] and consequently causes the release of free AA. This mechanism may play a particularly important role in lung carcinogenesis. While molecular oxygen in the atmosphere is unreactive, it presumably contains traces of activated forms [14]. We have studied the stimulation of arachidonic acid metabolism in lung epithelioid cells A549 by $O_2^-$ which was generated extracellularly by xanthine/xanthine-oxidase. The HPLC profile of ethyl-acetate extracts of the culture medium is shown in Figure 1. While the release of free AA increased only moderately, we observed a rapid and strong stimulation of the synthesis of prostaglandins $F_2\alpha$ and $E_2$ and of a mixture of metabolites which were not well resolved in our HPLC system and probably contains prostaglandins $A_2$ and $B_1$. No increase in lipoxygenase products was observed. Similarly, superoxide, $H_2O_2$, and tert-butylhydroperoxide stimulated prostaglandin synthesis in pig aorta endothelial cells [15] and lung fibroblasts [16], respectively. $H_2O_2$ formed by dismutation from $O_2^-$ represents an active species also for xanthine-xanthine oxidase. Active oxygen under certain conditions inactivates rather than activates cyclooxygenase, and it appears that this system is intricately tuned [17]. Prostaglandins and the products of the lipoxygenase pathway are formed from metastable hydroperoxy-precursors hydroperoxy-AA (HPETEs), $PGG_2$, and 15- hydroperoxy-$PGE_2$) and it has been suggested

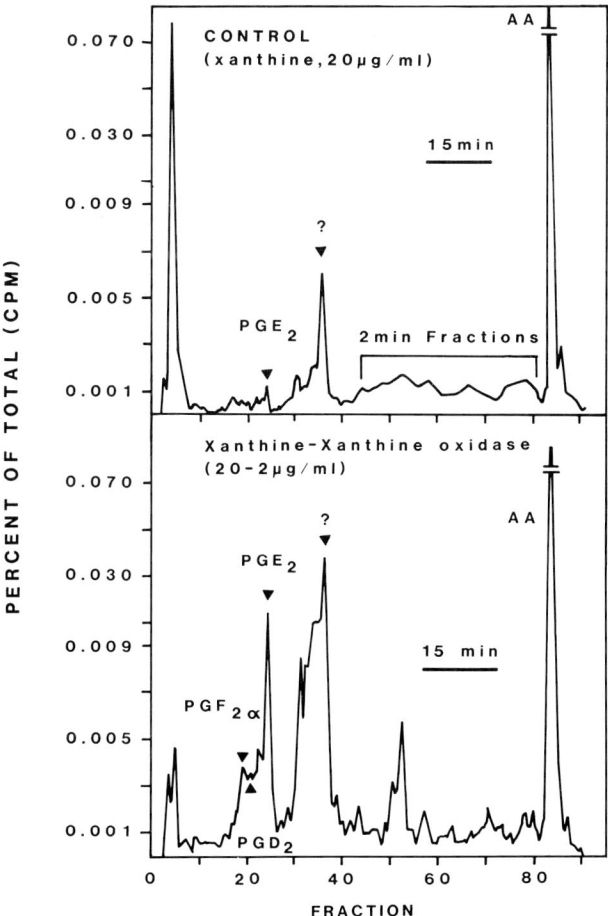

Fig. 1. HPLC analysis of ethylacetate extracts of the culture medium from $^3$H-arachidonic-acid-labeled human epithelioid lung cells A549. **Upper half**: Control (incubation with xanthine, no superoxide production). **Lower half**: Treatment with superoxide generated by xanthine-xanthine oxidase for 15 min.

that these intermediates, when present at low concentrations, may participate in a feedback loop which results in further amplification of AA-metabolism [18]. The activation of AA-metabolism is common to oxidant-and phorbolester promoters. For the latter the stimulation of the formation of $PGE_2$ has been adjudged to be necessary for the induction of hyperproliferation and of $PGF_2\alpha$ for its maintenance in mouse skin [19]. Furthermore, the overproduction of 8-HPETE appears to play a role in mouse skin promotion by PMA. On a molecular level the mechanism of action of these AA metabolites is largely unknown. As discussed below, our observation that HPETE's mobilize $Ca^{2+}$ from mitochondria and induce DNA breakage gives a first insight [20].

In general, oxidants with promotional activity are clastogenic. They induce DNA single-strand breaks with high efficiency, while they are usually only weak inducers of chromosomal aberrations. $H_2O_2$ causes DNA damage in part by Fenton-type reactions in cultured fibroblasts. This is documented by the inhibitory action of the iron chelators desferrioxamine and o-phenanthroline [21,22]. Extracellularly generated $O_2^-$ induces chromosomal aberrations and sister chromatid exchanges in human lymphocytes which can be suppressed by CuZn superoxide dismutase (SOD) [23]. In contrast, DNA strand breakage by $O_2^-$ in white blood cells is not affected by SOD [24]. The mouse skin promoter and progressor benzoylperoxide causes DNA breakage in mouse epidermal cells by an unknown mechanism [25].

The role of DNA damage induced by oxidant promoters in carcinogenesis remains unclear, although it has been demonstrated that DNA breaks per se, introduced by liposome-encapsulated DNAase I, sufficed to transform cultured Syrian hamster embryo cells [26]. DNA breaks may modulate the expression of neighboring genes or initiate genomic rearrangements. It has been suggested that they might represent a signal in cell differentiation [27]. This appears unlikely for cultured mouse epidermal cells. In these cells, PMA induces breaks only after several hours and preferentially in the differentiating rather than promoted subpopulation. In contrast, benzoylperoxide and x-rays rapidly induce DNA breakage in all cells but not differentiation [25]. The quality of the breaks and their distribution may determine their biological effects. Oxidative DNA base damage rather than strand breakage might play a role in the action of oxidant promoters [28]. The popularity of strand-break studies is explained by the ease with which they can be determined.

DNA containing breaks strongly stimulates ADP-ribosyl-transferase and, therefore, the poly ADP-ribosylation of chromosomal proteins. This suggests a genetic-epigenetic mechanism by which oxidant promoters might affect gene expression (see below).

## THE ROLE OF INFLAMMATION IN PROMOTION

As discussed in the introductory section, tumor promotion has to be understood as the composite reaction of a tissue rather than of specific target cells. Chronic inflammatory diseases with increased cancer incidence and lung carcinogenesis lend support to the notion that phagocytic leukocytes can participate in tumorigenesis. Particularly impressive are the inflammatory reactions in the bronchial tissue which follow the exposure to particulates, irritants in tobacco smoke and mineral fibers [6]. Mouse skin promotion by PMA is another example where the immigration into the dermis of neutrophils and later monocytes/macrophages represents an early obligatory reaction [4]. Macrophages stay in the tissue over prolonged periods, although their numbers diminish with repetitive PMA treatment. Supporting a probable role of inflammation in mouse skin carcinogenesis is the observation that the skin of an athymic nude mouse fails to demonstrate an inflammatory or hyperplastic response to PMA and, at the same time, is resistant to the PMA promotional protocol unless grafted with a normal thymus [29].

Upon stimulation phagocytic cells release large amounts of active oxygen, metastable clastogenic components and a host of biologically active molecules such as plasminogen activator, elastase, complement components, monokines, tumor necrosis factor, etc [30]. It appears likely that at least some of these inflammatory mediator molecules play a role in promotion. Troll and Goldstein and colleagues [31] were the first to speculate that active oxygen from stimulated polymorphonuclear leukocytes might play a role in mouse skin promotion by PMA, and this model has since gained in popularity. It is worth noting that the magnitude of the respiratory burst which is induced by various phorbolester-type promoters is proportional to the tumor-promoting activity [31] and that inhibitors of the oxidative burst are antipromoters [32,33]. We have expanded this hypothesis and suggested that not only short-lived active oxygen but also diffusible, metastable clastogenic factor (CF) and non-clastogenic paracrine mediators participate in the promotional process [5].

Support for a participation of inflammatory cells in malignant transformation derives from in vitro coculture experiments. Unstimulated and PMA-stimulated human neutrophils transformed C3H10T1/2 mouse embryo fibroblasts [34]. Transforming agents in this system presumably are active oxygen and metastable, diffusible CF. We had demonstrated earlier that $O_2^-$, produced extracellularly by xanthine-xanthine-oxidase, was a potent promoter and a weak complete carcinogen in this system [9]. The mechanism by which active oxygen, CF, and possibly other components which are released by phagocytic leukocytes exert their carcinogenic effect is unknown. The introduction of DNA and chromosomal damage which can be demonstrated in cocultured lymphocytes [35,36], erythroleukemia cells [37], fibroblasts [28], hamster ovary cells [39], and mouse epidermal cells [39] could play a role. Our finding that PMA-treated human monocytes release a lipophilic diffusible CF [36,40] offered the interesting possibility that long-range genetic damage may result from the formation of secondary oxidative products rather than directly from active oxygen itself. Under conditions of CF formation human monocytes were found to release $O_2^-$, $H_2O_2$, thromboxane $B_2$, $PGF_2\alpha$, $PGE_2$, 12-hydroxy-5,8,10-hepta-decatrienic acid (HHT), hydroxy- and hydroperoxy-derivatives of AA, free AA, and small amounts of mono- and diacylglycerol and phospholipids [41]. It is believed that in the presence of $O_2^-$ and $H_2O_2$ this mixture of lipids undergoes free-radical chain reactions, initiates autooxidation in the target cell, and ultimately induces chromosomal damage. The fact that both SOD and catalase independently inhibit CF formation [36,42] further suggests that superoxide-reduced iron, reacting with $H_2O_2$ in a superoxide-driven Fenton reaction to form HO·, plays a critical role in the process. Our present understanding of the clastogenic action of $O_2^-$ and its inhibition is visualized in Figure 2.

The hydroxy-AA metabolites released by PMA-treated monocytes derive from their corresponding hydroperoxy-precursors, ie, 5-, 11-, and 15-HETE from 5-, 11-, and 15-HPETE; thromboxane $B_2$ and $PGE_2$ from the corresponding peroxide intermediate $PGG_2$. In vivo these hydroperoxy-metabolites are thought to be sufficiently stable to exert their effects on nearby cells and perhaps to mediate DNA damage. Using alkaline elution analysis, we in fact recently demonstrated that several hydroperoxy-derivatives of AA and AA itself cause DNA strand breaks in mouse fibroblasts [43]. The relative order of potency was determined to be hydroperoxy-AA, hydroxy-AA, AA. For HPETES the formation of damage depended on the presence of calcium and Fe. While the Fe-dependent mechanism most likely implicates the formation of OH-radicals in Fenton-type reactions, $Ca^{2+}$-dependent DNA breakage might involve the activation of a $Ca^{2+}$-dependent endonuclease. Evidence for a disturbance of $Ca^{2+}$ homeostasis by HPETE's comes from our studies with isolated rat mitochondria. We found that HPETE and, less efficiently, AA stimulated $Ca^{2+}$ release in a reaction which was accompanied by intramitochondrial pyridine nucleotide oxidation and hydrolysis. Measurements of the mitochondrial membrane potential indicate that $Ca^{2+}$ release is not due to uncoupling of mitochondria [20]. It is suggested that HPETE's and AA in CF act as mitochondrial $CA^{2+}$ mobilizers in signal transduction

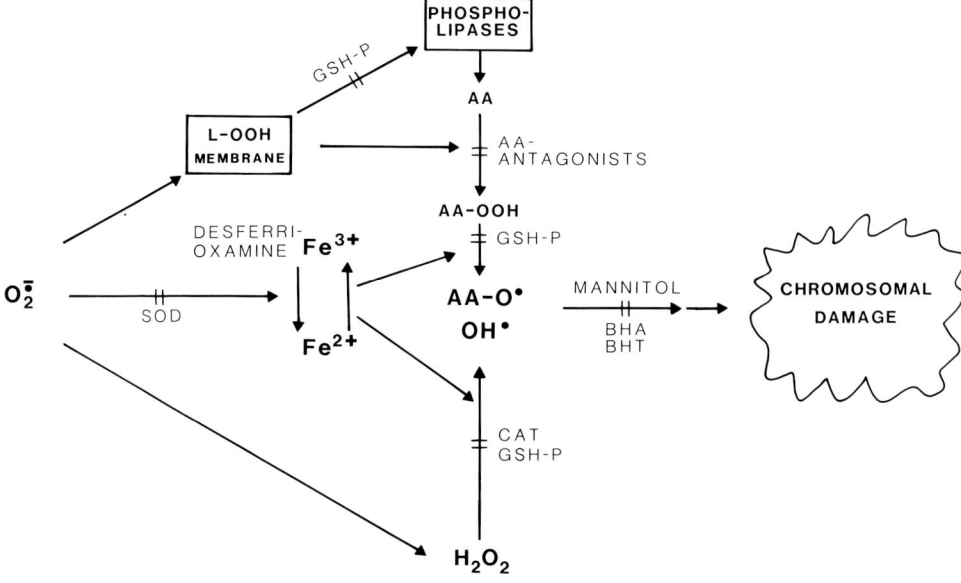

Fig. 2. Clastogenic action of superoxide.

related to proliferation and tumor promotion. This proposition is supported by the finding that free AA also triggered $Ca^{2+}$ release into the perfusate in isolated rat liver (H. Sies and P. Graf, personal communication).

As mentioned in the previous section, hydroperoxides are required to initiate and maintain the catalytic activity of fatty acid cyclooxygenase of the AA cascade. The involvement of inflammatory cells in creating such a "peroxide tone" [18] in target cells may be part of the carcinogenic effect of inflammation. It may be relevant to note that stable PG-endoperoxide analogs (U-46619 and U44069) possessed promotional activity on mouse skin [44] and that HPETE's (and HETE's) induced ornithine decarboxylase in rat colonic mucosa [45]. Of course, promoter-stimulated phagocytes also produce a host of more stable, active molecules which are likely to affect the growth and differentiation of surrounding cells. In summary, inflammation is bound to influence cell- and tissue- kinetics, but the degree of complexity is such that no general prediction can be made when the overall effect is promotion or possibly inhibition of tumorigenesis. The outcome will depend on the tissue, the promoter, the schedule of promoter application, etc.

## DIRECT PRO-OXIDANT ACTION OF PMA ON TARGET CELLS

Although PMA may facilitate tumor growth indirectly through its effects on leukocytes, it also has direct pro-oxidant effects on the target cells themselves. This is documented by the observation that PMA induces chemiluminescence (CL) in mouse epidermal cells [46]. CL reflects the formation of singlet oxygen. Further support for the involvement of a PMA-induced cellular pro-oxidant state comes from the antipromotional activity of a variety of antioxidants. The mechanism of the formation of a pro-oxidant state in nonphagocytic cells by PMA is unknown. It could be the consequence of PMA stimulation of the AA cascade, of the overproduction of active oxygen from mitochondria and the endoplasmic reticulum or of a weakening of the antioxidant defense owing to a transitory decrease in the cellular levels of SOD and catalase (CAT) [2,47]. The establishment of a cellular pro-oxidant state may have important consequences on the expression of a family of "pro-oxidant genes" [2]. We refer to recent reviews of this topic for further discussion [ 2,32,33]. We have already stressed that promotion has to be considered as a tissue phenomenon rather than the reaction of a single type of target cells. Therefore, we suggest that both direct interaction of promoters with target cells as well as indirect paracrine and clastogenic effects of surrounding cells contribute to the end result.

## POLY ADP-RIBOSYLATION OF CHROMOSOMAL PROTEINS BY OXIDANT PROMOTERS

The post-translational modification of proteins represents an epigenetic mechanism for the modulation of gene expression. PMA has already been shown to cause the phosphorylation of many proteins, including the receptors for EGF [48] and insulin [49], histones H2B and H4 and two 28-kilodalton "stress" proteins [51]. Another type of post-translational protein modification — namely, poly ADP-ribosylation (polyADPR)—is of particular interest because it specifically involves chromosomal proteins. It could provide the cell with a mechanism for modulating gene expression by oxidant promoters because poly ADPR is related to the cellular redox state and because it is stimulated by DNA containing strand breaks. It is intriguing to speculate that active oxygen and diffusible clastogenic components released by inflammatory cells might modulate gene expression in neighboring target cells via poly ADP-ribosylation.

We found that PMA induces poly ADP-ribosylation in mouse embryo fibroblasts C3H 10T1/2 as well as human monocytes [52,53]. Acceptor proteins in 10T1/2 cells included the core histones H2B, A24, H3d [54] and numerous medium and high molecular weight nonhistone proteins [55]. The major acceptor was ADPR-transferase and its proteolytic fragments of 20–25, 45, and 72–95 kilodaltons. The partial suppression of poly ADPR by the antioxidants SOD, catalase, glutathione-peroxidase, and butylated hydroxytoluene underscores an oxidative mechanism. The antipromotional protease inhibitor antipain also inhibited poly ADP-ribosylation of ADPR-transferase [55] . It should be noted that poly ADP-ribosylation by PMA in 10T1/2 cells was not accompanied by any detectable DNA strand breakage.

Recently we have concentrated on poly ADP-ribosylation of chromosomal proteins in mouse epidermal cells JB6 (Cl 41) by bona fide oxidant promoters. A single burst of extracellular $O_2^-$ (produced by xanthine-xanthine oxidase) stimulated poly ADPR in a dose-dependent manner. Superoxide was a moderate promoter of JB6 (Cl 41) cells [56]. The identification of acceptor proteins and mechanistic studies are presently being undertaken. Weak induction of poly ADP-ribosylation by benzoylperoxide has been observed by H. Hilz and colleagues in mouse 3T3 cells [57]. Exposure to ozone for several days increased poly ADP-ribose levels in mouse lung tissue [58].

## CONCLUDING REMARKS

While we have focussed on promotion by oxidants in mouse skin, recent observations suggest that oxidants play a role at several steps in tumorigenesis. Evidence for a role of oxidants in progression — a late step in carcinogenesis — has been obtained by Slaga and colleagues. They found that benzoylperoxide, which possesses promotional activity, also enhances the progression of benign mouse skin papillomas to carcinomas. A common denominator of these multiple carcinogenic activities of oxidants might be their capability to modulate gene expression in target cells. This could be due to gene rearrangements as a consequence of the clastogenic activity of oxidants or to epigenetic mechanisms such as the post-translational poly ADP-ribolysation of chromosomal proteins discussed in this article.

## ACKNOWLEDGMENTS

The original work reported in this article was supported by grants from the Swiss National Science Foundation and the Swiss Association of Cigarette Manufacturers.

## REFERENCES

1. Nishizuka Y: Nature 312:315, 1984.
2. Cerutti P: Science 227:375, 1985.
3. Levine L: Adv Cancer Res 35:49, 1981.
4. Slaga T, Fischer S, Viaje A, Berry D, Bracken W, Le Clerc S, Miller D: In Slaga T, Sivak A, Boutwell R (eds): "Mechanisms of Tumor Promotion and Cocarcinogenesis." Carcinogenesis 2. N.Y.: Raven Press, 1978, p 173.
5. Cerutti P: In Harris CC (ed): "Biochemical and Molecular Epidemiology of Human Cancer." N.Y.: Alan R. Liss, Inc., 1986, p167.
6. Mass M, Kaufman D, Siegfried J, Steele V, Nesnow S (eds): Carcinogenesis: A Comprehensive Survey, Vol. 8. N.Y.: Raven Press, 1985.
7. Slaga T, Solanki V, Logani M: In Nygaard O, Sinic M(eds): "Radioprotectors and Anticarcinogens." N.Y.: Academic Press, 1983.
8. Gindhart T, Nakamura Y, Stevens L, Hegameyer G, West M, Smith B, Colburn N: In Mass M, Kaufman D, Siegfried J, Steele V, Neshow S (eds): In "Cancer of the Respiratory Tract." Carcinogenesis— A Comprehensive Survey, Vol. 8. N.Y.: Raven Press, 1985, pp 341–367.
9. Zimmerman R, Cerutti P: Proc Natl Acad Sci USA 81:2085, 1984.
10. Dettwer C, Kramer S, Gottlieb S, Aponte G: J Natl Cancer Inst 41:751, 1968.
11. Srinivas L, Colburn N: Cancer Res 44:1510, 1984.
12. Slaga T, Klein-Szanto A, Triplett L, Jotti L, Trosko J: Science 213:1023, 1981.
13. Yasuda M, Fujita T: Jpn J Pharmacol 27:429, 1977.
14. Fridovich I: Science 221:875, 1978.
15. Ager A, Gordon J: J Exp Med 159:592, 1984.
16. Taylor L, Menconi M, Polgar P: J Biol Chem 258:6855, 1983.
17. Hemler M, Cook H, Lands W: Arch Biochem Biophys 193:340, 1979.
18. Lands W: Prostaglandins Leukotrienes Med 13:35, 1984.
19. Fürstenberger G, Gross M, Marks F: In Thaler-Dao H, de Paulet A, Paoletti R(eds): "Icosanoids and Cancer." N.Y.: Raven Press, 1984, p 101.
20. Richter C, Frei B, Cerutti P: Biochem Biophys Res Commun 143:609, 1987.
21. De Mello Filho A, Meneghini R: Biochem J 218:273, 1984.
22. De Mello Filho A, Meneghini R: Biochim Biophys Acta 847:82, 1985.
23. Emerit I, Keck M, Levy A, Feingold J, Michelson A: Mutat Res 103:165, 1982.
24. Birnboim H, Kanabus-Kaminska M: Proc Natl Acad Sci USA 82:6820, 1985.
25. Hartley J, Gibson N, Zwelling L, Yuspa S: Cancer Res 45:4864, 1985.

26. Zajac-Kaye M, Ts'o P: Cell 39:427, 1984.
27. Williams G, Johnstone A: Biosci Rep 3:815, 1983.
28. Lewis J, Adams D: Cancer Res 45:1270, 1985.
29. Holland J. Perkinds E, Gipson L: Proc Am Assoc Cancer Res 18:10, 1977.
30. E g, see in Volkman A (ed): "Mononuclear Phagocyte Biology." N.Y.: Marcel Dekker, Inc., 1984.
31. Goldstein B, Witz G, Amoruso M, Stone D, Troll W: Cancer Lett 11:257, 1981.
32. See in Kozumbo W, Cerutti P: In Holaender A, Hartman P, Brusick D, Shankel D (eds): "Mechanisms of Antimutagenesis and Anticarcinogenesis." N.Y.: Plenun Press, p491, 1986.
33. Kensler T, Trush M: Environ Mutagen 6:593, 1984.
34. Weitzman S, Weitberg A, Clark E, Stossel T: Science 227:1231, 1985.
35. Emerit I, Cerutti P: Nature 293:144, 1981.
36. Emerit I, Cerutti P: Carcinogenesis 4:1313, 1983.
37. Birnboim H: In Nygaard O, Simic M(eds): "Radioprotectors and Anticarcinogens." N.Y.: Academic Press, 1983, p539.
38. Weitberg A, Weitzman A, Destrempes M, Latt S, Stossel T: N. Engl J Med 308:26, 1983.
39. Dutton D, Bowden G: Carcinogenesis 6:1279, 1985.
40. Emerit I, Cerutti P: Proc Natl Acad Sci USA 79:7509, 1982.
41. Kozumbo W, Mühlematter D, Jorg A, Emerit I, Cerutti P: Carcinogenesis 8:5218, 1987.
42. Emerit I, Cerutti P: (unpublished).
43. Ochi T, Cerutti, P: Proc Natl Acad Sci USA 84:990, 1987.
44. Lupulescu A: Experientia 40:289, 1984.
45. Bull A, Nigro N, Golembieski W, Crissman J, Marnett L: Cancer Res 44:4924, 1984.
46. Fischer S, Adams L: Cancer Res 45:3130, 1985.
47. Solanki V, Rana R, Slaga T: Carcinogenesis 2:1141, 1981.
48. Iwashita S, Fox C: J Biol Chem 259:2559, 1984.
49. Takayama S, White M, Lauris V, Kuhn C: Proc Natl Acad Sci USA 81:7797, 1984.
50. Patskan G, Baxter C: Cancer Res 45:667, 1985.
51. Welch W: J Biol Chem 260:3058, 1985.
52. Singh N, Poirier G, Cerutti P: EMBO J 4:1491, 1985.
53. Singh N, Poirier G, Cerutti P: Biochem Biophys Res Commun 126:1208, 1985.
54. Singh N, Cerutti P: Biochem Biophys Res Commun 132:811, 1985.
55. Singh N, Poirier G, Cerutti P: Carcinogenesis 6:1489, 1985.
56. Mühlematter D, Cerutti P: (unpublished).
57. Hilz H, reported at 9th Int Symp on "ADP-Ribosylation Reactions,"Vitznau, 1984.
58. Hussein M, Mustafa M, Ghani Q, Bhataagar R: Arch Biochem Biophys 241:477, 1985.

# Signals and Sequences Involved in the Ultraviolet- and 12-O-tetradecanoylphorbol-13-acetate (TPA)-Dependent Induction of Genes

Peter Herrlich, Carsten Jonat, Hans Jobst Rahmsdorf, Peter Angel, Alois Haslinger, Masayoshi Imagawa, and Michael Karin

*Kernforschungszentrum Karlsruhe, Institute of Genetics and Toxicology, and Institute of Genetics, University of Karlsruhe, D-7500 Karlsruhe 1, Federal Republic of Germany (P.H., C.J., H.J.R., P.A.); Division of Pharmacology, School of Medicine, M-036, University of California, San Diego, La Jolla, California 92093 (A.H., M.I., M.K.)*

> Various conditions which elicit cellular stress cause the increased or reduced expression of genes. Several of these genes have been identified by gene cloning and sequencing. These genes permit the analysis of their regulation and of the signals sensed by their control elements. The cellular fos oncogene, the metallothionein IIA gene, and the gene for fibroblast collagenase serve to characterize the transcriptional responses to ultraviolet irradiation and to treatment of cells with the tumor promoter 12-O-tetradecanoylphorbol-13-acetate (TPA) and to identify the components of a novel type of positive gene regulation.

**Key words:** primary human fibroblasts, DNA damage, c-fos mRNA instability, protein kinase C, TPA-responsive sequences, metallothionein, collagenase

That genes respond to agents that interfere with cellular proliferation (eg, carcinogens), is not only a well-studied observation of bacterial geneticists [1] but has been observed in many eukaryotic cells [2–7]. Even part of the reaction of a multicellular organism to injury involves activation of specific genes [8].

Our research concerns the immediate changes in gene expression which follow treatment of cultured cells with carcinogenic or cocarcinogenic agents. The search for such changes was motivated by the apparent need for carcinogen-triggered endogenous reactions that would occur in the majority of a cellular population and would be responsible for the relatively high frequency and the tissue specificity of early molecular steps in carcinogenesis [see discussion on p. 500 of ref 9]. Using primary human skin fibroblasts, we found that a carcinogenic treatment with ultraviolet irradiation and with the tumor promoter 12-O-tetradecanoylphorbol-13-acetate (TPA)

Received May 28, 1986.

© 1988 Alan R. Liss, Inc.

induced the appearance of an overlapping set of gene products [10]. Many of the gene products are also induced by other toxic agents such as mitomycin C, hydroxyurea, or gamma irradiation [9]. Thus on the cellular level, a number of genes seems to respond to adverse agents in a coordinate uniform manner. This does not exclude that agent-specific gene activations may exist.

Much of the information on ultraviolet (UV) and TPA-inducible genes in primary human fibroblasts stems from cDNA cloning [11]. We will use three identified genes here to discuss the mechanism of their regulation and physiologic significance. It is likely that the principles derived will apply for other TPA- and UV-inducible genes. These human genes—the cellular fos oncogene, metallothionein IIA (hMT-IIA), and the gene encoding collagenase—all respond to more than one inducing factor, that is, other factors in addition to TPA and UV. c-fos has been shown to respond to a number of growth factors in many types of cells ([12–16] and review in [17]). hMT-IIA is ubiquitously expressed but is further induced upon either glucocorticoid hormone, heavy metal, interleukin-1, or interferon treatment ([6, 18] and see review in [19]), and collagenase is also synthesized upon interleukin-1 treatment [9]. Possibly these genes are activated through multiple regulatory elements and sequences [20]. We are studying the responses of these genes to UV and TPA using primary human fibroblasts and HeLa cells, both with their endogenous complement of genes and with transfected chimeric constructs.

## THE TRANSIENT NATURE OF THE RESPONSE

The response to TPA and UV is transient. The accumulation of mRNAs and proteins reaches a maximum, then ceases and the products are degraded [9, 10, 21]. Apparently the signal transduction chain(s) responsible for induction of these genes does not remain in an activated state for longer periods of time.

Within this generalization, the three genes we study follow different kinetics of activation. While c-fos mRNA [21] and hMT-IIA [11,22] mRNA start to appear within a few minutes of the inducing TPA or UV treatment, the major increase in collagenase mRNA level occurs only after 30 minutes [23]. There may be a minor immediate response. Collagenase mRNA induction is also partly inhibited by inhibitors of protein synthesis, suggesting that its accumulation is enhanced by prior synthesis of another protein.

As will be discussed below, the fates of the mRNA species after reaching maximum levels, also differ. c-fos mRNA (see below) and protein [14, 24] are particularly labile, while the other two gene products show long half-lives. Therefore the induction of hMT-IIA and collagenase is much longer-lasting than the induction of c-fos.

The transient activation of all three genes under study by UV and TPA is due to an enhanced transcriptional rate. This has been proven for TPA-mediated gene expression by both nuclear runoff and promoter fusion experiments [23,25, and see below]. Also chimeric constructs using 5′flanking DNA of the collagenase promoter fused to a foreign gene are directly TPA inducible. For the UV-induced transcription of this gene formal proof is still pending.

For all three genes conditions have been described which cause constitutive expression. Elevated fos RNA is maintained in cells transformed by the FBJ virus, which carries the fos gene in an altered form (see review in [17]). Also certain differentiated cells retain increased c-fos RNA levels [17,26]. Constitutive collagenase expression has been observed in cultured fibroblasts of patients with Bloom's syndrome [27,28] and several transformed cell lines [29].

Collagenase enzyme activity is correlated with tumor phenotypes [29,30]. Increased metallothionein expression can be stably established by prolonged treatment with cadmium [31]. Future studies of the genes and their flanking regions will identify the structural features leading to either transient or constitutive expression.

We will now expand briefly on the mechanism of c-fos RNA turnover, which will reveal one way of achieving elevated RNA levels. The high turnover leads to suggestions for the still unknown role of the fos gene. We will subsequently describe experiments on the regulation of transcription.

## c-fos mRNAase

Since transcription of c-fos is reduced to a low rate within 60 min of induction, the rate of c-fos RNA degradation can be measured directly without adding an inhibitor of transcription. We measured the half-lives of c-fos mRNA in the presence and absence of actinomycin D. Under both conditions, the half-life was low: in the absence of further transcription the $t\frac{1}{2}$ was 9 min (at 37°C), without actinomycin D about 10–15 min [32]. This high turnover is specific for c-fos RNA and perhaps for other species that are yet unknown. Many known RNA species, however, are not subjected to such high turnover in the very same cells, eg, the RNAs for actin, hMT-IIA, or collagenase. For example, the half-life of hMT mRNA is 2.5 hr [33]. Thus c-fos RNA must possess a property other RNA species do not share.

The specific RNA-degrading mechanism—we will tentatively call it c-fos mRNAase although we cannot distinguish at this time whether the enzyme is specific for c-fos mRNA or rather for all RNAs showing the same structural feature—depends on interaction with sequences in the 3′portion of the RNA. Replacing the promoter region of the gene to +70 by some other promoter does not stabilize the RNA. An insertion in the 3′untranslated region (M. Buescher, unpublished) or replacing the 3′untranslated sequence by long terminal repeat (LTR) as in the FBJ viral constructs (p 76/21 in [31]) increased the stability of the RNA. We tested these mutants by transfecting the constructs under metallothionein promoter control into 3T3 murine fibroblasts and SV40-transformed human fibroblasts [32]. Indeed, the changes of the 3′untranslated region stabilized the RNA. The structure recognized by c-fos mRNAase is thus formed under participation of the 3′untranslated portion of the RNA.

Addition of inhibitors of protein synthesis together with inducing agents causes tremendous superinduction of c-fos expression [12, 13, 21]. There has been some confusion as to whether transcriptional activation by the inhibitor contributed to the superinduction phenomenon. We and others observed increased "rates" of c-fos transcription in nuclear runoff experiments [32,35]. It turned out that protein synthesis inhibitors alone had only a marginal effect on the transcription of the c-fos gene, whereas in combination with growth factors they enhanced and prolonged transcriptional activation [35]. In addition, protein synthesis inhibitors have a major effect on RNA degradation. In an experiment as described above in which the amount of RNA is probed at various times after actinomycin D, the addition of cycloheximide together with TPA or UV led to complete stabilization of c-fos mRNA ($t\frac{1}{2} > 18$ hr; [32]).

Similar superinduction results were obtained with emetine, puromycin, or anisomycine.

The time period required for changing the slope of RNA degradation was extremely brief [32]. The accumulation of c-fos RNA after addition of TPA plus cycloheximide diverged almost instantaneously from the curve without cycloheximide. When cycloheximide was washed out of the culture, the 9-min half-life was reestablished within less than 15 min, regardless of the time spent in actinomycin D plus cycloheximide, eg, up to 3 hr. Two interpretations of these data seem possible. The c-fos mRNAase or a critical component of an enzyme complex is extremely labile, requires constant replenishment, and is synthesized from a stable mRNA. More likely, the RNAase is a stable enzyme, but the structure of c-fos mRNA recognized and formed with participation of the 3'end, is only maintained during ongoing translation of the c-fos mRNA. We imagine two mutually exclusive conformations of c-fos mRNA: a degraded structure formed by the free untranslated 3'end and a stable form in which the 3'end interacts with the protein coding sequence. This latter conformation is prevented by ribosomes moving along the coding sequence. The hypothesis leads to testable predictions: A deletion in the coding sequence would abolish the superinduction phenomenon and amino acid analogs which inactivate the function of newly synthesized protein such as the RNAase, would not affect the c-fos RNA-turnover (if c-fos mRNAase is stable).

## TPA AND ULTRAVIOLET IRRADIATION ACTIVATE c-fos AND METALLOTHIONEIN IIA TRANSCRIPTION THROUGH PROTEIN KINASE C— A NOVEL MECHANISM OF TRANSCRIPTIONAL REGULATION

How do genes sense that a carcinogenic agent such as ultraviolet irradiation or a tumor promoter has hit the cell? We have argued earlier that the progress through the cell cycle may be monitored and that an interruption of the cell cycle may be one of several possible mechanisms which cause overexpression of genes [9]. An interruption of the cell cycle could be caused by UV-induced DNA damage. Whether DNA damage is indeed involved in the induction of genes is a testable question. In cells from a patient with xeroderma pigmentosum group A, which cannot repair UV-induced DNA damage, a lower dose of UV is required to cause a certain number of lesions at any given time than for inducing the same number of lesions in wild-type cells. If gene induction in xeroderma cells would be accomplished at a lower dose than in healthy cells, this would be a strong indication for involvement of DNA damage in the induction process. The experimental results seem to divide the TPA- and UV-inducible genes into two subgroups. The collagenase and hMT-IIA genes represent the group of genes whose induction by UV irradiation involves DNA damage. Maximal RNA accumulation is achieved with 30 $J/m^2$ in wild-type cells, while in the xeroderma cells only 2 to 5 $J/m^2$ give the same result (Rahmsdorf et al, in preparation; see also [10]). In the same sets of cells, c-fos behaved totally differently. Maximal UV induction occurred at 50–60 $J/m^2$ in both xeroderma and wild-type cells. Thus c-fos induction possibly does not need DNA damage. What then could be the mediator of the UV induction of c-fos? We hypothesized that UV could act on the cellular membrane and perhaps transiently change ionic conditions. This hypothesis should be testable.

An important question is what could be the receiving structure responding to a signal whether generated by DNA damage or elsewhere. A suitable candidate for signal transduction is protein kinase C, since protein kinase C is modulated by membrane components and can also be activated by the other inducer examined—TPA (reviewed in [36]). Indeed, a fairly specific inhibitor of protein kinase C, H7 (1-[5-isoquinolinesulfonyl]-methylpiperazine) [37], blocked both the TPA- and UV-induced expression of c-fos while it did not affect glucocorticosteroid-induced transcription from the mouse mammary tumor virus promoter (Rahmsdorf et al, in preparation). An inhibitor with higher specificity for the cAMP-dependent protein kinases—N-(2-guanidinoethyl)-5-isoquinoline-sulfonamide—had no such effect. These experiments speak for a role of protein kinase C, although they cannot rule out the participation of other protein kinases.

The putative activation of protein kinase C by TPA and by UV and its involvement in c-fos induction may occur through different pathways: the actions of TPA and UV on c-fos are additive, and both pathways become independently refractory for a second dose of the same agent within 3 h but are restimulated by the other agent (Rahmsdorf et al, in preparation). TPA-responsive sequence elements may interact with protein kinase C or phosphorylated enhancer factors.

We made use of the large number of deletion and insertion mutants of the hMT-IIA promoter [20] to determine which sequence is required for inducibility by TPA. The behaviour of the mutants in both transient transfections using the chloramphenicol acetyltransferase (CAT) assay (examples shown in Figs. 1, 2) or in stable transfectants using S1 protection experiments, soon revealed that TPA, cadmium, and glucocorticosteroid hormone acted through different sequence elements of the promoter region. Several functional TPA-responsive sequences exist within the 5′ flanking region. The activated protein kinase C could possibly bind to these sequences directly. More likely the enzyme phosphorylates regulatory proteins that thereby gain increased affinity for the sequence or altered ability to interact with a protein neighbor. When HeLa cell nuclear extracts were used as a source of factors which bind to DNA in vitro, all sites became nuclease resistant. The footprints indicate that the sites mentioned are recognized by cellular DNA binding proteins. Using one of these binding motifs as an isolated DNA fragment, we have been able to prove its role as a TPA-responsive element and its binding protein as the signal receiving structure [38].

We envisage the active promoter as a stretch of DNA packed with regulatory proteins which need to interact with each other for full rate transcription. If one of the regulatory proteins is missing, the activity is reduced. The new feature detected here concerns the positive modulation of one or more of the regulatory proteins by protein kinase C (Fig. 3). This modification could affect not only the DNA binding ability but also the interaction between neighboring regulatory proteins.

## IS c-fos A STRESS GENE?

The UV- and TPA-induced genes identified so far raise testable suggestions for the physiology of the response. For example, collagenase expression appears to mediate part of a tumor-like phenotype [29]. The cloned gene will be transferred into various cells, and the consequence of its overexpression examined in tumor cell assays.

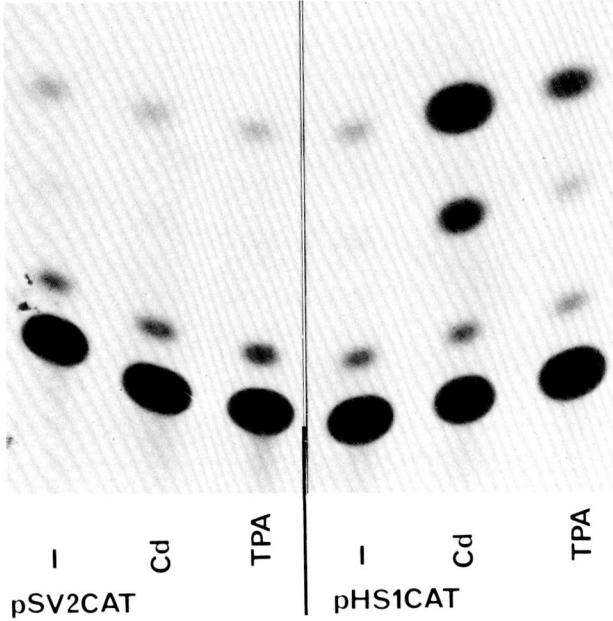

Fig. 1. TPA regulates the hMT-IIA promoter. It has been shown earlier that the endogenous hMT-IIA gene responds to TPA [11]. To show both that this induction is due to increased transcription and that the regulation is an in-built property of the hMT-IIA promoter which can be transferred to other cells and cells of other species, a series of transfection experiments were performed. One such transfection experiment is shown here. The bacterial gene for chloramphenicol acetyltransferase (CAT) was expressed in L-cells using chimeric constructs containing the bacterial gene either in conjunction with the SV40 promoter (pSV2CAT) or with the hMT-IIA promoter (pHS1CAT). The latter construct carries hMT-IIA promoter sequences from $-770$ to $+73$ [20] such that transcription into the CAT gene needed the start at the hMT-IIA promoter and all putative regulatory sequences would be provided by the hMT-IIA 5′flanking region. Eight hours after the transfection with 10 $\mu$g DNA/$10^6$ cells, $5 \times 10^{-6}$ M cadmium chloride (Cd) or 100 ng/ml TPA were added as indicated. At 40 hr after addition of the inducers, the cells were harvested, and equal amounts of cellular protein assayed for CAT activity. In the autoradiogram of the thin layer resolution shown, two of the acetylated products are seen at the top. While pSV2CAT induced only low levels of CAT activity, and the CAT activity was neither increased by cadmium nor by TPA in these cells, the hMT-IIA construct led to a similar basal level of CAT activity, but transcription was enhanced by both cadmium or TPA. Thus the construct contained cis-acting sequences that mediate the response to TPA and to cadmium. Using constructs with smaller portions of the 5′flanking DNA in a similar type of experiment, it could be shown that the TPA and cadmium responses depended on the presence of different elements.

On the basis of the observed c-fos induction within minutes of addition of agents that cause cell-cycle arrest we have proposed at the Cold Spring Harbor (CSH) meeting on Cancer in September 1984 that c-fos does in fact not function as a proliferation gene but rather as a stress gene [21]. We believe that more recent evidence further supports this interpretation. c-fos responds to any type of cellular injury, to various changes of culture conditions, and to growth factors, among others. As a nuclear short-lived protein the c-fos product may form a second or third messenger and perhaps act at an essential crosspoint for several nuclear pathways. The increase in cytoplasmic c-fos mRNA level may elicit pleiotropic changes, only one of which could be transformation. We hope to gain insight into the fascinating problem of c-fos function by using inducible chimeric c-fos constructs.

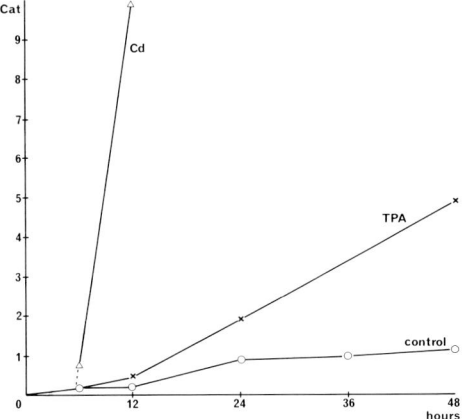

Fig. 2. Kinetics of CAT synthesis in HeLa cells transfected with pHS1CAT. An experiment similar to that in Figure 1 was performed in HeLa cells. Here, several time points were evaluated. CAT activity was quantitated. Twelve hours after transfecting 15 μg DNA/$10^6$ cells, $5 \times 10^{-6}$ M cadmium or 60 ng/ml TPA were added. Cells were harvested at various times thereafter. The scale of the ordinate expresses $10^{-3}$ units for the cadmium induction and $10^{-2}$ units for control cells and TPA induction. One unit is the amount of enzyme contained in 1 mg of cellular protein which converts 1 pmol of chloramphenicol per hour. The experiment shows that the rates of accumulation of chloramphenicol acetyltransferas were quite different whether cadmium or TPA was the inducer using the constructs with the full-length 5'flanking DNA of hMT-IIA. The lag periods of induction, however, were similar in that both curves extrapolated to the background level at about 6 hr after addition of the inducers.

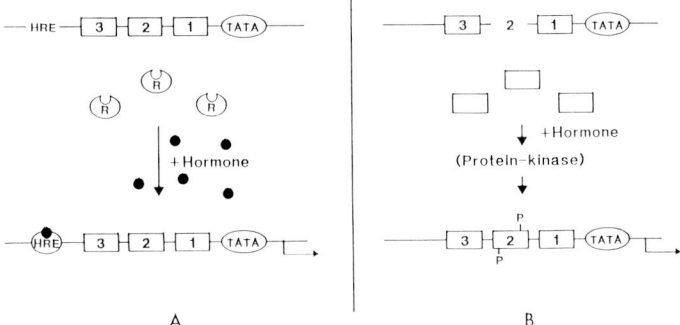

Fig. 3. Two pathways of hormone-mediated gene activation. In both schemes the external agent or its intracellular messenger is referred to as a hormone. In scheme **A**, which applies to steroid hormones, the hormone binds to a specific receptor protein which translocates to the nucleus and binds to a specific cis element, termed HRE [39], to increase the transcription of linked genes. The HRE acts as a hormone-dependent enhancer element; ie, it is relatively independent of position and orientation, and has no effect on the basal level of expression. In scheme **B**, which applies to various peptide hormones and tumor promoters, the exogenous agent binds to a cell surface receptor and leads to activation of an intracellular protein-kinase. The activated protein-kinase can increase the phosphorylation state of a certain transcription factor required for basal expression of the hormone-responsive promoter. This could increase either the affinity of the factor towards its binding site or its ability to interact with other components of the transcriptional machinery. Either one would lead to increased transcription. This model seems to apply for induction of hMT-IIA and collagenase by TPA and serum factors. The boxes labeled 1, 2, and 3 refer to binding sites occupied by various transcription factors responsible for the basal expression of the gene.

## ACKNOWLEDGMENTS

We thank Marga Litfin, who did part of the Northern blot determinations, and G. Kammerer, U. Baltzer, and C. Heinold, who organized our lab in Karlsruhe and provided the typescript.

## REFERENCES

1. Walker GC: Annu Rev Biochem 54:425, 1985.
2. Sachs L: Nature 274:535, 1978.
3. Griffin J, Munroe D, Major P, Kufe D: Exp Hematol 10:774, 1982.
4. Scher W, Friend C: Cancer Res 38:841, 1978.
5. Rahmsdorf HJ, Koch N, Mallick U, Herrlich P: EMBO J 2:811, 1983.
6. Rudd CF, Herschman HR: Toxicol Appl Pharmacol 47:273, 1979.
7. Bienz M: TIBS 4:157, 1985.
8. Kushner I: Ann NY Acad Sci 389:39, 1982.
9. Herrlich P, Angel P, Rahmsdorf HJ, Mallick U, Pöting A, Hieber L, Lücke-Huhle C, Schorpp M: Adv Enzyme Regul 25:485–504, 1986.
10. Schorpp M, Mallick U, Rahmsdorf HJ, Herrlich P: Cell 37:861, 1984.
11. Angel P, Pöting A, Mallick U, Rahmsdorf HJ, Schorpp M, Herrlich P: Mol Cell Biol 6:1760, 1986.
12. Cochran BH, Zullo J, Verma IM, Stiles CD: Science 226:1080, 1984.
13. Greenberg ME, Ziff EM: Nature 311:433, 1984.
14. Müller R, Bravo R, Burckhardt J, Curran T: Nature 312:716, 1984.
15. Curran T, Margan JI: Science 229:1265, 1985.
16. Kruijer W, Schubert D, Verma IM: Proc Natl Acad Sci USA 82:7330, 1985.
17. Müller R: BBA 823:207, 1986.
18. Karin M, Herschman HR: Science 204:176, 1979.
19. Karin M: Cell 41:9, 1985.
20. Karin M, Haslinger A, Heguy A, Dietlin T, Cooke T: Mol Cell Biol 7:606, 1987.
21. Angel P, Rahmsdorf HJ, Pöting A, Herrlich P: Cancer Cells 3:315, Cold Spring Harbor Laboratory, 1985.
22. Imbra RJ, Karin M: Mol Cell Biol 7:1358, 1987.
23. Angel P, Baumann L, Stein B, Delius H, Rahmsdorf HJ, Herrlich P: Mol Cell Biol 7:2256, 1987.
24. Curran T, Müller DA, Zokas L, Verma IM: Cell 36:259, 1984.
25. Imbra RJ, Karin M: Nature 323:555, 1986.
26. Müller R, Müller D, Guilbert L: EMBO J 3:1887, 1984.
27. Herrlich P, Angel P, Pöting A, Schorpp M, Mallick U, Rahmsdorf HJ: In Ebashis (ed): "Cellular Regulation and Malignant Growth." Lipmann Symposium. Berlin, N.Y., Tokyo: Springer Verlag, 1985, pp 50–57.
28. Angel P, Rahmsdorf HJ, Pöting A, Lücke-Huhle C, Herrlich P: J Cell Biochem 29:351, 1985.
29. Liotta LA, Cancer Res 46:1, 1986.
30. Sheela S, Kennedy AR: Carcinogenesis 7:201, 1986.
31. Beach LR, Palmiter RD: Proc Natl Acad Sci USA 78:2110, 1981.
32. Rahmsdorf HJ, Schönthal A, Angel P, Litfin M, Rüther U, Herrlich P: NAR 15:1643, 1987.
33. Karin M, Slater EP, Herschman HR: J Cell Physiol 106:63, 1981.
34. Rüther U, Wagner EF, Müller R: EMBO J 4:1775, 1985.
35. Greenberg ME, Hermanowski AL, Ziff EB: Mol Cell Biol 6:1050, 1986.
36. Nishizuka Y: Nature 308:693, 1984.
37. Kawamoto S, Hidaka H: Biochem Biophys Res Commun 125:258, 1984.
38. Angel P, Imagawa M, Chiu R, Stein B, Imbra RJ, Rahmsdorf HJ, Jonat C, Herrlich P, Karin M: Cell 49:729, 1987.
39. Hynes H, van Ooyen AYY, Kennedy N, Herrlich P, Ponta H, Groner B: Proc Natl Acad Sci USA 80:3637–3641, 1983.

# Oncogene Mutation and Amplification During Initiation and Progression Stages of Mouse Skin Carcinogenesis

**M. Quintanilla, K. Brown, F. Fee, S. Young, and A. Balmain**

The Beatson Institute for Cancer Research, Bearsden, Glasgow G61 1BD, Scotland

> Over 90% of mouse skin tumours initiated with dimethylbenzanthracene display a transversion mutation (A-T) at codon 61 of the $ras^H$ gene. The presence of the mutation depends upon the type of initiating carcinogen but is independent of the promoting agent, which strongly supports the idea that a $ras^H$ gene mutation is critically involved in initiation of mouse skin carcinogenesis. When activated $ras^H$ genes are introduced into epidermal cells, in vivo, by direct application of retroviruses to mouse skin, benign papillomas arise only after promotion with 12-O-tetradecanoylphorbol-13-acetate (TPA). Some of these progress to invasive carcinomas. The results show that activated *ras* genes can replace chemical carcinogens in initiation of mouse skin carcinogenesis. In the chemically induced tumours, progression from papilloma to carcinoma may involve the amplification of the mutated $ras^H$ allele or the development of homozygous mutations. In the case of virus-induced tumours, the carcinomas have particularly high levels of expression of the viral $ras^H$ gene. The role of chromosomal changes occurring at late stages of tumorigenesis is discussed.

**Key words:** chemical carcinogenesis, viruses

Mutated oncogenes of the *ras* family have been found in a wide variety of human and animal tumours, but their role in the multistage process of carcinogenesis remains the subject of much debate. In particular, it has been postulated that such mutations can occur as early events [1–3], late events [4,5], or even that they may be a consequence of tumour development with no causative role [6,7]. This laboratory has tried to address these questions by using the mouse skin carcinogenesis system to investigate the possibility that activation of specific genes may correlate with particular biological events. Skin tumorigenesis can be operationally divided into the separate stages of initiation, promotion, and progression, the main characteristics of which have been reviewed elsewhere [8]. Most of the tumours which arise are benign papillomas, a small proportion of which progress to malignancy.

Received July 16, 1986.

© 1988 Alan R. Liss, Inc.

Previous results had shown that both papillomas and carcinomas initiated with dimethylbenzanthracene (DMBA) and promoted with 12-O-tetradecanoylphorbol-13-acetate (TPA) had an activated Harvey-*ras* gene (*ras*$^H$) [2]. The mutation must therefore be an early event, but could occur either at the time of initiation, by direct interaction of the *ras*$^H$ gene with the carcinogen [3], or at some point after initiation but before the appearance of papillomas. In the latter case, the mutations could be considered spontaneous, arising during the many rounds of cell division, stimulated by promoter treatment required to generate a papilloma. In the former case they may be "targeted," and as such might be expected to depend upon the metabolism and nucleotide-binding characteristics of the particular carcinogens used. Moreover, if a targeted mutation in a *ras* gene is indeed responsible for initiation, it would be predicted that introduction of the activated gene itself into epidermal cells in vivo would substitute for the initiating chemical. Benign papillomas should arise only in conjunction with promoter treatment, and should progress eventually to form carcinomas.

We have used two strategies to test these hypotheses. First, the mutations in *ras*$^H$ genes of tumours initiated and promoted by different chemical agents were analysed in order to determine whether they could be correlated with the type of initiator or promoter used. Secondly, retroviruses containing activated *ras* genes were applied directly to mouse skin to see whether tumours could be induced either by retroviral infection alone or in combination with promoter treatment. The results of both approaches lend strong support to the idea that initiation of carcinogenesis in mouse skin involves direct mutation of the *ras*$^H$ gene.

## A CARCINOGEN-SPECIFIC MUTATION IN MOUSE SKIN TUMOURS

Determination of the mutations in a series of DMBA-initiated tumours necessitated first of all the cloning and sequencing of the normal mouse *ras*$^H$ gene (Ramsden et al, manuscript in preparation). This was carried out in order to permit a computer-based prediction of restriction fragment length polymorphisms (RFLPs) which might be introduced by point mutations in the *ras*$^H$ coding sequence. Table I shows a list of possible restriction sites introduced or eliminated by mutations at either of the activating positions around codons 12 and 61. Digestion of DNA from NIH/3T3 foci obtained by transfection with the activated *ras*$^H$ genes of both papillomas and carcinomas showed that the vast majority had a new XbaI site caused by an A-T transversion mutation at the second nucleotide of codon 61 [9] (Fig. 1). This mutation within the second coding exon of the *ras*$^H$ gene results in the appearance of two new bands of approximate sizes 8 and 4 kilobases (kb), which can be easily distinguished from the 12-kb fragment seen in normal DNA (Fig. 2). No other tumour-specific RFLPs have been detected in this series of NIH/3T3 foci, although one transformant (lane a, Fig. 1) which was negative for the XbaI polymorphism was subsequently shown by *ras*$^H$ P21 analysis to have a mutation around codon 12 (data not shown).

The detection of the XbaI polymorphism in the majority of foci obtained by transfection enabled us to investigate a wider series of primary tumours which were initiated or promoted by different chemicals. The results are shown in Table II. Initiation with DMBA had a very high association with the A-T transversion at codon 61, since over 90% of the papillomas and carcinomas exhibited the XbaI polymorphism. This high frequency was not altered when a completely different promoting

TABLE I. Single Base Mutations at c-*ras*H Codons 12 and 61 Specifying Changes of Amino Acid Residues

|  | Restriction site formed |
|---|---|
| 12 | |
| Normal GGGGCT GGA GGCGTG | MnlI |
| GGCGCT AGA GGCGTG | |
| GGCGCT TGA GGCGTG | STOP |
| GGCGCT CGA GGCGTG | XhoI, TaqI |
| GGCGCT GAA GGCGTG | |
| GGCGCT GTA GGCGTG | |
| GGCGCT GCA GGCGTG | PstI, Fnu4HI |
| 61 | |
| Normal GCAGGT CAA GAAGAG | |
| GCAGGT AAA GAAGAG | |
| GCAGGT TAA GAAGAG | STOP |
| GCAGGT GAA GAAGAG | HphI |
| GCAGGT CTA GAAGAG | XbaI |
| GCAGGT CCA GAAGAG | AvaII |
| GCAGGT CGA GAAGAG | TaqI |
| GCAGGT CAT GAAGAG | |
| GCAGGT CAC GAAGAG | |

agent was used (chryserobin) which does not bind to the cellular protein kinase C receptor thought to be utilised by TPA [10]. Thus, although these two promoters probably have different mechanisms of action, they do not influence the nature of the mutations in the *ras*H genes of tumours initiated with DMBA.

A somewhat different picture was seen in tumours initiated with other carcinogens. N-methyl-N'-nitro-N-nitrosoguanidine (MNNG) is a direct-acting alkylating agent which is thought to exert its carcinogenic effects by methylating the O-6 position of guanosine residues (11). In a series of 12 skin tumours, comprising both papillomas and carcinomas, none had the XbaI polymorphism seen after DMBA initiation. Preliminary experiments on cell lines derived from papillomas and carcinomas initiated with benzo[a]pyrene (BP) gave similar results. None had the codon 61 mutation, but at least two of five lines investigated appeared to have codon 12 mutations in the *ras*H gene, suggesting that with this carcinogen, mutation may occur at guanosine residues.

The carcinogens DMBA and BP are among the most commonly used initiators of mouse skin tumorigenesis, and consequently a great deal is known about the metabolism and binding of these compounds to epidermal DNA. These studies have shown that the potent initiating effects of DMBA, which is 30 times more effective on a molar basis than BP, correlate with the much higher propensity of DMBA to bind to deoxyadenosine (dA) residues. In fact, Dipple and co-workers have previously predicted that the critical adduct for the carcinogenic potency of DMBA would be deoxyadenosine [12]. This prediction appears to be borne out by the fact that over 90% of the mutations in the *ras*H genes of DMBA-initiated tumours occur at the first A of codon 61. Other initiating agents do not appear to induce this mutation at such high frequency.

The first evidence for carcinogenic-specific mutations in chemically induced tumours came from the results of Barbacid and colleagues on rat mammary carcinomas induced by a single treatment with nitrosomethylurea (NMU). A high proportion

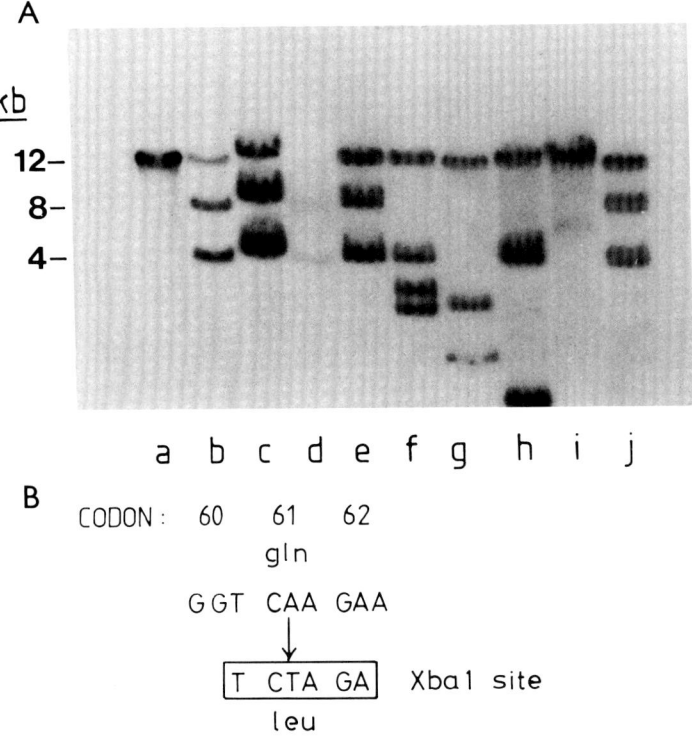

Fig. 1. XbaI polymorphism in NIH/3T3 transformants. **A:** Transformants obtained by transfection with papilloma (**lanes d–h,j**) or carcinoma (**lanes a–c**) DNAs. DNAs were digested with XbaI and hybridised on Southern blots with a $ras^H$-specific probe. Lane a contains DNA from a transformant with a $ras^H$ codon 12 mutation. **Lane i** contains DNA from a spontaneous NIH/3T3 transformant which has the same hybridisation pattern as normal DNA. **B:** Induction of an XbaI polymorphism by a specific mutation at codon 61. Digestion of normal mouse DNA with XbaI gives rise to a 12-kb fragment containing the mouse c-$ras^H$ gene. Mutation of the middle base of codon 61 from A to T would give rise to a new XbaI site, generating tumour-specific fragments of about 8 and 4 kb. (Reprinted from [9] with permission.)

of tumours exhibited the same G-A transition at the second position of codon 12 in the $ras^H$ gene [3]. Tumours induced by DMBA showed less frequent activation of the $ras^H$ gene, but those which did proved to have mutations at the deoxyadenosines of codon 61 [3] (M. Barbacid, personal communication).

Further evidence for carcinogen-specific mutations comes from the studies of Wiseman et al on mouse liver tumours induced by treatment with a variety of different chemical agents [13]. The mutations occurred predominantly in the $ras^H$ gene and at least in some cases could be correlated with the known mutational specificity of the carcinogen used [13]. These three independent sets of experiments involving skin, mammary gland, and liver carcinogenesis therefore lead to similar conclusions: the existence of carcinogen-specific mutations suggests that ras gene mutation can be a critical event in initiation.

## CAN ACTIVATED ras GENES INITIATE CARCINOGENESIS DIRECTLY?

A critical test for the putative role of activated ras genes in initiation would be to use the gene itself rather than a chemical carcinogen to initiate mouse skin

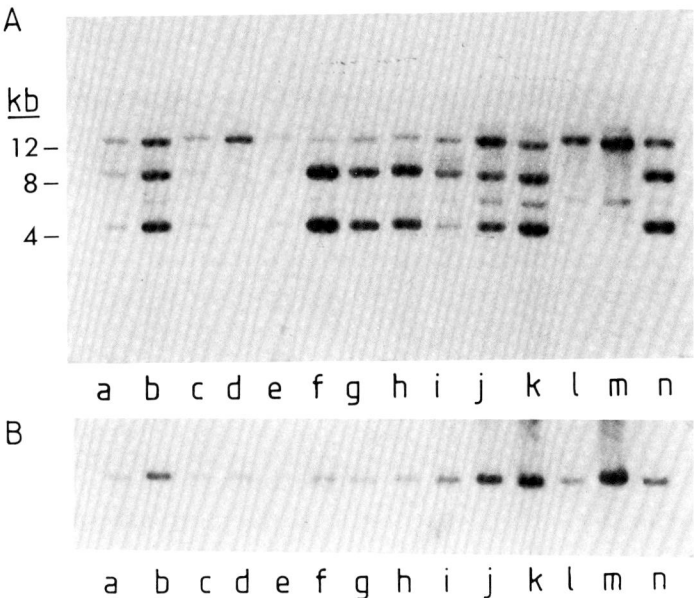

Fig. 2. Demonstration of the XbaI polymorphism in DMBA-initiated primary tumours. **A**: DNAs obtained from primary papillomas are in **lanes a, d,** and **e. Lanes b, c,** and **f–k** contain carcinoma DNAs. **Lane l** contains DNA from normal skin. **Lanes m** and **n** respectively contain DNAs from spontaneously transformed 3T3 cells and a NIH/3T3 transformant derived by transfection from a papilloma with the codon 61 polymorphism. DNAs were manipulated as described in the legend to Figure 1. **B**: Autoradiograph showing the 1-kb band obtained by rehybridisation of the blot shown in A with a cDNA probe for the I1-3 gene. (reprinted from [9] with permission.)

**TABLE II. An XbaI Polymorphism in the $ras^H$ Gene in Skin Tumours**

| Strain | Tumour | Initiator | Promoter[a] | No. positive/No. tested |
|---|---|---|---|---|
| Sencar | Papilloma | DMBA | TPA | 12/14 |
|  |  | DMBA | Chryserobin | 5/5 |
| NMRI | Papilloma | DMBA | TPA | 2/2 |
| NIH | Papilloma | DMBA | TPA | 3/3 |
| Sencar | Carcinoma | DMBA | TPA | 0/2[b] |
| NMRI | Carcinoma (transplanted) | DMBA | TPA | 3/3 |
| NIH | Carcinoma | DMBA | TPA | 8/8 |
| NIH | Papilloma | MNNG | TPA | 0/6 |
| NIH | Carcinoma | MNNG | TPA | 0/6 |

[a]Promotion was carried out twice weekly by treatment with an acetone solution of TPA or chryserobin.
[b]One tumour has a $ras^H$ mutation presumably at codon 12.

tumorigenesis. If the $ras^H$ gene is indeed mutated at the time of initiation, direct introduction of the gene into epidermal cells in vivo should give rise to a population of cells which, upon treatment with TPA, should develop into papillomas.

A series of experiments was therefore carried out with Harvey-and Balb-murine sarcoma viruses, both of which contain activated $ras^H$ genes. These viruses were applied to mouse skin by scarification under ether anaesthesia, and various groups of animals were subsequently treated with either TPA or acetone. The results in Table III show that papillomas appeared in the TPA-treated animals after only 4–5 wk,

TABLE III. Mouse Skin Papilloma Formation After Initiation With Harvey Murine Sarcoma Virus (HaMSV)

| Strain | No. mice | Initiator[a] | Latent period[b] (mo) | Promoter[c] | Total papillomas/ mice with papillomas |
|---|---|---|---|---|---|
| NIH | 27 | HaMSV + F | — | TPA | 151/26 |
| NIH | 7 | HaMSV + F | — | — | 0 |
| NIH | 5 | PBS | — | TPA | 0 |
| NIH | 5 | — | — | TPA | 0 |
| Sencar | 15 | HaMSV + F | — | TPA | 15/13 |
| Sencar | 12 | HaMSV + F | — | — | 0 |
| Sencar | 5 | PBS | — | TPA | 0 |
| Sencar | 4 | — | — | TPA | 0 |
| NIH | 10 | F | — | TPA | 0 |
| NIH | 10 | F | — | — | 0 |
| NIH | 9 | Balb MSV | — | TPA | 12/4 |
| NIH | 3 | HaMSV + F | 4 | TPA | 9/3 |
| Sencar | 4 | HaMSV + F | 4 | TPA | 8/4 |

[a]F indicates that the Friend murine leukemia virus helper virus was used.
[b]A dash indicates no latent period; ie, mice were treated twice weekly with TPA or acetone beginning the day after initiation.
[c]A dash indicates treatment twice weekly with acetone.

whereas no skin tumours were observed in the acetone group over a period exceeding 1 yr [14]. Some of the papillomas grew very rapidly, reaching a size of 8–10 mm in diameter after 7–8 wk of TPA treatment. Histologically, the papillomas were well-differentiated lesions very similar in appearance to chemically initiated tumours. Table III also shows that virus-infected cells can remain in a "dormant" state within the epidermis for several months, since promotion after this lag period leads to the appearance of papillomas within 3–4 wk. We conclude that viruses containing activated *ras* genes can mimic the effect of chemical carcinogens in initiation of two-stage carcinogenesis.

## ALTERNATIVE MECHANISMS OF INITIATION

The combined approaches outlined above, ie, the detection of initiator-specific mutations in mouse skin tumours and the demonstrations that activated *ras* genes can initiate, lend strong support to the idea that at least in some carcinogenesis systems, initiation can involve direct mutation of a *ras* gene. As might be expected, however, alternative mechanisms of initiation appear to take place in other systems. When C3H/10T1/2 cells in culture are irradiated, "initiation" of a high proportion of the treated cells is followed by a rare mutational event which eventually gives rise to transformed foci [15]. It seems unlikely that this early event involves a single mutation in a specific gene since the frequency with which it occurs suggests a much larger target size. The nature of this putative target and indeed the relevance of this mechanism to tumour development in vivo remain unclear.

Guerrero et al [16] have also shown that G-A transition is not always the route by which *ras* genes become activated in tumours induced by the carcinogen NMU. Thymic lymphomas induced by treatment with NMU frequently have activated N-*ras* genes, at least one of which exhibits a C-A mutation at codon 61. This is obviously not the same mutation as that observed by Zarbl et al in rat mammary carcinomas, indicating either that this carcinogen is capable of inducing mutations by different mechanisms or, more probably, that a different gene target may be involved in initiation of this system.

## CHROMOSOMAL CHANGES DURING TUMOUR PROGRESSION

It is well known that chromosomal aberrations involving amplification, deletion or translocation of DNA sequences are frequently observed in tumour cells [17]. In mouse skin tumours, aneuploidy develops during tumour progression, since papillomas are usually diploid, but most carcinomas have a hypertetraploid DNA content [18]. The XbaI polymorphism found in DMBA-initiated tumours enabled us to investigate whether these chromosomal changes included further alterations at the $ras^H$ locus.

The Southern blot in Figure 2 shows that papillomas in general have a heterozygous mutation in the $ras^H$ gene, since both the normal 12-kb band and those corresponding to the mutated allele can be seen (lanes a,e). Some carcinomas, on the other hand, showed clear evidence of amplification of the mutated allele (lanes f–h) or a change in the ratio of normal to mutated $ras^H$ genes (lanes b,k). These conclusions can be drawn from a comparison of the relative band intensities after hybridisation with a $ras^H$ probe (Fig. 2A) or a probe for the mouse interleukin-3 gene (Fig. 2B). We conclude that heterozygous mutations in $ras^H$ genes in papillomas can become amplified or homozygous during progression to carcinomas, although it is not yet possible to say exactly when these molecular changes take place.

Since only a relatively small proportion of papillomas progress to carcinomas, it is possible that only those papilloma cells which undergo additional chromosomal changes go through this transition. An alternative interpretation would be that these additional changes take place in a subset of papillomas at an early stage of their development with the consequent generation of a population of papillomas with a high probability of conversion to malignancy. The existence of such a population has recently been demonstrated [19]. In line with this possibility, from a series of 22 DMBA-initiated papillomas positive for the XbaI polymorphism, one appeared to have already amplified the mutated allele (data not shown). However, it should be stressed that no histological verification was obtained for this particular papilloma, and it remains possible that it had already partially converted to a carcinoma.

## PROGRESSION OF VIRUS-INDUCED PAPILLOMAS

Papillomas initiated with viral $ras^H$ genes rather than DMBA also progress to carcinomas, but appear to do at least in some cases by a different mechanism. Restriction of DNA from virus-induced tumours with enzymes which cut the viral genome only once leads to the conclusion that the papillomas are probably polyclonal or oligoclonal in origin, whereas carcinomas are clonal [14]. This is shown by the Southern blot in Figure 3. The carcinomas show 1 or 2 novel hybridising fragments with a $ras^H$ probe, indicative of single or at most two copies of the integrated provirus (lanes 2–6, 12–16). Papillomas, on the other hand, only rarely show discrete novel bands (lanes 1,8,11), suggesting that they arise from infection of several primary epidermal cells. It is interesting to note that no evidence of amplification of the

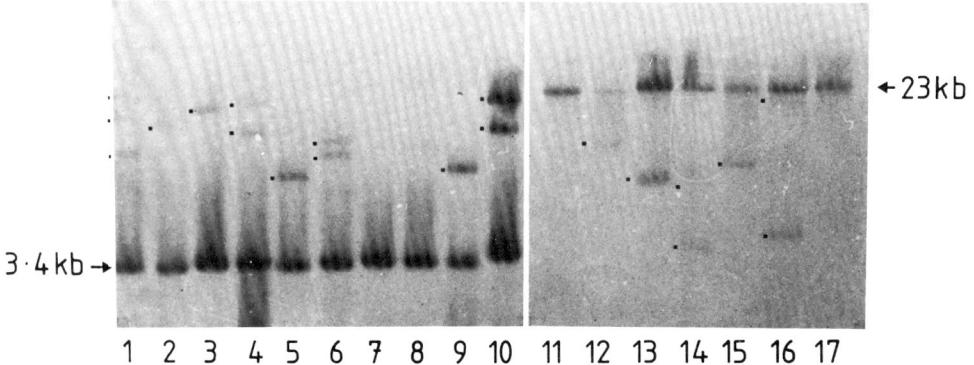

Fig. 3. Southern blot showing *ras*-sequences in HaMSV-initiated tumours. DNAs in **lanes 1–10** were digested with BamH1 and **11–17** with EcoRI. Arrows show the positions of the cellular $ras^H$ bands at 3.4 kb and 23 kb, respectively. Additional hybridising fragments in the tumours are indicated by dots. Lanes 7 and 17 contain normal mouse epidermal DNA. Lanes 1, 8, and 11 contain papilloma DNA. Lanes 2–6 and 12–16 contain carcinoma DNA. Lanes 9 and 10 contain DNA from carcinoma cell lines. The cell line in lane 10 was derived from the primary tumour in lane 4. (Reprinted from [14] with permission.)

mutated viral gene can be seen in the carcinomas, in contrast to the situation in some DMBA-initiated tumours. It could, however, be shown by Northern blotting using RNA's from virus-induced tumours that, in particular, the carcinomas express high levels of v-$ras^H$ mRNA. Figure 4 shows a Northern blot of *ras*-specific transcripts in a chemically induced papilloma (lanes 1,4) and in HaMSV-induced papilloma (lane 3) or carcinomas (lanes 5–9). All of the virus-induced tumours show transcripts ranging in size up to the full-length genomic RNA of 5.4 kb. The level of expression is particularly high in some carcinomas, as shown by rehybridisation of the same blot with an actin-specific probe (lanes 10–15). This could be confirmed at the protein level by immunocytochemical techniques using $ras^H$-specific monoclonal antibodies. The papillomas express fairly high P21 levels mainly in the basal cells, whereas in the carcinomas the staining is more intense and more uniformly distributed throughout the tumours [14]. Consequently, it is possible that elevated expression of the v-*ras* gene plays a role in progression to malignancy. In chemically induced tumours, the same effect may be achieved by amplification of the mutated gene or the development of homozygosity.

It is not possible at present to say whether changes involving the mutated allele which take place during tumour progression are accompanied by complete loss of the normal allele of the $ras^H$ gene. It is, however, intriguing that loss of the normal allele has been seen in a carcinogen-induced thymic lymphoma [16] and in some human tumour cell lines [20]; and loss of a $ras^H$ gene polymorphism on one human chromosome with reduplication of the other allele has been reported in a series of bladder carcinomas [21].Chromosomal changes of this kind may represent late events in tumour development in some, but by no means all, mouse skin carcinomas. At least four carcinomas in NIH or NMRI mice, including a transplantable squamous cell carcinoma, do not show any evidence of amplification of the mutated allele.

An exciting possibility is that chromosomal deletions occurring at late stages of carcinogenesis may involve loss of specific suppressor genes. Normal cells have been shown to contain suppressor genes which can reverse the transformed phenotype after

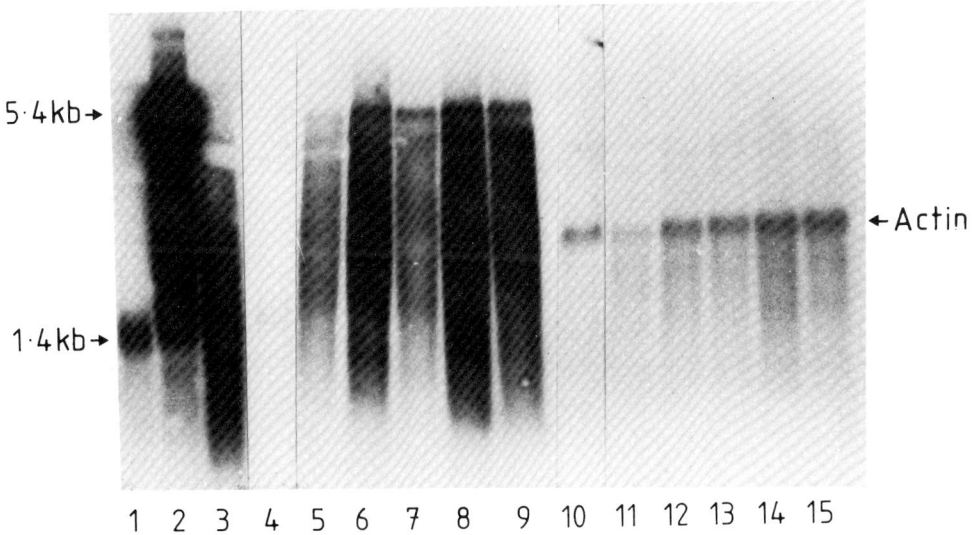

Fig. 4. Expression of viral $ras^H$ sequences in HaMSV-initiated tumours. The figure shows an autoradiograph of a Northern blot hybridised with $ras^H$-specific or actin-specific probes. 20 μg of total RNA was run in each lane unless otherwise stated. Sources of RNA were as follows. **Lanes 1,4,10:** DMBA-initiated papillomas. **Lane 2:** HaMSV-infected 3T3 cells (200 ng of poly (A)$^+$ RNA were run in this case). **Lane 3:** HaMSV-initiated papilloma. **Lanes 5–9,11–15:** HaMSV-initiated carcinomas. **Lanes 1–9** were hybridised with a $ras^H$ probe. **Lanes 10–15** show a rehybridisation of lanes 4–9 with an actin-specific probe. Exposure times are 2 days for **lanes 1–3** and about 12 hr for **lanes 4–9**. (Reprinted from [14], with permission.)

fusion with tumorigenic cells [22]. Loss of these genes is associated with tumour development in certain forms of hereditary cancer [23], but specific chromosome loss in sporadic human tumours [21] suggests that repressor sequences may play a more widespread role. It is interesting that human chromosome 11, which contains the c-$ras^H$ gene, is frequently lost during carcinoma development [21]. An interesting parallel is that mouse chromosome 7, which contains the mouse c-$ras^H$ gene, is also frequently underrepresented in cells derived from skin carcinomas [24]. The mouse skin carcinogenesis system may therefore be a suitable model to investigate the role of suppressor genes during tumour development in vivo.

## ACKNOWLEDGMENTS

M.Q. is supported by an EMBO long-term fellowship, and K.B. is the recipient of a Training Fellowship from the Medical Research Council. We are grateful to I.B. Kerr, M. Ramsden, and J. Paul for helpful discussions. Some tumours used in this study were kindly supplied by G.T. Bowden, T. Slaga, and J. Di Giovanni. The Beatson Institute is supported by the Cancer Research Campaign.

## REFERENCES

1. Balmain A: Br J Cancer 51:1, 1985.
2. Balmain A, Ramsden M, Bowden G, Smith J: Nature 307:658, 1984.
3. Zarbl H, Sukumar S, Arthur AV, Martin-Zanca D, Barbacid M: Nature 315:382, 1985.

4. Vousden KH, Marshall CJ: EMBO J 3:913, 1984.
5. Albino AP, LeStrange R, Oliff AI, Furth ME, Old LJ: Nature 308:69, 1985.
6. Duesberg PH: Science 228:669, 1985.
7. Cichutek K, Duesberg PH: Proc Natl Acad Sci USA 83:2340, 1986.
8. Hecker E, Fusenig NE, Kunz W, Marks F, Thielmann HW (eds): Carcinogenesis, Vol. 7. NY: Raven Press, 1982.
9. Quintanilla M, Brown K, Ramsden M, Balmain A: Nature 322:78, 1986.
10. Declos KB, Nagle DS, Blumberg PM: Cell 19:1025, 1980.
11. Margison GP, O'Connor PJ: In Grover P (ed): "Chemical Carcinogens and DNA," Vol. 1. Florida: CRC Press, 1978.
12. Dipple A, Sawicki JT, Moschel RC, Bigger CAH: In Rystrom J, Montelius J, Bentsson M (eds): "Extrahepatic Drug Metabolism and Chemical Carcinogenesis." Amsterdam: Elsevier, 1983, p 439.
13. Wiseman RW, Stowers SJ, Miller EC, Anderson MW, Miller JA: Proc Natl Acad Sci USA 83:5825, 1986.
14. Brown K, Quintanilla M, Ramsden M, Kerr IB, Young S, Balmain A: Cell 46:447, 1986.
15. Kennedy AR, Cairns J, Little JB: Nature 307:85, 1984.
16. Guerrero I, Villasante A, Corces V, Pellicer A: Proc Natl Acad Sci USA 82:7810, 1986.
17. Klein G, Klein E: Nature 315:190, 1985.
18. Balmain A, Sauerborn R, Ramsden M, Pragnell IB, Bowden GT, Smith J, Cole G: In Omenn G, Harris C, Gelboin H (eds): "Banbury Report 16." Cold Spring Harbor Laboratory, 1984, p 243.
19. Hennings H, Shores R, Mitchell P, Spangler EF, Yuspa SH: Carcinogenesis 6:1607, 1985.
20. Santos E, Martin-Zanca D, Reddy EP, Pierotti MA, Della Porta G, Barbacid M: Science 223:661, 1984.
21. Fearon ER, Feinberg AP, Hamilton SH, Vogelstein B: Nature 318:377, 1985.
22. Stanbridge EJ, Der CJ, Doersen CJ, Nishimi RY, Pechl DM, Weissman BE, Wilkinson J: Science 215:252, 1982.
23. Knudson A: Cancer Res 45:1437, 1985.
24. Fusenig N, Dzarlieva-Petrasevska RT, Breitkreuz D: In Barrett JC, Tennant RW (eds): Carcinogenesis, Vol. 9. New York: Raven Press, 1985, p 293.

# In Vivo and In Vitro Expression Pattern of Genes Activated During Multistage Carcinogenesis in the Mouse Skin

**Peter Krieg, Karl Melber, Gerhard Furstenberger, and G. Tim Bowden**

*Institutes for Virus Research (P.K., K.M.) and Biochemistry (G.F.), German Cancer Research Center, D-6900 Heidelberg, Federal Republic of Germany; Radiation Oncology Department, University of Arizona, Health Sciences Center, (G.T.B.) Tucson, Arizona 85724*

Studies on the expression of sequences activated during multistage carcinogenesis in mouse skin showed that 3 sequences (mal-1, -2, -3) were already activated at the benign papilloma stage. In contrast, overexpression of mal-4 appeared to be specific for the malignant state. In addition, a change in the transcript pattern of mal-3-related sequences was observed during progression from benign to malignant tumors. Tumor promoters were able to induce a transient expression of the mal sequences in the normal epidermis. A strong transcriptional activity of the mal sequences was also observed in different epidermal cell lines. Expression of mal-1 and mal-2 was shown to be dependent on the serum requirement in putatively initiated basal cells, but not in transformed tumorigenic keratinocytes.

**Key words: gene expression, tumor promoters, skin tumors, epidermal cell cultures**

To investigate molecular mechanisms involved in the process of carcinogenesis, it is useful to identify and characterize genes whose expression is altered during tumor development. It is advantageous to perform such studies using an in vivo system, because in vitro systems may only incompletely reflect the complicated in vivo situation of carcinogenesis which is recognized to be a multistep process, and because gene expression may undergo further alteration during in vitro cultivation which is not correlated to the process of carcinogenesis.

The two-stage model of skin carcinogenesis in the mouse is a suitable system for such studies [1]. There are operationally and mechanistically clearly defined and isolatable stages of tumor development; and, in addition, suitable in vitro systems of the different stages are available facilitating further mechanistic studies.

Using this model of tumor induction and utilizing molecular hybridization as well as cDNA cloning techniques, we recently isolated a number of sequences (mal-

Received June 10, 1986.

© 1988 Alan R. Liss, Inc.

1–mal-6) which were activated during tumor development [2]. To assess a possible functional role of these sequences in tumorigenesis, we studied their expression during tumor development and in different epidermal in vitro systems.

## RESULTS

### Expression of Mal Sequences in Benign and Malignant Mouse Skin Tumors

To investigate a potential role of the mal sequences in the process of carcinogenesis, we asked whether there was a correlation between the state of tumor development and the level of expression of the isolated mal sequences. Tumors were induced in the back skin of NMRI mice by initiation with 7,12-dimethylbenz [a]anthracene (DMBA) or with N-methyl-N'-nitro-N-nitrosoguanidine (MNNG) and promotion with 12-O-tetradecanoylphorbol-13-acetate (TPA).

Northern blots were performed with RNA isolated from normal mouse epidermis, from benign papillomas, from a benign keratoacanthoma, and from malignant squamous cell carcinomas using the mal cDNA clones as probes. As a control in all Northern blots the RNA bound to the filters was hybridized either simultaneously or in a second hybridization cycle with a probe specific for the 7S cytoplasmic RNA, present in the same abundance in normal epidermis as well as in tumors [3].

With one exception (mal-4), the transcripts corresponding to the mal cDNA clones were already activated in the papilloma stage of multistep carcinogenesis, and we did not observe further activation of these sequences during the progression from the benign papilloma to the malignant tumor. There were no detectable differences in the expression of mal-1-related sequences in tumor-promoter-dependent papillomas (Fig. 1h,i), those that had been isolated immediately after their appearance in the animals, compared to tumor-promoter-independent (so-called autonomous) papillomas (Fig. 1f,g), which were taken 12 wk after the end of the TPA treatment.

In contrast, mal-2 expression was slightly enhanced in tumor-promoter-independent papillomas compared to tumor-promoter-dependent tumors (Fig. 1f–i). A keratoacanthoma, another benign tumor sometimes arising during mouse skin carcinogenesis, showed only a very weak expression of mal-1 and 2 (Fig. 1a). The expression of these mal sequences was only slightly, if at all, enhanced compared to normal epidermis (Fig. 1j). In all malignant squamous cell carcinomas we observed a high expression level of mal-1- and mal-2-related sequences. There were only slight differences in the expression levels in several individual carcinomas induced by the two-step protocol with either DMBA/TPA (Fig. 1c–e) or with MNNG/TPA (Fig. 1b).

In contrast, there was a change in the patterns of mal-3-related transcripts during tumor development in the mouse skin. In normal epidermis, small amounts of three transcripts of 1.3, 2.3, and 2.9 kilobases (kb) were detectable (Fig. 2a). In benign tumors, in a keratoacanthoma and in six different papillomas tested so far, only the 1.3- and the 2.3-kb transcripts were overexpressed (sometimes to a greater extent than in carcinomas, see [2], whereas the largest 2.9-kb transcript was not detectable (Fig. 2b). In malignant squamous cell carcinomas, this largest transcript related to mal-3 was always overexpressed, whereas the 2.3-kb transcript in most of the carcinomas disappeared (Fig. 2d), detectable only in a few carcinomas in lower amounts (Fig. 2c). In all benign tumors tested so far, we only detected an overexpres-

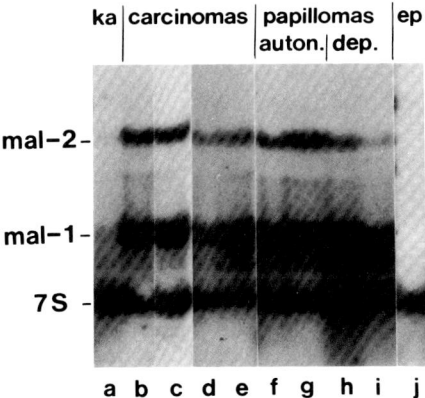

Fig. 1. Expression of mal-1 and mal-2 in different mouse skin tumors. 10 μg of total RNA from normal epidermis or tumors were applied to each lane, size fractionated in 1.4% agarose/2.2 M formaldehyde gels, blotted onto cellulose nitrate paper, and hybridized to a mixed probe of nick-translated plasmid DNA pmal-1, pmal-2, and pA6 [2,3]. Tumor induction, preparation of tumor and epidermal tissues, and RNA isolation and hybridization were performed as described elsewhere [2]. **a:** RNA from a keratoacanthoma. **b:** RNA from a squamous cell carcinoma induced by MNNG and TPA. **c–e:** Induced by DMBA and TPA **(f,g):** RNA from tumor-promoter, independent papillomas. **(h,i):** From tumor-promoter-dependent papillomas. **j:** RNA from normal epidermis.

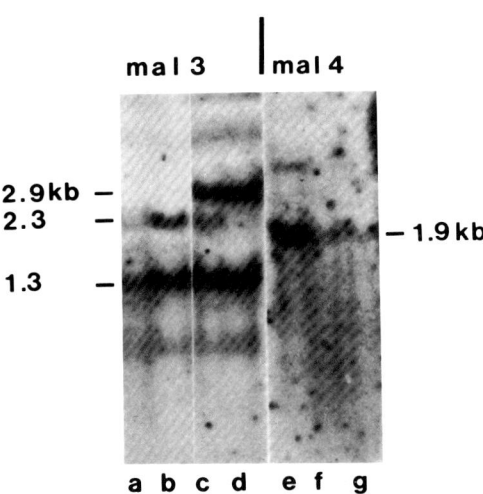

Fig. 2. Expression of mal-3 and mal-4 in different mouse skin tumors. Northern blots were performed as described in the legend to Figure 1 and elsewhere [2]. **a–d:** Hybridized with pmal-3. **e–g:** Hybridized with pmal-4 **a,g:** RNA from normal epidermis. **b,f:** RNA from papillomas, **c–e:** RNA from squamous cell carcinomas.

sion of the 1.3-kb and 2.3-kb transcripts. Thus, overexpression of the 2.9-kb transcript may be related to the malignant state.

In a similar way, overexpression of mal-4-related sequences appeared to be specific for the malignant state. In Northern blot analysis of mal-4, we detected only a slight overexpression of a 1.9-kb transcript in benign papillomas compared to normal epidermis (Fig. 2f,g). In malignant squamous cell carcinomas, however, this mal-4 transcript was present in high abundance (Fig. 2e). As estimated by densitometry of the autoradiograms, the factor of elevated transcription levels in carcinomas compared to papillomas and normal epidermis was greater than 10.

## Induction of Transient Expression of Mal-Related Sequences by Tumor Promoters

Although there is evidence that during tumor development tumor promoters act by mediating clonal expansion of previously initiated cells, it is unclear if genetic effects of tumor promoters are also necessary for the promotion process. It has been shown in vitro that tumor promoters are able to modulate the transcription of a number of cellular and viral genes [4].

To test such effects of tumor promoters in vivo, we asked whether the expression of mal-related sequences was affected by tumor promoters. Using Northern blot analysis, we detected a transcriptional activation of mal sequences in the mouse epidermis after treatment with tumor promoters. A single application of 10 nmol of TPA on the uninitiated back skin of NMRI mice stimulated a transient expression of the mal-related sequences. Interestingly, the different mal sequences showed different kinetics of stimulated expression (Fig. 3).

The expression of mal-1 and mal-2 was enhanced within 4 hr after TPA treatment and reached a maximum between 18 and 24 hr. The expression of these sequences returned to control levels between 48 and 72 hr after treatment. Mal-4 expression was enhanced earlier and for a shorter time. Enhanced expression was observed within 2 hr, reached a maximum within 4–18 hr and decreased within 24 hr after TPA treatment. A similar kinetic pattern was observed for the enhanced expression of mal-3-related transcripts (data not shown). Multiple treatments with TPA did not further enhance the expression of the mal sequences compared to that seen with a single dose of TPA (data not shown).

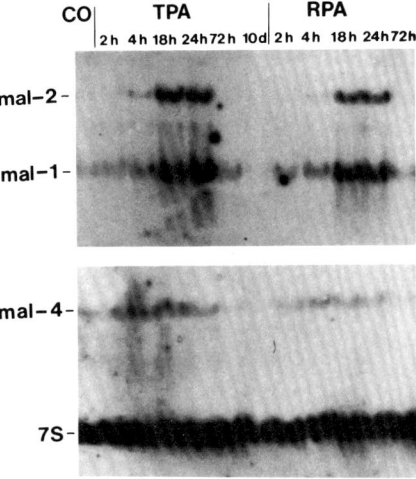

Fig. 3. Expression of mal sequences in mouse epidermis after treatment with tumor promoters. Shaved back skins of NMRI mice were treated with a single application of 10 nmol TPA or RPA (dissolved in 100 µl of acetone). At different times after application, as indicated, the animals were sacrificed and epidermal RNA was isolated to perform Northern blot analysis as described [2]. **Top:** Hybridized with a mixed probe of mal-1 and mal-2. **Bottom:** Hybridized with a mixed probe of mal-4 and 7S-specific pA6.

Treatments with the incomplete second-stage tumor-promoter retinoyl phorbol acetate (RPA) resulted in nearly the same responses except that the level of stimulated mal expression appeared to be lower (Fig. 3).

## Expression of Mal-Related Sequences in Different Epidermal In Vitro Systems

To obtain information concerning a possible function of the mal sequences, it is useful to study their expression in different epidermal cells and to investigate whether there is a correlation between the level of expression and phenotypic properties of the cells examined.

Primary cultures of basal cells isolated from adult and neonatal epidermis exhibited strong transcriptional activity of mal sequences, independent of the growth state of the cells (data not shown). High expression of mal-2, mal-3, and mal-4, but not of mal-1, was also observed in different immortalized keratinocytes (Fig. 4). HEL30 are spontaneously transformed mouse keratinocytes. PDV are in vitro transformed keratinocytes (transformed by DMBA and TPA) and HDII are in vitro cultivated carcinoma cells, derived from a mouse carcinoma which had been induced by DMBA and TPA [5]. Weak transcriptional activity of mal-1 was only detectable in growing HEL30 and in growing HDII cells. Mal-2, -3, and -4 were expressed in all these lines.

Whereas transcription of mal-4 was enhanced in growing cells compared to those which reached confluency, mal-2 and -3 transcription was independent of the cells' growth rate (Fig. 4). Interestingly, the 2.9-kb transcript related to mal-3 present in the other transformed keratinocytes as well as in vivo in all carcinomas was replaced in the in vitro cultivated carcinoma cells HDII by a smaller RNA species of 2.5 kb. A high transcriptional activity of mal sequences was also observed in the

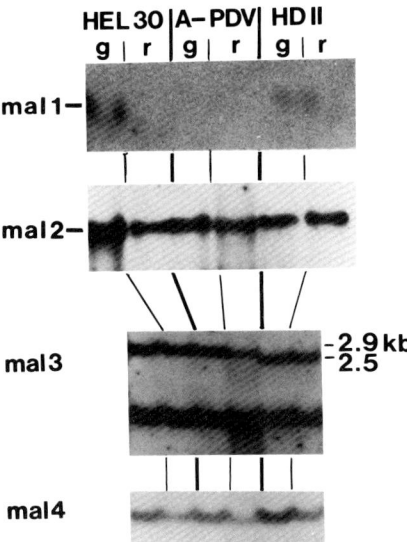

Fig. 4. Expression of mal sequences in different keratinocyte cell lines. Cell cultures (keratinocyte cell lines HEL30, PDV, or HDII [5]) growing logarithmically (g) or cultures which had reached confluence (r) were harvested. RNA was isolated, and Northern blot analysis was performed as described [2], using as probes the nick-translated pmal c-DNA clones as indicated.

putatively initiated basal cell lines MCA 3F and MCA 3D [6]. A difference between the untransformed putatively initiated cells and the transformed keratinocytes, however, was observed in respect to mal expression in serum-depleted cells. Whereas in the transformed cells mal expression was independent of serum, in the putatively initiated basal cells transcription of mal-1 and -2 decreased after removal of the serum and was turned on again within 4–6 hr after addition of serum or TPA to the medium (Fig. 5). In contrast, expression of mal-3 and -4 could not be influenced by serum or TPA. In addition, a different transcript pattern of mal-3-related sequences was observed in these putatively initiated cells. The 2.9-kb transcript, present in the other keratinocytes, was replaced by a larger one of 4.3 kb.

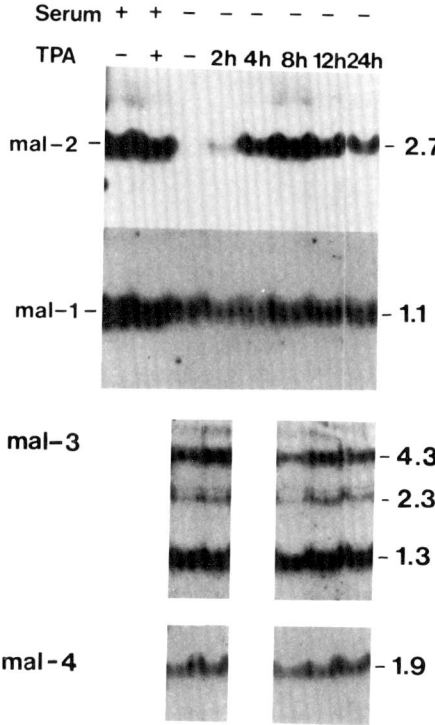

Fig. 5. Expression of mal sequences in serum-depleted MCA3/F cells and after TPA treatment. MCA3/F cells [6] were grown after trypsination in medium for 3 days (until confluent) in the presence of 10% fetal calf serum. Serum was removed and the cells were grown in serum-depleted medium for 24 hr. TPA ($10^{-7}$ M) was added to the medium, and the cells were harvested at different times after TPA application as indicated. RNA was isolated from cells grown 3 days in medium with serum (**1st lane**), from cells grown in medium with 10% FCS and treated for 12 hr with TPA (**2nd lane**), from cells grown for 48 hr in serum-depleted medium (**3rd lane**), or from cells treated with TPA after serum depletion for the time as indicated. Northern blots were performed as described [2] using the nick-translated cDNA clones as indicated.

## DISCUSSION

We have previously described the isolation of sequences activated during tumor development, using the well-defined in vivo system of multistage carcinogenesis in the mouse skin [2]. Molecular characterization of these mal-sequences showed that they are not homologous to 20 known retroviral oncogenes tested so far. They also did not show homology with actin- or keratin-specific probes. Preliminary data from DNA sequencing of mal-1 revealed that a stretch of 150 base pairs (bp) on the 3' end exhibits 65% DNA homology with the cDNA sequence of a mouse lipid binding protein [12] (data not shown). This suggests that the gene product of mal-1 may possess a lipid binding domain and, therefore, possibly is involved in lipid metabolism or is a membrane associated protein.

Overexpression of mal-4 was a marker for the malignant state, because this sequence was not overexpressed in the benign tumors and transcriptional activation occurred during progression from the benign to the malignant state. In contrast, overexpression of mal-1 and mal-2 was specific for neoplasia, in that they were transcriptionally activated in both papillomas and squamous cell carcinomas. Only weak transcriptional activity of the mal-sequences, however, was detectable in a keratoacanthoma. Benign keratoacanthoma differs from the more common benign papilloma and malignant carcinomas in that they arise from different cell types—namely, the cells of hair follicles [7]. It may be that tumor development in these cells occurs by other molecular mechanisms which can be characterized by differences in the level of mal-expression.

A change in the transcription pattern of mal-3-related sequences was observed during progression from the benign to the malignant state. We do not know whether these RNA species are different precursor molecules reflecting a change in RNA processing or whether these are different mature mRNA species. In different in vitro cultivated cell lines, we observed further transcripts of different sizes related to mal-3. It may be that these are transcribed from different promoters or that they are encoded by different genes sharing a common 3' end. Southern analysis of the genomic organization detected at least four copies of mal-3-related sequences (data not shown), which suggests that there exists a gene family of mal-3-related sequences. Nevertheless, in all benign tumors tested so far, we only detected the two transcripts of 1.3 and 2.3 kb, and the 2.9-kb transcript was only overexpressed in malignant tumors. Therefore, overexpression of this 2.9-kb transcript related to mal-3 may be used as a genetic marker to distinguish between the benign and malignant state of tumor progression in mouse skin.

The aim of our work was to investigate if transcriptional activation of genes is involved in the multistep carcinogenesis process. Initiation is thought to be a mutational event resulting in the production of initiated basal cells [8]. Indeed mutational activation of the c-Ha-ras oncogene has been reported to occur during tumor development in the mouse skin [9]. The development of papillomas during promotion may occur only by clonal expansion of initiated cells. However, if one considers the multiple effects of tumor promoters on cellular and viral genes in different in vitro systems [4] and that there exist genetic determinants for promotion sensitivity [10], it may be possible that in addition genetic effects of tumor promoters on the initiated cells play a role during the process of promotion.

Our data demonstrated that in vivo tumor promoters can turn on a transient expression of mal sequences in normal mouse epidermis which are normally repressed in vivo, but expressed in tumors or after in vitro cultivation of epidermal cells. Interestingly, the different mal sequences exhibited different kinetics of stimulated expression. Tumor-promoter-induced expression of mal-1 and -2 showed similar kinetics compared to promoter-induced DNA synthesis. It is known that the labeling index of epidermal cells reaches a maximum at 18 hr and 30 hr after promoter application [11]. Thus, we have correlative data suggesting that mal-1 and -2 are genes which are induced in those cells which are stimulated by promoters for replication. Expression of mal-3 and -4, induced earlier and for a shorter time by promoters, may be involved in processes preceding DNA synthesis. Strong transcriptional activity of the mal sequences was observed in different epidermal in vitro systems. Mal-1 and -4 expression appeared to be correlated with the growth rate in transformed epidermal cells. Expression of mal-2 and -3, in contrast, was constitutive in these cells. Whereas in transformed and tumorigenic keratinocytes transcriptional activity of mal sequences was independent of the serum requirement, in the immortalized but nontumorigenic, putatively initiated cells [6], expression of mal-1 and -2, but not of mal-3 and -4, could be influenced by serum or TPA. Thus, transformed keratinocytes do not need external stimuli like serum or TPA to maintain expression of mal-1 and -2. Possibly, expression of these genes is correlated with a certain growth or differentiation state of the cell, which in untransformed cells need external stimuli. In contrast, transformed cells are able to maintain this expression pattern without external stimuli, perhaps because they produce in an autocrine fashion enough stimulating factors by themselves.

The role of the mal sequences, however, can only be clarified by functional assays, because the observed transcriptional activation during tumor develpment could be related to epiphenomena. These functional assays, to be performed by transfecting full-length cDNA clones of the mal sequences in eukaryotic expression vectors into epidermal recipient cells, should tell us whether the mal sequences are able to induce phenotypic changes such as differentiation, immortalization, or malignant trans formation.

## ACKNOWLEDGMENTS

This work was supplied in-part by grant CA-40584 awarded to G.T.B. by the National Institutes of Health, Bethesda, Maryland.

## REFERENCES

1. Slaga TJ: Environ Health Perspect 50:3, 1983.
2. Melber K, Krieg P, Furstenberger G, Marks R:Carcinogenesis 7:317, 1986.
3. Balmain A, Krumlauf R, Vass JK, Birnie GD: Nucleic Acids Res 10:4259, 1982.
4. Hecker E, Fusenig NE, Kunz W, Marks F, Thielmann HW: Carcinogenesis—A Comprehensive Survey, Vol. 7. New York: Raven Press, 1982, pp 417–427, 617–625.
5. Fusenig NE, Breitkreutz D, Dzarlieva RT, Boukamp P, Herzmann E, Bohnert A, Pohlmann F, Rausch C, Schutz S, Hornung J: In Smith GJ, Stewart BW (eds): "*In Vitro* Epithelial Differentiation and Neoplasia." Sydney: Australian Cancer Society, 1982, p 209.
6. Kulesz-Martin M, Kilkenny AE, Holbrook KA, Digernes V, Yuspa SH: Carcinogenesis 4:1367, 1983.

7. Klein-Santo AFP: In Slaga TJ (ed) "Mechanisms of Tumor Promotion, Vol. II. Tumor Promotion and Skin Carcinogenesis." Boca Raton, FL: CRC Press, 1984, p 41.
8. Farber E: In Becker FF (ed) "Cancer: A Comprehensive Treatise," Vol. 1. New York: Plenum Press, 1982, p 485.
9. Balmain A, Ramsden M, Bowden GT, Smith J: Nature 307:658, 1984.
10. Colburn NH, Talmadge CB, Gindhart, TD: Mol Cell Biol 3:1182, 1983.
11. Krieg L, Kuhlmann I, Marks F: Cancer Res 34:3135, 1974.
12. Bernlohr DA, Angus WC, Lane MD, Bolanowski MA, Kelly TJ, Jr: Proc Natl Acad Sci USA 81:5468, 1984.

# Tumorigenic Transformation of Human Teratocarcinoma Cells by Activated *ras* Oncogene but Not the Homologous Photo-Oncogene

## Michael A. Tainsky

*Department of Tumor Biology, M.D. Anderson Hospital and Tumor Institute, Houston, Texas 77030*

*Ras* oncogenes have been found in approximately 15% of the human tumors analyzed. However, a causal role of these genes in the tumorigenesis of human cells has yet to be shown. Tumorigenic late passages PA-1 human teratocarcinoma cells (L-PA-1) contain an activated N-*ras* gene. In this report evidence is presented that nontumorigenic early-passage PA-1 cells (E-PA-1) contain only the germ line proto-oncogene. Introduction by gene transfer of the activated L-PA-1 oncogene induces E-PA-1 cells to form tumors, suggesting that the activated N-*ras* oncogene has a causal role in the tumorigenesis of these cells. These and other experiments indicate one possible mechanism by which activated *ras* oncogenes induces malignant transformation.

**Key words:** human cell transformation

Oncogenes in the *ras* family are forms of the germ-line photo-oncogenes with specific point mutations which when transfected onto NIH-3T3 murine fibroblasts induce foci of morphologically altered cells [1–13]. The *ras* genes code for proteins of molecular weight of approximately 21,000 daltons with guanine nucleotide binding activity which are able to hydrolyze GTP (GTPase) [14]. The oncogenic proteins with point mutations at amino acid positions 12 or 61 retain the ability to bind guanine nucleotides but have reduced GTPase activity [15–17]. Approximately 15% of human tumors contain activated transfectable oncogenes. However, the role in oncogenesis of these genes in human neoplasms is unclear. It has yet to be determined whether these genes have a causal role in tumor induction or are mutated as a consequence of their presence in tumor cells which may simply have increased mutation rates. Chemical carcinogenesis studies in rats have shown that specific oncogene activations can be detected following carcinogen treatment [18–21]. However, it has not been

Received May 15, 1986.

© 1988 Alan R. Liss, Inc.

demonstrated that these altered genes are responsible for tumor induction in these systems.

The present study addresses this issue using PA-1 human terato cells [22]. This cell line was isolated by culturing the ascites fluid from a 12-yr-old female with an ovarian teratocarcinoma. PA-1 cells at early passages do not form tumors in athymic nude mice, while late-passage cells ($>100$) readily from tumors in athymic nude mice. In a previous report [23] it was shown that the presence of an activated oncogene was correlated with the ability of PA-1 cells to form tumors in athymic nude mice, implying a causal role in the tumorigenic transformation of L-PA-1. Tumorigenic PA-1 cells contain an oncogene which has been activated by a point mutation at codon 12 of the ras p21 protein. In contrast, transfection of DNA isolated from E-PA-1 cells into NIH-3T3 murine fibroblasts failed to detect this activated N-*ras* which was readily detected using DNA from L-PA-1 cells. In this report I demonstrate a causal role for the activated N-*ras* in the tumorigenicity of PA-1 cells by introducing this oncogene into nontumorigenic early-passage cells and showing that the resulting cells exhibit a tumorigenic phenotype when injected into athymic nude mice.

## MATERIALS AND METHODS
### Cell Culture

PA-1 cells were grown in modified eagle's medium with 10% fetal calf serum. The cells were subcultured weekly by treatment with 0.25% trypsin in PBS with 2 mM EDTA. High molecular weight DNA was prepared from animal cells as previously described [19].

### Molecular Cloning

In order to be sure that the N-*ras* gene in DNA from passage 41 E-PA-1 cells did not contain an activating point mutation it was cloned. The N-*ras* gene was contained on R-I fragments of 9 and 7 kilobases [24]. Since the point mutation responsible for the activation of the N-*ras* L-PA-1 cells resides on the 9-kb 5' Eco RI fragment, I cloned this fragment from DNA isolated from E-PA-1 cells at passage 41. DNA which had been digested to completion with Eco RI endonuclease was electrophoresed on preparative agarose gels and fractions containing fragments of 9 kb which hybridized a 5'-specific N-*ras* probe were isolated. DNA contained in these fractions was ligated to purified Eco RI arms of the bacteriophage vector lambda GT WES B [25], packaged, and screened by plaque hybridization to the N-*ras* "R" probe [9]. The Eco RI fragments from eight independent phage were ligated to the 3' N-ras 7-kb fragment cloned from human placental DNA (a gift from M. Wigler) and tested for biological transforming activity by transfection in the NIH-3T3 focus formation assay.

### Transfection

DNA-mediated transfection was performed by the method of Graham and Vander Eb [7] as previously described [13]. Monolayers of NIH-3T3 cells were carried at confluency for 14–21 days after transfection in pre-selected calf serum in Dulbecco's modified Eagle's medium (DMEM) and observed for the appearance in foci of morphologically altered cells.

## Protoplast Fusion

Protoplast fusion [28] was a useful gene transfer method to introduce cloned genes into certain cells which like PA-1 human teratocarcinoma cells do not survive the standard transfection protocol. Human recipient cells received oncogene plasmid constructions by protoplast fusion, and the resulting cell lines were analyzed for tumor formation in athymic nude mice. By linking the genes of interest to the selectable plasmid pSV2-neo [26] for resistance to the neomycin analog G-418, oncogenes can be introduced and selected for in a single step. The resulting drug-resistant cells have been used as a pool or as individual colony derived clonal lines. Plasmids which are to be introduced into cells were freshly transformed into the *Escherichia coli* DH-1 prior to each experiment. A bacterial colony was picked 2 days before the actual fusion experiment, and an overnight culture was prepared in the L-broth with the appropriate antibiotic. Two milliliters of the overnight culture was added to 50 ml of L-broth with the appropriate antibiotic in 250-ml flasks, and this culture was grown to A600 = 0.5 when chloramphenicol was added to 0.2 mg/ml. This culture was then incubated for an additional 14–18 hours. The polyethylene glycol 1500 (PEG) (BDH Chemical Co.) was prepared the day before the fusion experiment. Fifty millileters of molten PEG was mixed with 50 ml of DMEM and filtered immediately through a Nalgene 0.45-$\mu$m sterilizing unit. This 50% PEG was kept overnight in a tissue culture incubator in sterile plastic 50-ml tubes.

The protoplasts were prepared as follows: Each 50-ml bacterial culture was centrifuged at 4,000 rpm for 10 min. The medium was decanted and the bacteria resuspended on ice in a 2.5 ml of a buffer containing 20% sucrose and 50 mM Tris-HCl, pH 8.0. This suspension was incubated on ice with 0.5 ml of 5 mg/ml lysozyme (freshly prepared) in 0.25 m Tris-HCl, pH 8.0, for 5 min. Then 1 ml of 0.25 M EDTA, pH 8.0, was added. After 5 min, 1 ml of 50 mM Tris-HCl, pH 8.0, was added, and this mixture was incubated at 37°C until the bacteria were converted to protoplasts (usually 15 min) as judged by phrase contrast microscopy. The protoplasts were then diulted very slowly and gently with 20 ml serum-free DMEM containing 10% sucrose and 10 mM $MgCl_2$ and kept at 37°C for 15 min.

Recipient cells for protoplast fusion were seeded at 300,000 per 60-mm dish 14–18 hr before the experiment. The cells were washed once with 5 ml of prewarmed serum-free DMEM. Then 6.25 ml of the protoplast suspension was added to each 60-mm dish and centrifuged in a Sorvall RT-6000 centrifuged at 2,000 rpm for 12 min at room temperature. At the inside edge of the tissue culture dish nearest to the center of rotation, few cells on protoplasts remained after centrifugation. Therefore, all subsequent additions were performed at this edge of the dish. The medium was removed by aspiration, and 2.5 ml of the 50% PEG was added. After 2 min the 50% PEG was removed and the cells were washed five times with prewarmed serum-free DMEM. Complete DMEM was then added to the cells, which were then returned to the incubator. After 24 hr the cells from one 60-mm dish were trypsinized and transferred to a T-75 flask. After an additional 24 hr, the selective medium was added. The culture medium was changed twice weekly until large colonies were apparent. No cells remained in control cultures receiving no plasmid. The G-418-resistance colonies were then subcultured, expanded, and inoculated into animals.

## Animal Inoculations

Athymic nude mice, 3–6 wk old, were inoculated with 1 million cells in 0.1 ml Hanks' solution. Subcutaneous injection will be performed at an intrascapular location. Mice were treated with x-rays at a dose of 500 rads. Mice were checked weekly for tumor growth and general health. Mice were sacrificed by cervical dislocation, and the tumors were excised under sterile conditions for the establishment of cell lines and the extraction of DNA.

## RESULTS

It was known from previous transfection studies that DNA from nontumorigenic E-PA-1 cells does transfect the oncogene that was activated N-*ras* found in DNA from L-PA-1 cells. At that time I correlated the presence of this oncogene with tumorigenicity. It was possible that the activated oncogene was present in these cells sequestered by some mechanism (such as methylation) which masked its biological activity in transfection assays. Therefore, the N-*ras* locus from DNA isolated from passage 41 E-PA-1 cells was molecularly cloned. The gene is contained in Eco RI fragments of 9 kb and 7 kb [24]. Since the point mutation responsible for the activation of the N-*ras* in L-PA-1 cells resides on the 9-kb 5' Eco RI fragment, this fragment was cloned from DNA isolated from E-PA-1 cells at passage 41. The 9-kb 5' N-*ras* R-I fragments from eight independent phage were ligated to the 3' N-*ras* 7-kb fragment cloned from human placental DNA (a gift from M. Wigler) and tested for biological trasnforming activity in the NIH-3T3 focus formation assay. None of the eight early-passage genes induced foci on transfection, while genes constructed utilizing the equivalent R-I fragment cloned from L-PA-1 DNA produced numerous foci. Therefore, the N-*ras* genes in E-PA-1 cells do not contain an activating mutation within the 5' portion of the gene.

In order to determine whether this activated N-*ras* oncogene has a causal role in the tumorigenesis of these cells, gene transfer studies were undertaken. The goal of this study was to determine if a nontumorigenic E-PA-1 cell line could form tumors in athymic nude mice after the addition of the activated oncogene. For these experiments a clonal PA-1 cell line was used to avoid potential complications caused by heterogeneity in the recipient cell population. A nontumorigenic clone of E-PA-1 cells (clone 1) was isolated at passage 63. The cloned cells were carried an additional 50 passages before these experiments but remained nontumorigenic (Table I). The activated N-*ras* was cloned onto the drug selectable plasmid pSV2-neo [26].

Introduction of this plasmid into recipient cells, followed by selection with the neomycin analogue G-418, should select for a population of cells carrying both the selected plasmid sequences and the N-*ras* oncogene. Since PA-1 cells do not survive the standard calcium phosphate transfection procedure [27], protoplast fusion [28]

**TABLE I. Tumorigenicity of N-*ras*-Transformed PA-1 Clone 1 Cells***

| Experiment | Fraction with tumors | Latent period (wk) |
|---|---|---|
| 1 | 1/3  2/3  2/3  0/3  0/3  0/3 | 12  13  21  22  25 |
| 2 | 5/8 | 15  18  19  21  22 |
| 3 | 1/3 | 7 |
| 4 | 2/3 | 8  13 |

*PA-1 cells cells with the pSV2-neo N-*ras* plasmid construction were inoculated subcutaneously into 3–6-wk-old athymic nude mice at 1 million cells per animal. Mice were observed for tumor growth. Experiment 1 was performed using cell lines derived from individual G-418-resistant colonies. Experiments 2–4 were performed using pools of 20–100 G-418-resistant colonies.

was employed to introduce the construct into the recipient cells. In an initial experiment six G-418-resistant colonies were isolated and expanded into cells lines. Three of these six cell lines also were able to form tumors in athymic nude mice (experiment 1; Table I).

Further, in multiple independent experiments, clone 1 E-PA-1 cells, into which the activated N-*ras* was introduced, acquired the ability to form tumors in nude mice (expt. 2, 3, 4; Table 1). Since I had found that as few as 100 L-PA-1 cells in a mixture with 1 million nontumorigenic E-PA-1 cells could form tumors (data not shown), the remaining three experiments were performed using pools of G-418-resistant colonies. These populations were able to form tumors when injected into the nude mice (Table I). Therefore, it appears that either method (individual colonies or pools of colonies) is capable of giving rise to tumorigenic PA-1 cells and that the presence of nontumorigenic cells does not inhibit tumor induction by the transformed cells. Introduction of the pSV2-neo vector alone, or a pSV2-neo construct containing the normal human N-*ras* proto-oncogene, into clone 1 cells did not result in tumor formation in athymic nude mice.

DNA from the clone 1 transformants prior to injection into mice, as well as DNA from tumors induced by these cells, was analyzed to determine the state of the newly acquired N-*ras* sequences. Approximately single-copy level of the gene transferred injection and in the tumors formed after injection, indicating that gene amplification of the oncogene sequences was not required for tumor formation by these clone 1 cells (data not shown). Chromosomal analyses of the cells before and after the pSV2-neo N-*ras* construct were indistinguishable (data not shown). Therefore, the induction of the tumorigenic phenotype of PA-1 cells was consistent with its being the result of the addition of a single copy of an activated oncogene.

## DISCUSSION

PA-1 cells were shown to be tumorigenic in athymic nude mice when they contained an activated N-*ras* oncogene and nontumorigenic when it was absent. In order to prove that the activated N-*ras* oncogene caused the tumorigenicity, the oncogene was introduced into a nontumorigenic E-PA-1 cell clone, and the resulting cells were tumorigenic. Introduction by gene transfer of the nonmutated proto-oncogene did not induce tumorigenesis. These data suggest that the activated *ras* oncogenes induce the ability to form tumors. The activated oncogenes may simply increase the growth rate of these cells or they may alter some other factor in the cell which affects the growth in the animal host. I suggest the latter. A few observations may bear on identifying a potential mechanism by which the oncogene induces tumorigenic transformation in these cells. Tumors in L-PA-1 cells as well as clone 1 cells transformed by the activated L-PA-1 N-*ras* oncogene can be classified histologically as neuroblastomas. In addition, PA-1 cells when induced to differentiate in culture, resemble cells of neural origin. The state of differentiation of a cell can affect the cell's sensitivity to host immune surveillance. While undifferentiated embryonal carcinoma cells are highly sensitive to natural-killer (NK) cells they become resistant to killing after they have been induced to differentiate [29]. This has been confirmed for PA-1 cells. Athymic nude mice are unable to produce mature T-cells and are high in NK activity. I postulate that the activated ras oncogene may induce some partial differentiation of PA-1 cells sufficient for them not to be recognized by host defenses (in this case the NK cells of the nude mouse) but not enough differentiation to shut down growth and cell division. Consistent with this hypothesis, recently two groups have found that *ras* oncogenes can induce differentiation of rat pheochromocytoma cells [31,32], and in one case the corresponding proto-oncogene was not able to induce differentiation [32]. In summary, I postulate that the mechanism by which *ras* oncogenes transform PA-1 cells may involve partial differentiation which allows the cells to become resistant host immune defenses. This immunological basis for the mechanism by which *ras* oncogenes tumorigenically transform cells may be different from the mode by which enhanced expression of the proto-oncogene acts. This may be mediated through signal transduction of an adenyl cyclase-like system [33,34] which affects the expression of factors responsible for the escape from immune defenses.

## REFERENCES

1. Shih C, Shelo BZ, Goldfarb MP, Dannenberg A, Weinberg RA: Proc Natl Acad Sci USA 76:5714, 1979.
2. Lane MA, Sainter G, Cooper M: Cell 28:273, 1982.
3. Parada LF, Tabin CJ, Shih C, Weinberg RA: Nature (Lond) 297:474, 1982.
4. Santos E, Tronick SR, Aaronson SA, Pulciani S, Barbacid M: Nature (Lond) 298:343, 1982.
5. Der CJ, Krontris TG, Cooper GM: Proc Natl Acad Sci USA 79:3637, 1982.
6. Der CJ, Cooper GM: Cell 32:201, 1983.
7. Goldfarb MP, Shimizu K, Perucho M, Wigler MH: Nature (Lond) 296:404, 1982.
8. Hall A, Marshall CJ, Spurr NK, Weiss RA: Nature (Lond) 303:396, 1983.
9. Shimizu K, Goldfarb M, Perucho M, Wigler M: Proc Natl Acad Sci USA 80:383, 1983.
10. Reddy EP, Reynolds RK, Santos E, Barbacid M: Nature (Lond) 300:149, 1982.
11. Tabin CJ, Bradley S, Bargman C, Weinberg R, Papageorge A, Scolnick E, Dhai R, Lowy D, Chang E: Nature (Lond) 300:143, 1982.
12. Taparowsky E, Suard Y, Fasano O, Shimizu K, Goldfarb M, Wigler M: Nature (Lond) 300:762, 1982.
13. Cooper CS, Blair DG, Oskarsson MK, Tainsky MA, Eader LA, Vande Woude GF: Cancer Res 44:1, 1984.
14. Popageorge A, Lowy D, Scolnick E: J Virol 44:509–519, 1982.
15. Sweet RW, Yokoyama S, Kamata T, Feramisco J, Rosenberg M, Gross M: Nature 311:273–275, 1984.
16. McGrath J, Capon D, Goeddel D, Levinson A: Nature 310:644–649, 1984.
17. Gibbs J, Sigal I, Poe M, Scolnick E: Proc Natl Acad Sci USA 81:5704–5709, 1984.
18. Sukamar S Notario SV, Martin-Zanca D, Barbacid M: Nature 306:658–661, 1983.
19. Balmain A, Pragnell IB: Nature 303:72–74, 1983.
20. Eva A, Aaronson SA: Science 220:955–956, 1983.
21. Guerrero I, Calzada P, Mayer A, Pellicer A: Proc Natl Acad Sci USa 81:202–205, 1984.
22. Zeuthen J, Norgaard JDR, Avner P, Fellows M, Wartiovaara J, Vaheri A, Rosen J, Giovannela BC: Int J Cancer 25:19, 1980.
23. Tainsky MA, Cooper CS, Giovanella BC, Vande Woude GF: Science 225:634–645, 1984.
24. Taparowsky E, Shimza K, Goldfarb M, Wigler M: Cell 34:586, 1983.
25. Tilghman SM, Tiemeier DC, Polsky F, Edgell M, Seidman JG, Leder A, Enquist IW, Norman B, Lee P: Proc Natl Acad Sci USA, 74:4406–4410, 1977.
26. Berg P, Southern PJ: Mol Appl Genet 1:327, 1982.
27. Graham FL, Vander Eb AJ: Virology 52:456, 1973.
28. Yoakum GH: Bio Techniques 24–29 (Jan/Feb), 1984.
29. Stern P, Gidlund M, Wigzell H: In Muramatsu T, Gachelin G, Moscona A, Ikawa Y (eds): "Teratocarcinoma and Embryonic Cell Interactions." Academic Press, Orlando Florida:1982, Chapter 6, pp 75–101.
30. Tainsky MA: J Virol 37:922–930, 1981.
31. Bar-Sagi D, Feramisco J: Cell, 42:841–848, 1985.
32. Noda M, Ko M, Ogura A, Liv D, Amano T, Takaro T, and Ikara Y: Nature 313:73–75 (1985).
33. Gilman AG: Cell 36:577–579, 1984.
34. Tamanoi F, Walsh M, Kataoka T, Wigler M: Proc Natl Acad Sci USA 81:6924–6928, 1984.

# Use of Cell Variants to Study the Molecular and Cellular Determinants of Tumor Promotion

H. Yamasaki, M. Hollstein, E. Hamel, L. Giroldi, E. Rivedal, T. Sanner, and T. Kakunaga

*International Agency for Research on Cancer, 69372 Lyon, France (H.Y., M.H., E.H., L.G.); Norwegian Radium Hospital, Oslo 3, Norway (E.R., T.S.); Osaka University, Osaka 565, Japan (T.K.)*

In order to study molecular and cellular determinants of tumor promotion and of cellular susceptibility to tumor-promoting agents, we used several genetic cell variants; they include (1) 12-O-tetradecanoylphorbol-13-acetate (TPA)-sensitive and -resistant murine erythroleukemia cells in which cell differentiation can or cannot be inhibited by TPA; (2) a murine erythroleukemia cell variant which is dependent on TPA for its growth (these cells are already committed to differentiate); (3) Syrian hamster embryo cell lines which are TPA-sensitive or -resistant with respect to enhancement of cell transformation; and (4) BALB/c 3T3 cell variants with high and low susceptibility to induction of cell transformation by chemical carcinogens or ultraviolet (UV) light. From the use of TPA-resistant cells, we conclude that the difference in receptor occupancy by TPA is not a mechanism by which cells become TPA-resistant. From these variant cells, we also concluded that diacylglycerol is a functional analogue of TPA, since diacylglycerol inhibits differentiation of TPA-sensitive, but not TPA-resistant, murine erythroleukemia cells, and since it also inhibits intercellular communication of TPA-sensitive Syrian hamster embryo cells, but not TPA-resistant variants. From the use of TPA-resistant Syrian hamster embryo cells and BALB/c 3T3 cell variants, we confirmed that intercellular communication blockage plays an important role in enhancement of cell transformation. Analysis of proto-oncogenes of several variants of murine erythroleukemia cells indicates that methylation patterns of certain oncogenes are different in these genetic variants, suggesting that expression of some proto-oncogenes may be associated with cell differentiation state and/or susceptibility to TPA-mediated inhibition of differentiation.

**Key words: TPA, erythroleukemia cells, intercellular communication, proto-oncogenes**

There are numerous reports on the effects of tumor-promoting agents on cells [1–3]. However, we are still far from understanding the mechanisms of action of tumor-promoting agents and the mechanisms of tumor promotion. It is now important

Received November 6, 1986.

© 1988 Alan R. Liss, Inc.

to know which of these reported effects or phenotypes are really essential in the process of tumor promotion. In this regard, development and analysis of genetic variants in terms of their susceptibility to tumor-promotion agents or their difference in expression of promotion-related phenotypes is very useful.

In the process of tumor promotion, it is postulated that there is selective clonal expansion of initiated cells, but not of surrounding normal cells; ie, there should be excess proliferation of initiated cells. However, proliferation alone may result only in hyperplasia and we consider that it is also important that there be inhibition of cell differentiation to obtain a tumor. In order to see how these events occur, we used two biological systems, in vitro cell transformation and cell differentiation. Based on results from these studies and from other laboratories, we present the working hypothesis that, in order for initiated cells to expand clonally, it is important to disturb intercellular communication between initiated cells and surrounding normal cells [4,5]. Since cell proliferation in itself is not enough, we believe that some gene which is involved in cell differentiation should be altered during the initiation phase, and the same gene may be expressed during the promotion phase.

In order to study gene expression during differentiation and the effect of tumor-promoting agents on gene expression, we used different genetic variants of murine erythroleukemia cell lines. These cells enter into terminal cell differentiation in the presence of different inducing agents, such as DMSO, hexamethylene bisacetamide (HMBA), butyric acid, etc [6]. When low concentrations of 12-O-tetradecanoylphorbol-13-acetate (TPA) are added to culture media, both spontaneous and induced cell differentiation is inhibited [7,8]. We have isolated TPA-resistant clonal variants from the parent lines [9]; we have also established a cell line which was incubated for a prolonged time with an inducer of differentiation (HMBA) and TPA; this cell line became TPA dependent for growth, since the removal of TPA results in terminal cell differentiation and terminal cell divisions [10]. These variants do not require any inducer for their differentiation upon removal of TPA. Therefore, we consider this cell line as a variant which is TPA-dependent for proliferation and committed to differentiate.

We used two different cell lines in which two-stage cell transformation has already been successfully carried out, ie, BALB/c 3T3 and Syrian hamster embryo cells. One of us (T.K.) has isolated a number of clonal variants of BALB/c 3T3 A-31 cells with different susceptibilities to ultraviolet (UV) light or chemicals with respect to induction of cell transformation [11,12]. Since these cell lines responded quite equally to known carcinogens such as benzo(a)pyrene and UV in terms of their effects on DNA, we speculated that these cell lines are different in their expression of certain phenotypes related to the tumor-promotion process [12]. Syrian hamster embryo cell variants were isolated with differing susceptibilities to TPA-induced enhancement of cell transformation [13]. These variants were used to see whether some of the TPA-induced phenotypes are related to TPA-induced cell-transformation enhancement.

## CELL LINES AND EXPERIMENTAL METHODS

TPA-sensitive and -resistant subclones of murine erythroleukemia cells (MELC) were isolated and characterized previously [9,14]. Another clonal variant, HT clone 2, is the line which had been incubated in the presence of HMBA and TPA over a 4-

yr period [10]. Isolation and preliminary characterization of BALB/c 3T3 clonal variants in terms of their susceptibility to chemical or UV-induced cell transformation have been described previously [11,12], as have isolation and characterization of TPA-sensitive and -resistant Syrian hamster embryo cell lines [13].

Detailed procedures for the phorbol 12,13-dibutyrate (PDBu) binding site assay [15], protein kinase C activity assay [16], and cell transformation assay [17,18] have been already published. Junctional communication capacity was measured with a micro-injection dye transfer assay in which Lucifer Yellow CH was microinjected into individual cells, after which the spread of the Lucifer Yellow molecules into neighbouring cells was monitored under a microscope [19].

The methylation state of cellular proto-oncogenes was assessed by parallel DNA restriction enzyme digestions with the isoschizomers HpaII and MspI. HpaII does not cleave if the internal cytosine of the CCGG recognition sequence is methylated. DNA was extracted from MELC variants, digested with restriction enzyme, resolved on agarose gels, and hybridised under stringent conditions following transfer to solid support, with $^{32}$P-labelled v-onc probes, according to published procedures [20].

## RESULTS

### Relationship Between TPA-Resistance, Phorbol Ester Binding Sites and Protein Kinase C Activity

Phorbol ester binding sites were measured in two sets of TPA-resistant and TPA-sensitive cell-type pairs. Both TPA-sensitive and TPA-resistant MELC had a similar number of binding sites with similar affinities for phorbol esters as measured with $^3$H-PDBu as a ligand. Similarly, TPA-resistant Syrian hamster embryo cells had a normal level of PDBu binding sites with normal affinity. However, when we analysed PDBu binding in TPA-sensitive Syrian hamster embryo cell lines, we were surprised to see that there were almost no binding sites. Further studies comparing TPA-sensitive and -resistant Syrian hamster embryo cells suggested that TPA-sensitive cells probably metabolize PDBu so quickly that we cannot detect PDBu binding sites. In fact, when we used $^3$H-TPA as a ligand, we could demonstrate that TPA-resistant and TPA-sensitive cells have high-affinity binding sites.

Consistent with these results, we also found that TPA-resistant murine erythroleukemia cells and TPA-resistant Syrian hamster embryo cells have protein kinase C activity similar to that of their TPA-sensitive counterparts. These results indicate that the absence of TPA binding sites or protein kinase C is not the mechanism of TPA resistance in these cell lines.

### TPA Resistant Variants Are Also Resistant to Diacylglycerols

Since TPA and other related tumor-promoting agents bind to and activate protein kinase C, replacing the role of diacylglycerols [21,22], it has been suggested that diacylglycerol is a functional analogue of phorbol esters [23]. There is now a long list of biological systems in which diacylglycerol indeed mimics TPA effects. In order to see whether TPA resistance also confers diacylglycerol resistance, we utilized our available genetic variants.

We used 1-oleoyl 2-acetyl glycerol (OAG), since this analogue can intercalate membranes and exert biological effects. When TPA-sensitive MELC were incubated with OAG, the induction of differentiation by HMBA was inhibited. However, as expected, OAG failed to inhibit differentiation of TPA-resistant MELC. In this experiment, it was necessary to add OAG many times a day, for 3 days, since OAG is rapidly metabolized, and continuous presence of OAG is necessary for inhibition of cell differentiation [16].

We have previously shown that OAG can inhibit intercellular communication between BALB/c 3T3 cells [24]. When we incubated TPA-sensitive Syrian hamster embryo cells with OAG, we also inhibited intercellular communication. However, when TPA-resistant variants of Syrian hamster embryo cells were incubated with OAG, there was no inhibition of intercellular communication.

Effects of OAG on these genetic variants are summarized in Table I. These results support the model that diacylglycerols are functional analogues of phorbol esters in exerting various biological effects.

**TABLE I. Effects of 1-Oleoyl 2-Acetyl Glycerol (OAG) on 12-O-Tetradecanoylphorbol-13-Acetate (TPA)-Sensitive and -Resistant Variant Cells***

| OAG treatment | TPA-sensitive cells | TPA-resistant cells |
|---|---|---|
| Effect on junctional intercellular communication in Syrian hamster embryo cell lines (% communication of control) | | |
| Treated for 1 hr (µg/ml) | BPNi | 83-106 |
| 10 | 59 | 86 |
| 100 | 34 | 106 |
| 200 | 9 | 80 |
| Effect on induced differentiation of Friend erythroleukemia cells (% differentiation of control) | | |
| Treated 7 times per day for 3 days (µg/ml for each treatment) | TS 19-101 | TR19-9 |
| 0.03 | 50 | 94 |
| 0.3 | 26 | 92 |
| 3 | 5 | 87 |

*See Cell Lines and Experimental Methods for measurement of intercellular communication and Friend cell differentiation. Some data are taken from [16].

**TABLE II. Characterization of TPA-Sensitive (BPNi) and TPA-Resistant (83-106, 74-100) Syrian Hamster Embryo Cells***

| Cell line | TPA effect on— | | Phorbol ester binding | | Protein kinase C activity | OAG inhibition of intercellular communication |
| | Enhancement of cell transformation | Inhibition of intercellular communication | PDBu | TPA | | |
|---|---|---|---|---|---|---|
| BPNi | + | + | −[a] | + | + | + |
| 83-106 | − | − | + | + | + | − |
| 74-100 | − | − | ND[b] | ND | ND | ND |

*Some data are from [13].
[a]Negative finding is probably due to rapid metabolism of PDBu.
[b]No data.

## The Relationship Between Enhancement of Cell Transformation and Blocked Junctional Intercellular Communication

In order to see whether TPA-mediated enhancement of cell transformation is related to TPA blockage of junctional intercellular communication, we have incubated TPA-sensitive or TPA-resistant Syrian hamster embryo cells with TPA. When a TPA-sensitive cell line was incubated with TPA, there was a rapid inhibition of junctional communication at low concentrations of TPA, but there was almost no inhibition of junctional communication by TPA in two TPA-resistant clones of Syrian hamster embryo cells [13]. Since TPA can enhance cell transformation of only TPA-sensitive clonal cells, this result suggests a good correlation between the block of junctional communication and enhancement of cell transformation by TPA. Characterization of Syrian hamster embryo cell lines is summarized in Table II.

A variant of BALB/c 3T3 cells was used to see whether blockage of junctional intercellular communication is also important in the process of enhancement of cell transformation when no TPA is present. Transformation-sensitive clonal cell variants can be transformed quite easily by chemicals or by UV, whereas resistant variants were unresponsive to these carcinogenic stimuli [11,12]. Detailed studies by Kakunaga's group have shown that they have a similar ability to metabolize and bind chemical carcinogens and they have a similar ability to repair DNA damage [25]. Moreover, they were both similarly susceptible to chemical- or UV-induced mutation [12]. When these cells were exposed to TPA, we found they have similar affinities and numbers of phorbol esters receptors, and they showed similar responses to TPA, such as the inhibition of junctional intercellular communication, stimulation of 2-deoxyglucose, and enhancement of cell transformation [26]. These preliminary results have indicated to us that these cell variants are different in their intrinsic ability to express some of phenotypes related to enhancement of cell transformation.

When these cell lines were cultured for 5 wks, and intercellular communication capacity was measured, we found a dramatic difference between these two cell lines. When cells were in their growing phase, both cell lines showed a similar capacity to communicate junctionally. However, when they reached confluence, only transformation-resistant cell lines maintained the same ability to communicate junctionally; transformation-sensitive cell lines lost their junctional communication capacity, as though we had added TPA into cell cultures. These results suggest that transformation-sensitive clonal cells can express a TPA-like effect, ie, intercellular communication blockage, by reaching confluency. We conclude that this transformation-sensitive clonal cell is a promotion-proficient cell variant of BALB/c 3T3 cells, and these results again support the idea that blocked communication is an important determinant of transformation enhancement [26].

Results from comparative studies of two clonal variants from BALB/c 3T3 A 31 cells are summarized in Table III.

## Methylation Pattern of Proto-Oncogenes and Murine Erythroleukemia Cell Differentiation

Our previous results on characterization of TPA-resistant Friend cells in comparison to TPA-sensitive variants are summarized in Table IV.

The methylation of cytosine in proto-oncogene-containing DNA of MELC was studied in order to see whether at these sequences methylation patterns are related to

TABLE III. Characterization of Transformation-Sensitive (1-13) and -Resistant (1-8) Variants of BALB-c 3T3 A 31 Cells*

| Transformation by MCA or UV | Response to benzo[a]pyrene (B[a]P) | | | | |
|---|---|---|---|---|---|
| | Transformation frequency ($\times 10^5$/surviving cells) | Metabolism $^3$H-B(a)P (%3-OH B[a]P) | B(a)P binding[a] ($\mu$mol/mol nucleotide) | Excision of B[a]P-DNA (% removal/ 48 hr) | $^3$H-PDBu binding KD (mM) |
| Cell line 1-8 Very low | < 0.5 | 43.2 | 3.8 ± 0.2 | 10-15 | 12.7 |
| Cell line 1-13 Very high | 20.8 ± 8.0 | 45.3 | 3.3 ± 0.1 | 10-15 | 16.1 |

*Data are taken from [11, 12, 26]. UV, ultraviolet light; MCA, 3-methylcholanthrene; B[a]P, benzo[a]pyrene.
[a]Type of DNA adduct: (±)-7, 8-dihydroxy-9, 10-epoxy-7,8,9,10-tetrahydrobenzo(a)pyrene-deoxyguanosine.

the expression of differentiation or to cell susceptibility to TPA-mediated inhibition of differentiation. We have probed the methylation state of three cell variants—namely, TPA-resistant, TPA-sensitive, and TPA-dependent cell lines—and also have looked for changes in DNA methylation as the cells differentiate. Methylation patterns of sequences at or near seven proto-oncogenes were studied using the restriction enzyme pair HpaII and MspI. We found that sequences surrounding abl, raf, myc, and K-ras were heavily methylated, and no differences were detected among cell variants tested (Table V).

Sequences surrounding the fos and H-ras genes were more heavily methylated in TPA-resistant cell lines than in TPA-sensitive cell lines. The methylation pattern of TPA-sensitive cells did not change when the cells were induced to differentiate. Cellular sequences hybridizing with the v-sis probe in TPA-sensitive or TPA-resistant cells are relatively unmethylated in comparison with DNA of the differentiation committed variant, and again the methylation pattern of sequences probed with this oncogene did not change when cells were induced to differentiate.

Methylation patterns of TPA-dependent variants which are differentiation committed were studied both in differentiating and undifferentiating cells. Sequences surrounding the fos gene were heavily methylated, and those of sis- and H-ras genes were partially methylated. All three oncogene methylation patterns were not changed following differentiation by removal of TPA.

## DISCUSSION

Our results suggest that TPA-induced biological effects do not depend on the receptor occupancy by TPA, since both TPA-sensitive and -resistant cells have similar receptors for phorbol esters. Moreover, we found protein kinase C activity both in TPA-resistant and -sensitive cells. Similar results have been reported using other TPA-resistant cell lines [27–31]. Recent studies suggest that for the activation of protein kinase C by phorbol esters, the protein kinase C should be translocated from

| | Response to 12-O-tetradecanoylphorbol-13-acetate (TPA) | | | | | |
|---|---|---|---|---|---|---|
| $^3$H-PDBu binding (No./cell) | TPA metabolism (T1/2, hr) | Enhancement of MCA-initiated transformation | Stimulation of 2-deoxy-glucose uptake (-fold) | Inhibition of junctional communication (% inhibition) | | Junctional intercellular communication at culture confluence |
| | | | | Electrical coupling | Dye transfer | |
| $2.9 \times 10^5$ | < 24 | + | 8.0 | 35 | 86 | High |
| $2.6 \times 10^5$ | < 24 | + | 6.2 | 23 | 89 | Low |

cytoplasm to the membrane [23]. We have not yet carried out the translocation experiments. On the other hand, PDBu binding usually measures the binding to already translocated protein kinase C, and the extent of the binding is similar in these variants. Therefore, it is not likely that translocation is the mechanism of this resistance [16].

Although our results described above suggest that receptor occupancy and protein kinase C activation are not the mechanisms for TPA susceptibility, our results do suggest that protein kinase C is involved in exerting various TPA effects. Thus the presumptive functional analogue of TPA, diacylglycerol, can inhibit cell differentiation of only TPA-sensitive MELC; and similarly, diacylglycerol can inhibit junctional intercellular communication of only TPA-sensitive Syrian hamster embryo cells. These results tend to suggest that the determining factors in TPA resistance come into play after phorbol ester binding to, and activation of, protein kinase C, and one possible mechanism for this TPA susceptibility difference is that these different genetic variants have different substrates for protein kinase C; some of these substrates which are essential to exert TPA effects may be missing in TPA-resistant cells. This hypothesis has been tested recently by Kramer and Sando [32] using EL 4 mouse thymoma cell variants, and they indeed found a difference in C kinase substrates between phorbol ester-sensitive and-resistant cells.

The results obtained from our genetic variants also reinforce our working hypothesis that blocked junctional intercellular communication is an important determinant in the process of tumor promotion. Although we used junctional intercellular communication as an end point to measure cell-to-cell interaction, it is still possible that junctional communication inhibition is secondary to more general cell-to-cell recognition. In accordance with this idea, we have recently found that when BALB/c 3T3 cells are transformed by chemicals or by transfection of activated oncogenes, these transformed cells communicate junctionally in a similar way as normal cells. However, the transformed cells did not communicate with normal cells [33; unpublished observations]. Since these transformed or normal cells have the ability to communicate with each other, it is reasonable to assume they have the ability to produce junctional structures. These results suggest that it is rather cell-to-cell recognition which inhibits the formation of junctions between transformed and nontrans-

**TABLE IV. Characterization of TPA-Sensitive (TS 19-101) and -Resistant (TR 19-9) Clonal Variants of Friend Erythroleukemia Cells***

| Cell line | Induction of differentiation by DMSO or HMBA | Inhibition of differentiation | | | TPA activity | | | | | | | |
|---|---|---|---|---|---|---|---|---|---|---|---|---|
| | | TPA | OAG | Dexamethasone | Local anesthesia | Cell adhesion induction | Plasminogen activator induction | Transient growth inhibition | Induction of prostaglandin synthesis | Changes in surface membrane protein | Metabolism of TPA | PDBu binding sites | Protein kinase C activity |
| TS 19-101 | + | + | + | + | + | + | + | + | + | + | + | + | + |
| TR 19-9 | + | – | – | + | + | – | – | – | – | – | + | + | + |

*Some data are taken from [9,14,15,37,38]. HMBA, hexamethylene bisacetamide.

TABLE V. Methylation Pattern of Various Proto-Oncogenes in TPA-Sensitive (TS-19-101), TPA-Resistant (TR 19-9), and TPA-Dependent (HT Cl 2) Variants of Friend Erythroleukemia Cells*

| Differentiation state | Phenotype | Methylation surrounding proto-oncogenes | | | |
|---|---|---|---|---|---|
| | | fos | sis | H-ras | abl myc raf K-ras |
| MELC TR 9 | Undifferentiated, TPA$^R$ | + | − | + | Sequences surrounding these genes are heavily methylated in all cell variants |
| TS 101 | Undifferentiated, TPA$^S$ | (+) | − | (+) | |
| TS 101-D | Differentiating | (+) | − | (+) | |
| HT 2 | Committed to differentiation | + | (+) | (+) | |
| HT 2-D | Differentiating | + | (+) | (+) | |

*+, heavily methylated; (+), partial demethylation; −, extensively demethylated.

formed cells. Therefore, cell-to-cell recognition may be a triggering event for formation of junctional communication. In tumor promotion, therefore, it is possible that the primary action of TPA is blockage of cell-to-cell recognition rather than a direct action of junctional communication. Blockage of junctional communication, however, may still be important. If there is junctional communication between cells, they can exchange physiological factors and thereby neutralize each other, which may be acting toward the inhibition of carcinogenesis. Stanbridge et al [34] have found that normal cells are dominant to transformed cells in expressing phenotypes when these two types are hybridized, suggesting the ability of normal cells to inhibit expression of the transformed phenotype. Therefore, for the formation of tumors, especially in the tumor-promotion process, we believe that blockage of junctional communication itself is also relevant.

Evidence is accumulating which suggests that normal functions of proto-oncogenes include production of growth factors, or their receptors, important in control of cell proliferation and differentiation [35,36]. Since tumor-promoting agents are potent modulators of a variety of cell differentiation programmes [37], we thought TPA might modulate the expression of certain proto-oncogenes; there are a few reports to support this hypothesis [38–41]. To investigate this possibility we have chosen the system of MELC in which we have previously shown that TPA can inhibit reversibly both induced and spontaneous cell differentiation [7,8]. We screened the degree of DNA methylation in sequences near or in seven proto-oncogenes in the hope that the extent of cytosine methylation could give a preliminary indication of the expression state of these genes. When we compared TPA-sensitive and -resistant cells, we found that DNA of TPA-resistant cells is more heavily methylated at sequences surrounding fos and H ras. These methylation patterns did not change when TPA-sensitive cells were induced to differentiate. Although we cannot make any conclusive remarks without having studied the messenger RNA accumulation in these cells, these results suggest that certain oncogene methylation patterns may be related to TPA susceptibility.

Using different sytems of TPA-sensitive and TPA-resistant cells, Colburn's group has recently shown that TPA-sensitive cells have a TPA-responsive gene which can be transfected to TPA-resistant cells to render TPA-resistant cells TPA sensitive [42]. These sequences, called pro sequences, did not have homology to known existing oncogenes, including H-ras. Homology between these pro genes and fos gene sequences has not been tested yet.

It is clear that the use of various genetic variants of TPA susceptibility or transformation sensitivity will provide valuable information on the mechanism of action of tumor promoters and the process of tumor promotion.

## ACKNOWLEDGMENTS

Part of the studies presented here was supported by NCI grant 1 RO1 CA 40534-01 to H.Y. The authors thank Dr. T. Enomoto, Dr. U. Frixen, and Ms N. Martel for their valuable contribution to the work, and Ms C. Fuchez for her secretarial aid.

## REFERENCES

1. Slaga TJ, Sivak A, Boutwell R (eds): Carcinogenesis, A Comprehensive Survey, Vol. 2: "Mechanisms of Tumor Promotion and Cocarcinogenesis." New York: Raven Press, 1978.
2. Hecker E, Kunz W, Fusenig NW, Marks F, Thielmann HW (eds): Carcinogenesis, A Comprehensive Survey, Vol. 7: "Cocarcinogenesis and Biological Effects of Tumor Promoters. New York: Raven Press, 1982.
3. Borzsonyi M, Lapis K, Day NE, Yamasaki H (eds): "Models, Mechanisms and Etiology of Tumor Promotion." IARC Scientific Publications No. 56. Lyon: 1984.
4. Yamasaki H, Enomoto T, Martel N: In Borzsonyi M, Lapis K, Day NE, Yamasaki H (eds): "Models, Mechanisms and Etiology of Tumor Promotion. IARC Scientific Publications No. 56. Lyon: IARC 1984, pp 217–238.
5. Yamasaki H: In Butterworth BE, Slaga TJ (eds): Banbury Report on Non-Genotoxic Mechanisms of Carcinogenesis. Cold Spring Harbor: Cold Spring Harbor Laboratories, 1987, in press.
6. Marks PA, Rifkind RA: Annu Rev Biochem 47:419–448, 1978.
7. Rovera G, O'Brien TG, Diamond L: Proc Natl Acad Sci USA 74:2894–2898, 1977.
8. Yamasaki H, Fibach E, Nudel U, Weinstein IB, Rifkind RA, Marks PA: Proc Natl Acad Sci USA 74:3451–3455, 1977.
9. Fibach E, Yamasaki H, Weinstein IB, Marks PA, Rifkind RA: Cancer Res 38:3685–3688, 1978.
10. Yamasaki H, Martel N, Fusco A, Ostertag W: Proc Natl Acad Sci USA 81:2075–2079.
11. Kakunaga T, Crow JD: Science 209:505–507, 1980.
12. Kakunaga T, Hamada H, Leavith J, Crow JD, Hirakawa T, Lo KY: In Harris CC, Cerutti PA (eds): "Mechanisms of Chemical Carcinogenesis." New York: Alan R. Liss, Inc., 1982, pp 517–529.
13. Rivedal E, Sanner T, Enomoto T, Yamasaki H: Carcinogenesis 6:899–902, 1985.
14. Yamasaki H, Drevon C, Martel N. In Hecker E, Kunz W, Fusenig NW, Marks F, Thielmann HW (eds): "Carcinogenesis, Cocarcinogenesis and Biological Effects of Tumor Promoters," Vol. VII. New York: Raven Press, 1982, pp 359–377.
15. Yamasaki H, Drevon C, Martel N: Carcinogenesis 3:905–910, 1982.
16. Giroldi L, Hamel E, Yamasaki H: Carcinogenesis 7:1183–1186, 1986.
17. Kakunaga T: Int J Cancer 12:463–473, 1973.
18. IARC/NCI/EPA Working Group: Cancer Res 45:2395–2399, 1985.
19. Enomoto T, Martel N, Kanno Y, Yamasaki H: J Cell Physiol 121:323–333, 1984.
20. Maniatis T, Fritsch EF, Sambrook J: "Molecular Cloning, a Laboratory Manual." Cold Spring Harbor: Cold Spring Harbor Laboratory, 1982.
21. Castagna M, Takai Y, Kaibuchi K, Sano K, Kikkawa U, Nishizuka Y: J Biol Chem 257:7847–7851, 1982.
22. Sharkey NA, Leach KL, Blumberg PM: Proc Natl Acad Sci USA 81:607–610, 1984.
23. Nishizuka Y: Nature 306:693–698, 1984.
24. Enomoto T, Yamasaki H: Cancer Res 45:3706–3710, 1985.
25. Lo KY, Kakunaga T: Cancer Res 42:2644–2650, 1985.
26. Yamasaki H, Enomoto T, Shiba Y, Kanno Y, Kakunaga T: Cancer Res 45:637–641, 1985.
27. Blumberg PM, Butler-Gralla E, Herschman HR: Biochem Biophys Res Commun 102:818–823, 1981.
28. Clarke PRH, Varani J: Cancer Res 44:4967–4971, 1984.

29. Colburn NH, Gindhart TD, Hegamyer GA, Blumberg PM, Delclos KB, Magun BE, Lockyer J: Cancer Res 42:3093–3097, 1982.
30. Lehrer RI, Cohen LE, Koeffler HP: Cancer Res 43:3563–3566, 1983.
31. Shupnik MA, Tashjian AHJ: J Biol Chem 257:12161–12164, 1982.
32. Kramer, CM, Sando JJ: Cancer Res 46:3040–3045, 1986.
33. Enomoto T, Yamasaki H: Cancer Res 44:5200–5203, 1984.
34. Stanbridge EJ, Der CJ, Doersen CJ, Nishimi RY, Peehl DM, Weissman BE, Wilkinson JE: Science 215:252–259, 1982.
35. Weinberg RA: Science 230:770–776, 1985.
36. Bishop JM: Annu Rev Biochem 52:301–304, 1983.
37. Yamasaki H: In Slaga TJ (ed): Mechanisms in Tumor Promotion, Vol. IV: Cellular Responses to Tumor Promotion. Boco Raton, FL: CRC Press, 1984, pp 1–26.
38. Kelly K, Cochran BH, Stiles CD, Leder P: Cell 35:603–610, 1983.
39. Muller R, Curran T, Muller T, Guilbert L: Nature 314:546–548, 1985.
40. Connan G, Rassoulzadegan M, Cuzin F: Nature 314:277–279, 1985.
41. Dotto GP, Parada LF, Weinberg RA: Nature 318:472–475, 1985.
42. Lerman MI, Hegamyer GA, Colburn NH: Int J Cancer 37:293–302, 1986.
43. Yamasaki H, Fibach E, Weinstein IB, Nudel U, Rifkind RA, Marks PA: In Ikawa Y, Odaka T (eds): "Oncogenic Viruses and Host Cell Genes," New York: Academic Press, 1979, pp 365–376.
44. Yamasaki H, Drevon C: In Letnansky K (ed): "Biology of the Cancer Cells." Amsterdam: Kugler Publications 1980, pp 317–325.

# Mitogen-Specific Nonproliferative Variants of Swiss 3T3 Cells

## Harvey R. Herschman

*Department of Biological Chemistry and Laboratory of Biomedical and Environmental Sciences, Center for the Health Sciences, University of California, Los Angeles, California 90024*

> A variety of chemically diverse agents (polypeptide growth factors, prostaglandins, tumor promoters) can stimulate quiescent, density-arrested 3T3 cells to reenter the cell cycle and divide. We have selected variant cell lines unable to respond to the mitogenic effect of epidermal growth factor. The epidermal growth factor (EGF)-nonresponsive 3T3 variants do not make any product cross-reactive with antiserum directed against the EGF receptor. They also fail to produce any detectable mRNA from the EGF receptor gene. We have also isolated variant cell lines unable to respond to the tumor promoter tetradecanoyl phorbol acetate (TPA). The TPA-nonproliferative variants retain TPA-stimulated protein kinase C activity and demonstrate a variety of TPA-mediated physiological and biochemical responses known to occur during the mitogenic response elicited by TPA in the parental 3T3 cells. Two biological responses—induction of ornithine decarboxylase activity and gene amplification—stimulated by TPA in 3T3 cells do not occur in a TPA-nonproliferative variant.

**Key words:** epidermal growth factor, tetradecanoyl phorbol acetate, growth factors, tumor promotors, mitogenesis

Many of the cells in an organism are in a resting, nonproliferating state termed G0. G0 cells respond to a mitogenic stimulus by reentering the cell cycle and, after passing through a period of DNA synthesis, divide. The transition from a resting, nonproliferative cell to one commited to divide is accompanied by extensive alterations in the metabolic, physiological, biochemical, and molecular properties of the responding cell. Distinguishing the causal cellular responses necessary for the G0-to-cycling commitment from those changes accompanying, but not necessary for, the mitogenic response is a problem of major magnitude. We have approached this problem by isolating variant cell lines that are unable to mount a proliferative response when challenged with a specific mitogen. The operative assumption in our work is that those cellular properties deficient in a mitogen-nonresponsive variant are likely to be properties whose modulation is necessary for the mitogenic response.

Received May 28, 1986.

© 1988 Alan R. Liss, Inc.

The model system we have chosen is the clonal murine embryo Swiss 3T3 cell line. 3T3 cells grow to a saturation density proportional to the concentration of serum in medium. The cells exit the cell cycle at cell densities proportional to the serum concentration, and enter a G0, nonproliferating state [1]. When additional serum is added to a quiescent population of 3T3 cells, the majority of the cells leave the G0, resting state and reenter the cell cycle. Since the cell population is clonal and the mitogenic response is synchronous, the 3T3 mitogenic response can be used to study the biochemistry, physiology, and metabolism of the G0-G1 transition.

There are a number of agents in addition to serum that can stimulate the proliferative response of quiescent 3T3 cells. These include polypeptide growth factors, prostaglandins, and tumor promoters (Table I). Thus, a wide variety of structurally dissimilar agents can elicit a common biological response in quiescent 3T3 cells. In our studies with mitogen-nonresponsive variants we have used the polypeptide mitogen epidermal growth factor (EGF) and the macrocyclic tumor promoter tetradecanoyl phorbol acetate (TPA).

## SELECTIONS OF MITOGEN NONRESPONSIVE VARIANTS

Our selection technique to isolate mitogen-nonresponsive variants takes advantage of the ability of certain drugs to prevent the polymerization of microtubules. The procedure for isolating mitogen-nonresponsive variants has recently been described in detail [2] and is illustrated in Figure 1. Briefly, 3T3 cells are grown to density arrest in medium containing 5% fetal calf serum. Entry of the cells into the G0, quiescent phase is monitored by cell count on successive days. After the cells have arrested in the G0 state, the selective mitogen (eg, EGF) is added, along with a drug such a colcemid, which prevents microtubule formation. The cells respond to the mitogen and reenter and traverse the cell cycle. Once the responding cells enter mitosis, however, they are unable to complete cell division, since their microtubules cannot repolymerize in the presence of colcemid. Thus, responsive cells are "frozen" in mitosis. Mitotic cells are rounded up and not tightly adherent to the culture dish. They can, therefore, be differentially removed from the culture dish with a shearing stream of medium. In practice, mitogen and drug are added for 3–4 days before mitotic cells are removed by the wash. An example of the cellular response to the selective process is shown in Figure 2. The remaining cells are grown again to density arrest and the selection is repeated. Individuals cells are cloned after several rounds of selection, and the cloned colonies are tested for mitogenic response to the "selecting" mitogen and a representative group of nonselective mitogens.

**TABLE I. Partial List of Mitogenic Agents for 3T3 Cells**

Polypeptide growth factors
   Epidermal growth factor
   Platelet-derived growth factor
   Fibroblast growth factor

Prostaglandins
   PGF$2\alpha$

Tumor promoters
   Tetradecanoyl phorbol acetate
   Teleocidin

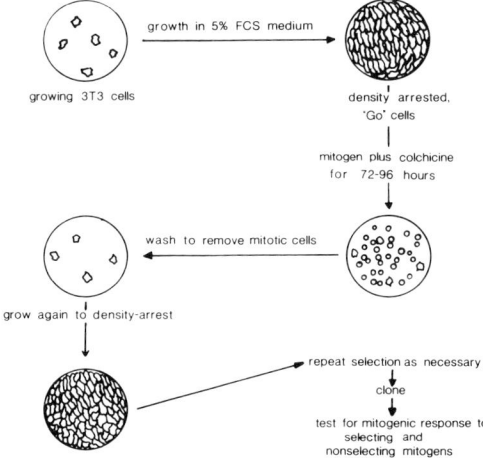

Fig. 1. Selection scheme for the isolation of mitogen-specific nonproliferative variants.

We have isolated and characterized four independent mitogen-nonresponsive 3T3 cell lines (Table II). Two of the cell lines, 3T3-NR6 and 3T3-ENR7, were isolated using EGF as the selective mitogen [3,4]. As expected, neither 3T3-NR6 nor 3T3-ENR7 can divide in response to EGF. Both cell lines can, however, respond to other mitogens; their mitogen response repertoire differs from the parental 3T3 cell only in their inability to divide when challenged with EGF [3,4]. The other two cell lines, 3T3-TNR2 and 3T3-TNR9, were isolated with TPA as the selective mitogen [5]. One, 3T3-TNR9, showed the expected phenotype. Of the mitogens tested, only TPA was unable to stimulate cell division of quiescent 3T3-TNR9 cells. The second variant, 3T3-TNR2, was also unable to respond to TPA. This cell line, however, was also unable to respond to EGF, although EGF was not used as the selective mitogen. 3T3-TNR2 cells are not generally defective in the mitogen response; both serum and fibroblast growth factor can elicit a substantial proliferative response [5].

## THE MOLECULAR BASIS FOR THE EGF-NONRESPONSIVE PHENOTYPE

The first interaction between EGF and 3T3 cells is the binding of EGF to a specific cell surface receptor for EGF [6]. Binding can be analyzed by using $^{125}$I-labeled EGF. There are approximately 50,000–100,000 EGF receptors present on a 3T3 cell. When the three EGF-nonresponsive variants were analyzed for the presence of EGF receptors, no significant binding of EGF could be demonstrated (Fig. 3). The three variants are, therefore, unable to respond to EGF because they do not possess functional EGF receptors.

To determine the biochemical basis for the deficit in EGF binding of the variants, we investigated the production of molecules serologically related to the EGF receptor, using an antiserum to the receptor obtained from Stanley Cohen. The cell surface proteins of 3T3 cells and the three EGF-nonresponsive variants (3T3-NR6, 3T3-ENR7, and 373-TNR2) were labeled with $^{125}$I-EGF, and the EGF receptor was immunoprecipitated from cell membrane extracts [7]. Although a substantial quantity of $^{125}$I-labeled EGF receptor could be precipitated from 3T3 cells, no labeled band could be precipitated from the variant cell lines that was not also precipitated with a

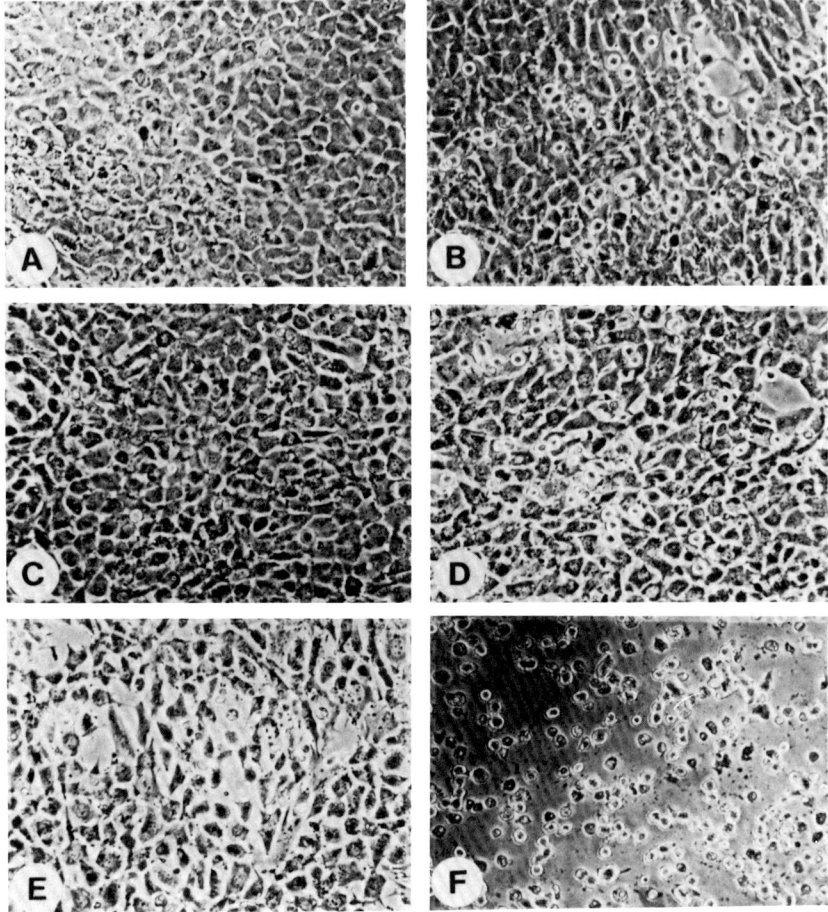

Fig. 2. Response of quiescent 3T3 cells to colchicine alone or EGF plus colchicine. 3T3 cells were treated with colchicine (1.5 $\mu$/ml) alone or colchicine and EGF (10 ng/ml). Cells treated with colchicine alone for (**A**) 1 day, (**C**) 2 days, and (**E**) 3 days. Cells treated with colchicine and EGF for (**B**) 1 day, (**D**) 2 days, and (**F**) 3 days. (Reproduced from [3].)

**TABLE II. Mitogen-Nonresponsive 3T3 Cell Lines**

| Cell line | Selective agent | Mitogenic response to— | | | | Reference |
| | | Serum | FGF | EGF | TPA | |
|---|---|---|---|---|---|---|
| 3T3 | — | + | + | + | + | — |
| 3T3-NR6 | EGF | + | + | — | + | [3] |
| 3T3-ENR7 | EGF | + | + | — | + | [4] |
| 3T3-TNR2 | TPA | + | + | — | — | [5] |
| 3T3-TNR9 | TPA | + | + | + | — | [5] |

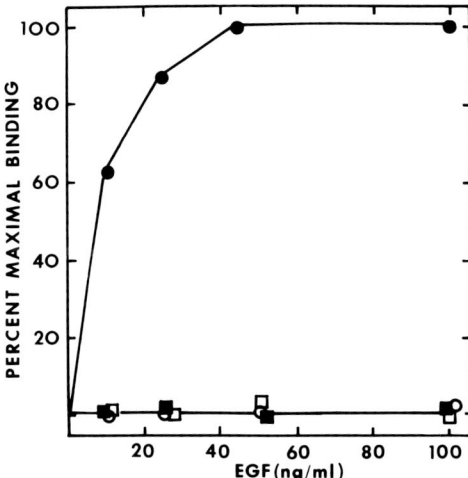

Fig. 3. Binding of $^{125}$I-labeled EGF to 3T3 cells and the EGF-nonresponsive variants. Data are expressed as the percent binding relative to the saturating value observed for 33T3 cells and are normalized for cell number. 3T3 (●), 3T3-NR6 (○), 3T3-TNR2 (□), 3T3-ENR7 (■). Reproduced from [7].

preimmune serum (Fig. 4). Immunoprecipitation of mixtures of extracts of 3T3 and 3T3-TNR2 cells demonstrated that, had a cross-reactive protein been present in the variants at a concentration of greater than 1% that found on 3T3 cells, our immunoprecipitation procedure would have identified the molecule (Fig. 4, right panel).

We thought it possible that an altered receptor might be synthesized, but rapidly degraded and/or not inserted into the membrane of the variant cells. Labeling of cell surface molecules might not detect such an altered receptor. To examine this possibility, the parental cell line and the EGF-nonproliferative variants were labeled with $^{35}$S-methionine for a time period substantially shorter than the half-life of the receptor, and total cellular extracts were used for immunoprecipitation with antiserum to the EGF receptor. Although extensive labeling of EGF receptor could be demonstrated in the parental 3T3 cells, no labeled molecules specifically precipitated with anti-EGF receptor antiserum were detectable in any of the EGF-nonproliferative variants (Fig. 5). The three EGF-nonproliferative variants do not synthesize detectable cross-reactive proteins related to the EGF receptor.

To determine the molecular basis for the lack of receptor cross-reactive antigens, we turned to measurement of EGF receptor mRNA. PolyA-plus mRNA was prepared from human A431 cells (which contain large numbers of EGF receptors) and from 3T3 cells and the EGF-nonproliferative variants. After electrophoresis and transfer, the blots were probed with a nick-translated cDNA probe for the human EGF receptor, provided by Drs. M.G. Rosenfeld and R. Evans. While transcripts for the EGF receptor are clearly present in 3T3 cells, no mRNA for EGF receptor was demonstrable in any of the EGF-nonproliferative variants (Fig. 6). This surprising result suggests that the EGF receptor genes in all three EGF-nonresponsive cell lines are silent, at least at the level of cytoplasmic message. The simplest explanation of these data is that the gene for the EGF receptor has been deleted in the EGF-nonresponsive variants. However, genomic Southern blot analysis of DNA from the parental and EGF-nonresponsive variants demonstrated that the receptor gene in the

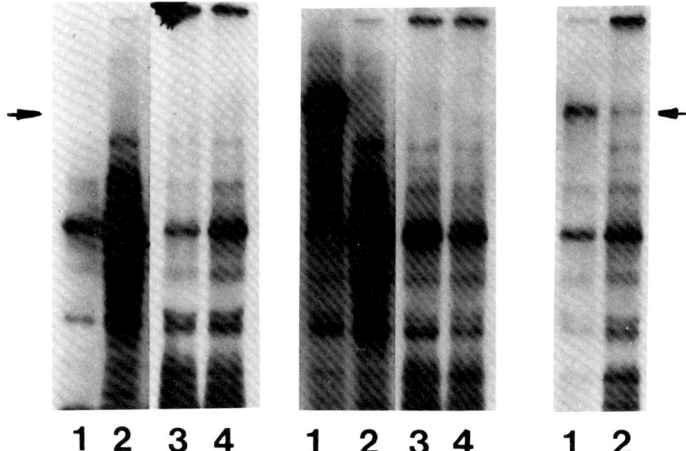

Fig. 4. Immunoprecipitation of $^{125}$I-labeled EGF receptors from 3T3 cells and the EGF-nonresponsive variants. A double-immunoprecipitation procedure was used to reduce nonspecific background. Immunoprecipitates using the control antiserum are shown in the **left panel**; immunoprecipitates using antireceptor antiserum are in the **center panel**. Arrows indicate the position of the EGF receptor. **Lanes 1**, 3T3; **2**, 3T3-TNR2; **3**, 3T3-NR6; **4**, 3T3-ENR7. To analyze the sensitivity of the immunoprecipitation assay $^{125}$I-labeled 3T3 and 3T3-TNR2 extracts were mixed together in varying proportion, while keeping the total amount of radioactivity constant (**right panel**). Immunoprecipitate from a 14:86 mixture of 3T3:3T3-TNR2 extracts is shown in **lane 1**; immunoprecipitate from a 1.4:98.6 mixture of 3T3:3T3-TNR2 extracts is shown in **lane 2**. (Reproduced from [7].)

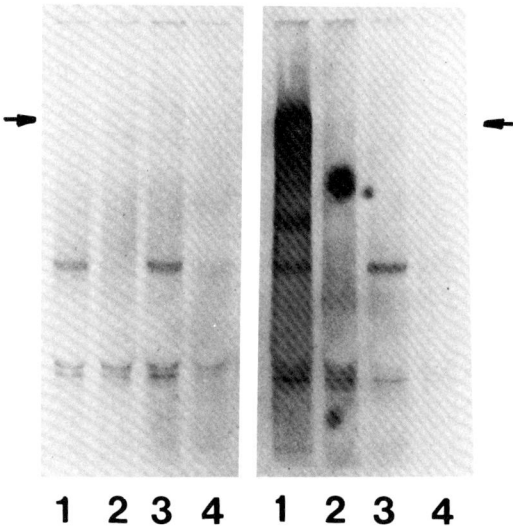

Fig. 5. Immunoprecipitates of $^{35}$S-methionine-labeled cell extracts. Immunoprecipitates using control antiserum are shown in the **left panel**, immunoprecipitates using antireceptor serum are in the **right panel**. Arrows indicate the position of the EGF receptor. **Lane 1**, 3T3; **2**, 3T3-TNR2; **3**, 3T3-NR6; **4**, 3T3-ENR7. (Reproduced from [7].)

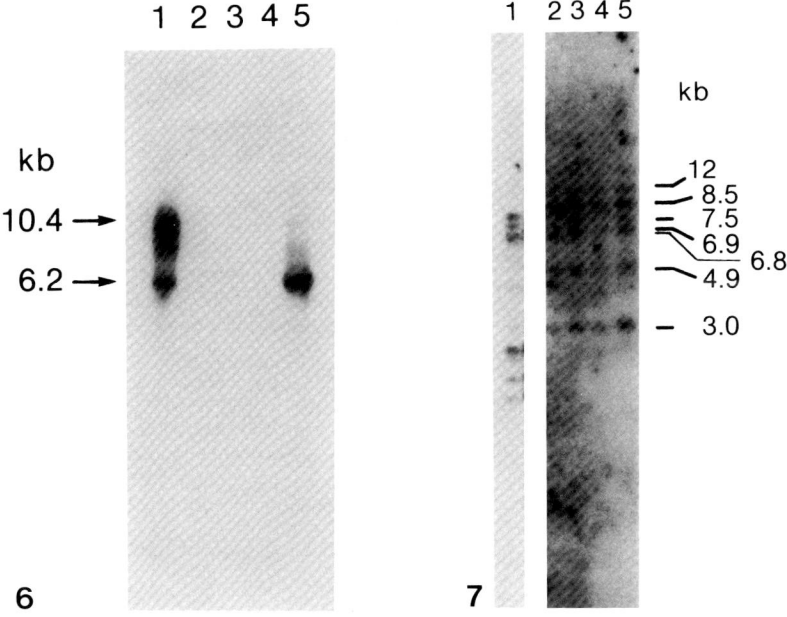

Fig. 6. RNA blot analysis of EGF-receptor mRNA. Each lane contains 10 μg of poly(A)$^+$ RNA. **Lane 1**, A431; **2**, 3T3-NR6; **3**, 3T3-TNR2; **4**, 3T3-ENR7; **5**, 3T3. Lane 1 was autoradiographed for 2 hr; lanes 2–5, for 28 hr. The molecular size in kilobases is indicated. (Reproduced from [7].)

Fig. 7. Genomic Southern blot of the EGF-receptor gene. Each lane contains 10 μg of EcoRI-digested DNA. **Lane 1**, A431; **2**, 3T3-NR6; **3**, 3T3-TNR2; **4**, 3T3-ENR7; **5**, 3T3. Lane 1 was autoradiographed for 2 hr; lanes 2–5, for 28 hr. The molecular size in kilobases is indicated. (Reproduced from [7].)

variants is not completely deleted or, indeed, even grossly rearranged (Fig. 7). Transcription rate experiments will be required to determine if the defect lies in the transcriptional vs. processing mechanisms.

We have also attempted to approach by somatic-cell genetic procedures the nature of the mutation(s) leading to the EGF-receptorless phenotype [8]. The parental 3T3 cell line and the three variants were each used to form hybrids with 3T3-R5, a cell line that has EGF receptors and is both ouabain resistant and thymidine kinase deficient. Hybrids of 3T3-NR6, 3T3-ENR7, and 3T3-TNR2 with 3T3R5 were selected in HAT-ouabain medium. The parental cells and three hybrid clones for each variant were analyzed for the presence of EGF receptors by $^{125}$I-EGF binding. The two variants selected with EGF as mitogen were both recessive; hybrids of 3T3-NR6 and 3T3-ENR7 with 3T3R5 have EGF receptors (Fig. 8). In contrast, the pleiotypic mutant 3T3-TNR2, which is unable to respond to either EGF or TPA, was dominant with regard to the EGF receptor phenotype; hybrids of 3T3-TNR2 and R5 did not demonstrate statistically significant levels of EGF receptor binding activity.

The recent cloning of the complete EGF receptor of cDNA [9–11] will permit an analysis of structure and function of the EGF receptor by site-directed mutagenesis. The EGF-nonproliferative variants should be ideal recipients to study the effects of transfected mutant EGF-receptor expression constructs, since these cells make no detectable endogenous EGF receptor message or protein, but respond to mitogenic signals.

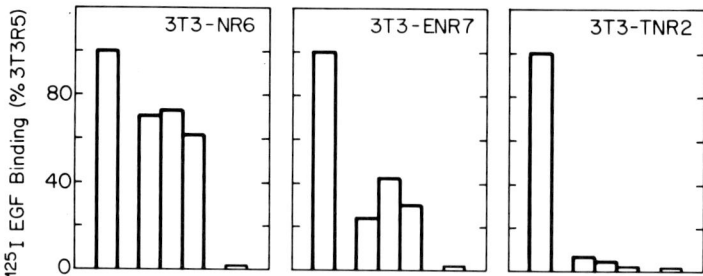

Fig. 8. Binding of EGF to hybrids between 3T3R5 and the "EGF-receptorless" variants. Data are presented relative to binding the 3T3R5 parent (ie, cpm bound per $10^6$ hybrid cells/cpm bound per $10^6$ 3T3R5 cells). In each panel the columns are, from left to right, binding to 3T3R5 cells, binding to three hybrid clones, and binding to the EGF-nonproliferative variant. **Left panel**, 3T3R5×3T3NR6 hybrids; **center panel**, 3T3R5×3T3ENR7 hybrids; **right panel**, 3T3R5×3T3NR2 hybrids. (Reproduced from [8].)

## THE MOLECULAR BASIS FOR THE TPA-NONRESPONSIVE PHENOTYPE

The two TPA-nonresponsive variants, 3T3-TNR2 and 3T3-TNR9, were first tested for the presence of TPA "receptors," using labeled phorbol dibutyrate, in a collaborative study with Peter Blumberg [12]. The TPA-nonproliferative variants and the parental 3T3 cells were indistinguishable with regard to either total binding or affinity for phorbol dibutyrate. These early data suggested that, unlike the EGF-nonresponsive variants, the TPA-nonproliferative variants had functional TPA "receptors." We also tested several of the early biochemical and metabolic responses known to occur when 3T3 cells are exposed to TPA. The TPA-nonproliferative variants, when treated with TPA, also released arachidonic acid and derivative prostaglandins [13]. These data suggested that the early responses to TPA were intact in the two TPA-nonproliferative variants. Thus the TPA-specific alteration causing TPA mitogenic nonresponsiveness must lie distal to these early initial responses.

The 3T3-TNR2 variant is unable to respond either to EGF or TPA. Moreover, the EGF receptorless phenotype is dominant in somatic cell hybrids (vida supra). We prepared additional hybrids between 3T3-TNR9 and 3T3-R5, and tested 3T3-TNR9×3T3R5 and 373-TNR2×3T3R5 hybrids for TPA mitogen responses. The 3T3-TNR2 phenotype was also dominant for TPA-stimulated mitogenesis; 3T3-TNR2×3T3R5 hybrids did not respond to TPA. In contrast, 3T3-TNR9×3T3R5 hybrids respond to TPA; the 3T3-TNR9 phenotype was recessive [8]. Although the pleiotypic dominance of the 3T3-TNR2 phenotype is a fascinating question, we have emphasized characterization of the recessive, TPA-restricted 3T3-TNR9 cell line in most of our subsequent studies.

The galvanizing observation that TPA activates protein kinase C [14], and the subsequent data that suggest that most, if not all, of the cellular actions of TPA are mediated by protein kinase C lead us to examine the level and activity of this enzyme in the 3T3-TNR9 TPA-nonproliferative variant. Both phospholipid/$Ca^{++}$-dependent TPA binding and phospholipid/$Ca^{++}$-dependent protein kinase C activity (using histone as substrate) were analyzed [15]. The TPA-nonproliferative 3T3-TNR9 variant and the parental 3T3 cells had essentially identical levels of TPA binding activity (Table III). If anything, the variant cell line has higher protein kinase C activity. At the level of in vitro protein kinase C analyses, the defect leading to the TPA-

TABLE III. Phospholipid/Ca$^{++}$-Dependent TPA Binding and Protein Kinase C Activity in 3T3-TNR9 and 3T3 Cells

| Cells | $^3$H-TPA binding | | Protein kinase C (pmol/min per mg protein) |
|---|---|---|---|
| | pm/mg protein | % total binding | |
| 3T3 | | | |
| 100K supernatant | 1.1 ± 0.2 | 57 | 148 ± 19 |
| 100K pellet | 0.52 ± 0.17 | 18 | — |
| 1K pellet | 0.48 ± 0.10 | 25 | — |
| 3T3-TNR9 | | | |
| 100K supernatant | 0.99 ± 0.15 | 61 | 235 ± 42 |
| 100K pellet | 0.60 ± 0.17 | 24 | — |
| 1K pellet | 0.28 ± 0.07 | 15 | — |

nonproliferative phenotype in 3T3-TNR9 cells is not due to a reduction in this enzyme.

We next investigated whether the protein kinase C molecule was activated in a similar fashion in the 3T3-TNR9 TPA-nonproliferative variant and 3T3 cells. In collaboration with the laboratories of Michael Weber and Perry Blackshear, we examined the phosphorylation of intracellular substrates in the two cell lines in response to TPA. The major quantitative substrate for phosphorylation in 3T3 cells following treatment with TPA is an as yet unidentified 80-kilodalton (kDa) substrate [16]. Phosphorylation of this 80-kDa protein in 3T3-TNR9 and 3T3 cells was stimulated to equivalent levels (Fig. 9). TPA treatment of cells is also known to cause phosphorylation of the EGF receptor by protein kinase C [17]. When 3T3 cells and the TPA-nonproliferative 3T3-TNR9 variant were exposed to TPA, both cell types demonstrated phosphorylation of the EGF receptor, despite a lack of TPA-dependent DNA synthesis in 3T3-TNR9 cells (Fig. 10). By these criteria the TPA activation of protein kinase C appears to be intact in 3T3-TNR9 cells. The deficit leading to the TPA-nonproliferative phenotype must be distal to TPA activation of protein kinase C; we are inside the "black box" of signal transduction in mitogenesis.

The phosphorylation, on tyrosine, of a 42-kDa protein substrate occurs in response to a variety of mitogenic and transforming agents [18]. TPA also stimulates phosphorylation of this substrate in 3T3 cells. Since protein kinase C is a threonine/serine protein kinase, the tyrosine phosphylation of the 42-kDa protein in response to TPA must be the consequence of an activated protein kinase cascade. TPA treatment of 3T3-TNR9 cells causes tyrosine phosphorylation of the 42-kDa protein, just as it does in 3T3 cells [15]. Thus, we conclude that (i) the protein kinase cascade activated by TPA in 3T3 cells is intact in the TPA-nonproliferative variant, and (ii) activation of the cascade and phosphorylation of 42-kDa substrate is not sufficient to cause a commitment to DNA synthesis.

We have described, in 3T3-TNR9 cells exposed to TPA, a variety of biochemical and metabolic changes that occur in a fashion indistinguishable from responses observed in the parental 3T3 cells. These processes *may* be necessary for a TPA-stimulated mitogenic response. They cannot, however, be sufficient for TPA-stimulated cell division, since 3T3-TNR9 cells do not proliferate when exposed to TPA. What TPA-induced events observable in quiescent 3T3 cells do not occur in 3T3-TNR9 cells? Increased levels of ornithine decarboxylase (ODC) frequently precede

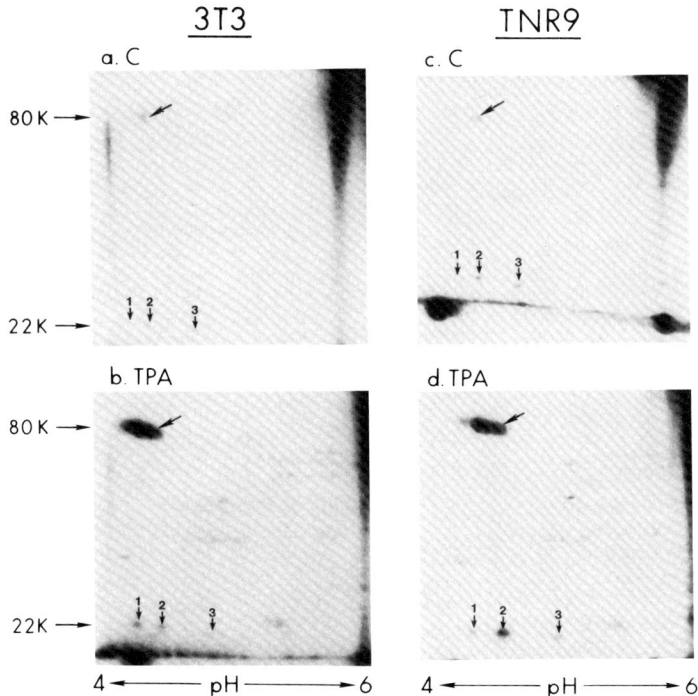

Fig. 9. Two-dimensional electrophoresis of acidic phosphoproteins in 3T3 or 3T3-TNR9 cells treated with TPA. Quiescent fibroblasts were labeled with $^{32}$P, and exposed to either dimethyl sulfoxide (0.1% for 15 min 37°C) or TPA (1.6 μM for 15 min), then were harvested and subjected to two-dimensional electrophoresis. The position of the 80-kDa protein is indicated by the arrows: the three components of a 22-kDA protein whose phosphorylation is stimulated by TPA also are indicated. (Reproduced from [15].)

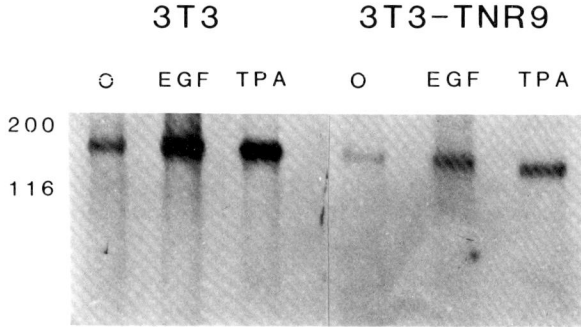

Fig. 10. Phosphorylation of the EGF receptor in 3T3 and 3T3-TNR9 cells. Cell cultures were labeled to equilibrium with $^{32}$PO$_4$ and then stimulated with EGF (100 ng/ml) or TPA (50 ng/ml) for 10 min. The cultures were lysed and the EGF receptor was immunoprecipitated and electrophoresed. (Reproduced from [15].)

the initiation of DNA synthesis in the proliferative responses to a variety of mitogens [9]. Serum, a competent mitogen for 3T3 cells, stimulates a substantial elevation of ODC activity. Serum stimulation of quiescent 3T3-TNR9 (or 3T3-TNR2) cells also induces an increase in both DNA synthesis and in the level of ODC activity in these cells (Fig. 11). TPA also induces an increase of ODC activity in quiescent 3T3 cells, as well as a concomitant elevation of DNA synthesis and cell division. In contrast, TPA is unable to stimulate either DNA synthesis or ODC activity in the TPA-nonproliferative 3T3-TNR2 and 3T3-TNR9 variants. The ODC gene must not be altered in any substantive fashion in these variants, since serum—a competent mitogen—can stimulate an increased level of the enzyme in 3T3-TNR9 and 3T3-TNR2 cells [20]. The inability to stimulate ODC activity in response to TPA administration appears to be a biochemical correlate of the TPA-nonproliferative phenotype. We conclude that either (i) the induction of ODC activity by TPA is a necessary event for TPA-induced mitogenesis or (ii) that a step common to the stimulation of ODC and DNA synthesis by TPA is altered in the TPA-nonproliferative variants. One of our major immediate goals is to determine the molecular basis for the lack of TPA-induced ODC expression in the TPA-nonproliferative variants.

Varshavsky demonstrated that TPA can enhance gene amplification in 3T6 cells [21]. TPA-enhanced resistance to methotrexate toxicity occurred as a consequence of an increased copy number of the dihydrofolate reductase gene. Barsoum and Varshavsky [22] later showed that polypeptide mitogens such as EGF or insulin could also enhance gene amplification. Barsoum and Varshavsky suggested that there exists "a qualitative correlation between the ability of a substance to act as a mitogen for 3T6 cells and its ability to increase methotrexate resistant 3T6 colonies." If this hypothesis is correct, then one would expect that TPA-nonproliferative variant cell lines should be resistant to the TPA-stimulated amplification described by Varshavsky [21,22].

We first wished to demonstrate that the parental 3T3 cell line was subject to TPA-induced gene amplification. Bojan et al [23] reported that the frequency of resistance to cadmium toxicity could be increased by TPA, as a consequence of

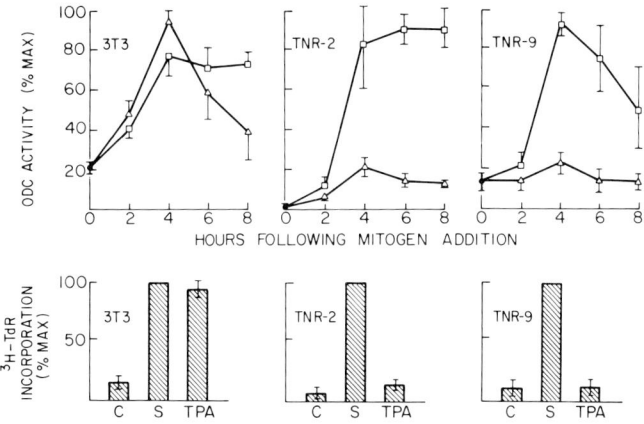

Fig. 11. **Top**:Ornithine decarboxylase induction following serum (10%) or TPA (100 ng/ml) stimulation. □, serum stimulated; △, TPA stimulated; ○, unstimulated. **Bottom**: 3H-thymidine incorporation at 20 hr following serum or TPA stimulation. C, no stimulation; S, serum stimulation; TPA, TPA stimulation. Bars show standard error of the mean. (Reproduced from [20].)

amplification of the metallothionein (MT) genes. We exposed 3T3 cells to cadmium in the presence and absence of TPA [24]. A large increase in cadmium-resistant colonies occurred in the presence of TPA (Fig. 12). Five cadmium-resistant clones were picked, grown out, and tested for increased cadmium resistance. All the cell lines exhibited substantially increased cadmium resistance (Fig. 13). To demonstrate that the increased resistance to cadmium was due to amplification of the metallothionein gene, DNA was isolated from the parental 3T3 cell line; and the five cadmium-resistant variants, digested with Hind III, and the presence of an MT-1-containing restriction fragment was analyzed by Southern analysis. All five lines have amplified

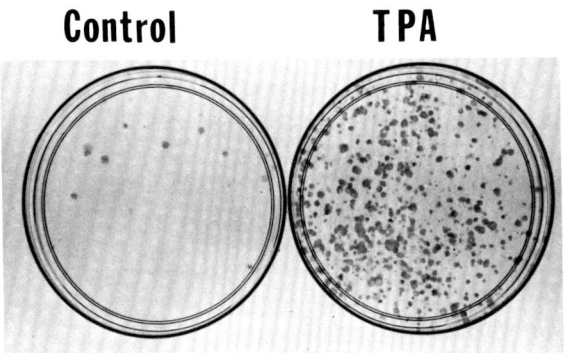

Fig. 12. 3T3 colony formation in the presence of cadmium or cadmium plus TPA. 3T3 cells ($5 \times 10^5$ cells per dish) were plated on 10-cm plates. TPA, when present, was at 50 ng/ml. After 1 day, $CdCl_2$ was added to the medium to a concentration of 18 $\mu$M. After visible colonies were observed, the plates were washed and stained.

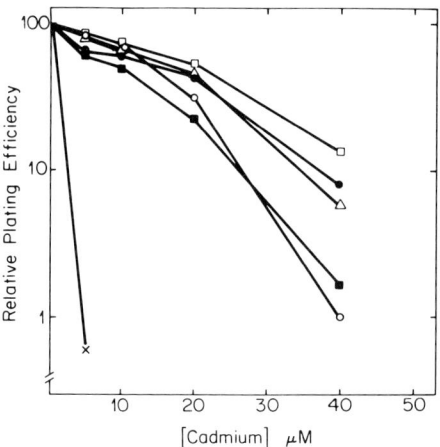

Fig. 13. Colony-forming ability of 3T3 cells and cadmium-resistant ($Cd^R$) 3T3 isolates in the presence of cadmium. Approximately 200 cells of each line were plated in 100-mm dishes. Cadmium was added after 1 day. The plates were washed and stained after 10 days, and the colonies were counted. The relative plating efficiency of each cell line (ie, normalized to the number of colonies on the control, cadmium-free plate) is plotted as a function of cadmium concentration. Data are the average of three plates per point. Symbols: ×, 3T3; ●, 3T3-$Cd^{R1}$; ○, 3T3-$Cd^{R2}$; □, 3T3-$Cd^{R3}$; ■, 3T3-$Cd^{R4}$; ▲, 3T3-$Cd^{R5}$. Data reproduced from [24].

copies of the MT-1 gene (Fig. 14). We concur with the conclusion of Varshavsky [22] that TPA can stimulate gene amplification.

We next asked whether TPA could stimulate gene amplification in the 3T3-TNR9 TPA-nonproliferative variant. Development of cadmium-resistant clones in 3T3 and 3T3-TNR9 cells was tested as a function of TPA concentration. While a tenfold increase in cadmium-resistant colonies occurred in the presence of TPA for the parental 3T3 cells, TPA was unable to stimulate gene amplification in 3T3-TNR9 cells (Fig. 15). It is possible that 3T3-TNR9 cells are simply less responsive to all stimuli in the gene amplification response. To test this possibility the frequency of cadmium-resistant colonies in response to a powerful mitogen active on both 3T3 cells and the TPA-nonproliferative 3T3-TNR9 cells was examined. Fibroblast growth factor increased the frequency of cadmium-resistant colonies for both 3T3 and 3T3-TNR9 cells (Fig. 16) Representative plates are shown in Figure 17. We conclude that (i) 3T3-TNR9 cells are, like 3T3 cells, subject to mitogen-induced gene amplification, and (ii) that the pathway of TPA-induced mitogenesis and gene amplification share a common step, a step blocked in the 3T3-TNR9 TPA-nonproliferative variant.

## CONCLUSIONS AND DIRECTIONS

Studies of EGF-receptorless mutants have defined both an interesting problem in gene expression and potentially valuable tool for analysis of EGF action. Why is the EGF gene silent in all three EGF-nonresponsive cell lines? Expression of hybrid genes utilizing the promoter region of the EGF receptor gene may help to clarify this issue. If vectors containing EGF receptor expression constructs can be successfully

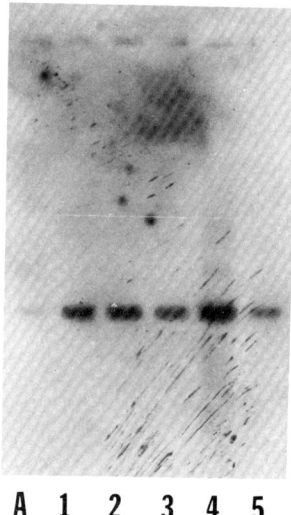

Fig. 14. Amplification of the MT-1 gene in cadmium-resistant ($Cd^R$) 3T3 isolates. DNa isolated from 3T3 cells and five cadmium-resistant clonal isolates was digested with HindIII. Equal amounts (10 μg) of each DNA were subjected to electrophoresis. After transfer to nitrocellulose, the filter was hybridized with a $^{32}P$-labeled probe for murine MT-I and analyzed by autoradiography. **Lane A**, 3T3 cells; **1**, 3T3 $Cd^{R1}$ cells; **2**, 3T3 $Cd^{R2}$ cells; **3**, 3T3 $Cd^{R3}$ cells; **4**, 3T3 $Cd^{R4}$ cells; **5**, 3T3 $Cd^{R5}$ cells. (Data reproduced from [24].

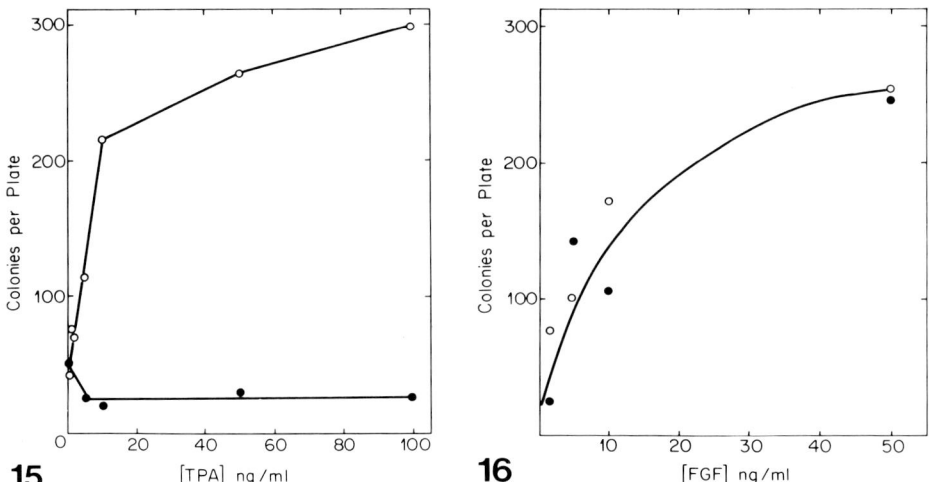

Fig. 15. Quantitation of the TPA effect on the number of cadmium-resistant colonies of 3T3 and 3T3-TNR9 cells. 3T3 and 3T3-TNR9 cells were plated at $5 \times 10^5$ cells per dish with the indicated concentrations of TPA. After 1 day, CdCl$_2$ (18 M) was added. After 3 wk the plates were washed and stained, and the colonies were counted. Data are averages of duplicate plates at each point. Symbols, ○, 3T3 cadmium-resistant colonies; ●, 3T3-TNR9 cadmium resistant colonies. (Reproduced from [24].)

Fig. 16 Fibroblast growth factors increases the frequency of cadmium-resistant colonies of 3T3 and 3T3-TNR9 cells. 3T3 and 3T3-TNR9 cells ($5 \times 10^5$ per dish) were plated and treated as described in the legend to Figure 15. Fibroblast growth factor concentrations are indicated on the figure; data are the average of duplicate plates at each point. Symbols: ○, 3T3 cadmium-resistant colonies; ●; 3T3-TNR9 cadmium-resistant colonies. (Reproduced from [24].)

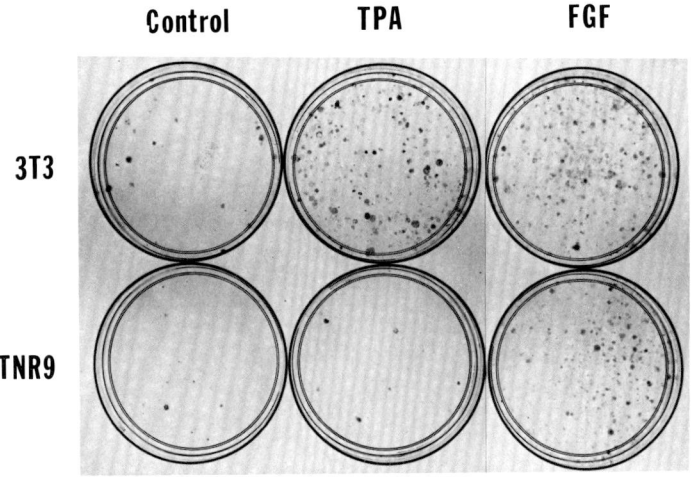

Fig. 17. Representative plates used to develop the data shown in Figures 15 and 16.

transfected into the EGF-nonresponsive cells, this system should provide an ideal model to study structure function relationships for the EGF receptor.

For TPA-nonproliferative variants we apparently have a defect distal to post-protein kinase C activation. Why is ODC not elevated by TPA in these cells? We are currently examining expression of the ODC gene in 3T3 and 3T3-TNR9 cells in response to serum, TPA, and other mitogens. TPA, like other mitogens, is an inducer of several "early response" or "primary response" genes in quiescent cells. We are currently identifying members of the "primary response" family of genes induced by to TPA in quiescent 3T3 cells (ie, those genes whose level of mRNA is induced in TPA-treated cells in the absence of any intervening protein synthesis). We will then examine TPA-stimulated 3T3-TNR9 cells, to determine if any of the genes induced by TPA in the parental 3T3 cells are not induced in 3T3-TNR9 cells. Such genes should be likely candidates for the synthesis of proteins whose functions are causal in the TPA-induced mitogenic response.

## ACKNOWLEDGMENTS

I thank those graduate students and postdoctoral fellows in my laboratory—Rebecca Pruss, Edith Butler-Gralla, Ernest Terwilliger, Aharon Aharonov, Robert Lim, Susan Beatty, and Carol Schneider—who have participated in these experiments. I also thank Drs. Peter Blumberg, Michael Weber, Richard Bishop, and Perry Blackshear for their collaborations in various portions of this work. These studies were supported by contract DE AC03 76 SF00012 between the Regents of the University of California and the Department of Energy, and by grant GM 24797 from the National Institutes of Health.

## REFERENCES

1. Holley RW, Kiernan JA: Proc Natl Acad Sci USA 60:300, 1986.
2. Herschman HR: In Barnes D, Sirbasqu D (eds): "Peptide Growth Factors: Methods in Enzymology." New York: Academic Press, in Press.
3. Pruss RM, Herschman HR: Proc Natl Acad Sci USA 74:3918, 1977.
4. Terwilliger E, Herschman HR: Biochem Biophys Res Commun 118:60, 1984.
5. Butler-Gralla E, Herschman HR: J Cell Physiol 107:59, 1981.
6. Aharonov A, Pruss RM, Herschman HR: J Biol Chem 253:3970, 1978.
7. Schneider CA, Lim RW, Terwilliger E, Herschman HR: Proc Natl Acad Sci USA 83:333, 1986.
8. Terwilliger E, Herschman HR: J Cell Physiol 123:321, 1985.
9. Lin CR, Chen WS, Kruiger W, Stolarsky LS, Weber W, Evans RM, Verma IM, Gill GN, Rosenfeld MG: Science 224: 843, 1984.
10. Xu Y-H, Ishii S, Clark AJL, Sullivan M, Wilson RK, Ma DP, Roe BA, Merlino GT, Pastan I: Nature (Lond) 309:806, 1984.
11. Ullrich A, Coussens L, Hayflick JS, Dull TJ, Gray A, Tam AW, Lee J, Yarden Y, Libermann TA, Schlessinger J, Mayes ELV, Whittle N, Waterfield MD, Seeburg PH: Nature (Lond) 309:418, 1984.
12. Blumberg PM, Butler-Gralla E, Herschman HR: Biochem Biophys Res Commun 102:818, 1981.
13. Butler-Gralla E, Taplitz S, Herschman HR: Biochem Biophys Res Commun 111:194, 1983.
14. Castagna M, Takai Y, Kaibuchi K, Sano K, Kikkawa U, Nishizuka Y: J Biol Chem 257:7847, 1982.
15. Bishop R, Martinez R, Weber MJ, Blackshear PJ, Beatty S, Lim R, Herschman HR: Mol Cell Biol 5:2231, 1985.
16. Rozengurt E, Rodriguez-Pena M, Smith KA: Proc Natl Acad Sci USA 80:7244, 1983.
17. Cochet C, Gill GN, Meisenhelder J, Cooper JA, Hunter T: J Biol Chem 259:2553, 1984.

18. Bishop R, Martinez R, Nakamura KD, Weber MJS: Biochem Biophys Res Commun 115:536, 1983.
19. O'Brien TG, Diamond L: Cancer Res 37:3895, 1977.
20. Butler-Gralla E, Herschman HR: J Cell Physiol 114:317, 1983.
21. Varshavsky A: Cell 25:561, 1981.
22. Barsoum J, Varshavsky A: Proc Natl Acad Sci USA 80:5330, 1983.
23. Bojan F, Kinsella AR, Fox M: Cancer Res 43:5217, 1983.
24. Herschman HR: Mol Cell Biol 5:1130, 1985.

# Workshop: "The Use of Genetic Variants to Find Genes for Growth and Transformation"

### H. Herschman, M. Gottesman, and L. Cantley

*Department of Biological Chemistry and Laboratory of Biomedical and Environmental Sciences, University of California, Los Angeles, California 90024 (H.H.); Laboratory of Molecular Biology, National Cancer Institute, National Institutes of Health, Bethesda, Maryland 20892 (M.G.); and Department of Physiology, Tufts University School of Medicine, Boston, MA 02111 (L.C.)*

The workshop opened with an overview, presented by Dr. Gottesman, of a somatic-cell genetic approach to growth control and transformation. He described several criteria that would increase confidence in a conclusion that phenotype is related to a biochemical/genetic change in somatic variants. These include (i) isolation of multiple independent mutants, (ii) revertants that simultaneously lose phenotype and biochemical effect, (iii) isolation of conditional mutants, and (iv) transfer of the gene responsible for the defect. The workshop presentations and ensuing discussions then centered around three topics: the selection and characterization of variants, the use of "natural mutations" in growth and transformation pathways, and the use of "reverse genetics"—the manipulation and subsequent reinsertion of known genes involved in proliferation, transformation, or promotion pathways.

Michael Shepard described the isolation of variant cell lines resistant to the cytoxic action of tumor necrosis factor. Emma Moore reported on the isolation of F9 teratocarcinoma variants unable to differentiate in response to retinoic acid. These cells retained the retinoic acid binding protein, suggesting they are altered in a distal function. Causal genes are being sought by genomic transfection studies. Len Benade described the use of differential ouabain sensitivity as a selective means to isolate revertant cell lines carrying multiple copies of a transfected oncogene. The system is designed to facilitate the subsequent isolation of dominant "anti-oncogenes." A spirited discussion followed on the issue of distinctions between tumor suppressor genes and anti-oncogenes, with active participation by a variety of speakers. Cori Bargman followed with a presentation that defined the molecular basis for the oncogenicity of the *neu* oncogene—one amino acid substitution in the single mem-

Received May 28, 1986.

© 1988 Alan R. Liss, Inc.

brane-spanning region of the presumptive growth factor/hormone receptor coded for by this gene. The same mutation has been observed in a number of independently isolated brain tumors induced by ethylnitrosourea. Tom O'Brien has isolated $Na^+K^+Cl^-$ cotransport mutants that fail to respond to TPA. The working hypothesis is that these early ionic changes may be important second messengers mediating postreceptor events in mitogenesis. These cell lines are unable to respond to tetradecanoyl phorbol acetate (TPA) challenge with either an elevation of ornithine decarboxylase activity or with a mitogenic response. The cells have normal protein kinase C levels, and their TPA-stimulated phosphorylation responses are similar to those of the parental cells. Barbara Walner reported the cloning of two forms of lipocortin. Analyses of the sequences revealed that one of the proteins was the major substrate of the tyrosine kinase activity of the epidermal growth factor (EGF) receptor, while the other was the well-known 35-kilodalton (kDa) substrate of the src gene product $pp60^{src}$. A discussion of the potential role of phospholipase $A_2$ and lipocortin in mitogenic response followed, with Tony Hunter sounding a note of caution by pointing out the large amount of 35-kDa $pp60^{src}$ substrate in cells and its association with cytoskeletal structures.

The "natural mutations" portion of the workshop began with a presentation by Dan Rossen that compared the molecular structures of the c-myb and v-myb gene. The deletions and substitutions necessary for the establishment of oncogenicity were considered. Susan Naylor described the genetics of small cell lung cancer. A recessive deletion on chromosome 3 appears to be implicated in the disease. Richard Gatti illustrated a collection of restriction fragment length polymorphisms that track with genes that are involved in the differentiation of HL60 genes.

Etta Sivnek began the "reverse genetics" portion of the program with a characterization of deletion mutants of the EGF receptor. Truncations of the c-terminal end of the molecule destroyed internalization and mitogenicity. Perhaps the most surprising result of the workshop was presented at this point; a small deletion within the receptor kinase region eliminated detectable kinase activity, but did not concomitantly eliminate EGF-mediated stimulation of DNA synthesis. Stella Clark described transfection experiments with EGF receptor constructs in which successful expression in Chinese hamster ovary (CHO) cells was achieved, but not in the EGF receptorless 3T3-NR6 cells. Promoter construct mutants with the chloramphenicol acetyl transferase gene were used to compare promoter efficacy in CHO and 3T3-NR6 cells. David Kaplan reported on the activation of both tyrosine kinases and PI kinases by wild-type and mutant polyoma viruses, in an attempt to identify the role of the polyoma T-antigen in transformation. The evening concluded, after three hours of intense discussion, with the remaining aficionados of somatic-cell genetics retiring to a nearby pub to evaluate the night's work.

# c-fos Proto-Oncogene Expression Is Necessary for Normal Growth of Mouse 3T3 Cells

## J.T. Holt and A.W. Nienhuis

*Clinical Hematology Branch, National Heart, Lung, and Blood Institute, National Institutes of Health, Bethesda, Maryland 20892*

Inhibition of expression of proto-oncogenes in intact cells may provide information about the roles of these genes in vivo. We have produced anti-sense RNA complementary to c-fos in mouse 3T3 cells employing gene transfer techniques. Transcriptional units were constructed which produced steroid-inducible RNAs that contained sequences complementary to the endogenous mouse c-fos mRNA. The c-fos sequences were positioned between the mouse mammary tumor virus (MMTV) promoter and the 3' portion of the human beta globin gene. Appropriate controls were constructed that produced steroid-inducible RNAs that were identical except that the c-fos sequences were in the sense orientation. A neomycin-resistance gene was included in the constructs to allow isolation of stable transformants. There was a tenfold decrease in the number of stable transformants obtained with both a mouse (pM84) and a human anti-sense construct in the presence of steroid. A second mouse anti-sense construct (pM301) produced considerably smaller clones but only a 40% decrease in the number of clones. There was no difference in the size or number of clones obtained from transfection of control DNAs that would not produce anti-sense c-fos RNA. These results indicate that anti-sense RNA may be useful in elucidating proto-oncogene function in normal cells, and provides evidence that the c-fos gene product has a required role in normal cell division.

**Key words:** anti-sense RNA, cell proliferation

Proto-oncogenes are presumed to serve essential roles in normal growth because their mutant forms are associated with malignancy [1,2]. Attempts to study the function of proto-oncogenes in nonmalignant cells have employed two different approaches: (1) overexpression of the gene of interest by injection of the protein [3] or gene transfer into appropriate cells [4] and (2) inhibition of gene expression by antibodies [5,6] or anti-sense RNA. The second approach produces mutant cells that

Received July 28, 1986.

© 1988 Alan R. Liss, Inc.

are deficient in the expression of the gene of interest, demonstrating which cellular processes require expression of the proto-oncogene.

c-fos is a nuclear phosphoprotein that binds DNA [7]. It represents the cellular homologue of the Finkel-Bistis-Jenkins (FBJ) osteosarcoma virus which was isolated from a murine tumor [8]. Induction of c-fos mRNA and protein occurs following the addition of growth factors to fibroblasts, hematopoietic cells, and pheochromocytoma cells [7,9,10]. Although rapid induction of c-fos mRNA is intriguing, this does not establish a regulatory role for c-fos in cellular proliferation.

The use of anti-sense RNA has been proposed as a general method for inhibiting gene expression [11–14]. RNA complementary to the endogenous mRNA for a specific gene is introduced into cells and inhibits the expression of the gene product. Inhibition results from the formation of RNA:RNA duplexes between the anti-sense RNA and the specific cellular mRNA. Transcriptional units that generate anti-sense RNAs have been employed to inhibit expression of thymidine kinase, actin, and heat-shock protein [12–14].

We introduced plasmids into mouse 3T3 cells which produce inducible anti-sense RNA complementary to c-fos. Induction of the anti-sense c-fos RNA results in inhibition of cell growth. These results suggest that c-fos expression is essential for normal cell proliferation.

## MATERIALS AND METHODS
### Cells and Culture Conditions

Mouse Swiss 3T3 cells were grown in improved modified essential media with 10% fetal calf serum. Transfections were performed by standard technique [15]. Following transfection, the cells were split into two aliquots. Both aliquots were grown in 1 mg/ml of the antibiotic G418, but only one aliquot was grown in 1 $\mu$M dexamethasone. Stable transformants were isolated from clones that grew on plates with G418 but not dexamethasone. Growth curves were performed by counting cells at various time points as previously described [16].

### Plasmid Constructs and Nucleic Acid Analysis

The details of construction of these plasmids have been described previously [16]. The 84- and 301-base pair (bp) anti-sense plasmids contain sequences complementary to mouse c-fos exon I. The 196-bp anti-sense plasmid contains sequences complementary to the human c-fos exon I. Each of the anti-sense plasmids has a corresponding sense plasmid which is identical except for the orientation of the c-fos fragment within the transcriptional unit. DNA and RNA were isolated from clones as described [17]. Southern blots, Northern blots, and S1 nuclease protection assays were performed by established methods [18–20]. The c-fos exon II M13 probe was constructed as described in the legend to Figure 2.

## RESULTS

The presence of the plasmid constructs in the neomycin-resistant clones was confirmed by Southern blot analysis with a probe for the neomycin-resistance gene. The copy number of the plasmids in the clones was variable ranging from one to fifty copies per cell (Fig. 1). Analyses of RNA from the clones demonstrated that the anti-

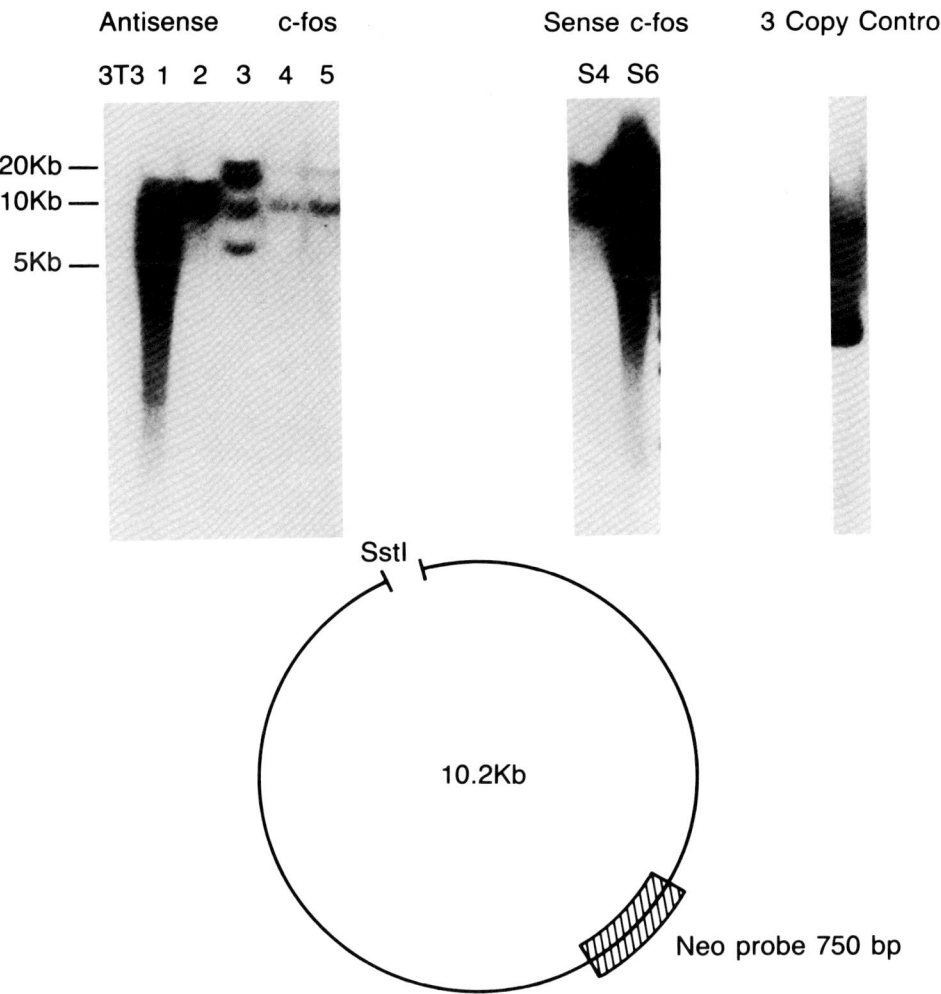

Fig. 1. Southern blot analysis of DNA derived from 3T3 clones transfected with the human 196bp antisense and sense plasmids. The probe is a 750-bp Pvu II fragment derived from the neomycin phosphotransferase gene. This probe was nick-translated to a specific activity of $4 \times 10^8$ cpm per $\mu$g. The lane labeled "3T3" represents DNA from untransfected 3T3 cells. The 3 copy control represents approximately three copies per cell of plasmid DNA which was electrophoresed on the same gel with the samples as a control to assess the copy number of the constructs in each clone. All DNAs were digested with Sst I, which digests the construct DNA in only one place (within the MMTV promoter). Shorter exposures of the filter reveal the expected 10.2-kilobase band in each sample. This photograph is from a 4-hr exposure of the autoradiograph.

sense and control RNAs were steroid inducible and of the expected size and structure [16]. S1 nuclease protection assays employing a probe for the endogenous mouse exon II demonstrated that c-fos mRNA is a rare species in logarithmically growing 3T3 cells (less than 10 copies per cell). The induction of anti-sense c-fos RNA does not affect the steady-state concentration of the endogenous c-fos mRNA (Fig. 2). These constructs produce a vast excess of anti-sense RNA in comparison to the endogenous c-fos mRNA.

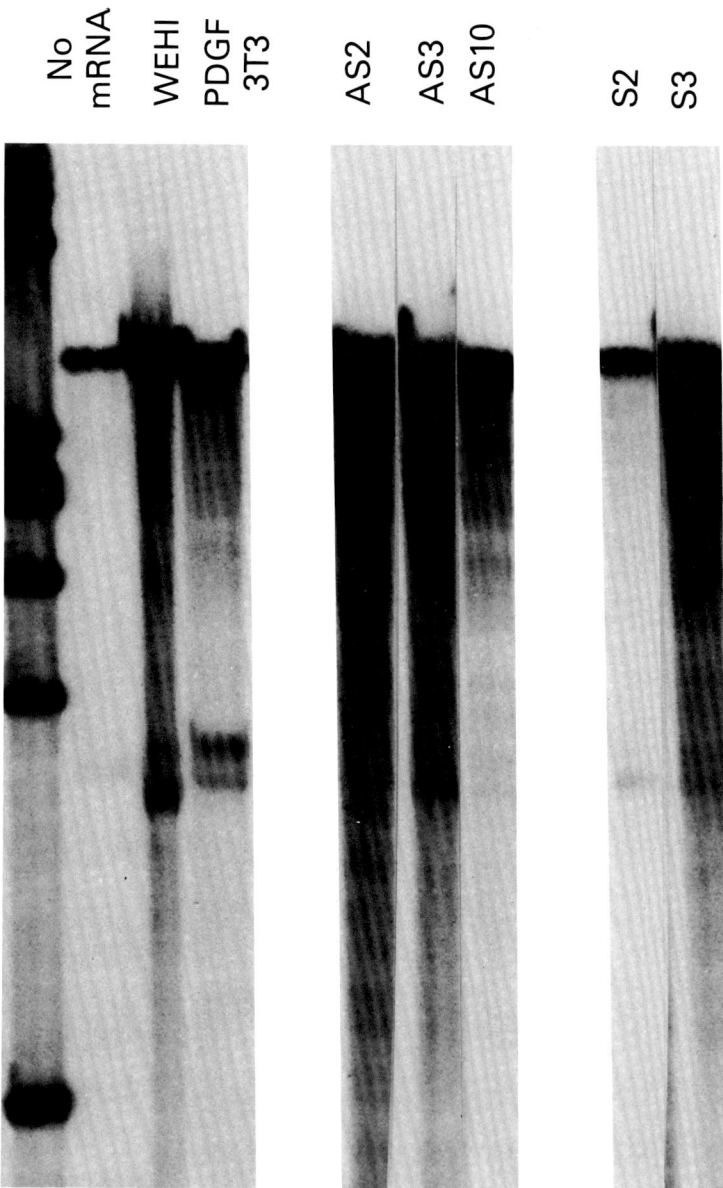

Fig. 2. This figure illustrates the presence of endogenous c-fos mRNA in the anti-sense and sense human 196-bp clones. The lane labeled "No mRNA" represents the S1 analysis of the probe only, without incubation with cellular RNA samples. The lane labeled "WEHI" represents RNA from WEHI cells. The lane labeled "PDGF 3T3" represents RNA from growth-arrested 3T3 cells which were exposed to platelet-derived growth factor for 1 hr. Lanes AS2, AS3, and AS10 represent anti-sense clones. Lanes S2 and S3 represent sense clones. All RNAs obtained from clones were incubated in 1 μg/ml of dexamethasone for 4 hr prior to RNA preparation. This autoradiograph was exposed for 14 days. The probe was constructed by cloning a 450-bp Pst I-Xba I fragment (from exon II and intron II of mouse c-fos) into M13mp19. The probe will protect a 176-bp fragment at the 3' end of exon II (this represents the lower band in the autoradiographs. The origin of the upper band is unknown and may represent an artifact of the analysis).

Transfection of the sense and anti-sense plasmids into 3T3 cells in the presence of dexamethasone demonstrates the effect of anti-sense c-fos RNA on cell growth. Induction of the anti-sense RNA produced by the mouse 84-bp and the human 196-bp constructs resulted in a 90% reduction in the number of G418-resistant clones (Fig. 3). Induction of the mouse 301-bp anti-sense RNA resulted in only a 40% reduction in the number of G418-resistant clones [16], but the clones that did appear were much smaller (Fig. 4). Dexamethasone induction of the sense clones did not result in a decrease in cither the number or the size of G418-resistant clones. Analysis of the growth rates of the anti-sense clones in the presence and absence of the inducer (dexamethasone) confirms the growth inhibitory effect of c-fos anti-sense RNA.

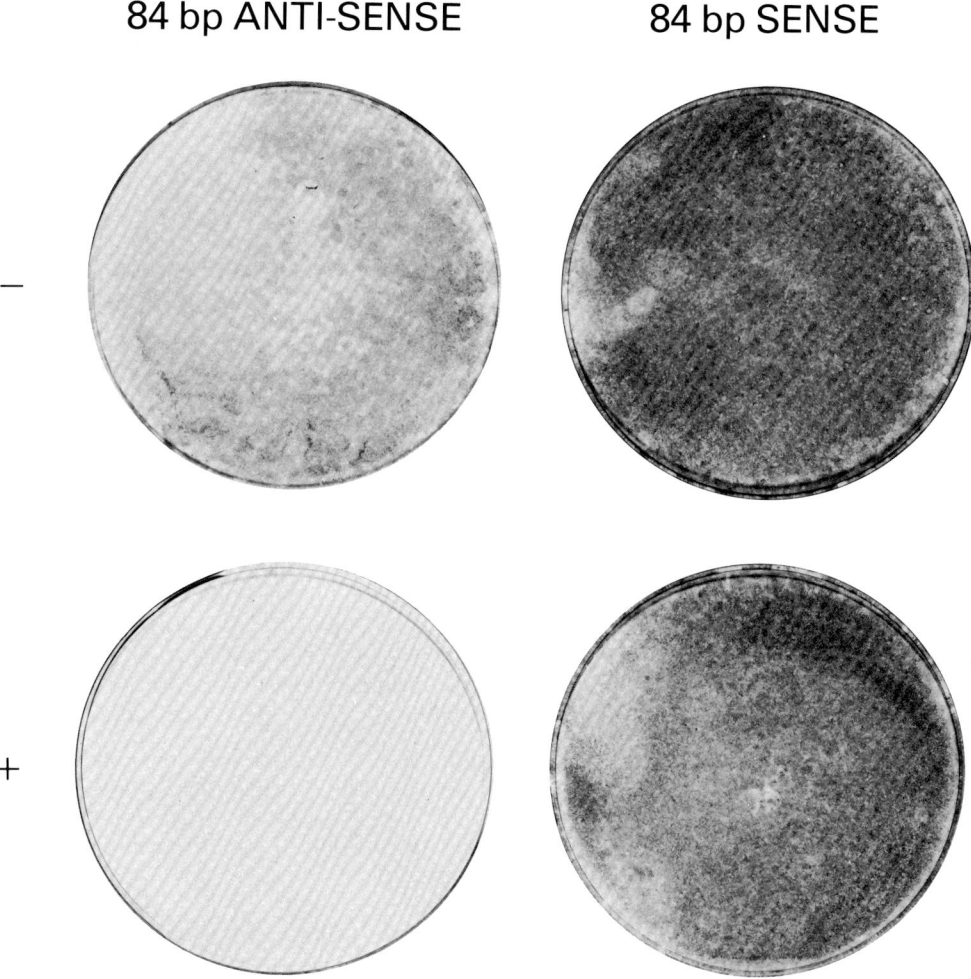

Fig. 3. This figure illustrates the growth inhibition that results from the induction of anti-sense c-fos RNA. The 84-bp mouse anti-sense or sense constructs were transfected into 3T3 clones as described in the Materials and Methods. Plates labeled with a minus sign are those that were grown in media with 10% fetal calf serum and 1 mg/ml G418. Plates labeled with a plus sign are those grown in the above media plus 1 μg/ml of G418. After growth for 14 days, the plates were washed with phosphate buffered saline, fixed in buffered 10% formaldehyde, washed with phosphate buffered saline, and stained with 5% Giemsa.

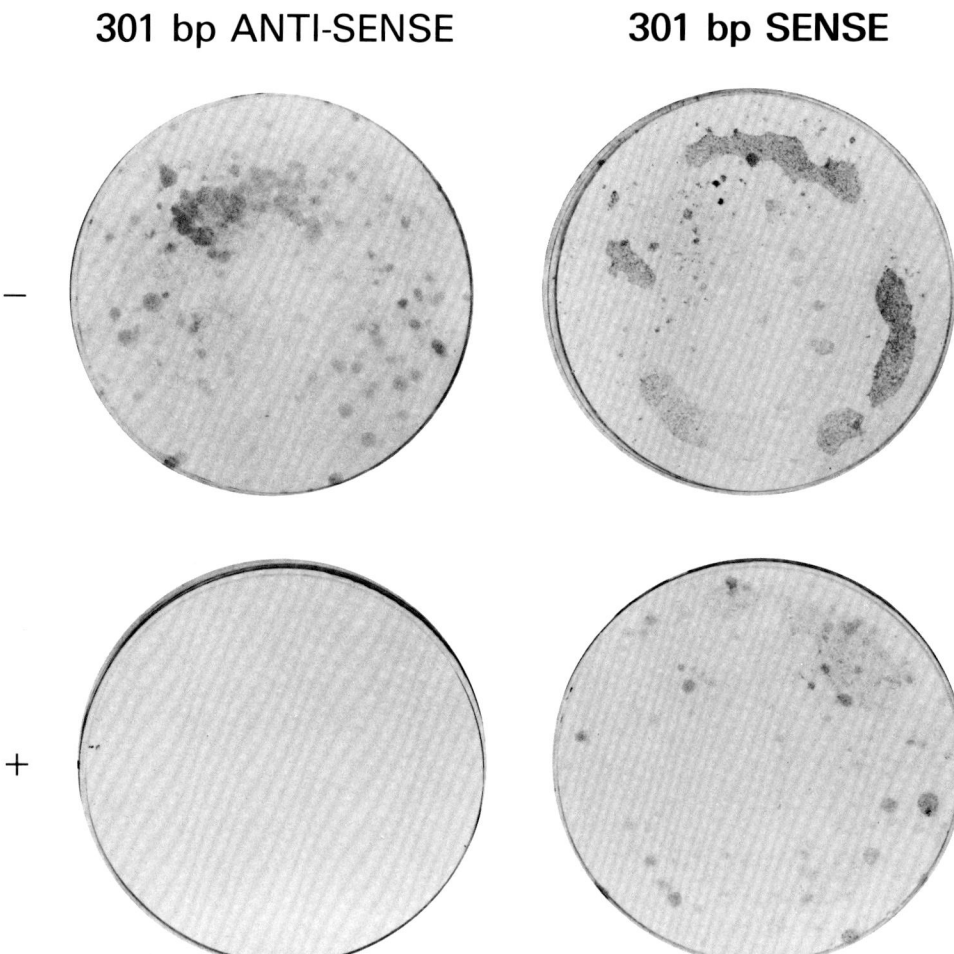

Fig. 4. This figure illustrates the growth inhibition that results from the induction of the 301-bp mouse anti-sense RNA construct. Plates are labeled and prepared as described in Figure 2. Examination of the plates reveals the presence of small G418-resistant clones following induction of the 301-bp anti-sense, although these are difficult to photograph.

There was a 50–80% reduction in the growth rate of clones when anti-sense RNA was induced, whereas control RNA induction has no significant effect on growth rate [16].

## DISCUSSION

These results suggest that the small amount of basal c-fos expression present in 3T3 cells is required for normal fibroblast proliferation. The S1 nuclease protection assay for mouse c-fos mRNA demonstrates that the presence of a vast excess of anti-sense RNA does not result in a rebound increase in c-fos transcription. This implies that the c-fos protein product may not directly regulate c-fos transcription, as had been proposed [21].

The growth inhibition which results from the inducible production of c-fos anti-sense RNA provides direct support for the hypothesis that c-fos expression is essential

## MOUSE-HUMAN c-fos HOMOLOGY

```
                           *  **  **           ***
Human      CTGCAGCGAGCAAACTGAGAAGCCAAGACTGAGCCGGCGGCCGCGGCGCA

Mouse      CTGCAGCGAGCAAACTGAGAAGACTGGATAGAGCCGGCGGTTCC******


                              *    *
Human      GCGAACGAGCAGTGACCGTGCTCCTACCCAGCTCCTACCCAGCTCTGCTT

Mouse      GCGAACGAGCAGTGACCGCGCTCCCACCCAGCTCCTACCCAGCTCTGCTT


                  ** *   *  * **              *   *     *  *
Human      TCACAGCGCCCACCTGTCTCCG*******CCCCTCGCCCGGCTTTGCCT

Mouse      TCAACGCTCCCACCAGTGTCTACCCCTGGACCCCTTGCCGGGCTTTCCCC


               * **    *              *        *
Human      AACCGCC*ACGATGATGTTCTCGGGCTTCAACGCAGACTACGAGGCGTCA

Mouse      AAACTTCGACCATGATGTTCTCGGGTTTCAACGCCGACTACGAGGCGTCA



Human      TCCTCCCGCTGCAG

Mouse      TCCTCCCGCTGCAG
```

Fig. 5. This figure demonstrates the homology between mouse and human c-fos in the 196-bp region present in the human anti-sense RNA construct. Areas of sequence divergence are marked by an asterisk. The coding region is underlined beginning at the ATG.

for cell growth. These findings also indicate that anti-sense RNA inhibition of proto-oncogenes may require the use of inducible expression vectors. If the presence of the anti-sense RNA inhibits cell proliferation, then clones must be obtained in the absence of this effect.

Our studies demonstrate that a human anti-sense construct can inhibit the proliferation of mouse cells. Previous studies have emphasized the specificity of anti-sense RNA [11–14], and appear to be in conflict with our results. However, the sequence conservation between mouse and human DNA at the 5' end of c-fos is striking. The homology is greater than 80% with several stretches of complete homology. A comparison of mouse and human c-fos sequence in this region is presented in Figure 5.

c-fos mRNA and protein levels are difficult to measure in normal growing cells but are induced by growth factors. Other investigators have detected c-fos mRNA or protein in cells that have not been stimulated with growth factors [22,23]. Our studies suggest that that this small amount of basal c-fos expression is required for cell division. Inhibition of gene expression provides a technique to establish the importance of low level gene expression, since important genes may not always be transcribed at high levels. This study establishes the feasibility of employing anti-sense RNA methodology to study the role of proto-oncogene expression in cellular biochemistry.

## ACKNOWLEDGMENTS

The authors acknowledge the excellent technical assistance of Patricia Turner and Austine Davis Moulton.

## REFERENCES

1. Bishop JM: Cell 42:23, 1985.
2. Holt JT, Morton CC, Nienhuis AW, Leder P: "Molecular Basis of Blood Diseases." Philadelphia: W.B. Saunders, p 347, 1987.
3. Kaczmarek L, Hyland JK, Watt R, Rosenberg M, Baserga R: Science 228:1313, 1985.
4. Ruther U, Wagner EF, Muller R: Embo J 4:1775, 1985.
5. Huang JS, Huang SS, Deuel TF: Cell 39:79, 1984.
6. Mulcahy LS, Smith MR, Stacey DW: Nature 313:241, 1985.
7. Muller R: Biochim Biophys Acta 823:207, 1986.
8. Curran T, Teich NM: Virology 116:221, 1982.
9. Gonda TJ, Metcalf D: Nature 310:249, 1984.
10. Curran T, Morgan JI: Science 229:1265, 1985.
11. Izant JG, Weintraub H: Cell 36:1007, 1984.
12. Izant JG, Weintraub H: Science 229:345, 1985.
13. Kim SK, Wold BJ: Cell 42:129, 1985.
14. McGarry TJ, Lindquist S: Proc Natl Acad Sci USA 83:399, 1986.
15. Wigler M, Sweet R, Sim GK, Wold B, Pellicer A, Lacy E, Maniatis T, Silverstein S, Axel R: Cell 16:777, 1979.
16. Holt JT, Gopal TV, Moulton AD, Nienhuis AW: Proc Natl Acad Sci USA 83:4794, 1986.
17. Humphries RK, Ley TJ, Turner P, Moulton AD, Nienhuis AW: Cell 30:173, 1982.
18. Southern SM: J Mol Biol 98:503, 1975.
19. Alwine JC, Kemp DJ, Stark GR: Proc Natl Acad Sci USA 74:5350, 1977.
20. Ley TJ, Anagnou NP, Pepe G, Nienhuis AW: Proc Natl Acad Sci USA 79:4775, 1982.
21. Meulink F, Curran T, Miller AD, Verma IM: Proc Natl Acad Sci USA 82:4987, 1985.
22. Curran T, Van Beveren C, Ling N, Verma IM: Mol Cell Biol 5:167, 1985.
23. Adamson ED, Meek J, Edwards SA: Embo J 4:941, 1985.

# Structure and Function of p21 *ras* Proteins: Biochemical, Immunochemical, and Site-Directed Mutagenesis Studies

**Thomas Y. Shih, David J. Clanton, Seisuke Hattori, Linda S. Ulsh, and Zhang-qun Chen**

*Laboratory of Molecular Oncology, Division of Cancer Etiology, National Cancer Institute, National Institutes of Health, Frederick, Maryland 21701*

p21 *ras* proteins contain at least two functional domains. The membrane binding domain apparently involves a short sequence at the C-terminus of p21. We have determined by chemical methods that the cysteine-186, four amino acid residues from the end, is the palmitylation site. We have studied a highly purified p21 protein overproduced in *Escherichia coli*. Scatchard analysis and probing with monoclonal antibodies against p21 indicate a single site per molecule for binding GTP or GDP with a Kd of approximately $1 \times 10^{-8}$ M. The same site apparently is involved with the autokinase and GTPase activities associated with p21. We have constructed point mutations in order to dissect the structure-function relationship of the GTP binding domain. Mutations at asparagine-116 of p21 to lysine or tyrosine, but not the adjacent 117th or 118th positions, abolish GTP binding activity. Mutation of the autophosphorylation site at threonine-59 to a serine or a alanine residue reduces or abolishes the autokinase activity, consistent with juxtaposition of this residue to the GTP gamma-phosphate. Mutations of glycine residues to valine of the consensus sequence, GXXXXGK, of the p21 N-terminal region, also greatly affect the GTP binding activity. These studies suggest that the basic structure of the GTP binding domain is very similar between p21 and EF-Tu, the *E. coli* elongation factor, of which a three-dimensional structure has recently been determined by other investigators. Transfection studies of the mutant *ras* genes indicate that the GTP binding domain is crucial for p21 cellular function.

**Key words:** *ras* oncogene, G-proteins

The *ras* gene family encodes a group of closely related 21,000-dalton p21 proteins [1]. In the human genome, there are three distinct c-*ras* genes; c-*ras*[H], which is homologous to the *ras* gene of Harvey murine sarcoma virus (Ha-MuSV); c-*ras*[K],

Received May 7, 1986.

Seisuke Hattori's present address is Department of Pure and Applied Sciences, University of Tokyo, Tokyo 153, Japan.

© 1988 Alan R. Liss, Inc.

which is homologous to the closely related Kirsten murine sarcoma virus (Ki-MuSV); and c-$ras^N$, which was originally identified in a neuroblastoma cell line and has, as yet, no known viral counterpart. These proteins of 188–189 amino acid residues are highly conserved in sequence, which suggests not only that most of their structure is functionally significant but also that cellular *ras* genes must play an essential role in growth and development. Activated oncogenes have been detected in numerous human and experimentally induced animal tumors and most of these oncogenes are activated by single point mutations replacing either glycine-12 or -13, or glutamine-61 with different amino acid residues.

There are three known biochemical activities associated with p21. The p21 of all *ras* genes binds GTP or GDP with high affinity and displays a low GTPase activity, which is further reduced by some mutations activating the proto-oncogenes [2–7]. Additionally, the p21 of Ha-MuSV and Ki-MuSV exhibits an autokinase activity owing to substitution of a threonine residue for the alanine-59 of c-*ras* genes at the autophosphorylation site [8]. Like many other GTP binding proteins, p21 is localized on the inner surface of the plasma membrane, which apparently requires palmitylation at cysteine-186 [9–12]. This has led to the suggestion that p21 may function as a coupling protein relaying extracellular messages to the interior cellular effectors in the signal transduction pathways controlling cell growth and proliferation.

In this paper, we wish to report our studies on the structural domains for GTP binding and membrane association of p21 molecules, and to relate the structural significance of the GTP binding domain to the cellular function of p21 proteins.

## MATERIALS AND METHODS
### Purification of p21 From *E. coli*

*E. coli* carrying plasmid, pJLcIIrasI [13], was induced by raising the temperature from 31°C to 41°C to overproduce p21, which constituted approximately 10% of total proteins. p21 was purified to over 95% purity, without denaturation, by high-speed centrifugation of lysates, ammonium sulphate precipitation, DEAE-Sephacel, and Sephadex G-150 as previously described [6]. Some mutant p21s, however, were less soluble when overproduced in *E. coli*. In these cases, an alternative procedure was used which involved cell lysis with lysozyme and Nonidet p-40 (NP40) and, after centrifugation, extraction of p21 from pellets with 8 M urea [14].

### Guanine Nucleotide Binding Assays

The standard assay involves incubation of p21 with 1–10 $\mu$M $^3$H-GDP in a buffer containing 90 mM Tris-HCl (pH 8.0), 5 mM $MgCl_2$, 100 mM NaCl, 10% (v/v) glycerol, and 500 $\mu$g bovine serum albumin at 37°C for 10–30 min. The binary complexes of p21 and $^3$H-GDP were retained by nitrocellulose filters [6]. An alternative Western-blot GTP binding assay was performed as previously described by McGrath et al [3], in which binding was assayed after Western blot transfer of p21 from SDS-PAGE.

### Autokinase and GTPase Assays

Autokinase activity was assayed by incubation of p21 with gamma-32p-GTP, and the phosphorylated p21 was detected by autoradiography after SDS-PAGE [15]. GTPase activity was assayed by incubation of p21 with gamma-$^{32}$P-GTP, and the inorganic phosphate was determined by the Fiske-Subbarow method or by PEI thin layer chromatography (TLC) plates [15].

## Monoclonal Antibodies to p21

The well-characterized rat monoclonal antibodies to p21 were previously developed by Furth et al [16]. Mouse monoclonal antibodies were prepared by immunizing Balb/c mice with purified p21 isolated from *E. coli*, and hybridoma clones were isolated after fusion of myeloma cells with spleen cells of immunized mice.

## Site-Directed Mutagenesis

The 1.1–kilobase (kb) SstII/XbaI fragment containing the entire *ras* oncogene from the pH-1 DNA clone of Ha-MuSV [17] was cloned into the M13 phage. Single-stranded DNA from this phage was used as the template for mutagenesis. Oligonucleotides of 17 bases containing single base changes from the v-$ras^H$ gene were used to construct mutants substituting a single amino acid residue of the wild-type v-$ras^H$ oncogene at specific amino acid residues. Phage containing mutations were identified by plague hybridization with $^{32}$p-labeled oligonucleotides, and DNA was sequenced to confirm the mutations. The SstII/XbaI fragment containing the mutations was used to reconstruct the pH-1. Alternatively, the internal 0.88-kb HindIII fragment from the mutated phage DNA was cloned into the bacterial expression vector, pJL6, for producing p21 in *E. coli* [13]. The mutagenesis scheme was previously described in detail [14].

## Transfection of NIH3T3 Cells

Focus-forming assays were performed as described by Lowy et al using plasmid DNA coprecipitated with calcium phosphate [18]. Foci were scored by trypan blue staining 30 days post-transfection. Cotransfection with the pSV-neo gene [19] was performed with a mixture of 500 ng mutant pH-1 DNA and 50 ng pSV-neo DNA. G418-resistant cell colonies were cloned after 14 days [14].

## Immunoprecipitation of p21

Cells were labeled with $^{35}$S-methionine or cysteine, or with $^{32}$P-orthophosphate, and p21 was immunoprecipitated by monoclonal antibodies as previously described [16, 20].

## RESULTS AND DISCUSSION

Figure 1 illustrates the pathway of p21 biosynthesis. We demonstrated previously, in pulse-chase experiments, that p21 is synthesized as a precursor form in the free cytosol, and shortly after synthesis, processed and associated with the plasma membrane. This is accompanied by a slight increase (approximately 1,000 daltons) in electrophoretic mobility on SDS-polyacrylamide gels [10]. p21 proteins of the v-*ras* and c-*ras* oncogenes, and their proto-oncogenes all appear to be processed by a common pathway [21]. Sefton et al first demonstrated post-translational acylation of p21 [22], and Willumsen et al, using a series of genetic mutations near the p21 C-terminus, demonstrated that the cysteine-186 residue is required for acylation to occur, which is essential for the transforming activity of the *ras* oncogene [11]. Recently, we have chemically identified the acylation site by comparing the peptide

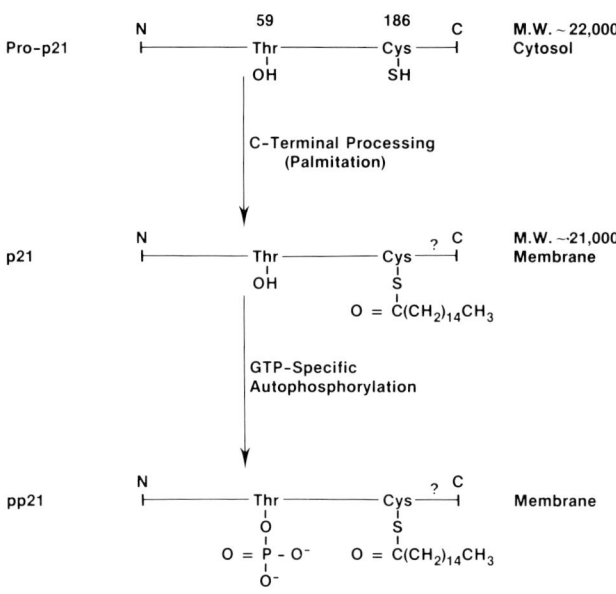

Fig. 1 Biosynthetic pathway of p21 ras proteins. Shortly after its synthesis in the free cytosol, the pro-p21 is palmitylated at cysteine-186, and phosphorylated at threonine-59. These processed products are associated with the plasma membrane. Question marks indicate the precise C-terminus and N-terminus have not yet been determined (Reproduced from Shih et al [46].

map of p21 labeled with $^3$H-palmitate in a Ha-MuSV-transformed cell to that of p21 overproduced in *E. coli*, which is not acylated and is an equivalent of pro-p21 [12]. The acylated hydrophobic tryptic peptide was isolated by high performance liquid chromatography (HPLC), and was found to be the C-terminal tetrapeptide of p21 containing cysteine-186. Interestingly, the fatty acid associated with this peptide is palmitic acid rather than myristic acid, which has been found to be associated with the *src* and many other cellular proteins on their N-termini [23,24]. Since p21 does not contain any long stretch of hydrophobic amino acid residues, the palmitic moiety may serve as the membrane-anchoring domain for *ras* proteins. The precise mechanism of this is still obscure, since only a small fraction of p21 appears to contain fatty acid [25], while most of the cellular p21 is membrane associated [9,10]. Also, our preliminary attempts to remove fatty acid from p21 do not restore electrophoretic mobility of p21 to that of pro-p21 on SDS-PAGE [12].

## The GTP Binding Domain

The p21 protein we have studied in detail is the gene product of the v-*ras* oncogene of Ha-MuSV, and is overproduced in *E. coli* as an N-terminal fusion protein as depicted in Figure 2, with the four p21 amino acid residues replaced by a stretch of 14 amino acids derived from the plasmid expression vector under the control of the phage lambda $P_L$ promoter [13]. About 50% of p21 of the wild-type v-*ras* gene overproduced in this system can be recovered in the high-speed supernatant of bacterial lysates without denaturation. p21 protein of over 95% in purity was obtained by ammonium sulphate precipitation, DEAE-Sephacel chromatography, and Sepha-

```
c-H-ras       ( 1 - 22)                    M T E Y |K L V V V G A G|G V|G K S A L|T I|Q
v-H-ras       ( 1 - 22)                    M T E Y |K L V V V G A R|G V|G K S A L|T I|Q
v-H-ras      (-10 - 22)  MVRANKRNEAL R I D |K L V V V G A R|G V|G K S A L|T I|Q
(E. coli)
α Transducin (27 - 48)              A K T V |K L L L L G A G|E S|G K S T I|V K|Q
α G₀                                A K D V |K L L L L G A G|E S|G K S T I|V K|Q
EF-Tu         ( 9 - 30)             K P N V |N|V|G T|I|G|H V D H|G K T T L|T A A
```

Fig. 2. Comparison of N-terminal sequences of p21 *ras* with G-proteins. The junction point of the chimeric p21 in *E. coli* between residues 4 and 5 precedes the boxed homologous regions of G-proteins. The following Dayhoff conservative categories are used: C; S, T, P, A, G: N, D, E, O; H, R, K; M, I, L, V; and F, Y, W [47]. Amino acid symbols are A, Ala; R, Arg; N, Asn; D, Asp; C, Cys; E, Glu; Q, Gln; G, Gly; H, His; I, Ile; L, Leu; K, Lys; M, Met; F, Phe; P, Pro; S, Ser; T, Thr; W, Trp; Y, Tyr; V, Val. Sequences are from transducin α, Medynski et al [32]; α-subunit of G₀ from brain, Hurley et al [33]; EF-Tu, Arai et al [48].

dex G-150 gel filtration in 25% glycerol [6]. Three biochemical activities with specificity for guanine nucleotides are copurified with p21. The dissociation constants ($K_d$s) for binding GTP or GDP are approximately the same, on the order of $2 \times 10^{-8}$ M. p21 also possesses a GTPase activity with a very low turnover number (0.2 mmol GTP hydrolyzed per mol p21 per min). This activity was found to be at least tenfold higher in p21 of the proto-oncogenes [2–5]. In addition, the p21 of Ha-MuSV and Ki-MuSV exhibits an autokinase activity owing to a threonine-59 mutation, which generates the phosphoryl acceptor site [8]. This autokinase activity may contribute to an additional activation step for creating the highly transforming v-*ras* oncogenes, and it also serves as a marker for evaluating the p21 biochemical activities in vitro or in vivo. The specific activities of these three biochemical properties of p21 isolated from *E. coli* are very similar to p21 purified form Ha-MuSV-transformed NIH3T3 cells [15]. Scatchard analysis on the highly purified protein indicates that p21 has a single site per molecule for binding either GTP or GDP with a $K_d$ on the order of 10 nM [6]. Significantly, using a panel of monoclonal antibodies against p21, we found that both GTP/GDP binding and autokinase activities of this v-*ras*[H] p21 protein were partially, but specifically, inhibited by the Y13-259 monoclonal antibody previously developed by Furth et al [16]. The same antibody, when microinjected into cells transformed by *ras* oncogene, was observed by Mulcahy et al to cause transient reversion to the normal phenotype [26]. Reduced ³H-GDP binding has also been observed in immunoprecipitates of p21 from lysates of *ras*-transformed cells by Y13-259 as compared to Y13-238 or YA6-173 monoclonal antibody [16] (Weeks and Fattorossi, personal communication). Lacal and Aaronson, however, observed no difference in GTP binding in their p21 preparation obtained from a different construct [27]. One significant difference among p21 preparations obtained from proteins overproduced in *E. coli* is their method of isolation. Figure 3 shows UV absorption spectra of the purified p21. p21 isolated without a denaturation step, such as 3 M guanidine-HCl or 8 M urea, demonstrates a higher spectral ratio of minimum at 253 nm to maximum at 277 nm and a shoulder at 260 nm, spectrum characteristic of p21 complexed with an equimolar amount of GDP [28]. In these "native" preparations, ³H-GDP binding assays presumably measure the exchange of labeled GDP with the prebound GDP, whereas in p21 prepared in a denatured form, nucleotide binding probably occurred upon renaturation of the protein. The capability of p21 to renature is most clearly seen in the Western-blot GTP binding assay [3], in which the pro-

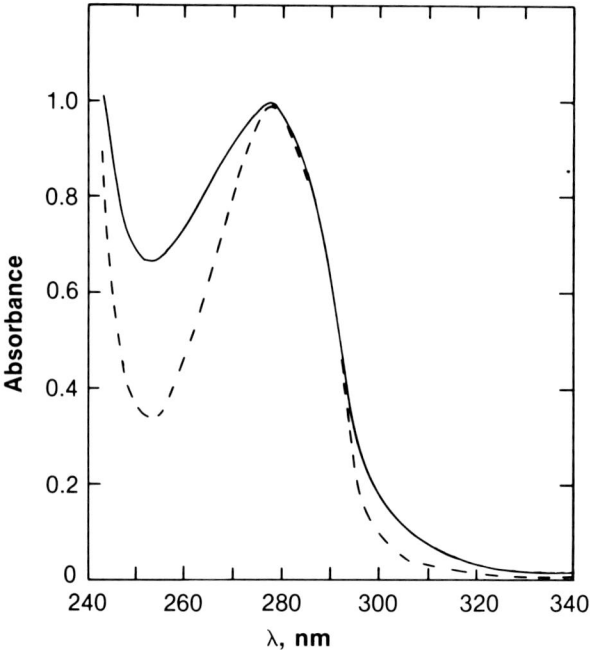

Fig. 3. Comparison of UV absorption spectra of p21 isolated with and without denaturation steps. p21 of over 95% in purity was isolated from *E. coli* without any denaturation step, as described by Hattori et al [6], or by extraction of pellet fractions with 3 M guanidine-HCl, as described by Lacal et al [7,27]. p21 preparations at a concentration of 0.5 $OD_{280}$ were extensively dialized against 20 mM MES buffer, pH 7.0, and 25% (v/v) glycerol. UV spectra were measured against the same buffer with a Cary 17 recording spectrophotometer. The normalized spectra were presented. The dashed line is for p21 isolated with denaturation steps, and the solid line for p21 without any denaturation step. For the latter preparation, the ratios of $A_{min}/A_{max}$ of 0.67 and $A_{260}/A_{277}$ of 0.75 are very close to those reported by Poe et al [28] for p21·GDP equimolar complex of 0.68 and 0.81, respectively.

tein has been completely denatured by boiling in SDS buffer before Western transfer (Fig. 4B).

The p21 proteins of the *ras* gene family have a highly conserved primary amino acid sequence. Of particular significance in defining the GTP binding domain is the extensive sequence homology found between p21 and the family of G-proteins which also bind GTP/GDP [29–35]. There are at least three regions of high homology, i.e., amino acids 5–27, 58–73, and 109–120 [Figs. 2, 6]. Recent studies by x-ray crystallography of EF-Tu·GDP reveal the structure of the GDP binding site [36,37]. The high homology region of G-proteins are within the peptide loops, connecting α-helices and β-sheets of the GDP binding domain, and a molecular model for p21 has been proposed based on the analogy to EF-Tu [38]. This p21 structure (Fig. 7) is remarkably consistent with the results of our site-directed mutagenesis studies in which we made many point mutations with single amino acid substitutions at the high homology regions of p21 (Table I). In the three-dimensional structure of EF-Tu·GDP, the homologous asparagine-116 and aspartate-119 of p21 have several critical interactions with the guanine ring perhaps by hydrogen-bonding (Figs. 6, 7). Our mutations, by drastically changing the side chain of asparagine-116 into that of lysine (116K) or tyrosine (116Y), result in p21 proteins apparently devoid of GTP/GDP binding and

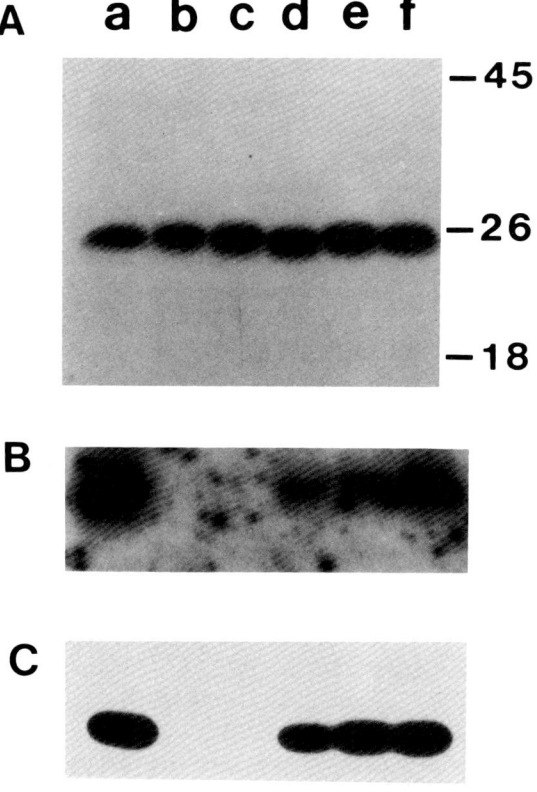

Fig. 4. Mutant p21 proteins overproduced in *E. Coli*. **A:** Western blotting with Y13-259 anti-21 monoclonal antibody. 1.25-μl of aliquots of 8 M urea extracts of pellet fractions of lysates were loaded on each lane. **Lane a,** wild type; **b,** clone 116K; **c,** clone 116Y; **d,** clone 117Q; **e,** clone 118R; and **f,** clone 118S. **B:** Western GTP binding assay. p21 (1.5 μg) were transferred onto a nitrocellulose filter and then reacted with 1 μM α-$^{32}$P-GTP. **C:** Autokinase activities. Extracts containing 0.35 μg p21s were incubated with 10 μM γ-$^{32}$P-GTP and the phosphorylated proteins were visualized by autoradiography after SDS-PAGE. (Reproduced from Clanton et al [14].

autokinase activities (Fig. 4), whereas mutations at the adjacent positions of 117 or 118 have little effect [14]. These results, together with the recent findings by Sigal et al [39] on mutations of aspartate-119, strongly suggest that the basic structure of the GTP binding site is very similar between p21 and EF-Tu.

We also made mutations changing the threonine-59 at the autophosphorilation site into alanine and serine. As expected, the alanine mutant cannot be phosphorylated, and the serine mutant exhibits a reduced autokinase activity (Fig. 5), which may reflect a 10–100-fold decrease in the GTP binding affinity of the serine mutant (Table I). Again these results are consistent with juxtaposition of this threonine residue with the gamma-phosphate of GTP (Fig. 7).

Figure 2 shows boxed regions of high homology of p21 N-terminal sequences to other G-proteins. It is interesting to note that in our current construct of the p21 fusion proteins, the fusion point precedes the homologous sequences starting from lysine-5, which may explain why this fusion protein has biochemical activities very close to the intact p21 in mammalian cells. This region of p21 also contains the consensus sequence of GXXXXGK(Gly$^{10}$XXXXGly$^{15}$Lys$^{16}$ of p21) found in most

**TABLE I. $^3$H-GDP Binding Activities of Mutant p21s**[*]

| Clone | Mutation(s) | Maximum binding (pmol GDP/pmol p21) | Km |
|---|---|---|---|
| R/T (wild-type) | — | 1.0 | 20 nM |
| G/T | Gly$^{12}$ | 0.4 | 75 nM |
| G/A | Gly$^{12}$, Ala$^{59}$ | 0.6 | 12 nM |
| R/A | Ala$^{59}$ | 0.8 | 31 nM |
| R/S | Ser$^{59}$ | 0.6 | 945 nM |
| G/S | Gly$^{12}$, Ser$^{59}$ | 0.6 | 7918 nM |
| 22K | Lys$^{22}$ | 0.3 | 97 nM |
| 10V | Val$^{10}$ | 0.001 | ca. 500,000 nM |
| 15V | Val$^{15}$ | 0.03 | ca. 500,000 nM |
| 116K | Lys$^{116}$ | 0 | >1,000,000 nM |
| 116Y | Tyr$^{116}$ | 0 | >1,000,000 nM |

[*]The $^3$H-GDP binding assay was performed as previously described [6,14] by incubating p21 extracted with 8 M urea from pellet fractions of bacterial lysates with $^3$H-GDP with varying specific radioactivities up to a concentration of 1 mM. Radioactivity associated with p21 was determined after retention of binary complexes on nitrocellulose filters. Km is the GDP concentration for half-maximal binding.
[a]The wild-type p21 has Arg$^{12}$, Thr$^{59}$, Gln$^{22}$, Gly$^{10}$, Gly$^{15}$, and Asn$^{116}$ at the respective mutation sites.

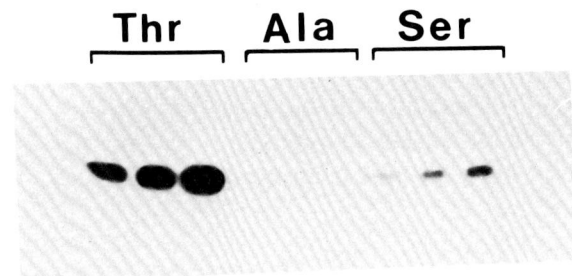

Fig. 5. Autokinase activities of p21 with mutations at threonine-59 to either an alanine or a serine residue. Extracts were assayed as in Figure 4. **Lanes 1–3:** 0.3, 0.5, and 1 μl of wild-type p21. **Lanes 4–6:** 0.3, 0.5, and 1 μl of the alanine mutant. **Lanes 7–9:** 0.3, 0.5, and 1 μl of the serine mutant.

SEQUENCE HOMOLOGY AT THE GUANINE BASE BINDING SITE

```
                                           G SITES
                                           v        v
H-ras        (109-120)    V P M V L V G - N K C D - L
K-ras        (109-120)    V P M V L V G - N K C D - L
N-ras        (109-120)    V P M V L V G - N K C D - L
RAS 1        (116-127)    I P V V V V G - N K L D - L
RAS 2        (116-127)    V P I V V V G - N K S D - L
rho          (111-122)    V P I L V G - N K K D - L
Transducin α (258-269)    T S - I V L F L N K K D - V
EF-Tu        (127-139)    V P Y I I V F L N K C D - M
EF-G         (134-146)    V P R I A - F V N K M D R M
```

Fig. 6. Comparison of amino acid sequences of p21 and G-proteins at the guanine base binding site. In the three-dimensional structure of EF-Tu, asparagine-135 and aspartate-138 have critical interactions with the guanine moiety of GDP. Mutations at the homologous sites of p21 critically affect its GTP/GDP binding activities. Amino acid symbols are the same as in Figure 2.

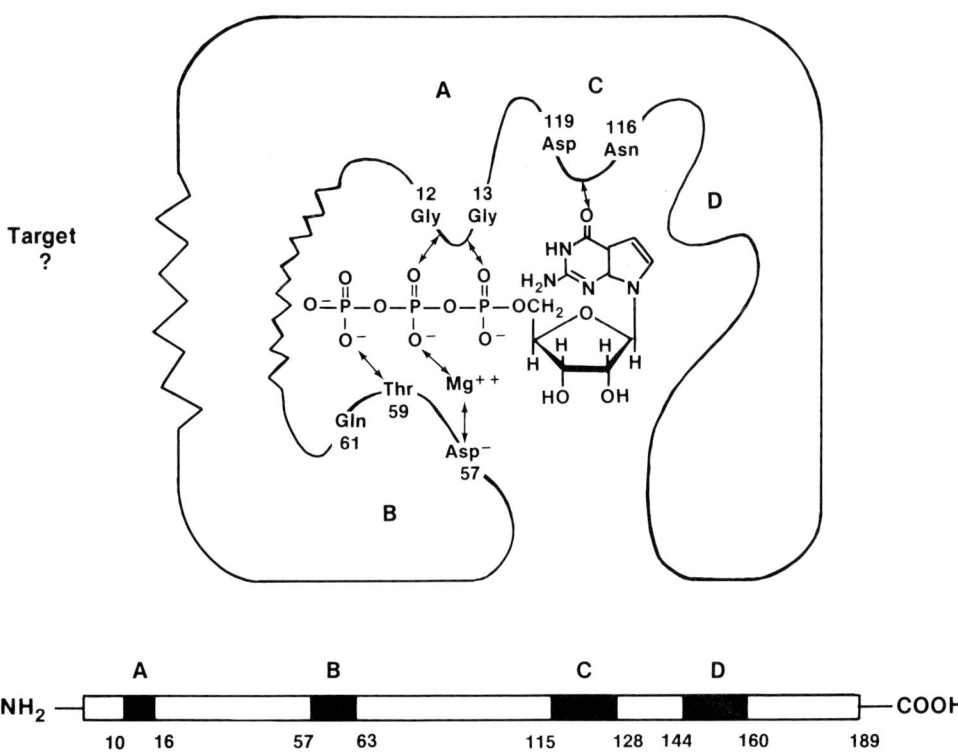

Figure 7. The GTP binding domain of p21 *ras* proteins. The detailed structure of the GTP binding site is the result of many studies described in the text. The four peptide loops which constitute the GTP binding site are shown in the p21 linear structure below. Arrows indicate interactions of specific amino acid residues with GTP. The zigzag regions suggest possible conformational changes of p21 upon binding with either GTP or following its hydrolysis to GDP, functioning as an on/off switch for its interaction with putative targets. (Reproduced from Shih et al [46].

ATP or GTP binding proteins [40]. As shown in Table I, our mutations changing these two glycine residues into valine greatly affect $^3$H-GDP binding activity of p21, suggesting the critical role of this sequence. Proximity of this peptide loop 10–16 to the GTP binding site is also suggested by the recent study by Clark et al using antisera directed against this peptide [41]. Figure 7 summarizes our current knowledge on the structure of the GTP binding domain of p21 proteins.

## Crucial Roles of the GTP Binding Domain to *ras* Cellular Function

Although it is clear from a previous study of p21 in a temperature-sensitive mutant of Ki-MuSV that p21 is required for virus-induced transformation [42] and perhaps it also plays a role in cellular proliferation in normal and abnormal growth of most eukaryotic organisms, crucial p21 properties required for its cellular function are not unequivocally established. Recently, we and others have identified antibodies that affect the GTP binding domain of p21 in vitro [6,41]. Upon microinjection of these antibodies into cells, Stacey, Feramisco and their colleagues were able to demonstrate transient reversion of cells transformed by *ras* oncogenes to their normal phenotypes [26,42]. These studies suggest that p21 in vitro activities may be responsible for its in vivo function. However, antibodies are large molecules as compared to p21 of 21,000 daltons. It is possible that antibodies may affect other unknown p21 activities apart from the epitopes (Y13-259 monoclonal antibody, which in our hands affects GDP binding, recognizes an epitope at residues 63–73 of p21, which is not at the GTP binding pocket [27]) (J.B. Gibbs, personal communication). We have reconstructed mutant DNAs into the proviral pH-1 DNA clone of Ha-MuSV. Upon transfection into NIH3T3 cells, mutants which retain GTP binding activity are all transforming, whereas both 116K and 116Y, which have lost their GTP binding ability, have reduced transforming efficiency (Table II). By cotransfection with pSV-neo as a positive selection marker, cell clones resistant to G418 have been isolated. Cells transfected with 116K or 116Y mutants are contact-inhibited. These studies suggest that GTP binding is crucial for *ras* gene function [14].

The precise cellular role of *ras* proteins, however, is still one of the major unresolved problems. The similarity in biochemical properties and subcellular localization between p21 and the group of G-proteins would suggest that p21 may function in an analogous manner in coupling the transduction of growth control signals, as depicted in Figure 8 [43]. This coupling function is turned on and off by the switch of binding GTP ligand and its hydrolysis to GDP. In our recent study, reconstitution, with purified p21 of a membrane preparation of S49 lymphoma cells lacking the $N_s$

**TABLE II. Transformation and GTP Binding of Proviral DNA Clones**[*]

| DNA | Mutation | Transformation | GTP binding |
|---|---|---|---|
| pH-1[a] | Wild Type | + | + |
| 116K | Lys[116] | − | − |
| 116Y | Tyr[116] | − | − |
| 117Q | Gln[117] | + | + |
| 118R | Arg[118] | + | + |
| 118S | Ser[118] | + | + |
| 59A | Ala[59] | + | + |
| 59S | Ser[59] | + | + |

[*]Transformation activities were evaluated by a focus-forming assay upon transfection of NIH3T3 cells. Positive clones scored more than 200 to 600 trypan-blue-stainable foci per µg DNA. Negative clones were less than 20 foci per µg DNA. For details see Clanton et al [14].

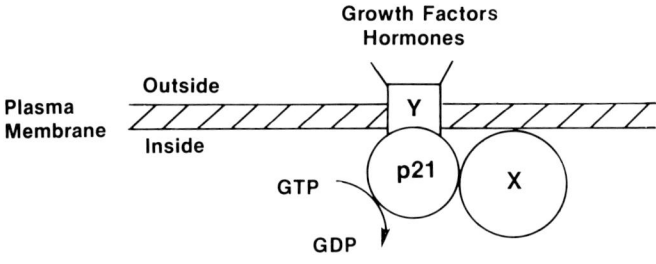

p21 : Membrane Signal Transducer
X : Cellular Effector
Y : Transmembrane Receptor

Fig. 8. A working model for the function of p21 ras proteins. p21 is shown as a coupling factor for transduction of growth control signals from extracellular factors across the plasma membrane to intracellular effectors. The binding of GTP/GDP serves as an on/off switch for these interactions. The cellular effector (X) and the transmembrane receptor (Y) are hypothetical at present.

protein of adenylate cyclase, did not restore adenylate cyclase activity. This suggests that p21 does not function directly on adenylate cyclase [44], an observation different from the marked effect seen in the *Saccharomyces cervisiae* yeast strain [45]. However, if indeed p21 does function as a switch in the signal transduction pathway, understanding of the GTP binding domain will help to delineate this function, and it may provide a starting point in testing and designing chemical agents to interfere with this swtiching function.

## REFERENCES

1. Shih TY, Weeks MO: Cancer Invest 2:109, 1984.
2. Gibbs JB, Sigal IS, Poe M, Scolnick EM: Proc Natl Acad Sci USA 81:5704, 1984.
3. McGrath JP, Capon DL, Goeddel DV, Levinson AD: Nature (Lond) 310:644, 1984.
4. Sweet RW, Yokoyama S, Kamata T, Feramisco JR, Rosenberg M, Gross M : Nature (Lond) 311:273, 1984.
5. Manne V, Bekesi E, Kung HF: Proc Natl Acad Sci USA 82:376, 1984.
6. Hattori S, Ulsh LS, Halliday K, Shih TY: Mol Cell Biol 5:1449, 1985.
7. Lacal JC, Santos E, Notario V, Barbacid M, Yamazaki S, Kung HF, Seamans C, McAndrew S, Crowl R: Proc Natl Acad Sci USA 81:5305, 1984.
8. Shih TY, Stokes PE, Smythers GW, Dhar R, Oroszlan S: J Biol Chem 257:11767, 1982.
9. Willingham MC, Pastan I, Shih TY, Scolnick EM: Cell 19:1005, 1980.
10. Shih TY, Weeks MO, Gruss P, Dhar R, Oroszlan S, Scolnick EM: J Virol 42:253, 1982.
11. Willumsen BM, Norris K, Papageorge AG, Hubbert NL, Lowy DR: EMBO J 3:2581, 1984.
12. Chen ZQ, Ulsh LS, DuBois G, Shih TY: J Virol 56:607, 1985.
13. Lautenberger JA, Ulsh LS, Shih TY, Papas TS: Science 221:858, 1983.
14. Clanton DJ, Hattori S, Shih TY: Proc Natl Acad Sci USA: 83:5076, 1986.
15. Shih TY, Papageorge AG, Stokes PE, Weeks MO, Scolnick EM: Nature (Lond) 287:686, 1980.
16. Furth ME, Davis LJ, Fleurdelys B, Scolnick EM: J Virol 43:294, 1982.
17. Chang EH, Maryak JM, Wei CM, Shih TY, Shober R, Cheung H, Ellis RW, Hager GL, Scolnick EM, Lowy DR: J Virol 35:76, 1980.
18. Lowy DR, Rands E, Scolnick EM: J Virol 26:291, 1978.
19. Southern PJ, Berg P: J Mol Appl Genet 1:327, 1982.
20. Shih TY, Weeks MO, Young HA, Scolnick EM: Virol 96:64, 1979.
21. Ulsh LS, Shih TY: Mol Cell Biol 4:1467, 1984.
22. Sefton B, Trowbridge IS, Cooper JA, Scolnick EM: Cell 31:465, 1982.
23. Schultz AM, Henderson LE, Oroszlan S, Garber EA, Hanafusa H: Science 227:427, 1985.
24. Marchildon GA, Casnellie JE, Walsh KA, Krebs EG: Proc Natl Acad Sci USA 81: 7679, 1984.
25. Buss JE, Sefton BM: Mol Cell Biol 6:116, 1986.
26. Mulcahy LS, Smith MR, Stacey DW: Nature (Lond) 313:241, 1985.
27. Lacal JC, Aaronson SA: Mol Cell Biol 6:1002, 1986.
28. Poe M, Scolnick EM, Stein RB: J Biol Chem 260: 3906, 1985.
29. Halliday KR: J Cyclic Nucleotide Protein Phosphor Res 9:435, 1984.
30. Leberman R, Egner U: EMBO 3:339, 1984.
31. Tanabe T, Nukada T, Nishikawa Y, Sugimoto K, Suzuki H, Takahashi H, Noda M, Haga T, Ichiyama A, Kangawa K, Minamino N, Matsu H, Numa S: Nature (Lond) 315:242, 1985.
32. Medynski DC, Sullivan K, Smith D, Van Dop C, Chang FH, Kung BKK, Seeburg PH, Bourne HR: Proc Natl Acad Sci USA 82: 4311, 1985.
33. Hurley JB, Simon MI, Teplow DR, Robinshaw JD, Gilman AG: Science 226:860, 1984.
34. Yatsunami K, Khorana HG: Proc Natr Acad Sci USA 82:436, 1985.
35. Madaule P, Axel R: Cell 41: 31, 1985.
36. la Cour TFM, Nyborg J, Thirup S, Clark BFC: EMBO 4:2385, 1985.
37. Jurnak F: Science 230:32, 1985.
38. McCormick F, Clark BFC, la Cour TFM, Kjeldgaard M, Norskov-Lauritsen L, Nyborg J: Science 230:78, 1985.
39. Sigal IS, Gibbs JB, D'Alonzo JS, Temeles GL, Wolanski BS, Socher SH, Scolnick EM: Proc Natl Acad Sci USA 83:952, 1986.

40. Moeller W, Amons R: FEBS Lett 186:1, 1985.
41. Clark R, Wong G, Arnheim N, Nitecki D, McCormick F: Proc Natl Acad Sci USA 82:5280, 1985.
42. Feramisco JR, Clark R, Wong G, Arnheim N, Milley R, McCormick F: Nature (Lond) 314:639, 1985.
43. Gilman AG: Cell 36:577, 1984.
44. Beckner SK, Hattori S, Shih TY: Nature (Lond) 317:71, 1985.
45. Toda T, Uno I, Ishikawa T, Powers S, Kataoka T, Broek D, Cameron S, Broach J, Matsumoto K, Wigler M: Cell 40:27, 1985.
46. Shih TY, Hattori S, Clanton DJ, Ulsh LS, Chen ZQ, Lautenberger JA, Papas TS: In Papas TS, Vande Woude GF (ed): "Oncogenes." New York: Elsevier, 1986.
47. Dayhoff MO: Atlas Protein Sequence Struct 5:345, 1978.
48. Arai K, Clark BFC, Duffy L, Jones MD, Kaziro Y, Laursen RA, L'Italien J, Miller DL, Nagarkatti S, Nakamura S, Nielsen KM, Petersen TE, Takahashi K, Wade M: Proc Natl Acad Sci USA 77:1326, 1980.

# Trans-Activation by the Adenovirus E1A Gene Product

## Joseph R. Nevins, Imre Kovesdi, and Ronald Reichel

*Howard Hughes Medical Institute, The Rockefeller University, New York, New York 10021*

The adenovirus E1A gene product *trans*-activates a number of viral and cellular promoters. The mechanism for this transcriptional induction has been investigated with an in vivo exoIII mapping technique to assay for proteins that interact with an E1A-inducible promoter. A protein bound to the early E2 promoter was detected in wild-type infected cells. In the absence of E1A induction, specific interactions at the promoter could not be detected, as indicated by the absence of an exoIII-protected fragment. These results suggest a model in which the efficient utilization of the E2 promoter is mediated by a cellular transcription factor. In the absence of E1A, the interaction can take place, but slowly and inefficiently in comparison with the interaction in the presence of E1A. We have employed a gel-retardation assay for the detection of a transcription factor in nuclear extracts from adenovirus-infected cells that interacts with the E2 promoter. Footprint analysis revealed a protection from DNase cleavage between $-33$ and $-74$. The factor could be detected in extracts of uninfected cells, although at greatly reduced levels, as assayed by a sensitive exoIII mapping technique. The increase in active factor required the E1A gene. These results suggest that the E2 binding activity is a cellular transcriptional factor whose concentration or binding activity increases as a result of the action of the E1A gene product.

**Key words:** transcription control, E2 promoter, E2 factor

The elucidation of the mechanism of transcriptional regulatory events is crucial to an understanding of the basis for phenotypic differences among cell types, including the oncogenic phenotype. One system of transcription control that is amenable to detailed study is the set of early adenovirus genes regulated by the E1A gene product. A functional E1A gene is required for the normal activation of early viral transcription [1–3]. In addition, at least two cellular genes are induced by E1A in a lytic infection [4–6]. Considerable effort by numerous laboratories has established many of the details of this process. Various lines of evidence indicate that the activation is indirect and does not involve recognition of promoter sequences by the E1A protein [7]. Rather, it appears that cellular factors may recognize these promoters. Essential sequences in each of the early viral promoters have been identified [8–15]. A key

Received May 7, 1986.

© 1988 Alan R. Liss, Inc.

remaining aspect in the elucidation of this system of gene control is the identification and isolation of the factor(s) that recognize the inducible promoters. Once this is accomplished, experiments to probe the basis for E1A action are possible.

Of added interest is the fact that the E1A gene is an oncogene that, together with the E1B gene [16] or an activated *ras* gene [17], can transform cells to an oncogenic state. The E1A gene alone can immortalize primary cells in culture to generate permanent cell lines [18]. Thus, the study of E1A function goes beyond the activation of genes in a lytic infection. The critical question relates to whether these defined activities of the E1A gene, transcription activation and transformation, are indeed linked. Whether they are or not, the study of transcription control by E1A should illuminate the details of eukaryotic transcription regulation.

## RESULTS AND DISCUSSION

Previous experiments have established several characteristics of the process of E1A stimulation of early viral transcription. First, E1A can be efficiently replaced by various other viral *trans*-activating genes [19] and inefficiently replaced by cellular activities [20]. These results suggest that the E1A protein may not act directly but rather may employ cellular factors in the process. Second, there appears to be in most cases no unique regulatory sequence required for E1A stimulation [9]. That is, the same sequences necessary for basal, uninduced transcription are also required for induced transcription. This result suggests that the same factor may be involved in both events. How could this be? If indeed a cellular transcription factor were utilized by an E1A-inducible promoter and if this factor were limiting in the cell at the start of a vrius infection such that there were not sufficient quantities of the factor to interact with each of the viral promoters, then the viral promoters would be unoccupied by factor and transcription would therefore be inefficient. A function of E1A might be to increase the supply of such a factor so that it was no longer limiting in the cell. Under these conditions the viral promoters would be occupied by the factor and transcription would be efficient. Such a change would be scored as a stimulation of transcription mediated through the action of E1A. A situation as just described would predict that an E1A-inducible viral promoter would be unoccupied or void of transcription factors in the absence of E1A but would be occupied in the presence of E1A. We have tested for such a possibility by analyzing the interaction of proteins with the adenovirus E2 promoter in virus-infected cells. For these experiments we employed the exoIII mapping technique described by Wu [21] for defining in vivo interactions at the *Drosophila* heat-shock promoters. A protein bound to the E2 promoter was detected in nuclei of wild-type infected cells, whereas in cells infected with an E1A deletion mutant (dl 312) no specific interaction was detected [22]. The position of the protection from exoIII digestion was consistent with a protein bound to sequences which have previously been described as critical for transcription of the E2 gene and induction by E1A. Specifically, an exoIII stop was generated that mapped to a position of $-85$ relative to the E2 transcription initiation site. These results would therefore suggest that in the absence of E1A, the E2 promoter is essentially unoccupied by specific factors, whereas in the presence of E1A there is a protein or proteins bound to the promoter.

These results suggest the possibility that E1A activation of viral transcription, in this case of the early E2 promoter, is mediated through the enhanced binding of a specific factor. To understand the basis for this activation, it is essential to identify and eventually isolate the factor. Once this is accomplished, then the origin of the factor can be determined, that is, is the factor of cellular or viral origin, and one can then ask what is the basis for the change in the factor as a function of E1A. To this end we have used various assays to identify a factor in extracts of adenovirus infected cells that interacts with the adenovirus E2 promoter. Using a gel assay for binding of proteins to specific DNA sequences, we have identified a protein in nuclear extracts of infected cells that interacts with the E2 promoter [23]. That the interaction is specific is indicated by the fact that it is eliminated by an excess of DNA containing the promoter sequence but not by an excess of nonspecific DNA. Furthermore, the interaction appears to involve critical promoter sequences since promoters deleted of these sequences do not compete.

The precise site for binding of the factor to the promoter was defined by an exoIII mapping assay as well as by DNase I footprinting. Both procedures defined the binding of the factor to E2 promoter sequences between $-33$ and $-74$ relative to the E2 initiation site at $+1$. This then places a factor binding to the E2 promoter to sequences that have previously been defined as critical for transcription.

All of the initial assays were performed using extracts from virus-infected cells, thus leaving open the question as to the origin of the factor. Therefore, we compared binding activities from extracts of uninfected cells versus infected cells. The results indicated that the binding activity was significantly enhanced in extracts of infected cells. Furthermore, using in vitro exoIII mapping, which provides a sensitive and accurate analysis of binding, we found that the binding activity was present in extracts of uninfected cells but at a level much reduced compared to that in infected cells. Furthermore, the binding activity in uninfected cells protects exactly the same sequence as the activity from infected cells. We thus conclude that the same protein from both uninfected and infected cells is responsible for the protection, thereby indicating that the activity is a cellular protein and that the activity increases as a result of infection. Furthermore, by comparing extracts from wild-type infected cells or E1A mutant infected cells, it was shown that the increase depended upon a functional E1A gene.

All of these results are consistent with a model for E1A action that involves binding of a cellular transcriptional factor to critical sequences in the E2 promoter. In the uninfected cell, it appears that this transcriptional factor is limiting, such that the promoters utilizing the factor are largely unoccupied and therefore inefficiently transcribed. We would suggest that the increase in the factor, dependent upon the action of E1A, as seen in extracts of wild-type infected cells, is responsible for the increased transcription of the E2 gene. The factor is no longer limiting for the viral promoters; they become fully occupied by the factor and transcription proceeds efficiently. Clearly the next step in these studies is the purification of the factor. By so doing, we can achieve an understanding for the basis of the control mediated by E1A. Does the factor increase in number as a result of E1A action, or is there an E1A-induced modification?

Finally, is the activation of this factor by the action of the E1A gene product involved in transformation? Recent results might suggest that it is. We previously found that mouse F9 teratocarcinoma cells exhibited an E1A-like activity that could complement an E1A mutant [20]. This E1A-like activity disappeared when F9 cells were induced to differentiate. In light of this result, we have analyzed nuclear extracts of F9 cells for the presence of the E2 factor. Using a gel-shift assay for binding, we found a high level of E2 binding factor, which appears to be the same factor that is induced in HeLa cells as judged by footprint analysis. Strikingly, the factor disappears when F9 cells are induced to differentiate. Furthermore, infection of the differentiated F9 cells results in the reappearance of the factor, suggesting that it is indeed subject to E1A control. As F9 cells are tumor cells, and the differentiated cells have lost this property, we suggest that there is a direct correlation between the level of active factor and the transformed state of these cells. Since this factor is also subject to E1A control, we would propose that E1A transcriptional control, mediated through this cellular transcription factor, is indeed an aspect of transformation.

## REFERENCES

1. Berk AJ, Lee F, Harrison T, Williams J, Sharp PA: Cell 17:9935, 1979.
2. Jones N, Shenk T: Proc Natl Acad Sci USA 76:3665, 1979.
3. Nevins JR: Cell 26:213, 1981.
4. Nevins JR: Cell 29:913, 1982.
5. Kao H-T, Nevins JR: Mol Cell Biol 3:2058, 1983.
6. Stein R, Ziff EB: Mol Cell Biol 4:2792, 1984.
7. Nevins JR: CRC Crit Rev Biochem 19:307, 1986.
8. Bos JL, ten Wolde-Kraamwinkel HC: EMBO J 2:73, 1983.
9. Imperiale MJ, Nevins JR: Mol Cell Biol 4:875, 1984.
10. Gilardi P, Perricaudet M: Nucleic Acids Res 12:7877, 1984.
11. Kingston RE, Kaufman RJ, Sharp PA: Mol Cell Biol 4:1970, 1984.
12. Elkaim R, Goding C, Kedinger C: Nucleic Acids Res 11:7105, 1983.
13. Weeks DL, Jones NC: Mol Cell Biol 3:1222, 1983.
14. Leff T, Corden J, Elkaim R, Sassone-Corsi P: Nucleic Acids Res 13:1209, 1985.
15. Murthy SCS, Bhat GP, Thimmappaya B: Proc Natl Acad Sci USA 82:2230, 1985.
16. Tooze J: "DNA Tumor Viruses." Cold Spring Harbor, NY: Cold Spring Harbor Laboratory, 1981.
17. Ruley HE: Nature 304, 602, 1983.
18. Houweling A, van den Elsen PJ, van der Eb AJ: Virology 105:537, 1980.
19. Feldman LT, Imperiale MJ, Nevins JR: Proc Natl Acad Sci USA 79:4952, 1982.
20. Imperiale MJ, Kao H-T, Feldman LT, Nevins JR, Strickland S: Mol Cell Biol 4:867, 1984.
21. Wu C: Nature 309:229, 1984.
22. Kovesdi I, Reichel R, Nevins JR: Science 231:719, 1986.
23. Kovesdi I, Reichel R, Nevins JR: Cell 45:219, 1986.

# Transcriptionally Active Domains in the 5′ Flanking Sequence of Human c-*myc*

Bruce Whitelaw, Neil M. Wilkie, Katherine A. Jones, James T. Kadonaga, Robert Tjian, and Jas C. Lang

*Beatson Institute for Cancer Research, Bearsden, Glasgow, Scotland (B.W., N.M.W., J.C.L.); Department of Biochemistry, University of California, Berkeley, California 94720 (K.A.J., J.T.K., R.T.)*

Transcriptionally active domains have been identified and located within the 5′ flanking sequences of the human c-*myc* oncogene. Regions important for c-*myc* promoter function in vivo have been identified by linkage to the coding sequences of the bacterial chloramphenicol acetyltransferase (CAT) gene. Promoter deletion studies and in vivo competition assays for c-*myc*/CAT recombinant plasmids have allowed the identification of a negative regulatory element (NRE2) located 5′ to the c-*myc* gene which functions in an orientation-independent manner by interaction with a trans-acting factor(s) and which is capable of repression of CAT gene expression directed by heterologous promoter sequences. An upstream promoter element (UPE) located between NRE2 and the c-*myc* mRNA cap sites which is responsible for activation of high levels of CAT gene expression has also been identified. Preliminary data further suggests the tentative involvement of two other distal regulatory domains (NRE1, and a putative activator or enhancer-type region; PRE,E) in control of c-*myc* gene expression. In vitro footprint analysis has allowed the identification of sequence-specific DNA binding proteins which interact with the NRE2 and UPE domains. The region encompassing NRE2 contains binding sites for transcription factors Sp1 and CTF. The promoter region, UPE, appears to be a highly complex domain involving several Sp1 binding sites.

Key words: transcriptional control, promoters, negative regulation, transcription factors, oncogenes

Aberrant expression of the c-*myc* oncogene is intimately associated with the genesis of lymphoid neoplasia in several animal systems. The most extensively analysed (human Burkitt lymphomas [BL] and the murine equivalent plasmacytomas [MPC]) are characterised by a chromosomal translocation between one of the immunoglobulin loci and the c-*myc* oncogene (reviewed by Klein [1]). In nearly all cases— the Raji cell line is one exception [2]—the normal nontranslocated allele is transcriptionally silent. This has led Siebenlist et al [3] to propose that c-*myc* expression may be controlled via a trans-acting repressor which regulates transcription by interaction

Received July 15, 1986.

© 1988 Alan R. Liss, Inc.

within the 5' noncoding exon, flanking sequences or both. Thus translocations observed in BL- and MPC-derived cell lines with breakpoints located within the region of putative negative regulation would allow escape from such controlling mechanisms. Indeed, the majority of BL translocation breakpoints are clustered around exon 1 [4] and those observed in mouse plasmacytomas predominantly either remove the dual promoter region or are clustered around a region approximately 350 base pairs (bp) 5' to the location of the c-*myc* mRNA cap site (P1) [5]. This clustering suggests the interruption of a regulatory region within the 5' domain of c-*myc*. However, translocation breakpoints located far from the c-*myc* coding regions (both 5' and 3') have been described and may represent the disruption of other as yet uncharacterised regions involved in the control of c-*myc* gene expression [6–8]. Alternatively, or in addition, cis-acting transcriptional regulatory sequences associated with the immunoglobulin gene locus may exert a dominant effect in altering the control of c-*myc* expression, even at a distance. In this respect, activation of c-*myc* in retroviral-induced lymphoma in chickens [9,10], cats [11], and mice [12,13] by proviral insertion may represent a similar situation. Additional mutations or small deletions occurring within the regulatory regions close to the gene cannot as yet be ruled out. Other possible levels of regulation of c-*myc* expression include both turnover of c-*myc* transcripts [14] and blockage of transcript elongation [15]. Indeed, altered half-lives of c-*myc* mRNA have been associated with translocation breakpoints which disrupt putative RNA target sequences located in the first exon [16].

For a repressor to regulate c-*myc* transcription a *cis*-responsive domain (negative regulatory element, NRE) located within the control region of c-*myc* may be postulated. With the localisation of such *cis*-regulatory domains in mind, DNAseI-hypersensitive sites have been mapped to the c-*myc* gene in vivo [3]. DNAseI-hypersensitive sites have previously been associated with *cis*-acting regulatory elements in several other gene systems [17–19]. Five such sites have been identified 5' to the c-*myc* coding region [3] and as such may indicate the presence of several regulatory elements in this gene.

We have undertaken transient transfection experiments using the bacterial chloramphenicol acetyltransferase (CAT) gene as an assayable marker in an attempt to identify and locate cis-acting transcriptionally active sequences in the human c-*myc* oncogene. Such an assay system has previously been used to identify transcriptional regulatory elements for both viral and cellular genes [20,21]. We report the localisation of a negative regulatory element (NRE) 5' to the human c-*myc* gene which functions in an orientation-independent manner by interaction with a *trans*-acting factor(s). In vitro footprint analysis has allowed a tentative identification of some sequence-specific DNA binding proteins which interact with this domain. In addition, the promoter region appears to be a highly complex domain involving several Sp1 binding sites. An additional NRE and a positive regulatory element (PRE) have been tentatively identified. We propose that removal or mutation of the NRE and/or other regulatory elements may be involved in the deregulation of c-*myc* expression observed in some BL and MPC cells.

## MATERIALS AND METHODS
### Construction of Recombinant Plasmids

Recombinant vectors were constructed by modifying parental donor plasmid molecules as described in the legends to appropriate figures. Ligations were carried out for 5 hr at room temperature or overnight in the presence of T4 DNA ligase (Bethesda Research Laboratories, USA) with restriction-digested plasmids and the insert DNA fragments at a 1:5 molar ratio. Transformations of HB101 or JM83 cells were performed as described [22]. Molecular linkers were used as suggested by the suppliers (Collaborative Research, Inc., USA).

### DNA-Mediated Gene Transfer and Assay of CAT Activity

Transfection of mouse LATK$^-$ and human HeLa cells were carried out using the calcium phosphate technique [23] with the following modifications. The DNA-calcium phosphate coprecipitate was added to the culture medium at a ratio of 0.5 ml coprecipitate/5 ml medium/$1 \times 10^6$ exponentially growing cells in a 25-cm$^2$ flask. After 16 hr the medium (SLM, 10% FCS, Flow Laboratories, UK) was replaced, and the cells were incubated for a further 32 hr. Cells were then harvested for determination of CAT activity as previously described [24].

### DNAseI Footprinting Analysis

DNA footprint probes were prepared by restriction enzyme digestion of recombinant plasmids, treated where necessary with calf intestinal alkaline phsophatase, and radiolabelled with polynucleotide kinase and $\gamma^{32}$P-ATP (Amersham, UK) (5' end label) or the large fragment of *Escherichia coli* DNA polymerase I (Klenow—Bethesda Research Laboratory, USA) and $\alpha^{32}$P-dNTP (3' end label). Radiolabelled DNA was then recleaved with a second restriction enzyme to generate fragments labelled at one end only, and purified by electrophoresis on a 4% acrylamide gel. DNA fragments were then isolated by isotachophoresis [25], precipitated with ethanol, and stored in 1 mM Tris, 1 mM NaCl, 0.1 mM EDTA (pH 7.5) at 4°C.

DNA binding reactions and DNAse I (Boehinger Newheim, UK) digestions were carried out as previously described [26]. Samples were then subjected to electrophoresis on 6% polyacrylamide/urea sequencing gels and dried, and autoradiograms were exposed to x-ray film (XAR5, Kodak, USA).

## RESULTS

In order to test the transcriptional activity of the c-*myc* 5' flanking sequences, the entire 2.8-kilobase (kb) region from the PvuII site located in the first exon (Fig. 1) was removed from the human c-*myc* containing recombinant plasmid pMC41 (provided by Dr. T. Papas) and ligated to a promoterless bacterial chloramphenicol acetyltransferase (CAT) gene. Also constructed were a series of 5' terminal deletions which progressively removed 5' flanking c-*myc* sequences. These plasmids were transfected into LATK$^-$ or HeLa cells, and the levels of CAT activity were determined. The results are given in Figure 1. As expected, the intact 5' c-*myc* domain restores expression of CAT activity to the inactive, promoterless CAT gene. Very little change in activity is observed in deletion mutants lacking DNA sequences which include the previously reported in vivo DNAseI-hypersensitive sites I and II$_2$ [3]. However, further deletions which remove site II$_2$ result in a marked increase in gene expression (plasmids pB38, pB14, and pB36). Similar results are obtained using fibroblastic LATK$^-$ cells or human epithelial HeLa cells, although the general level of enzyme activity obtained with the latter is lower. Similar results are also obtained by transfection of human erythroid K562 and murine NIH 3T3 cells (data not shown).

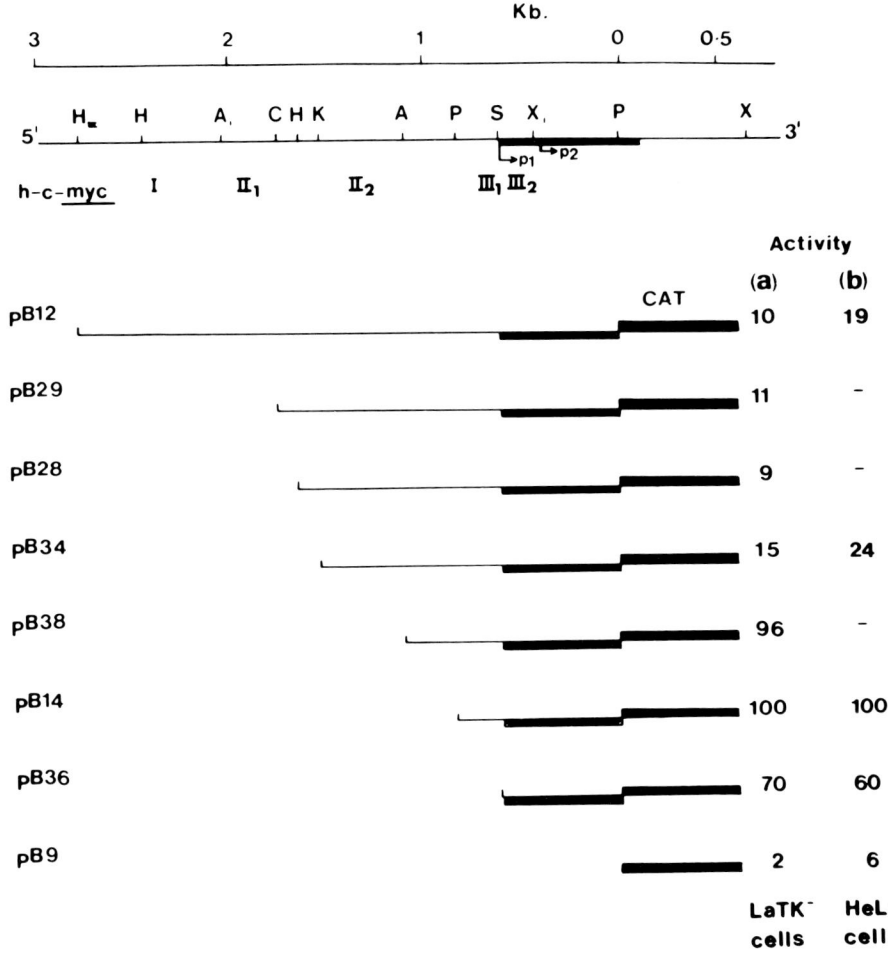

Fig. 1. Chloramphenicol acetyltransferase activity of human c-*myc*/CAT recombinant plasmids in mouse LATK$^-$ (**a**) and human HeLa (**b**) cells. Exon 1 of the c-*myc* gene and 2.3 kilobases (kb) of 5′ flanking DNA sequences are shown together with relevant restriction sites. The PvuII site within exon 1 was converted to a HindIII site and the resulting 2.8-kb HindIII fragment (position −2,318 to +513) containing the 5′ flanking sequences and majority of exon 1 was inserted into a promoterless CAT vector (pB9, constructed by removal of the HSV-2 IE5 promoter sequences from plasmid pLW2) [30] to generate plasmid pB12. Progressive 5′-3′ deletions of parental plasmid pB12 were then constructed by removal of successive restriction fragments as shown. Transfection conditions were as described in Materials and Methods. CAT activity is shown as a percentage of the value for pB14. Activities represent the average of two flasks from five separate LATK$^-$ and three separate HeLa experiments. Previously reported DNAseI-hypersensitive sites I–III$_2$ and position of mRNA cap sites P$_1$ and P$_2$ [3] are shown. A, AccI; A$_1$, AvaI; C, ClaI; H, HincII, H$_{III}$, HindIII; K, KpnI; P, PvuII; S, SmaI; X, XbaI; X$_1$, XhoI.

We tentatively interpret this result to indicate the presence of a cis-acting NRE in this domain which includes DNAse-hypersensitive site II$_2$.

In many independently repeated experiments a similar pattern of results was obtained, but the magnitude of the effect obtained by deleting the domain which includes site II$_2$ varied considerably. It was our impression that this might be related to the amount of DNA used to transfect cell cultures. To test this directly, the ratio of

CAT activities obtained after transfection with DNA containing the upstream NRE region in either orientation (pB12 and pB19) to that obtained with the upstream region deleted (pB14) was measured at different DNA concentrations. Figure 2 shows that the activation of CAT gene expression upon deletion of the upstream domain is obtained only at low concentrations of DNA. At higher concentrations (10 $\mu$g per dish) the undeleted constructions are almost as active as pB14. The results could be explained if the *cis*-acting NRE binds cellular factors, probably trans-acting DNA binding proteins, present in limiting amounts. The use of high DNA levels in transfection would then titrate out such factors, leading to "activation" or "derepression" of remaining unbound DNA. In addition, this experiment shows that the NRE appears to act in an orientation-independent manner (see Fig. 5).

To test this hypothesis more directly, a plasmid competition assay was carried out using pB18 as competitor, which is plasmid pUC12 carrying only the upstream element containing the putative NRE. Figure 3 shows CAT levels obtained when LATK$^-$ cells are transfected with 1 $\mu$g of the promoterless CAT gene (pB9), the CAT gene controlled by the entire c-*myc* domain alone (pB12), and the CAT gene controlled by the c-*myc* domain from which the putative NRE is deleted (pB14). Figure 3 also shows the CAT levels when cells are transfected with 1 $\mu$g of pB12 plus increasing concentrations of pB18. Clearly, pB18 (the competitor) can titrate some factor(s) in the cells which results in activation of pB12.

Figure 4 presents results which serve to narrow down the NRE domain within the c-*myc* flanking sequences. A 445-base pair (bp) fragment encompassing in vivo DNAse-hypersensitive site II$_2$ is shown. An internal deletion of this sequence from pB29, in which the CAT gene is controlled by c-*myc* sequences containing sites II$_2$, III$_1$, and III$_2$, results in activation of CAT gene expression (compare pB29 and pB29 $\Delta$NRE). Figure 5 shows that transfer of the c-*myc* domain containing sites I, II$_1$, and II$_2$ to other promoters (the herpes simplex virus [HSV]-2 1E5 and human $\epsilon$-globin promoters) also results in down-regulation of gene expression. Thus the NRE activity is not specific for the *myc* gene but also affects other well-characterised promoters. This experiment also confirms that this relatively large domain suppresses the activity of promoters in an orientation independent manner.

Figure 1 shows that terminal deletion of sequences containing DNAse-hypersensitive sites I and II$_2$ has little effect on CAT expression. However, transfer of DNA fragments from this domain to other promoters, or to a promoterless CAT gene, gives different results. Figure 6 shows that a domain encompassing sites I and II$_2$ (FUE in Fig. 6) behaves as a *cis*-acting NRE to the HSV-2 IE5 and $\epsilon$-globin promoters, in an orientation-independent manner. Furthermore, a short fragment encompassing only site II$_1$, (II$_1$ in Fig. 6) suppresses the HSV-2 IE5 promoter. We tentatively interpret these results to indicate a further *cis*-acting NRE, distinct from that encompassing site II$_2$. Figure 6 also shows that ligation of a short 5' fragment (E) to the promoterless CAT gene results in activation of gene expression. This element has not yet been characterised in as much detail as the others, but results from preliminary experiments (data not shown) lead us to suggest that this element is a *cis*-acting positive regulatory element (PRE), possibly of an enhancer type. We also note that this putative PRE does not encompass DNAse-hypersensitive site I.

An obvious question is whether and which specific DNA binding proteins interact with the regulatory domains of c-*myc* identified so far. Our analysis is as yet far from complete, but NRE2 and the "promoter" region (UPE, Fig. 9) have been

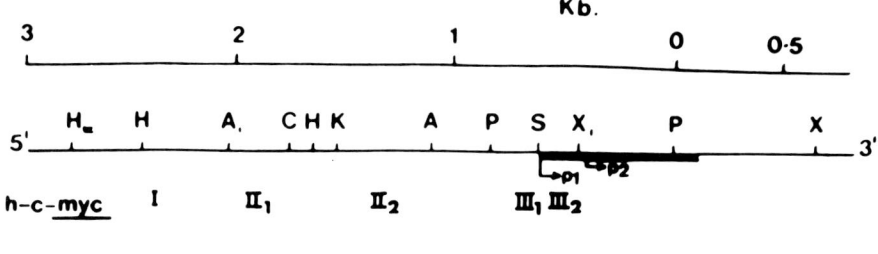

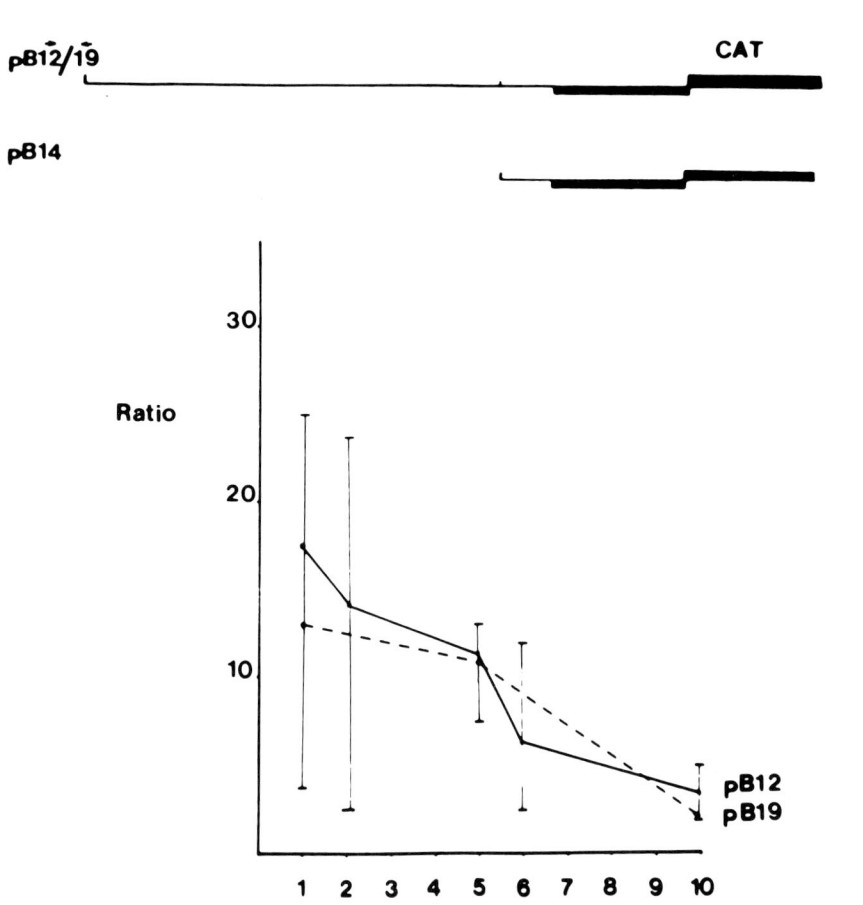

Fig. 2. "Self-titration" of c-*myc*/CAT recombinant plasmids. pB12 contains the intact c-*myc* regulatory region from −2,318 to +513 linked to the coding sequences of the CAT gene. pB19 contains the sequences from −2,318 (HindIII) to −346 (PvuII) inverted in 3'-5' orientation. Graph shows the relative activity of plasmids pB12 (solid line) and pB19 (dotted line) compared with pB14 after transfection into separate cell cultures and expressed as a ratio for molar plasmid concentrations of 1, 2, 5, 6, and 10 relative to pB14. Each point represents the average of two values from ten separate experiments using different plasmid preparations. Vertical bars indicate range of values for each sample point.

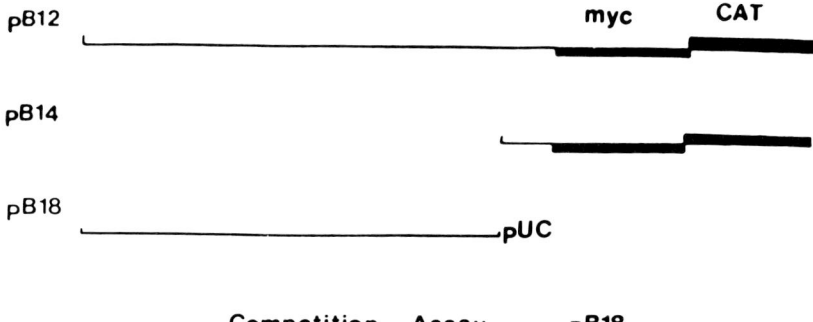
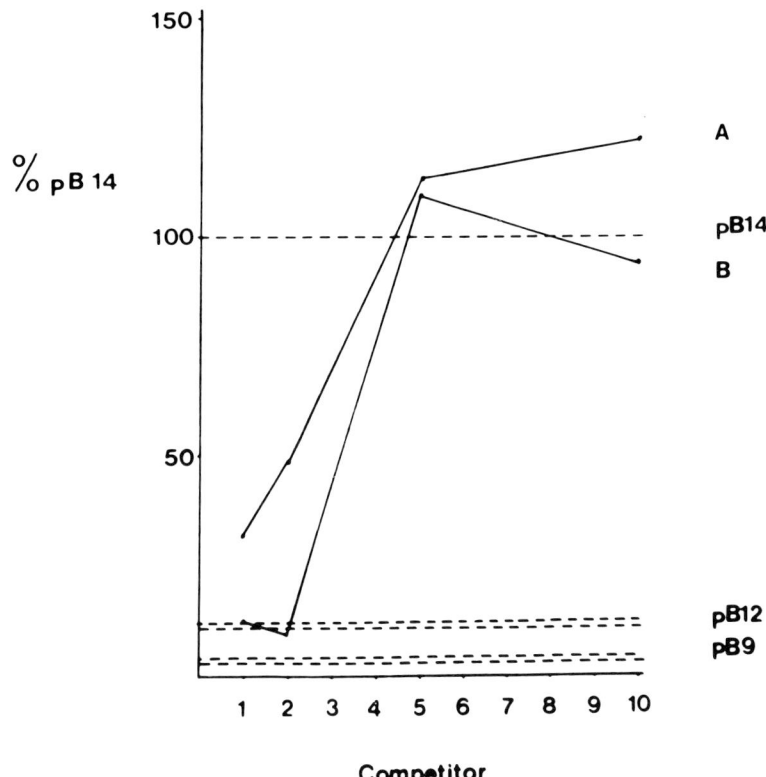

Fig. 3. Competition assay of c-*myc*/CAT recombinant plasmids. pB18 containing the c-*myc* sequences from −2,318 to −346 alone, cloned into the plasmid vector pUC12, was used in a competition assay against pB12 transfected into the same cells at molar plasmid concentrations of 1, 2, 5, and 10 competitor-to-parental plasmic pB12. Results represent the CAT activity of 1 μg pB9 and molar equivalents of pB12 and pB14 and are expressed as a percentage of pB14 for two flasks from two separate experiments (A and B).

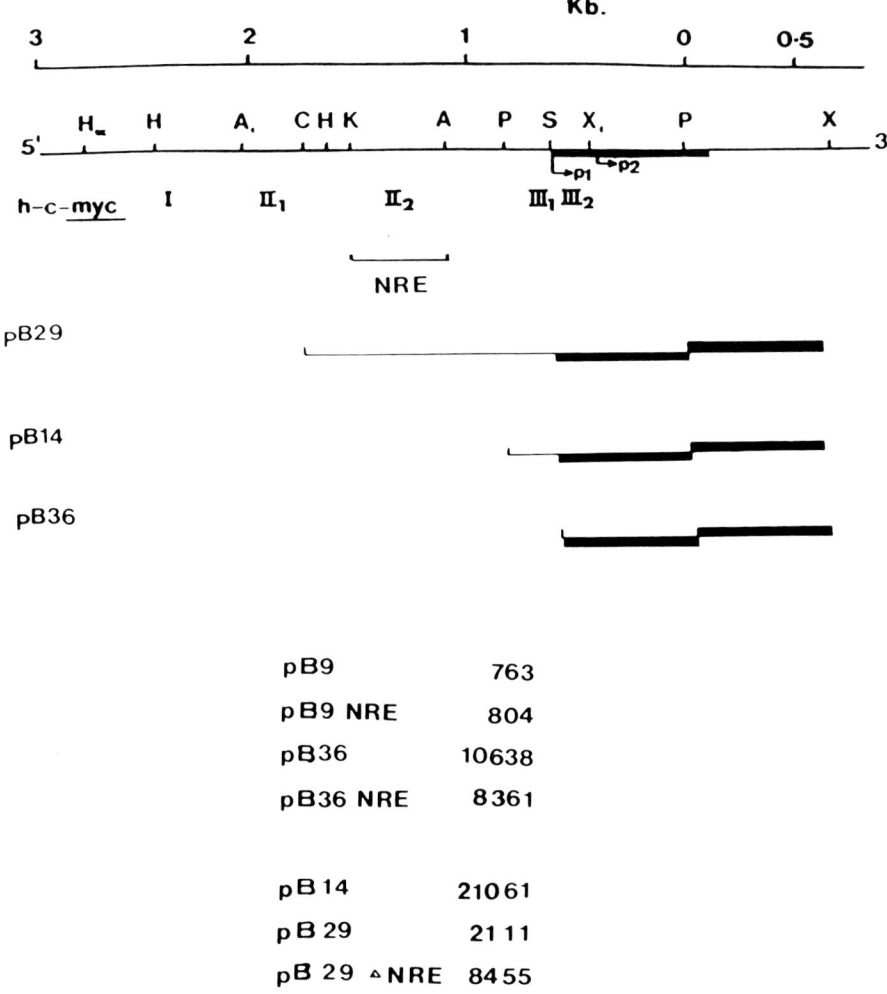

Fig. 4. Internal deletion of NRE2. Results show the CAT activity for recombinant plasmid pB29 ΔNRE constructed by deleting the internal KpnI/AccI fragment harbouring the NRE2 region. Also shown are recombinant plasmids pB9 NRE and pB36 NRE, which contain the NRE2 element placed in 5'-3' orientation 5' to the c-myc regulatory sequences of pB36 and the CAT coding sequences of pB9, respectively. CAT activity is shown as combined $^{14}$C CPM for the mono- and diacetylated forms of chloramphenicol and represents the average for two flasks from one experiment.

partially characterised DNA footprinting was carried out using a partially purified DNA binding protein preparation (H0.3) and two highly purified preparations prepared by specific DNA affinity chromatography from HeLa cells, highly enriched for transcription factors Sp1 and CAAT binding transcription factor (CTF), respectively [27]. Figure 7a shows the DNA footprints obtained with the 445-bp NRE2. Four distinct footprints are obtained, one of which can be accounted for by Sp1 binding, and two by CTF binding. In addition, several in vitro DNAse-hypersensitive sites are observed. The limits of the footprints and the location of in vivo and in vitro DNAse-hypersensitive sites are shown relative to the DNA sequence in Figure 7b. We interpret the results as indicating (from 5'-3') one Sp1 site (5'-GCCCCTCCCA-3'),

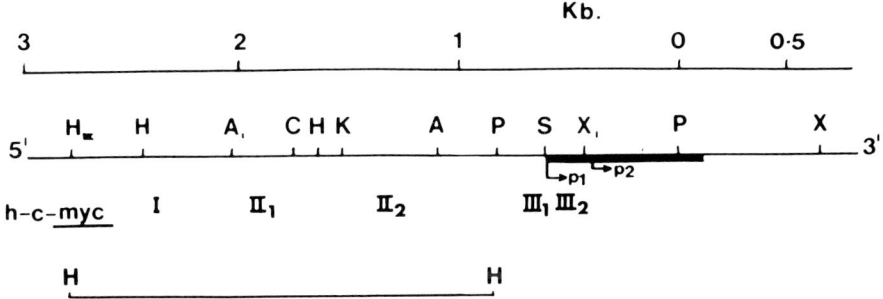

| Plasmid | Activity | Counts | Promoter |
|---|---|---|---|
| pB14 | 1·0 | 95041 | h-c-myc |
| pB12 | 0·10 | 9025 | |
| pB19 | 0·09 | 8645 | |
| pLW2 | 1·0 | 123034 | HSV-2 IE 5 |
| pB16 | 0·38 | 47215 | |
| pB17 | 0·27 | 33149 | |
| pB30 | 1·0 | 28087 | h-ε-globin |
| pB37 | 0·16 | 4498 | |

Fig. 5. Effect of c-*myc* regulatory sequences on CAT activity directed by heterologous promoters. Recombinant plasmids pB14, pLW2, and pB30 contain the c-*myc* promoter ($-346$ to $+513$), the herpes simplex virus type 2 immediate early mRNA 5 promoter (HSV-2 IE5) [30], and the human ε-globin promoter [31] sequences, respectively, linked to the CAT gene. The c-*myc* regulatory region from $-2,318$ to $-345$ was placed 5' to the promoter sequences of these recombinants in 5'-3' orientation, generating plasmids pB12, pB16, and pB37, respectively, and in 3'-5' orientation generating plasmids pB19 and pB17. CAT activity is shown as combined $^{14}$C CPM for the mono- and diacetylated forms of chloramphenicol or as a ratio relative to plasmid pB14. Equimolar ratios for 2 µg of pB14 or pLW2 were used for derived plasmids, while equivalent ratios for 20 µg were used for plasmids derived from pB30.

one Sp1-like site (5'-CAGGAGGGGC-30'), and two CTF sites in opposite orientations, respectively, (5'-TTTGG-3') and (5'-CCAAT-3'). Footprints of the upstream promoter element (UPE) domain using H0.3 are shown in Figure 8a. These comprise a long, partially overlapping region of protection. Using purified Sp1 preparation, and different electrophoresis conditions (data not shown), this region was better resolved, and some sites were identified as Sp1 binding sites. Figure 8b shows the limits of the footprints and the DNA sequences, as well as the cap sites for the two known c-*myc* mRNAs (P1 an P2) and the DNAse-hypersensitive sites. We interpret the results to indicate an upstream region of partially overlapping Sp1-like sites, and an Sp1 site adjacent to P1. On the basis of H0.3 titration experiments, the upstream

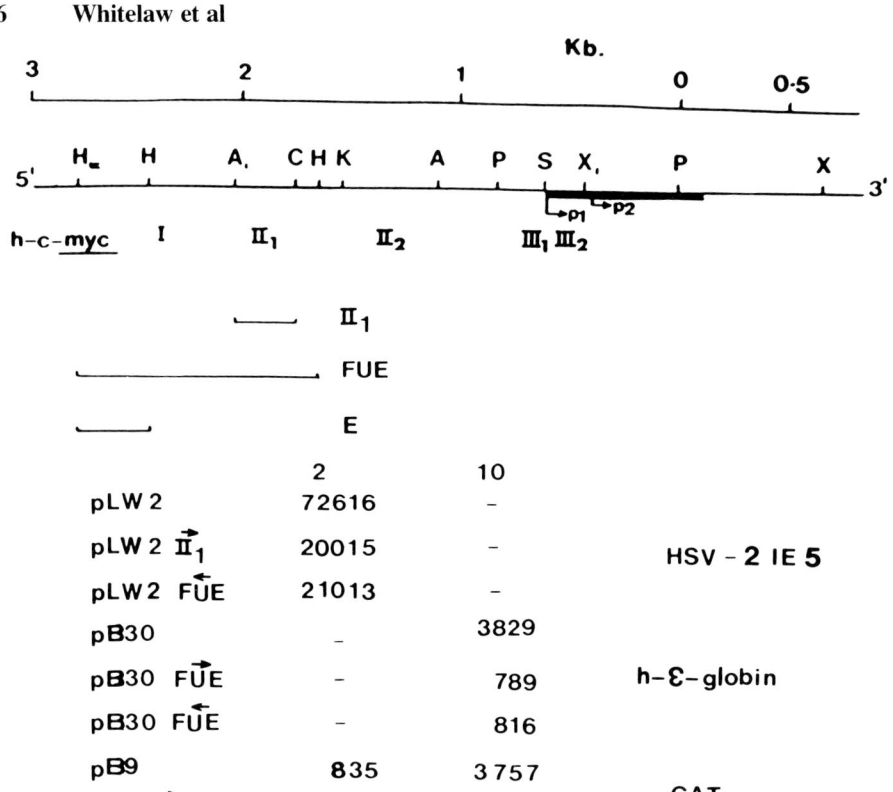

Fig. 6. Transfer of c-*myc* regulatory regions to recombinant CAT genes under the control of heterologous promoters. The AvaI/ClaI fragment from c-*myc* spanning DNAse-hypersensitive site II$_1$ [3], the HindIII/HincII fragment designated "far upstream element" (FUE), and a HindIII/HincII fragment containing a putative enhancerlike element (E) possessing four copies of the enhancer core consensus described by Weiher et al [32], were placed 5' to the promoter sequences in plasmids pLW2 and pB30 and 5' to the CAT coding sequences in pB9, with orientation as indicated. CAT activity is shown as combined $^{14}$C CPM for the mono- and diacetylated forms of chloramphenicol for molar equivalents of both 2 and 10 μg of pB9. Arrows indicate orientation of transferred c-*myc* sequences for each recombinant.

sites appear to be weaker targets for Sp1 than the proximal Sp1 site, or targets for a different DNA binding protein(s) (data not shown). No footprints are observed adjacent to P2, although DNase-hypersensitive sites are seen after protein binding. We are currently extending the footprinting to the PRR and NRE1 domains.

## DISCUSSION

On the basis of the data shown in this paper, we tentatively propose the incomplete model shown in Figure 9. We suggest that at least four distinct domains can be identified in the 5' flanking sequences of c-*myc* which regulate transcriptional activity. These include a 5' PRE and two *cis*-acting NREs which are located between −2,100 to −750 relative to the first known cap site of the gene. NRE2 is the better characterised and appears to function in an orientation-independent manner by interaction with a trans-acting factor(s) present both within an epithelial cell line (HeLa) and a murine fibroblastic cell line (LATK$^-$). Closer to the cap sites is a complex

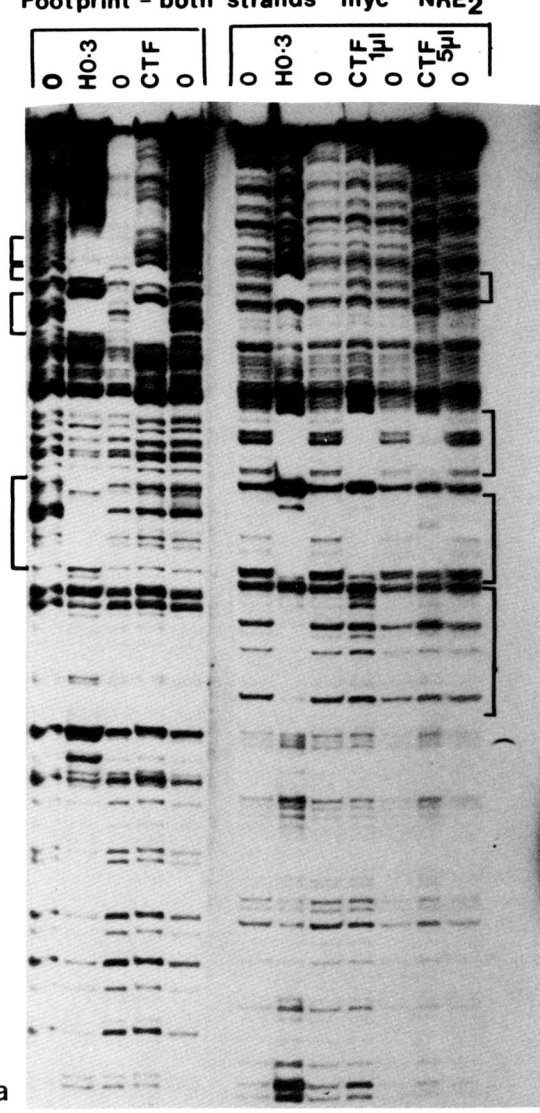

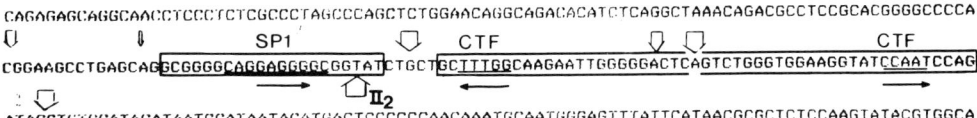

Fig. 7. **a:** DNAse footprint analysis of transcription factor binding specificities within the c-*myc* NRE2 region. DNA was 5'-end-labelled on either strand and incubated with H0.3 fraction or purified CTF, subjected to partial digestion with DNAseI, and analysed on a 6% polyacrylamide sequencing gel. **Lanes 2,7:** 10-$\mu$l H0.3 fraction. **Lanes 4,9,11:** Purified CTAF. **Lanes 1,3,5,6,8,10,12:** Unprotected. Regions of protection from DNAse digestion are indicated in brackets. **b:** DNA sequence of the c-*myc* upstream region encompassing NRE2. Regions of DNAseI protection are indicated by an open box, with consensus sequences for Sp1 and CTF underscored. Precise limits of footprinting were determined by comparison with the products of a "G" track reaction prepared by the method of Maxam and Gilbert [33] and run concurrently (data not shown). Solid arrows show 5'-3' orientation of binding sites, open arrows the location and magnitude of in vitro DNAseI-hypersensitive sites and II$_2$ the position of an in vivo hypersensitive site previously reported [3].

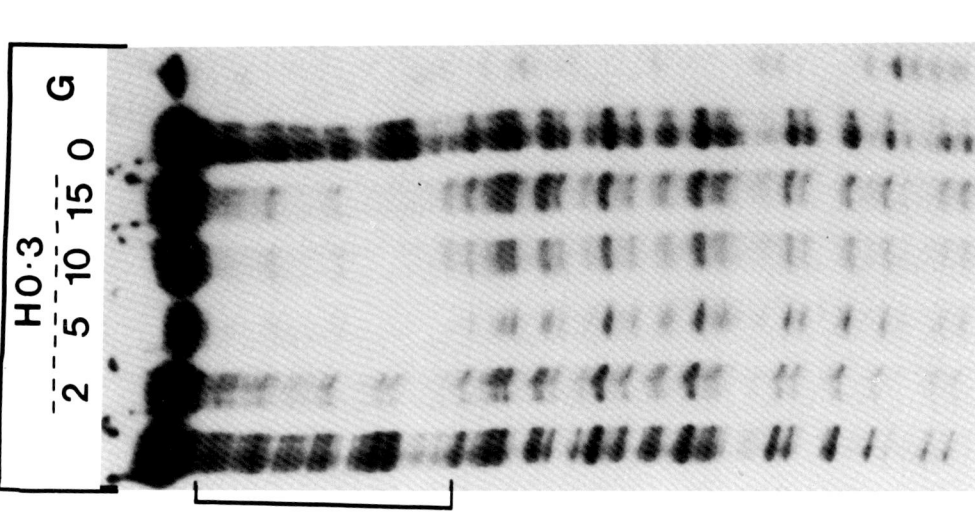

Fig. 8. **a:** DNAseI footprint analysis of DNA-binding protein interactions within the c-myc UPE region. DNA ws 3' labelled and incubated with varying concentrations of H0.3 fraction, partially digested with DNAseI, and analysed on a 6% polyacrylamide sequencing gel. **Lanes 2–5:** 2-, 5-, 10-, and 15-μl H0.3 fraction, respectively. **Lanes 1,6:** Unprotected. **Lane 7:** G track sequencing reaction. Regions of DNAseI protection are indicated in brackets. **b:** DNA sequence of the c-myc promoter region encompassing UPE and mRNA cap sites. Regions of DNAseI protection are indicated by an open box with consensus sequences for Sp1 underscored. Solid arrows show 5'-3' orientation of binding sites; and open arrows the location and magnitude of in vitro DNAseI-hypersensitive sites. III$_1$ and III$_2$ indicate previously reported in vivo hypersensitive sites [3]. The TATAA boxes (double underscore) and position of P1 and P2 cap sites as previously reported [3] are indicated.

**Model for cis-acting transcriptional control domains.**

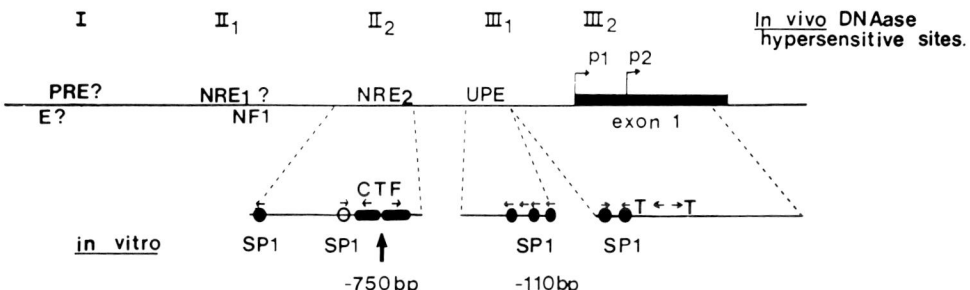

Fig. 9. Model for cis-acting transcriptional control domains within the regulatory region of the human c-*myc* oncogene. The location of functional domains within the c-*myc* regulatory region as defined by CAT expression assays are indicated. The position of transcription factor (SP1 and CTF) DNA binding sites within these functional domains as obtained by in vitro DNAseI footprinting studies are also shown. I–III$_2$ indicate previously described DNAseI hypersensitive sites [3]. PRE, positive regulatory element; NRE1 and 2, negative regulatory elements 1 and 2; UPE, upstream promoter element; E, putative enhancer element; NF1, nuclear factor 1 binding site [3]; SP1 and CTF, previously described transcription factor DNA binding proteins [28]. Arrows represent the 5'-3' orientation of SP1 and CTF binding sites.

promoterlike region we temporarily term UPE. NRE1 and 2 encompass the previously determined hypersensitive sites II$_1$ and II$_2$, while UPE encompasses hypersensitive sites III$_1$ and III$_2$. The DNAse-hypersensitive sites obtained by in vitro binding of proteins are in close proximity to the in vivo sites, but are more complex and do not match in position. The in vitro sites tend to be located at the edges of protected regions, but internal sites are also seen. We presume this reflects differences between the artificial situation in vitro and the more relevant in vivo situation where undoubtedly more complex interactions occur. At present we do not know whether or not sites II$_1$ and II$_2$ reflect chromatin interactions involved with regulation of transcription.

Our analysis of the interaction of DNA binding proteins with UPE in vitro suggest a complex region. There is an upstream cluster of Sp1-like sites which behave as somewhat weak target sites for DNA binding protein(s). We do not know that these are sites for Sp1 itself or a related or different protein(s). This region can be deleted with only a relatively small effect on gene expression (a 2–3-fold effect; eg, compare pB14 and pB36 in Fig. 1, and several experiments not shown). Closer to P1 is a stronger target site which binds purified Sp1, whose associated footprint overlaps the adjacent TATA box (see Fig. 7b). Surprisingly, no footprints were identified adjacent to P2, although the presence of in vitro DNAse-hypersensitive sites may suggest weak binding of some protein.

In NRE2, we have identified one strong 5' Sp1 site, one weak Sp1-like site, and two strong CTF sites (see Fig. 9). All the protected regions obtained with partially purified DNA binding proteins (H0.3) could be accounted for by those four footprints. We suggest that either or both CTF sites and possibly the weak Sp1-like site modulate the negative regulatory effect. If true, this is of considerable interest, since previously Sp1 and CTF have only been associated with up-regulation of promoter- and/or enhancerlike domains [28]. Of course we cannot exclude the binding of some other factor in this region, either to a specific target sequence or to the Sp1/CTF proteins bound to the DNA. At the present time we are undertaking site-directed mutagenesis studies to obtain evidence that CTF and/or Sp1-like proteins modulate the negative regulatory activity.

Associated with the 3' CTF footprint is the sequence 5'-GTGGAAGGTATC-CAAT-3'. This is highly homologous to the reported target sequence for the cellular protein nuclear factor 1 (NF1), which has been implicated in adenovirus DNA replication [29]. A similar sequence, 5'-CTGGAAGGCAGCCAAA-3', is located adjacent to DNAse-hypersensitive site $II_1$ in NRE1. Previous studies suggested that the NRE1 site is a target for NF1, although no evidence for NF1 binding to NRE2 was obtained [3]. Further studies are required to distinguish between NF1 and CTF binding and whether these two DNA-binding protein preparations contain common factors. Further studies are also required to determine whether any of the c-*myc* regulatory domains so far identified are implicated in the regulation of c-*myc* transcription associated with signal transduction during serum stimulation of cells and in differentiation and development.

## ACKNOWLEDGMENTS

We would like to thank Drs. J.L. Whitton and J.B. Clements for generously providing plasmid pLW2 and Dr. T. Papas for pMC41. We are indebted to the Cancer Research Campaign of Great Britain (B.W. and N.M.W.) and the Leukaemia Research Fund (J.L.) for support.

## REFERENCES

1. Klein G: Cell 32:311, 1983.
2. Nishikura K, Erikson J, AR-Rushdi A, Huebner K, Croce C M: Proc Natl Acad Sci USA 82:2900, 1985.
3. Siebenlist U, Hennighausen L, Battey J, Leder P: Cell 37:381, 1984.
4. Taub R, Moulding C, Battey J, Murphy W, Vasicek T, Lenoir GM, Leder P: Cell 36:339, 1984.
5. Yang J, Bauer SR, Mushinski JF, Marcu KB: EMBO J 4:1441, 1985.
6. Webb E, Adams JM, Cory S: Nature 312:777, 1984.
7. Cory S, Graham M, Webb E, Corcoran L, Adams JM: EMBO J 4:675, 1985.
8. Davis M, Malcolm S, Rabbitts TH:Nature 308:286, 1984.
9. Hayward WS, Neel BG, Astrin SM: Nature 290:475, 1981.
10. Payne GS, Bishop JM, Varmus HE: Nature 295:209, 1982.
11. Neil JC, Hughes D, McFarlane R, Wilkie NM, Onions DE, Lees G, Garrett O: Nature 308:814, 1984.
12. O'Donnell PV, Fleissner E, Lonial H, Koehne CF, Reicin A: J Virol 55:500, 1985.
13. Corcoran LM, Adams JM, Dunn AR, Cory S: Cell 37:113, 1984.
14. Blanchared JM, Piechaczyk M, Dani C, Chambard J-C, Franchi A, Pouyssegur J, Jeanteur P: Nature 317:443, 1985.
15. Bentley DL, Groudine M: Nature 321:702, 1986.
16. Piechaczyk M, Yang J-Q, Blanchard J-M, Jeanteur P, Marcu KB: Cell 42:589, 1985.
17. McGhee JD, Wood WI, Dolan M, Engel JD, Felsenfeld G: Cell 27:45, 1981.
18. Mills F, Fisher M, Kuroda R, Ford A, Gould H: Nature 306:809, 1983.
19. Dynan WS, Tjian R: Cell 35:79, 1983.
20. Gorman CM, Merlino GT, Willingham MC, Pastan I, Howard BH: Proc Natl Acad Sci USA 79:6777, 1982.
21. Ishii S, Merlino GT, Pastan I: Science 230:1378, 1985.
22. Norgard MV, Keen K, Monohan JJ: Gene 3:279, 1978.
23. Graham FL, van der Ebn AJ: Virology 52:456, 1973.
24. Gorman CM, Moffat LF, Howard BH: Mol Cell Biol 2:1044, 1982.

25. Ofverstedt L-G, Hannarstrom K, Balgobin N, Hjerten S, Pettersson U, Chattopadhyaya J: BBA 782:120, 1984.
26. Jones KA, Yamamoto KR, Tjian R: Cell 42:559, 1985.
27. Kadonaga JT, Tjian R: Proc Natl Acad Sci USA 83:5889, 1986.
28. Dynan WS, Tjian R: Nature 316:774, 1985.
29. Nagata K, Guggenheimer RA, Hurwitz J: Proc Natl Acad Sci USA 80:6777, 1983.
30. Gaffney DF, McLauchlan J, Whitton JL, Clements JB: Nucleic Acids Res 13:7847, 1985.
31. Lang JC, Wilkie NM, Spandidos DA: J Gen Virol 64:2679, 1983.
32. Weiher H, Konig M, Gruss P: Science 219:626, 1983.
33. Maxam AM, Gilbert GW: Methods Enzymol 65:497, 1980.

# Tumor Suppressor Genes

## Ruth Sager

*Division of Cancer Genetics, Dana-Farber Cancer Institute, Boston, Massachusetts 02115*

### THE YIN-YANG THEORY OF CANCER

This theory proposes that two classes of genetic changes determine the transformation of normal cells into neoplastic cells: genes that encode products which elevate the probability of transformation (ie, oncogenes), and genes whose products reduce that probability (ie, tumor suppressor genes). While numerous oncogenes have been identified and cloned over the past few years, tumor suppressor genes have only been recognized indirectly as yet.

Evidence for the existence of tumor suppressor genes comes principally from two lines of investigation which have been recently reviewed [1,2,3] and will be discussed by other speakers at this symposium. In brief, clinical studies of certain hereditary cancers, of which retinoblastoma is the paradigm, have shown that both copies of a single locus have to be lost or inactivated for tumors to arise, and thus by inference that this region encodes a suppressor of tumor formation. Expression of other hereditary cancers are also being shown to involve specific deletions or gene inactivations. In a second line of evidence based on experimental studies using cells in culture, stable tumor suppressor gene activity has been found in hybrids from fusions between normal and tumor cells [1,2]. In both clinical and cell culture systems, suppressor genes have been mapped to chromosomal locations.

The existence of suppressor genes that need to be inactivated for tumors to arise adds to the large body of evidence that neoplasia is a multistep genetic process and further confounds the oncogene evidence of one or two genes with decisive effects on tumorigenicity [4,5]. I call this conflict in concepts and in experimental evidence *the oncogene paradox*. The evidence that mutational or regulatory changes in one or two genes is sufficient to induce tumorigenesis is largely based on the use of chromosomally destabilized and partially transformed recipients. The importance of identifying all of the relevant genetic changes in the neoplastic transformation has become increasingly evident.

At the root of the problem in the past has been a widespread misunderstanding of the central importance of chromosomal changes in neoplasia. Cancer is fundamen-

Received May 20, 1986.

© 1988 Alan R. Liss, Inc.

tally a disease of the genome itself, in which a diverse set of genomic rearrangements arise. Cancer differs from all other diseases in the nature of the genomic changes: they are very *complex*, and they are *progressive*, not static. An example of genomic progression occurs in chronic myelogenous leukemia. In the chronic state one sees only a single translocation, associated with the oncogene *c-abl*, but the acute phase, blast crisis, is preceded by further changes elsewhere in the genome [6].

Genomic changes, including translocation, localized amplification, aneuploidy, deletions, and point mutations all contribute to the genetic heterogeneity of evolving tumors. The genetic heterogeneity in turn provides the phenotypic variability that is the fuel for rapid selection of increasingly successful tumor cell variants. Both activation of oncogenes and inactivation of suppressor genes play their roles on this changing genetic background.

## FUNCTIONS OF SUPPRESSOR GENES

It is appropriate now to consider what might be the functions of suppressor genes. Here one can only speculate on the basis of current knowledge. For purposes of discussion, several categories of suppressor function will be briefly considered.

### Growth Inhibitors

Inhibitors of tumor cell growth have been reported. Those discussed at this symposium include the interferons, tumor necrosis factor, and tumor growth factor beta (TGF-$\beta$) . TGF-$\beta$ is of particular interest since it has a growth promoting effect on various fibroblastic cell lines but an inhibitory effect on epithelial cell growth. In addition, inhibitory activities associated with quiescence or senescence are known [7,8], as well as a number of other growth inhibitory factors in various stages of purification (eg, [9,10]).

The existence of growth inhibitors of cellular origin bespeaks the importance of "Yin-Yang" regulation as a central feature of normal growth control. As the biochemistry of growth regulation becomes increasingly understood, the tight interaction of growth promotors and inhibitors is becoming more and more evident. Cancer clearly involves a disorganization of these regulatory controls. Transforming genes seem to act autonomously in some cell systems to drive growth; and suppressors identified in other systems appear to block tumor growth autonomously. Most likely activation of oncogenes as well as inactivation of suppressor genes will be found in all tumor types. Nonetheless, the retinoblastoma paradigm [3] as well as dominance of suppression in cell hybrids [1,2] encourages the view that transcriptionally activated suppressor genes may exert a permanent braking effect on tumor growth.

### Regulation of Tumor Angiogenesis Factor (TAF) Production

Recent identification of specific factors with TAF activity as measured by various assays has raised the hope that methods may be developed to regulate this activity. Folkman demonstrated many years ago that tumor cells produce one or more substances that attract capillary blood vessels [11]. Without new vascularization, tumors do not exceed about 1 mm in diameter. Thus, blocking this process would totally inhibit the development of solid tumors. With the cloning of genes that encode at least two factors with TAF activity, an important step has been taken towards understanding the molecular events that underlie TAF production [12–14]. In this laboratory, cell hybrids have been found that segregate for the ability to overproduce some TAFs, leading to formation of highly vascularized tumors in the nude mouse assay (unpublished). Other hybrid clones from the same fusion remain tumor suppressed+show not vascularization. These hybrids represent material from which a regulatory gene product might be identified.

**Suppression by Terminal Differentiation**

Since terminal differentiated cells no longer divide and thus cannot give rise to tumors, one mode of suppression consists of driving partially differentiated cells into the terminal state. Many laboratories work on aspects of this approach, and a large literature is available. At this symposium, a long-term ongoing experimental study was presented by Dr. Robert Scott.

**Suppression of Chromosomal Instability**

Since chromosomal instability appears to lie at the root of oncogenesis, an understanding of the mechanism underlying the instability phenomenon might provide a basis to engineer genetically determined inhibition of this instability. Transposable genetic elements provide a model both for the generation of instability and for its regulation [15]. Since transposition requires activation of a site-specific endonuclease, one mechanism of inhibition would be to block endonuclease expression. In *Drosophila,* this mechanism is used to confine P-element mobility to the germ line [16]. Identification of elements involved in genome rearrangements occurring in oncogenesis could lead to the development of methods to block their activity.

**Tumor Suppression by Senescence in Human Cells**

Normal human cells, whether embryonic, neonatal, or adult, are resistant to experimentally induced tumorigenesis in contrast to rodent cells treated in the same manner [18]. This resistance includes treatments with x-irradiation [19], chemicals [20], and viruses [21,22]. We have recently shown that presence of the SV40-encoded T-antigen has cellular transforming effects but does not induce tumor forming ability in the nude mouse assay [18] nor in the subrenal capsule assay (unpublished). Transfection of normal human fibroblasts (FS-2 cells) with the mutant human c-H-ras DNA, which transforms NIH/3T3 cells, neither induced morphological transformation nor tumorigenicity despite elevated expression of the p21 H-ras gene product [23]. In subsequent double-transfection experiments with mutant c-H-ras+myc, c-H-ras+SV40 DNA, and v-sis + SV40 DNA, no tumor formation resulted in assays of transfectants with large numbers of nude mice (unpubl.).

Recently we have transformed FS-2 cells first with SV40 DNA (defective in the late region), and subsequently, using SV40-transformed clones expressing high levels of T-antigen, we infected them with Kirsten murine sarcoma virus (KiMSV) pseudotyped with baboon endogenous virus (BaEV) to improve infectivity (prepared by Dr. Johng Rhim) [24,25]. The resulting FSVK cells expressed both T-antigen and viral Kirsten-encoded p21 protein but formed only transient tumors in nude mice [17]. All of the tumor-derived human cells senesced at various population doublings, all lower than that of the SV40-transformed cells of origin. FS-2 cells infected with the KiMSV preparation were transformed but soon senesced. Thus, prior transformation with SV40 DNA facilitated KiMSV transformation and led to transient tumorigenicity but not to immortalization. The clear separation of immortalization from loss of growth control with human fibroblasts is very different from the behavior of rodent cells in which spontaneous immortalization is a dependable outcome of cell culture. Thus, the block to immortalization may represent an important natural mechanism of tumor suppression in human cells.

## CONCLUDING REMARKS

Our studies of the effects of cloned oncogenes and tumor viruses on normal human neonatal fibroblasts have illustrated the resistance of human cells compared with rodent cells to experimentally induced oncogenesis. In the past, this resistance has often been ascribed to the stability of the human genome in contrast to that of rodents. Spontaneous transformation is a consistent feature of rodent cells grown in culture, while it has never been seen with human fibroblasts or epithelial cells. While genomic stability surely plays an important role, our recent studies have uncovered a second mechanism,—namely, resistance to immortalization, of importance in protecting human cells from oncogenesis. Both mechanisms are presumed to have evolved during human evolution, being very much weaker in rodents. Perhaps both mechanisms contribute to the fact that our long-lived species remains resistant to oncogenesis during most of its life span, cancer being primarily a disease of old age.

In general, it would appear that a number of different mechanisms may contribute to tumor suppression. Those mechanisms that have come to light to date may represent only a fraction of those that are active and important. Genetics provides a relatively unbiased methodology, analogous to a mutant hunt, for identifying genes with tumor suppressor activity, so long as they are expressed. The methodology is DNA transfer, using transfection or electroporation, of DNA from normal cells into tumor cells, followed by selection for phenotypic suppression with suitable assays. This methodology provides the means to clone out suppressor genes before their mode of action is known. With the application of this approach, it should be possible to identify tumor suppressor genes, and thereby to recover the proteins they encode, proteins with potential prophylactic or therapeutic activity against neoplastic cells.

## REFERENCES

1. Sager R: Adv Cancer Res 44:43–68, 1985.
2. Sager R: Cancer Res 46:1573–1580, 1986.
3. Knudson AG, Jr: Cancer Res 45:1437–1443, 1985.
4. Bishop JM: Sci Am 246:90–92, 1983.
5. Klein G, Klein E: Nature 315:190–195, 1985.
6. Rowley J: Cancer Res 44:3159–3168, 1984.
7. Stein G, Atkins L: J Cell Biochem [Suppl] 10C: (abs).
8. Lumpkin CK, Jr, McClung JK, Pereira-Smith OM, Smith JR: Science 232:393–395, 1986.
9. Hsu Y-M, Barry JM, Wang JL: Proc Natl Acad Sci USA 81:2107–2111, 1984.
10. McMahon JB, Farrelly JG, Iype PT: Proc Natl Acad Sci USA 79:456–460, 1982.
11. Folkman J: In Jaffe EA (ed): "Biology of Endothelial Cells." The Netherlands: Martinus Nijhoff, 1984, pp 412–420.
12. Shing Y, Folkman J, Sullivan R, Butterfield C, Murray J, Klagsbrun M: Science 223:1296–1299, 1984.
13. Maciag T: Prog Hemost Thromb 7:167–182, 1984.
14. Fett JW, Strydom DJ, Lobb RR, Alderman EM, Bethune JL, Riordan JF, Vallee BL: Biochemistry 24:5480–5486, 1985.

15. Shapiro JA (ed): "Mobile Genetic Elements." New York: Academic Press, 1983, p 688.
16. Laski FA, Rio DC, Rubin GM: Cell 44:7–19, 1986.
17. O'Brien W, G Stenman, Sager R: Proc Natl Acad Sci USA 83:8659–8663, 1986.
18. Sager R: Cancer Cell 2:487–493, 1984.
19. Borek C: Nature 283:776–778, 1980.
20. Kakunaga T: In Mishra N (ed): "Advances in Modern Environmental Toxicology," Vol. 1. Princeton Junction, New Jersey: Senate Press, 1980, p 355.
21. Tooze J (ed): "DNA Tumor Viruses." Cold Spring Harbor, N.Y.: Cold Spring Harbor Laboratory, 1980, p. 958.
22. Weiss R, Teich N, Varmus H, Coffin J (eds): "RNA Tumor Viruses." Cold Spring Harbor, N.Y.: Cold Spring Harbor Laboratory, 1982, p 1396.
23. Sager R, Tanaka K, Lau CC, Ebina Y, Anisowicz A: Proc Natl Acad Sci USA 80:7601–7605, 1983.
24. Rhim JS, Jay G, Arnstein P, Price FM, Sanford KK, Aaronson SA: Science 227:1250–1252, 1985.
25. Rhim JS, Fujita J, Arnstein P, Aaronson SA: Science 232:385–388, 1986.

# Oncogene and Chemical-Induced Neoplastic Progression: Role of Tumor Suppression

**J. Carl Barrett, Minoru Koi, Tona M. Gilmer, and Mitsuo Oshimura**

*Environmental Carcinogenesis Group, Laboratory of Pulmonary Pathobiology, National Institute of Environmental Health Sciences, Research Triangle Park, North Carolina 27709*

We previously demonstrated that carcinogen-induced neoplastic transformation of Syrian hamster embryo (SHE) cells requires multiple steps. In order to probe the mechanisms of these steps, normal, diploid SHE cells and carcinogen-induced preneoplastic cells were transfected with different oncogenes. The normal, early-passage SHE cells were not neoplastically transformed by either the v-Ha-*ras* or v-*myc* oncogenes alone; however, transfection of SHE cells by the two oncogenes in combination resulted in tumors in nude mice and syngeneic hamsters. In order to determine whether additional changes occurred in the *ras*-plus-*myc*-induced tumors, cytogenetic analyses of the tumors were performed. A nonrandom chromosome loss (monosomy of chromosome 15) was observed in the ras/myc tumors. In order to understand the biological role of this chromosome loss, hybrids between ras/myc tumor cells and normal SHE cells were prepared in vitro. Tumorigenicity of the ras/myc tumor cells was suppressed following hybridization with normal cells; reexpression of tumorigenicity at later passages correlated with the loss of chromosome 15, suggesting that this chromosome plays a role in suppressing tumorigenicity. The hybrid cells which were suppressed for tumorigenicity still expressed the *ras* and *myc* oncogenes.

An early change in carcinogen-induced neoplastic progression of SHE cells is induction of immortality. Carcinogen-induced immortal cells at early passages still retain the ability to suppress tumorigenicity in cell hybrids. This ability decreases with passaging of immortal cell lines, and subclones are heterogeneous in their ability to suppress tumorigenicity. The susceptibility of immortal cell lines to neoplastic transformation by DNA transfection with v-Ha-*ras* oncogene or tumor DNA inversely correlated with the tumor-suppressive ability of the cells in cell hybrids. Taken together these observations indicate that neoplastic transformation of Syrian hamster embryo cells involves at least three steps: (1) induction of immortality; (2) activation of a transforming gene or oncogene; and (3) loss or inactivation of a tumor suppression function.

Key words: tumorigenicity, immortality, cell hybrids, chromosome loss, aneuploidy

Received June 16, 1986.

© 1988 Alan R. Liss, Inc.

There is considerable evidence that neoplastic development occurs as a progressive process through qualitatively different stages (see [1] for review). Some of the steps in this process can be studied in culture by monitoring transformation of normal cells to the malignant state. Using cell culture systems, it has been possible to study and quantitate carcinogen-induced early, preneoplastic changes in cells as well as the progression of preneoplastic cells to the neoplastic state [1]. Different stages in the neoplastic development of Syrian hamster embyro (SHE) cells have been described [1–3] and some of the advantages of this system are listed in Table I. Following exposure to chemical carcinogens, normal, diploid SHE cells give rise to altered cells which are nontumorigenic but have an increased propensity to become neoplastic. These intermediate cells, termed "preneoplastic," can be isolated and cloned, allowing characterization of the cellular and molecular differences between normal, preneoplastic, and neoplastic populations [1].

The most common pathway of carcinogen-induced neoplastic development in these cells in culture is shown in Figure 1. Morphological alterations in colonies of cells can be observed within a week after carcinogen exposure, and these provide a quantitative measure of carcinogen-induced effects [1]. Some, but not all, of these morphologically transformed colonies can be isolated and grown indefinitely. The normal cells senesce after several passages, whereas following carcinogen exposure, immortal cells, which escape senescence and grow indefinitely, can be isolated. Whether these two early changes, morphological transformation and immortality, are induced at the same time or represent two steps in this process is unclear, and this point will be addressed further in the discussion of this paper. A nonrandom chromosome change (trisomy of chromosome 11) is observed in some immortal cell lines. Since not all immortal cells have an alteration in chromosome 11, multiple pathways for immortality may exist [4]. Morphological transformation and immortality are induced by chemicals which induce numerical chromosome changes in the absence of

**TABLE I. Advantages of Syrian Hamster Embryo Cell Transformation System**

1. Uses early-passage, diploid cells with stable karyotype
2. Low frequency of spontaneous transformation as measured by morphological transformation, immortalization or escape from cellular senescence, and neoplastic transformation
3. Neoplastic development of the cells in culture has been demonstrated to be a progressive, multistep process
4. Phenotypic markers for different stages in the neoplastic process have been identified
5. Quantitative assay for carcinogen-induced early preneoplastic changes has been developed; this assay detects a wide variety of chemical carcinogens but not structurally related noncarcinogens
6. Clones of preneoplastic cells have been isolated and can be used to study intermediate stages in neoplastic development; the cellular and molecular basis of preneoplastic cells and the influence of carcinogens on their progression can be studied

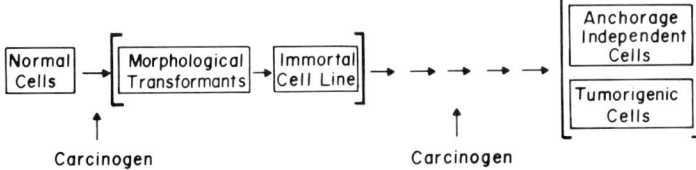

Fig. 1. Most common pathway of neoplastic progression of Syrian hamster embryo cells following carcinogen treatment.

detectable gene mutations [1,5,6] which is consistent with an important role for aneuploidy in this step.

Carcinogen-induced immortal cell lines are initially nontumorigenic and have to be passaged several times before neoplastic conversion occurs. In this system, the phenotypes of anchorage-independent growth and tumorigenicity are closely associated [1]. Preliminary data indicate that neoplastic SHE cells transformed in vitro by chemicals have a mutated Ha-*ras* oncogene [7]. Although this change is not observed in every tumor cell line, this finding does indicate a role for the *ras* oncogenes in the later, postimmortalization stages of neoplastic development in this model.

The multistep nature of neoplastic transformation of cells in culture, which was first shown following carcinogen treatment [1,2], is also observed when cells are transfected by activated oncogenes. Normal, primary rodent cells are generally not neoplastically transformed by a single oncogene, but certain combinations of oncogenes—for example, *ras* plus *myc*—can convert normal cells to the tumorigenic state [8–12]. Weinberg and colleagues [8,13] have proposed a two-step model of carcinogenesis based on the cooperation between two oncogenes. They suggest that one oncogene (eg, *myc*) is involved in the immortalization process, and the second oncogene (eg, *ras*) results in the expression of various transformed phenotypes, such as focus formation and anchorage-independent growth. The transforming oncogene alone is sufficient for tumorigenic conversion of immortal cells, whereas neoplastic transformation of primary cells requires both an immortalizing gene and a transforming gene. In general, the results of chemically induced neoplastic transformation are consistent with this model. As shown in Figure 1, the progression of Syrian hamster embryo cells following chemical carcinogen treatment results in immortal cells which later acquire anchorage-independence and tumorigenicity. Activation of the *ras* oncogenes may play a role in the later stages. To date, the oncogenes presumed to play a role in the early, immortalization stage of chemically induced neoplastic progression have not been identified.

Although the two steps of immortalization and transformation are involved in carcinogen-induced neoplastic development, certain observations suggest that changes in addition to these two steps are also needed. Not all immortal cell lines are the same. Some readily progress to the tumorigenic state either spontaneously or following treatment with carcinogens or oncogenes, whereas other cell lines are refractory to induction of neoplastic transformation [1,14]. The evidence that more than one step is involved in the neoplastic transformation of certain immortal cell lines is summarized in Table II.

Possible insight in the nature of the additional step in neoplastic transformation comes from experiments involving hybridization of normal and malignant cells [1,14–20]. Many, but not all, of these experiments indicate that tumorigenicity is a recessive

**TABLE II. Evidence for More Than One Step in the Neoplastic Transformation of Certain Immortal Cells**

1. The passage levels at which immortal cells undergo neoplastic transformation vary considerably
2. The rates at which anchorage-independent variants arise vary by three orders of magnitude between cell lines (from $10^{-7}$ to $10^{-4}$ variants per cell generation)
3. Some immortal cell lines are refractory to carcinogen/mutagen-induced neoplastic transformation
4. Some immortal cell lines are refractory to oncogene-induced neoplastic transformation
5. Tumorigenicity and anchorage-independent growth are recessive traits in hybrids between tumorigenic and some immortal, nontumorigenic cells

trait in cell hybrids. A major paradox in cancer biology, therefore, exists: DNA transfection experiments have identified dominantly acting cancer genes (oncogenes), whereas cell hybridization experiments suggest that tumorigenicity is recessive in nature.

In this report we will present results from two lines of research which bear on this problem. During our studies on the genetic changes involved in different steps of carcinogen- and oncogene-induced neoplastic progression of Syrian hamster embryo cells, we observed that two classes of genes are involved—oncogenes and tumor suppressor genes. The evidence for the existence of the latter and their role in neoplastic progression will be presented.

## MATERIALS AND METHODS

Syrian hamster embyro (SHE) cell cultures were established from 13-day gestation fetuses as described previously [2]. Secondary cultures were initiated from the frozen stocks and tertiary or later cultures were used for all experiments. The immortalized cells used in this study were DES-4 cells (passage 34–35 and 58–60) which were isolated as a clone after treatment of SHE cells with diethylstilbestrol (0.1 $\mu$g/ml) [21] and 10W cells (passage 5–6 and 15–17), a clone isolated after treatment of SHE cells with asbestos [22,23]. Two highly tumorigenic hamster cell tumor lines induced with benzo(a)pyrene (BP6T) or by transfection with v-H-*ras* plus v-*myc* (*ras*/*myc*-T) were chosen for study. Mutant clones of these cells resistant to both 1.5 mM ouabain and 10 $\mu$g/ml 6-thioguanine were isolated and designated BP6T M3 and *ras*/*myc* T-1m. The cell culture medium, growth methods, cell hybridization procedures, transfection methods, cytogenetic methods, and assays for growth in agar and tumorigenicity have been described [2,5,11,14].

## RESULTS

### Nonrandom Loss of Chromosome 15 in Syrian Hamster Tumors Induced by v-Ha-*ras* Plus v-*myc* Oncogenes

We previously reported that normal, diploid Syrian hamster embryo cells were neoplastically transformed following transfection with the v-Ha-*ras* plus v-*myc* oncogenes; but neither oncogene alone was sufficient for neoplastic conversion [11]. These findings confirm the results of Land et al [8], Ruley [9], and Newbold and Overell [10].

Neoplastic transformation of SHE cells by two cooperating oncogenes is consistent with a multistep model of carcinogenesis. However, the number of steps necessary to convert a normal cell into a malignant cell is unknown. If activation of two oncogenes is sufficient for tumorigenicity, tumors derived from diploid cells transformed by the transfected oncogenes may remain diploid or have only random chromosome alterations. Therefore, we examined the karyotypes of tumors formed after transfection of Syrian hamster embryo cells with v-Ha-*ras* plus v-*myc* oncogenes [24]. We observed that tumors induced by v-Ha-*ras* plus v-*myc* were monoclonal and had a nonrandom chromosome change, monosomy of chromosome 15 (Fig. 2). This chromosome loss was found in six out of six *ras*/*myc* tumors examined but not in polyoma-induced tumors [24]. Thus, an additional change, loss of chromosome 15,

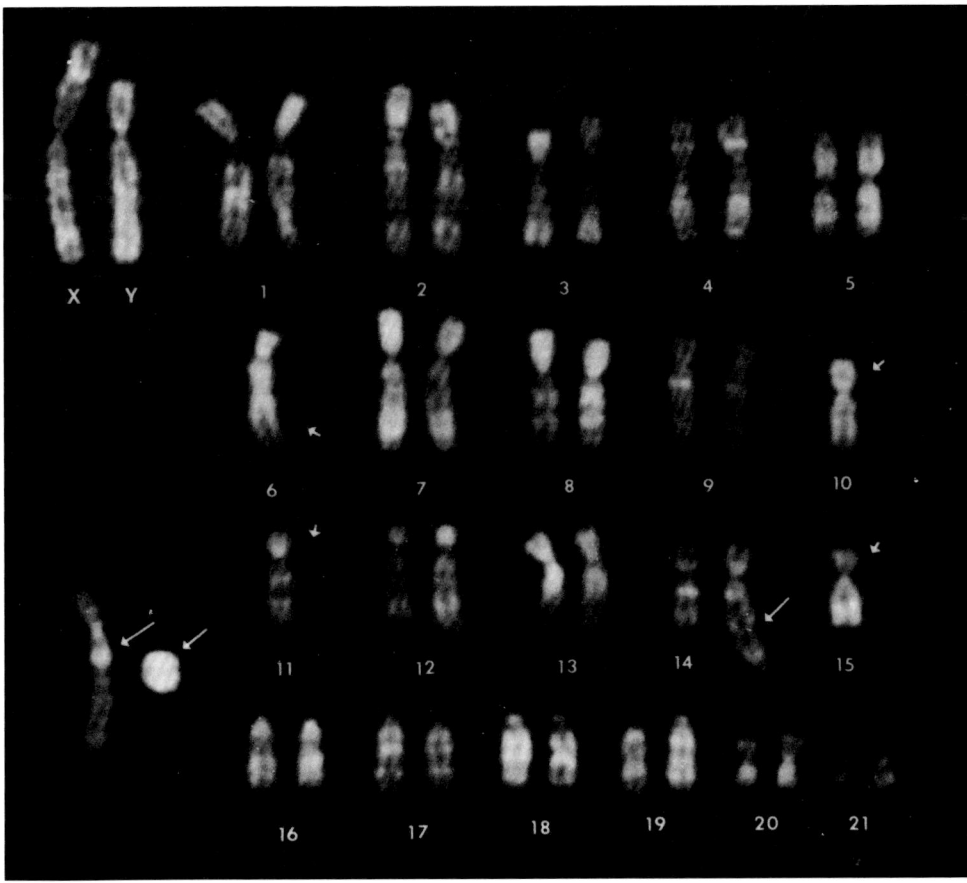

Fig. 2. Karyotype of tumor cells from Syrian hamster embryo cells transfected with v-Ha-*ras* and v-*myc* oncogenes. Alterations in chromosomes 10, 11, and 15 were observed in all the metaphases of these tumor cells. Loss of chromosome 15 was the only chromosome change common to all six *ras/myc* tumors examined.

is advantageous or required for tumorigenicity induced by v-Ha-*ras* plus v-*myc* oncogenes.

In order to clarify the role of monosomy 15 in the expression of the tumorigenicity of *ras/myc* cells, we made cell hybrids between these cells and normal SHE cells. We found that the tumorigenicity of the hybrid cells was totally lost or greatly reduced when compared to the *ras/myc* tumor cells. The hybrid cells also failed to grow in agar. The nontumorigenic *ras/myc* SHE hybrids still expressed the *ras* and *myc* RNA and high levels of the mutated form of the $p21^{ras}$ protein. Thus, the expression of tumorigenicity was not due to the loss or lack of expression of the oncogenes.

When the hybrid cells were passaged, anchorage-independent and tumorigenic variant cells arose at a low frequency in the population. These cells were cloned and comparisons were made of the karyotypes of the parental cells, hybrids which were suppressed for tumorigenicity and anchorage-independence, and hybrids which reexpressed these phenotypes. Banded karyotypes revealed that the suppressed hybrids contained the chromosome complement of both parental cells as anticipated. The hybrids which reexpressed tumorigenicity had lost only a few chromosomes; a nonrandom loss of chromosome 15, but no other chromosome, was observed. These results suggest that loss of chromosome 15 results in the loss of a cellular gene which effects a phenotype necessary for neoplastic transformation.

## Loss of Tumor Suppressor Function During Chemically Induced Neoplastic Progression of Syrian Hamster Embryo Cells

The cell hybrid experiments described above were performed with tumor cells induced by oncogene transfection. We next examined the suppression of tumorigenicity of chemically transformed cells by normal SHE in cell hybrids and determined the stage in the multistep neoplastic process at which this tumor suppression function was lost [14].

Cell hybrids between normal, early-passage Syrian hamster embryo cells and a highly tumorigenic, chemically transformed hamster cell line, BP6T-M3, were formed, selected, and analyzed [14]. Tumorigenicity and anchorage-independent growth were suppressed in the hybrid cells compared to the tumorigenic BP6T-M3 cells (Table III). These two phenotypes segregated coordinately in these cells.

To determine at what stage in the neoplastic process this tumor suppression observed in the normal SHE cells was lost, two chemically induced immortal cell lines (DES-4 and 10W) were examined at different passages. When DES-4 × BP6T hybrids were assayed for growth in agar, a significant suppression (500–5,000-fold) of anchorage-independent growth was observed [14]. Likewise, the tumorigenicity or tumor latency of the hybrids was suppressed. However, a different pattern was observed in 10W × BP6T hybrids compared to DES-4 × BP6T hybrids. All the 10W × BP6T hybrids grew in agar with a frequency from >1% up to 46% (only a 2–50-fold reduction). These hybrids were also much more tumorigenic; the latency period

**TABLE III. Suppression/Expression of Anchorage-Independent and Tumorigenic Phenotypes in Hamster-Hamster Hybrids***

| Clone | Plating efficiency on plastic[a] (%) | Plating efficiency in agar[b] (%) | Ratio of anchorage-independent to anchorage-dependent growth | Tumorigenicity[c] (average latency period in days) | Modal chromosome No. (range)[d] |
|---|---|---|---|---|---|
| Parental cells | | | | | |
| SHE | 21.2 | <.00005 | <.0001 | >30 | 44 (43–44) |
| BP6T-M3 | 70.7 | 58.0 | 0.82 | 7.0 | 43 (42–46) |
| SHE × BP6T-M3 | | | | | |
| clone 1 | 29.7 | 0.015 | 0.0005 | >30 | 116 (95–124) |
| clone 2 | 42.0 | <.0033 | <.0001 | >30 | 83 (69–85) |
| clone 3 | 44.0 | 0.057 | 0.0013 | >30 | 81 (76–84) |
| clone 4 | 46.4 | 0.01 | 0.0002 | 25[e] | 100 (57–105) |
| clone 5 | 18.7 | >.0033 | <.0002 | >30 | 120 (117–129) |

*Data from Koi and Barrett [14].
[a]$10^2$ hybrid cells per dish were inoculated onto 6-cm dishes and 7 days later the average PE was determined from five dishes.
[b]$10^2$ to $10^4$ hybrid cells per dish were inoculated into soft agar medium; 2 wk later, the average PE was determined from three dishes.
[c]$10^4$ hybrid cells were mixed with $10^6$ normal SHE cells and injected subcutaneously into nude mice. Average latency period was determined from three sites when first nodule was detected.
[d]Chromosome No. was determined from 20 well-spread metaphases.

was in some cases equal to BP6T parental cells and was increased to 13 days at most [14].

To confirm the difference between 10W cells and DES-4 cells to suppress anchorage-independent growth of BP6T cells in hybrids, a direct assay of anchorage-independent growth of hybrids was performed. In this assay the cells of interest were fused and after a 24-hr recovery period, the hybrids were selected in HAT/ouabain medium either on plastic or directly in agar with selective medium. The ratio of hybrids growing in agar ($Ag^+$) to total hybrids is therefore a measure of the cells' ability to suppress anchorage-independent growth. Control experiments with BP6T × BP6T hybrids resulted in the same number of hybrids in agar and on plastic (ratio = 1.076) (Table IV), while the ratio was reduced to 0.0009 for SHE × BP6T hybrids. Passage 15 10W cells reduced the ratio of $Ag^+$ hybrids to total hybrids to only 0.711. However, at an earlier passage (p5) the 10W cells were still effective in suppressing anchorage-independent growth (ratio = 0.0003). Hybrids of 10W (p5) × BP6T cells growing on plastic were isolated and analyzed for anchorage-independent growth and all were suppressed.

DES-4 cells at passage 35 and passage 58 were fused with BP6T cells and the ratio of $Ag^+$ hybrids to total hybrids was 0.0021 and 0.062, respectively. Therefore, at the later passage the ability of the cells to suppress anchorage-independence decreased but not to the degree observed with 10W cells. Subclones of 10W cells (p15) and DES-4 cells (p58) were randomly isolated and tested for the ability to suppress anchorage-independent growth of BP6T cells. These subclones were heterogeneous in their ability to suppress tumorigenicity in cell hybrids; some clones retained the tumor suppression ability and others lost this function.

In order to determine if the cells with differing abilities to suppress anchorage-independent growth differed in their susceptibility to transformation by DNA transfection, the cells were treated with BP6T cellular DNA or v-Ha-*ras* plasmid DNA by the calcium phosphate transfection method and after 5 days assayed for anchorage-independent growth or tumorigenicity.

When transfected with v-Ha-*ras* DNA, 10W cells, which are not effective in suppressing tumorigenicity and anchorage-independent growth ($sup^-$) grew in soft agar with a 1,000-fold higher frequency then DES-4 ($sup^+$) cells (Table V). This

**TABLE IV. Suppression/Expression of Anchorage-Independent Phenotype in Hybrids of BP6T-M3 Cells and Various Other Cells***

| Cell line fused to BP6T-M3 cells | Hybrid frequency[a] ($\times 10^{-4}$) | $Ag^+$ hybrid frequency[b] ($\times 10^{-4}$) | Ratio of $Ag^+$ frequency to total hybrid frequency |
|---|---|---|---|
| BP6T | 35.2 | 37.9 | 1.076 |
| SHE (p5) | 7.9 | 0.007 | 0.0009 |
| 10W (p5) | 17.6 | 0.005 | 0.0003 |
| 10W (p15) | 9.9 | 7.08 | 0.715 |
| DES-4 (p35) | 25.6 | 0.054 | 0.0021 |
| DES-4 (p58) | 22.4 | 1.39 | 0.062 |

*Data from Koi and Barrett [14].
[a]Hybrid frequency was determined from the number of colonies growing on plastic dishes in HAT/ouabain medium at two weeks after selection.
[b]$Ag^+$ hybrid frequency was determined from the number of colonies growing in agar containing selective (HAT/ouabain) medium at 3 wk after selection.

**TABLE V. Susceptibility of Immortal Cells to Transfection by Different DNAs**

| Recipient cells[a] | Phenotype[b] | G418$^r$ colonies[c] with pSV2neo | Agar colonies[d] with v-Ha-ras | Agar colonies[d] with BP6T DNA |
|---|---|---|---|---|
| DES-4 | sup$^+$ | $3 \times 10^{-3}$ | $2.5 \times 10^{-6}$ | $<1 \times 10^{-6}$ |
| 10W | sup$^-$ | $9 \times 10^{-4}$ | $3.8 \times 10^{-3}$ | $54 \times 10^{-6}$ |

[a]DES-4 p55; 10W p15.
[b]Cells either suppress (sup$^+$) or fail to suppress (sup$^-$) tumorigenicity and anchorage-independent growth of BP6T cells following cell hybridization.
[c]Cells transfected with pSV2neo as described [11]. Data expressed colonies per $\mu$g of DNA transfected.
[d]Cells were transfected with v-Ha-ras [11] or BP6T DNA [14] selected 3–5 days later for growth in agar.

**TABLE VI. Evidence for a Three-Step Model of Neoplastic Transformation of Syrian Hamster Embryo Cells**

1. Cells are not neoplastically transformed by a v-Ha-ras or v-*myc* oncogene alone
2. Cells neoplastically transformed by v-Ha-ras plus v-*myc* have a nonrandom loss of chromosome 15
3. Immortal cell lines are neoplastically transformed by v-Ha-ras oncogene alone, but the susceptibilities of cell lines vary
4. Tumorigenicity and anchorage-independent growth are recessive traits in hybrids between tumorigenic cells and normal Syrian hamster embryo cells
5. Some, but not all, immortal cells can suppress tumorigenicity; this ability decreases with passaging of immortal cells lines, and subclones are heterogeneous in their ability to suppress transformation
6. Susceptibility of immortal cell lines to neoplastic transformation by DNA transfection with viral oncogenes or tumor DNA is inversely correlated with suppressive ability in cell-cell hybrids

difference was unrelated to the efficiency of uptake and expression of DNA since the frequency of G418$^r$ colonies following transfection with pSV2-neo was greater for DES-4 cells (Table V).

## DISCUSSION

Based on our results we would like to propose the hypothesis that chemically induced neoplastic transformation of Syrian hamster embryo cells involves at least three steps: (1) induction of immortality; (2) activation of a dominantly acting, transforming oncogene; and (3) loss of a tumor suppression function. The lines of evidence supporting this model are summarized in Table VI.

The first two changes, immortality and transformation, are consistent with the model of Land et al [8] and Ruley [9] and with findings that carcinogen-induced immortality of these cells is an important step in the neoplastic progression of these and other cells [1,2,10,25]. However, our results indicate that in addition to activation of the oncogenes involved in immortality and transformation, inactivation of genes which suppress tumorigenicity is a necessary step in neoplastic transformation.

Genes analogous to the putative tumor suppression genes identified by cell culture studies of hybrids between normal and tumor cells are involved in the in vivo development of tumors in humans and nonhuman experimental models (Table VII). At least 50 dominantly inherited cancer susceptibilities have been identified in humans [26]. In none of these human cancer susceptibilities has an oncogene been implicated; rather, mutations in a different class of genes, tumor suppression genes, are inherited in these individuals [27]. Individuals who inherit a heterozygous, germ-line mutation

**TABLE VII. Evidence for Tumor Suppression Genes**

1. Cell-hybrids between normal and tumorigenic cells often show that tumorigenicity is suppressed and reexpression of the tumorigenic phenotype is related to the loss of specific chromosomes from the hybrid cells
2. Dominantly inherited cancer susceptibilities in humans (eg, Wilms' tumor and retinoblastoma and related tumors) involve a germ-line, heterozygous mutation which becomes homozygous or hemizygous in tumor tissue
3. Dominant inheritance of susceptibility of renal carcinoma in the rat may involve tumor-suppression gene
4. Recessive-lethal mutants of *Drosophilia melanogaster* are associated with predisposition to malignant neoplasms
5. Hybrid fish between ornamental platyfish (*Xiphophorus maculatis*) and ornamental swordtail (*Xiphophorus helleri*) have increased incidence of tumors which appear to result from inheritance of a tumor gene (oncogene) and elimination of regulatory (tumor suppressor) gene

in one of these genes are predisposed to develop specific cancers (eg, Wilms' tumor and retinoblastoma). In these cancers the locus of the inherited mutation becomes homozygous or hemizygous by a variety of secondary chromosomal mutations [28]. The implication of these findings is that these genes suppress the expression of the cancer unless a mutation and/or a deletion of both alleles of the gene arise.

Nonhuman models for this class of genes also exist. Eker and colleagues [29] have reported the dominant inheritance of susceptibility to development of renal carcinoma in the rat which is consistent with the involvement of a tumor suppression gene. In *Drosopohila melanogaster* a number of recessive-lethal mutants have been identified which are also associated with a predisposition to malignant neoplasms [30]. Hybrid fish between the ornamental platyfish (*Xiphophorus maculatis*) and the ornamental swordtail (*Xiphophorus helleri*) have an increased incidence of tumors which appears to result from inheritance of a tumor gene (oncogene) and elimination of regulatory (tumor suppressor) genes [31].

For a number of years, the recessive nature of tumorigenicity has been indicated by cell-cell hybridization experiments [14–20]. Yet the discovery of transforming genes (oncogenes) is predicated on the dominantly acting nature of these genes in DNA transfection experiments. Several possible explanations (which are not mutually exclusive) may exist for these seemingly disparate findings: (1) The ability to suppress the tumorigenic phenotype may depend on the genes activated. For example, *ras*-transformed cells are suppressed in hybrids with normal cells [14,16,17], whereas cells transformed by DNA viruses are sometimes not [15]. (2) The expression of the neoplastic phenotype may depend on the dosage of the putative tumor suppression and transforming genes. Results of certain studies are consistent with this hypothesis [32]. (3) The putative tumor suppression gene is lost either during the transfection process or during the selection for the transformed cells. The nonrandom loss of chromosome 15 in *ras*-plus-*myc*-induced Syrian hamster tumor cells may be an example of this mechanism [24]; and (4) the ability of a cell to suppress the tumorigenic phenotype may be dependent on the stage of progression of the cell. Our results [14] also illustrate this possibility.

In conclusion, the expression of neoplastic potential requires alterations in two classes of cellular genes—oncogenes and tumor suppression genes. Further studies on the nature and mechanisms of this tumor suppression function and isolation of the genes responsible for this function are needed.

## REFERENCES

1. Barrett JC, Fletcher WF: In Barrett JC (ed): "Mechanisms of Environmental Carcinogenesis, Vol II—Multistep Models of Carcinogenesis." Boca Raton: CRC Press, 1987, (in press).
2. Barrett JC, Ts'o POP: Proc Natl Acad Sci USA 75:3761, 1978.
3. Barrett JC: Cancer Res 40:91, 1980.
4. Oshimura M, Barrett JC: Environ Mutagen 8:129, 1986.
5. Tsutsui T, Maizumi H, McLachlan JA, Barrett JC: Cancer Res 43:3814, 1983.
6. Oshimura M, Hesterberg TW, Tsutsui T, Barrett JC: Cancer Res 44:5017, 1984.
7. Gilmer T, Annab L, Barrett JC: (in preparation).
8. Land H, Parada LF, Weinberg RA: Nature 304:596, 1983.
9. Ruley HE: Nature 304:602, 1983.
10. Newbold RF, Overell RW: Nature 304:648, 1983.
11. Thomassen DG, Gilmer TG, Annab LA, Barrett JC: Cancer Res 45:716, 1985.
12. Rassoulzadegan M, Cowie A, Carr A, Glaichenhaus N, Kamen R, Cuzin F: Nature 300:713, 1982.
13. Weinberg RA: Science 230:720, 1985.
14. Koi M, Barrett JC: Proc Natl Acad Sci USA 83:5992, 1986.
15. Weissman BE: In Barrett JC (ed): "Mechanisms of Environmental Carcinogenesis, Vol I—Epigenetic and Genetic Mechanisms." Boca Raton: CRC Press, 1987, in press.
16. Sager R: Cancer Res 46:1573, 1986.
17. Craig RW, Sager R: Proc Natl Acad Sci USA 82:2062, 1985.
18. Stanbridge EJ: Nature 260:17, 1976.
19. Harris H, Miller OJ, Klein G, Worst P, Tachibana T: Nature 223:363, 1969.
20. Bouck N, DiMayorca G: Mol Cell Biol 2:97, 1982.
21. McLachlan JA, Wong A, Degen GH, Barrett JC: Cancer Res 42:3040, 1982.
22. Hesterberg TW, Barrett JC: Cancer Res 44:2170, 1984.
23. Oshimura M, Hesterberg TW, Barrett JC: Cancer Genet Cytogenet 22:225, 1986.
24. Oshimura M, Gilmer TM, Barrett JC: Nature 316:636, 1985.
25. Newbold RF, Overell RW, Connell JR: Nature 299:633, 1982.
26. Mulvihill JJ: In Mulvihill JJ, Miller RW, Fraumeni JF (eds): "Genetics of Human Cancer." New York: Raven Press, 1971, pp 137–143.
27. Knudson AG: Cancer Res 45:1437, 1985.
28. Cavenee WK, Cryja TP, Phillips RA, Benedict WF, Godbout R, Gallie BL, Murphee AL, Strong LC, White RL: Nature 305:79, 1983.
29. Eker R, Mossige J: Nature 189:858, 1961.
30. Gateff E: Science 200:1448, 1978.
31. Anders F, Schartl M, Barnekow A, Anders A: Adv Cancer Res 42:191, 1984.
32. Benedict WF, Weissman BE, Mark C, Stanbridge EJ: Cancer Res 44:3471, 1984.

# Inhibition of HeLa Cell Growth Following Transfection With Genomic DNA From Human Embryo Fibroblasts

## Raji Padmanabhan, Tazuko Howard, and Bruce H. Howard

*Laboratory of Molecular Biology, Division of Cancer Biology and Diagnosis, National Cancer Institute, Bethesda, Maryland 20892*

We investigated whether gene transfer can be used to identify genomic DNA sequences that negatively regulate mammalian cell growth. WI38 human embryo fibroblast DNA mixed with a G-418 resistance plasmid (pRSVneo) was transfected into HeLa S3 cells. A double selection was then carried out for transfection-mediated growth inhibition: 1) transiently nonreplicating cells were enriched for by transfer into spinner medium and selection in bromodeoxyuridine/Hoechst 33258 medium; 2) surviving cells were replated into monolayer culture and selected for G-418 resistance. Parallel control transfections were carried out with *E. coli*/pRSVneo and HeLa/pRSVneo DNA mixtures. Transfection with the WI38/pRSVneo DNA combination yielded 15-fold more G-418-resistant colonies than control transfections, indicating the presence of specific growth-inhibitory sequences in WI38 DNA. Non-clonal WI38/pRSVneo-transfected cell populations exhibited twofold slower growth and greater sensitivity to growth restriction by low-serum conditions than control cells. Potential relationships between these results and genetic mechanisms involved in negative regulation of growth will be discussed.

**Key words: tumor suppression, growth inhibition, gene transfer**

Mammalian cell growth appears to be regulated both by genes that stimulate cell proliferation and genes that function to inhibit this process. There is a large body of information on oncogene-mediated stimulation of cell growth [1]. There are also multiple lines of evidence to support the functional importance of genes that negatively regulate cell growth; these include purification of growth-inhibitory proteins [2–5], studies on inhibition of DNA replication in heterokaryons formed by fusion of quiescent and growing cells [6,7], and studies on suppression of malignant transformation in somatic cell hybrids [8–10].

Relatively little is known about the interrelationships between stimulatory and inhibitory gene sets. Mitogenic stimulation of mouse fibroblasts with platelet-derived

Received June 2, 1986.

© 1988 Alan R. Liss, Inc.

growth factor induces increased expression of both cellular proto-oncogenes and the growth-inhibitory protein beta-fibroblast interferon [11]. It has been speculated that interferon expression under these circumstances provides negative feedback regulation of replication, but this remains to be established. The potential complexity of growth control is suggested by the finding that one gene product, human transforming growth factor-$\beta$ (TGF-$\beta$), can either stimulate or inhibit cell growth depending on the presence of other growth-regulatory factors [12]. It seems likely that regulation of cell proliferation will involve multiple mechanisms in which members of one set act on or modulate expression of the opposite set. Moreover, because of the quite different requirements placed on growth stimulatory and inhibitory mechanisms, the corresponding gene sets may be quite different in terms of structure and mode of action.

We decided to investigate whether DNA-mediated gene transfer, a valuable method for the isolation of oncogenes, could be used to isolate and characterize growth-inhibitory genes. In view of the usefulness of the NIH/3T3 cell line as a relatively nonspecific "indicator" for oncogene activity, we decided to search for a cell line that would serve as an indicator for the activity of growth-inhibitory sequences. For several reasons, the widely used HeLa cell line appeared to be a good candidate for this purpose. First, HeLa cells have not been (to our knowledge) reported to contain dominantly acting oncogenes and thus may represent a different transformation mechanism than that mediated by the known cellular oncogenes. Second, HeLa cells can be induced to arrest in the G1 phase of the cell cycle under anchorage-independent, but not monolayer, growth conditions [13]. This observation suggested that an assay for growth inhibitory genes could be devised in which HeLa replication was conditionally and reversibly inhibited. Finally, it has been demonstrated that a gene(s) active in human embryo fibroblasts can suppress HeLa tumorigenicity in stable somatic cell hybrids formed by fusion of those cells with HeLa cells [8,9].

As a source of DNA for transfection, we chose genomic DNA from quiescent WI38 human embyro fibroblasts. WI38 DNA (or control genomic DNA from *Escherichia coli* or HeLa cells) was transfected, together with marker plasmid pRSVneo [14], into rapidly growing HeLa cells. A selection was then carried out for transient inhibition of growth in the transfected cell population. We report here the surprising result that high levels of specific growth-inhibitory activity can be detected in WI38 genomic DNA. This growth inhibitory activity appears to be distinct from (and may mask the activity of) tumor suppression genes detected in other experimental systems.

## MATERIALS AND METHODS
### Materials

Bromodeoxyuridine and Hoechst 33258 were purchased from Sigma Biochemicals, St. Louis, MO; Dulbecco's modified Eagle's medium (DME) and the antibiotic G-418 were obtained from GIBCO, Grand Island, NY. MEM spinner medium without glutamine with bicarbonate was from Quality Biological, Gaithersburg, MD.

### Preparation of DNA

Plasmid DNA was purified as previously described from lysozyme/Triton X-100 clear lysates by two cesium chloride/ethidium bromide equilibrium centrifugation steps [14]. Genomic DNA was prepared by proteinase K digestion, followed by two phenol extractions, extensive dialysis in 10 mM Tris HCl (pH 7.9)/5 mM EDTA/10 mM NaCl, two cycles of cesium chloride/ethidium bromide equilibrium centrifugation, isobutanol extraction, and dialysis into HE (10 mM HEPES, 1 mM EDTA, pH 7.1).

### Cell Culture

HeLa S3 and WI38 cells, obtained from the American Type Culture Collection, were maintained as monolayer cultures in DME supplemented with penicillin (50 units/ml), streptomycin (50 µg/ml), glutamine (2 mM), and 10% fetal bovine serum. HeLa cell spinner medium was supplemented 10% filtered horse serum and 2% fetal bovine serum. WI38 and Rev-2 cells were checked following culture in our laboratory to exclude mycoplasma contamination.

### DNA-Mediated Transfection

Transfection conditions were as described elsewhere [14], except that HEPES-buffered saline (HBS) was adjusted to pH 7.05 just before use and $CaPO_4$/DNA coprecipitates were not allowed to stand more than 10 min. HeLa S3 cells, plated at $7.5 \times 10^5$ cells/75-cm$^2$ flask, were refed the following day 3–4 hr prior to transfection. Calcium phosphate precipitates containing pRSVneo (5µg) and either WI38, HeLa S3 or *E. coli* DNA (15µg) were added directly to the culture medium; after 4 hr cells were washed with DME, treated with 15% (w/v) glycerol in HBS for 30 sec, washed with DME, and refed. Forty-eight hours were allowed for phenotypic expression of exogenous DNA, after which the cells were trypsinized, counted, and placed in spinner medium at a density of $2.5–3 \times 10^5$ cells/ml. To determine relative pRSVneo transfection efficiencies, about $10^5$ cells were removed after 4 hr for plating at a density of 200 cells/cm$^2$ in G-418 medium (standard monolayer medium supplemented with 800–1,000 µg G-418). Bromodeoxyuridine(BrdUrd) and freshly prepared Hoechst 33258 were then added (final concentrations 100 µg/ml and 1 µg/ml) and the foil-wrapped spinner cultures were incubated for an additional 44 hr. Cells were exposed for 5 min to fluorescent light (25 W/m$^2$), counted, and plated at $10^4$ cells/cm$^2$ in G-418 medium. Selection for the pRSVneo marker was maintained by refeeding with G-418 medium at 4-day intervals.

## RESULTS

A protocol for detection of growth-inhibitory sequences in genomic DNA from quiescent WI38 fibroblasts is shown in Figure 1. A mixture of WI38 and pRSVneo DNAs were introduced into HeLa S3 cells in monolayer culture by the $CaPO_4$ method. As controls, *E. coli*/pRSVneo, and HeLa/pRSVneo mixtures were transfected in parallel. After an expression period of 48 hr, the transfected cultures were transferred into spinner culture in the presence of BrdUrd and Hoechst 33258. The rationale for this suspension culture step was to place transfected cells in a semipermissive growth environment during the BrdUrd/Hoechst selection for nondividing cells [15]. After 48 hr in spinner culture, cells were exposed to fluorescent light; surviving cells were then replated in monolayer culture in the presence of the antibiotic G-418 to select for the subpopulation that stably expressed the pRSVneo marker [14,16].

Transfect monolayer HeLa cells

a. Experimental DNA + pRSVneo
b. Control DNAs (E. coli or HeLa) + pRSVneo

↓

Grow in monolayer culture for phenotypic expression
(48 hrs) -> transfer to suspension (spinner) culture ->
remove cell aliquots to determine transfection efficiency

↓

Select against cells that continue to replicate in
suspension culture: add BrdUrd + Hoechst 33258
(48 hrs) -> expose to fluorescent light

↓

Select surviving cells for expression of pRSVneo
marker: plate in monolayer culture, G-418 medium

Fig. 1. Experimental protocol for detecting DNA sequences that inhibit HeLa cell growth. Details are given in Materials and Methods.

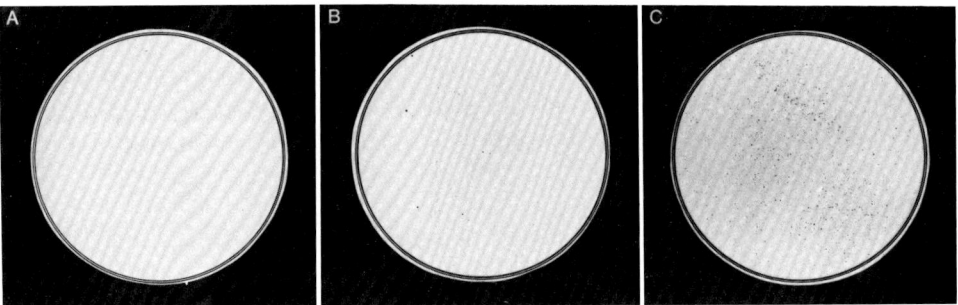

Fig. 2. Colonies surviving BrdUrd/Hoechst 33258 treatment followed by G-418 selection. HeLa S3 monolayer cultures were transfected with (**A**) pRSVneo and *E. coli* DNA, (**B**) pRSVneo and HeLa S3 DNA, (**C**) pRSVneo and WI38 DNA. Plates show colonies remaining after selection for 8 days in medium containing 800 μg/ml G-418.

The logic of this protocol is that cells which take up exogenous DNA and fail to divide in spinner culture will be scored as G-418-resistant colonies; thus, the number of colonies serves as an indirect measure of the growth-inhibitory activity of the transfected DNA. As a measure of nonspecific growth inhibition, we used the number of colonies following control transfections with *E. coli* or HeLa DNA. The increase in colonies relative to these controls provides a measure of specific growth-inhibitory activity in WI38 genomic DNA.

Results of such an experiment are shown in Figure 2. The number of colonies following WI38/pRSVneo transfection is about 15-fold higher than following transfection with either *E. coli*/pRSVneo or HeLa/pRSVneo controls. This difference does not reflect a disparity in the overall uptake of DNA, since aliquots of cells taken from spinner culture just prior to addition of BrdUrd/Hoechst indicated that transfection frequencies for the neo marker were similar (data not shown). The higher number of colonies after transfection with the WI38/pRSVneo mixture rather indicates specific growth-inhibitory activity in this genomic DNA.

To examine whether transfection with WI38 DNA leads to a persistent alteration in HeLa cell growth, more slowly growing colonies were pooled from a plate of WI38/pRSVneo-transfected cells similar to that shown in Figure 2. The population thereby obtained, termed Rev-2, was compared to a pool of HeLa S3 cells transfected with homologous DNA with respect to growth in either standard or low serum conditions. As shown in Figure 3, the Rev-2 population exhibited a longer doubling time (about 1.5–2-fold) than control HeLa cells in standard medium. Furthermore, unlike the control, the growth of Rev-2 cells was partially inhibited in low serum.

Since the growth of Rev-2 cells was not completely inhibited in low serum medium, it seemed unlikely that this population had lost its transformed phenotype. This question was examined by two experiments. First, Rev-2 cells were compared to HeLa cells with respect to suspension growth in soft agar. Under these conditions, Rev-2 growth was heterogeneously slowed but clearly demonstrable (data not shown). Second, Rev-2 cells were tested for growth in nude mice. Following subcutaneous injection of $10^5$ HeLa cells into nude mice, tumors formed in 5/5 animals. By comparison, following injection of $10^5$ Rev-2 cells, tumors formed in 4/5 animals. Although growth of the Rev-2 cell tumors was initially slower than the HeLa tumors, this result indicated that at least a subpopulation of cells within the Rev-2 population maintained its malignantly transformed phenotype.

## DISCUSSION

We have developed an assay that is designed to measure inhibition of HeLa cell replication following transfection with DNA from heterologous cell types. Although

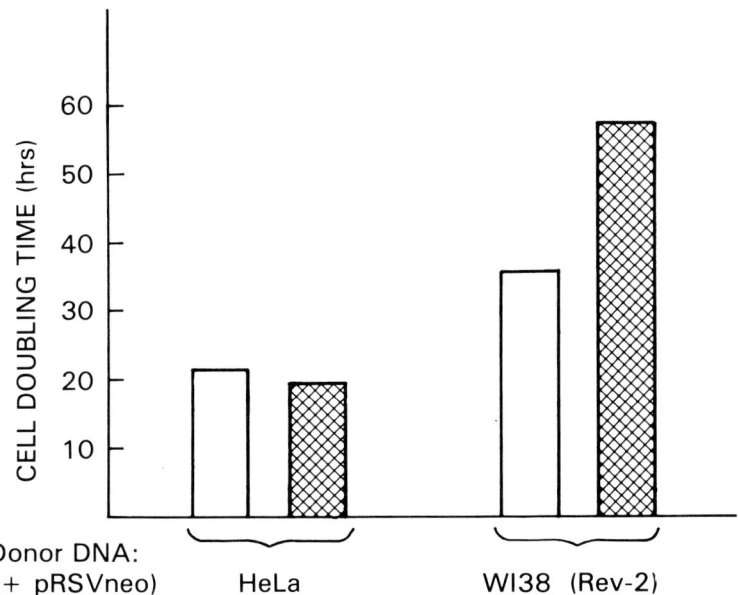

Fig. 3. Effect of serum concentration on growth of cells transfected with the WI38/pRSVneo or HeLa/pRSVneo DNA mixture. Growth rate was measured over an interval of 24–72 hr after plating into monolayer culture. Open bars indicate growth of cells in medium containing 10% fetal fetal calf serum; cross-hatched bars indicate growth in medium containing 0.5% fetal calf serum.

this assay detects specific inhibitory sequences in DNA from quiescent human embryo fibroblasts, it remains to be determined whether these sequences possess tumor suppression activity.

The failure to demonstrate loss of tumorigenicity in cell populations selected by our assay suggests that our experimental system does not detect the malignancy suppressing gene(s) extensively studied by Stanbridge and coworkers [8,9]. This would be consistent with the observation of those investigators that suppression of tumorigenicity in WI38×HeLa hybrids is separable from in vitro transformation as measured by anchorage-independent growth. It seems premature, however, to conclude that the sequences detected in our assay have no effect on tumorigenicity, since other transformed cell lines may be more sensitive to this type of negative growth regulation than HeLa.

Growth-inhibitory activity in human embryo fibroblasts has been demonstrated, in an independent line of investigation, by heterokaryon experiments in which those cells are fused with dividing cells. The inhibitory activity can be induced in fibroblasts by forcing the cells into a deep (G0) state of quiescence [6,7,17,18] or by passaging such cells until they become senescent [7,19–21]. The results obtained using this heterokaryon assay exhibit both similarities to and differences from the results that we report here.

The induction of growth-inhibitory activity by quiescence appears to parallel our findings. We observed substantially higher levels of inhibitory activity if genomic DNAs were harvested from fibroblasts that had been forced in G0 quiescence (by a combination of density and low-serum arrest) rather than harvested at subconfluence or confluence in high serum (unpublished results). The increased activity, which was on the order of twofold for WI38 embryo fibroblasts and tenfold for MRC-5 embryo fibroblasts, implies that epigenetic modification(s) and/or DNA rearrangements can strongly modulate the activity of genomic DNA preparations in our DNA-mediated transfection assay.

HeLa cells have been reported to exhibit a "dominant" phenotype in the heterokaryon assay, i.e., they induce replicative DNA synthesis in nuclei of quiescent cells [6] and senescent cells [20]. This dominance would appear to conflict with our findings; however, it is possible that the interactions between HeLa cells and nonreplicating fibroblasts are complex, involving both stimulation of and inhibition by the quiescent partner. Our assay would measure only the inhibitory component of such interactions. The observation that HeLa cells arrest more readily in the G1 phase of the cell cycle when grown in spinner culture [13] implies that HeLa cells might be more sensitive to growth inhibition when grown in anchorage-independent conditions. Interestingly, we have observed that the suspension-culture step in our assay potentiates the growth-inhibitiory activity that we detect.

One aspect of our findings that deserves comment is the level of growth-inhibitory activity. This level appears to be substantially above that which would be expected from one or a small number of single copy genes. It is possible that a large number of genes can mediate the arrest of HeLa growth in suspension culture. The alternative is that the sequences active in this assay are moderately to highly reiterated in the mammalian genome. We currently favor the latter possibility, since it is more easily reconciled with the reproducible and substantial variations in activity between DNAs from different sources. Another possibility is that a small number of growth-inhibitory gene copies can be activated or amplified to mediate an unexpectedly high level of activity in our assay.

The interpretation that abundantly represented sequences can mediate growth inhibition is supported by results obtained with cosmid libraries. Libraries constructed from PstI partial digestion of *E. coli*, HeLa, or WI38 DNA exhibited activities in our assay that were qualitatively similar to those of the genomic DNAs from which each library was derived; furthermore, fractionation of the WI38 library (sib analysis) indicated the presence of moderately repeated growth inhibitory sequences in this library (R. Padmanabhan, M. McCormick, T. Howard, M. Fordis, B. Howard, unpublished data).

The possibility that mammalian cells contain repeated sequences with the potential to inhibit growth has very interesting implications. Whereas activation of proto-oncogenes occurs rarely, activation of growth-inhibitory genes belonging to a family of repetitive sequence elements could occur with a probability several orders of magnitude higher than that expected for single copy genes. An epigenetic control mechanism, similar to that suggested by our data, could serve to modulate the frequency of this step. Especially if activation were to occur with a high (and/or modulated) frequency, it might function as a stochastic mechanism controlling basic processes such as terminal differentiation and cellular senescence. These processes have in fact been observed to possess stochastic features [22–24].

Transfection of tumor cell genomic DNA into NIH3T3 "indicator" cells has proven to be an effective approach for the isolation and subsequent characterization of cellular oncogenes. Our results suggest that repeated sequences with the potential to inhibit growth may obscure the detection of single copy "tumor suppression" genes in assays similar to the one that we have devised. On the other hand, the sequences that we detect may act to suppress malignant transformation in tumor cells that are less abnormal than HeLa, eg, tumor lines with minimal aneuploidy or those that are recessive in heterokaryon experiments [6,21,25]. Furthermore, we anticipate that it will be possible to identify the unexpected class of genes detected by our assay and to define roles that this class may have in cellular growth regulation.

## ACKNOWLEDGMENTS

We thank Ira Pastan, Mary McCormick, Michael Fordis, Rick Stead, Michael Gottesman, George Khoury, Radha Padmanabhan, Don Court, Mark Willingham, James Rose, and Katherine Sanford for support and helpful discussions. We are grateful to Cori Gorman and Irene Abraham for gifts of reagents.

## REFERENCES

1. Bishop JM: Annu Rev Biochem 52:301, 1983.
2. Holley RW, Böhlen P, Fava, R, Baldwin JH, Kleeman G, Armour R: Proc Natl Acad Sci USA 77:5989, 1980.
3. Raben D, Lieberman MA, Glaser L: J Cell Physiol 108:35, 1981.
4. Hsu YM, Barry JM, Wang JL: Proc Natl Acad Sci USA 81:2107, 1984.
5. Iwata KK, Fryling CM, Knott WB, Todaro GJ: Cancer Res 45:2689, 1985.
6. Stein GH, Yanishevsky RM: Proc Natl Acad Sci USA 78:3025, 1981.
7. Burmer GC, Rabinovitch PS, Norwood TH: J Cell Physiol 118:97, 1984.
8. Stanbridge EJ: Nature 260:17, 1976.
9. Stanbridge EJ, Der CJ, Doersen CJ, Nishimi RY, Peehl DM, Weissman BE, Wilkinson JE: Science 215:252, 1982.
10. Klinger HP, Shows TB: J Natl Cancer Inst 71:559, 1983.

11. Zullo JN, Cochran BH, Huang AS, Stiles CD: Cell 43:793, 1985.
12. Roberts AB, Anzano MA, Wakefield LM, Roche NS, Stern DF, Sporn MB: Proc Natl Acad Sci USA 82:119, 1985.
13. Paul D: Exp Cell Res 114:434, 1978.
14. Gorman C, Padmanabhan R, Howard BH: Science 221:551, 1983.
15. Stetten G, Davidson RL, Latt SA: Exp Cell Res 108:447, 1977.
16. Southern PJ, Berg, P: J Mol Appl Genet 1:327, 1982.
17. Augenlicht LH, Baserga R: Exp Cell Res 89:255, 1974.
18. Rossini M, Lin JC, Baserga R: J Cell Physiol 88:1, 1976.
19. Norwood TH, Pendergrass WR, Sprague CA, Martin GM: Proc Natl Acad Sci USA 71:2231, 1974.
20. Pendergrass WR, Saulewicz AC, Burmer GC, Rabinovitch PS, Norwood TH, Martin GM: J Cell Physiol 113:141, 1982.
21. Stein GH, Yanishevsky RM, Gordon L, Beeson M: Proc Natl Acad Sci USA 79:5287, 1982.
22. Holliday R, Huschtscha LI, Tarrant GM, Kirkwood TBL: Science 198:366, 1977.
23. Smith JR, Whitney RG: Science 207:82, 1980.
24. Bennett DC: Cell 34:445, 1983.
25. Stein GH, Yanishevsky RM: Exp Cell Res 120:155, 1979.

# Index

Acidic phosphoproteins, 3T3 cells, TPA, 304
$\beta$-Actin, 161
Actin, inhibition by anti-sense RNAs, 314
Actinomycin D, 251, 252
Activation of TGF-$\beta$, effect of cell's inability, 20
Acylation site, *ras* oncogene, 323–324
Adenovirus genes, mechanism of transformation induction, 37
Adenovirus genes, system of transcription control, 333
Adenovirus transformed CREF cells, wild-type, 38
Adenylate cyclase, and p21, 331
Adenylate cyclase pathway, 46–47
Ad5 DNA integration, 36–37
Ad5 E1a gene, mutated, 39
Ad5 mutant genes, tumorigenicity, 38
A431 cells, uptake of calcium, 235
A431 epidermal carcinoma cells, culture, 234
AIDS virus (HIV-1), 149
AKR-2b cells, stimulation by TGF-$\beta$, 18–19
Amino acid sequence
  human TGF-$\alpha$, 2–3
  p21, 326–327
  v-*fms* gene product, 28–29
Anchorage independence, role of TGF-$\alpha$, 6–7, 8
Anchorage-independent growth
  hamster-hamster hybrids, 365
  stimulation
    by growth factors, 14
    TGF-$\beta$, 17–18
Aneuploidy, 363
Angiogenesis effect of TGF-$\alpha$, 10
Anisomycine, superinduction of c-*fos* expression, 252
Anticancer activity, stages, 179–180
Anticarcinogenesis, 178–180
Anti-oncogenes, evidence for, 53
Antiproteases, JB6 model, 129
Anti-sense and sense plasmids, 3T3 cells, Southern blot analysis, 315
Anti-sense RNA, 314

inhibition of cell proliferation, 319
Aproliferin, 174
Arachidonic acid metabolism stimulation, oxidant promoters, 240–241
2ar gene(s), transcriptional analysis in JB6 C122 nuclei, 107
2ar isolation, 103–104
2ar/mRNA
  induction, 104–106
  isolation, 104
  in proliferating and nonproliferating cultures, 108–109
Arrest states, characteristics, 176
Atherosclerosis role of TGF-$\beta$, 22
ATP concentrations, effect on kinase activities, 62
Autocrine growth factors, and oncogenes, 191–192
Autocrine mechanisms
  lung carcinoma, 198
  malignant transformation, 7
Autokinase activities, p21, 328

BALB/c-3T3 cells, protein requirement to pass restriction point, 159
BALB/c-3T3 T mesenchymal stem cells, cellular differentiation and proliferation, 172–173
BCNS cells, human, lifespan extension, *pro* genes, 130–131
Benzoylperoxide, role in progression, 246
Beta-adrenergic system, signal transduction, 214–215
Binding characteristics, TGF-$\alpha$, human vs. rat, 8
Biochemical activities, p21, 322
Biochemical responses to carcinogenesis, 136–137
Biological effects
  TGF-$\beta$, 18
  TPA, and receptor occupancy, 288
Bithorax complex (BX-C), 203
Bone resorption, role of TGF-$\beta$, 22

Cadmium resistance, quantitation of TPA effect, 308

# Index

Cadmium toxicity, effect of TPA on 3T3 cells, 305–307
Calcium
  requirement of *pro* genes, 136
  as second messenger, 94
Calcium channel, role in c-*fos* induction, 218
Calcium concentration, mouse epidermal cells, 184–185
Calcium flux mechanism of vanadate activation, 238
Calcium influx, A431 cells, activation and inhibition, 236
Calcium ions
  and carcinogens, changes in keratinocyte differentiation program, 185–187
  role in protein kinase C activation, 224–225
  role of growth factors in release, 9
Calcium-mediated processes, modulation, 229
Calcium-regulated events, lanthanides in study, 134–136
Calcium transport systems, activation in A431 cells, 235–236
Calmodulin, 161
  induction of c-*fos*, 218
Cancer development, 171–172
Cancer prevention and treatment, agents mimicking growth inhibitory genes, 54
Carcinogen(s)
  and altered differentiation of keratinocytes, 185
  and differentiation defects, 178
Carcinogenesis
  direct initiation by *ras* genes, 260–262
  role of differentiation and proliferation defects, 177–178
Carcinogenic agents, inhibition of differentiation and proliferation control, 178
Carcinogenic chemical and physical agents, effects on NHBE cells, 196–197
Carcinogen induced neoplastic development, pathways, 360–361
CAT activity, competition assay, 343
cDNAs, expression in normal cells, 163–164
cDNA synthesis, 103
Cell-cycle-dependent genes, vs. oncogenes, 162
Cell-cycle-dependent mechanisms, suppression of tumorigenicity, 177
Cell cycle progression, 161–162
Cell differentiation, role of TGF-$\beta$, 22
Cell growth control, role of proto-oncogenes, 215–216
Cell hybridization, rat 208F cells, 112
Cell hybrids, Syrian hamster embryo and BP6T-M3, 364–365
Cell proliferation, normal cells, 157–158
Cell ruffling, TGF-$\alpha$ and EGF role, 9
Cell transformation enhancement, relation to blocked junctional intercellular communication, 287
Cell types and growth factors, 48–49
Cell-type specificity of signal transducing pathways, 98
Cellular hybrids, 112
Cellular receptor affinities, human and porcine TGF-$\beta$, 17
Cellular responsiveness to TGF-$\beta$, 20–21
Cellular TGF-$\beta$ sensitivity, mechanism, 20
c-*fms* gene product, relationship to mouse PDGF receptor, 32
c-*fms* locus rearrangements, 30
c-*fms* oncogene product, function, 29–30
c-*fms* transcripts, note in trophoblasts and choriocarcinoma, 30–31
c-*fos*
  activation role of protein kinase C, 219
  anti-sense RNA, growth inhibition, 3T3 cells, 317
  expression, and cell growth, 318–319
  induction, 218
    ability, 250
    pathways, 216
  as stress gene, 253–255
c-*fos* mRNA, anti-sense and sense human 196 bp clones, 316
c-fos mRNAase, 251–252
c-fos mRNA instability, 250–252
c-*fos* protein
  changes, 219–220
  detection, 216
c-*fos* transcription, regulation mechanism, 252–253
Chinese hamster cell lines, 120, 121–125
Chinese hamster $\times$ mouse hybrids
  loss and retention of mouse chromosomes with c-*onc* gene loci, 125
  relative oncogene mRNA levels, 124
CHO cell growth, influence of phorbol esters and cAMP, 78–79
CHO cells
  changes induced by db-cAMP, 75
  influence of phorbol esters and DAG on morphologic reversion, 77–78

levels of phosphoinositide after cAMP and phorbol ester, 74
Chloramphenicol acetyltransferase activity, 340
Chloramphenicol acetyltransferase (CAT) gene, 312
  assayable marker, 338
Cholera toxin
  cAMP enhancer, 194–195, 198
  enhancement of TPA inhibition of EGF binding, 145, 146
  in growth factor binding, 142
Chromosomal changes, tumor progression, 263
Chromosomal instability, suppression, 355
Chromosomal proteins, poly ADP-ribosylation, oxidant promoters, 245
Chromosome 15, loss in Syrian hamster tumors induced by v-Ha-*ras* plus v-*myc* oncogenes, 362–365
Chromosome location
  TGF-α gene, 14
  TGF-β gene, 15
Chronic myelogenous leukemia genomic progression, 354
Cigarette smoke condensate, 196–197
*Cis*-acting transcriptional control domains, c-*myc*, 349
C kinase activation. *See* Protein kinase C, activation
Clastogenic action $O_2$, 243, 244
c-*myb*, human leukemias, 166
c-*myc*
  domains in the 5' flanking sequences, 346–347
  human leukemias, 166
  marker for susceptibility to proliferative stimuli, 168
  regulatory domains, 341–346
  regulatory regions, transfer to recombinant CAT genes, 346
  regulatory sequences, effect on CAT activity directed by heterologous promoters, 345
  UPE region, footprint analysis, 348
  WI38 cells, 164
c-*myc*/CAT recombinant plasmids, self-titration, 342
Colcemid, 296
Colchicine, response of 3T3 cells, 298
Collagenase mRNA induction, inhibition, 250

Colony formation, TGF, 1–2
Colony screening, 108–109
Competence formation, role of PDGF, 142
Competence genes, 216
c-*onc* transcripts, size in hybrids, 122
Connecting tissue and wound healing, effects of TGF-β, 21–22
Control point, cell division. *See* Restriction point
CREF cell line(s)
  in adenovirus transformation studies, 37
  mutant adenovirus-transformed, DNA analysis, 40
CSF-1 receptor, 30
CTF footprint, 350
Cyclic AMP (cAMP)
  agents elevating, 78, 80
  agents that elevate or mimic effect on phosphoinositide levels, CHO cells, 74
  effect on morphology of H-*ras*-transformed NIH3T3 cells, 80
  increased effect of PDGF on EGF binding, 142, 146
  induced morphologic phenotype, CHO cells, 77
  interactions, 228
  and PMA, synergistic phosphorylation changes, 75
  as second messenger, 69–70
Cycloheximide, 145, 146, 159
  effect on c-*fos* gene, 251, 252
Cytoplasmic dot hybridization, 123

Defects, differentiation and proliferation, transformation requirement, 177
Density-dependent protein expression, 109
Depolarization, effects on vanadate-stimulated calcium influx, A431 cells, 235–236
Desmoplastic reaction, TGF-β-induced, 21
Dexamethasone induction of sense clones, 3T3 cells, 317
1,2-Diacylglycerol, activation of protein kinase C, 224
Diacylglycerol
  and increased calcium ion affinity of protein kinase, C, 225–226
  protein phosphorylation, 74–77
  and TPA resistance, 285–286
Differentiation
  biological mechanism regulating, 172–176

cancer, 172
NHBE cells, antagonists, 194–196
steps involved, 173–174
terminal characteristics, 175
Differentiation inducers, NHBE cells, 192–194
DMBA-initiated tumors, XbaI polymorphism, 261
DNA analysis, E1a 13S minus Ad2- and Ad5-transformed CREF cell lines, 41
DNA damage, oxidant promoters, 242
DNA synthesis, 157–158
inhibition, TGF-$\beta$, 193
thymidine kinase in, 160
Disulphide bridges, cysteines, TGF-$\alpha$ and EGF, 5
*Drosophila*, homeo boxes, 204, 205

E1A
activation of viral transcription, 334–335
effects, 36
gene products, 334
proteins, role, 39–40
E1A gene, transformation, 334
*E. coli*, mutant p21 proteins, 327
Embryonic development, 203–204
Emetine, superinduction of c-*fos* expression, 252
Enzyme in detergent-solubilized kinase activities, 61–62
Epidermal cell lines, differing mal transcriptional activity, 274
Epidermal differentiation, molecular control, 185
Epidermal growth factor (EGF)
binding
after freeze-thawing, 93
capacity loss, dependence on protein synthesis, 145
effects of PDGF and TPA, 143–144
hybrids and parental cells, 115
3T3 cells, 299
definition, 1
induction of cytosolic calcium ion increase, 233
induction of precocious murine eyelid opening, 9
labeled molecules, 299–300
nonresponsive cell lines, 307–309
nonresponsive phenotype, 3T3 cells, molecular basis, 297–302
stimulation of calcium uptake, 234
treatment, effects on binding, 97
Epidermal growth factor (EGF) receptor
antivaccinia effects, mechanism, 153
modulation effects, 89
phosphorylation, 3T3 cells, 304
and vaccinia virus, 149–150
Epidermal in vitro systems, expression of mal-related sequences, 271–272
Epinephrine, 194, 195, 198
Epithelial cell transformation, TGF-$\beta$ role, 21
Epstein-Barr virus, 149
Erythrocyte membranes, effects of vanadate, 65
Erythrocytes, PI and tyrosyl kinase activities, 60–62
Erythroleukemia cell lines, murine, 284
E2 factor, binding, 335, 336
E2 promoter, activation, 334–335
Expression
during embryogenesis, mouse M6-containing sequences, 208
murine homeo box genes, developmental, 212

Feedback control, role of protein kinase C, 226–227
Fibroblast growth factors, cadmium resistance, 3T3 cells, 308
Fibroblastic proliferation, TGF-$\beta$ role, 21
Fibronectin production, enhancement by TGF-$\beta$, 17
Footprinting, DNAase, c-*myc* NRE2 region, 347
*Fos*-related proteins, immunoprecipitation, 217
Friend erythroleukemia cells, methylation pattern, protooncogenes, 291

$G_D$ arrestor, 174
$G_D$ state of differentiation, nonterminal characteristics, 175
Gastrin-releasing peptide, 191–192
Gene amplification, enhancement, 3T6 cells, 305
Gene expression
alteration by TPA, 102
importance of low level, 320
Gene products, toxic agents, 250
Genes specifying sensitivity, 127–128, 129–130
Genetic heterogeneity, tumors, 354

Genetic loci, preneoplastic progression, 137–138
Genetic variants, responses to tumor promoters, 128
Genomic changes, 354
Genomic progression, chronic myelogenous leukemia, 354
GO cells, response to mitogenic stimulus, 295
G proteins
 binding, 322
 N-terminal sequences compared to p21 ras, 325
 and p21 amino acid sequences, 328
Growth-dependent genes, expression ratio, S phase gene, 167
Growth factor receptors, 142
 coupling with calcium permeability, 233
Growth factors
 induction of 2ar, 106–108
 induction of cell ruffling, 9
 mechanism of signal transduction, 46–47
 multiple pathways of EGF receptor modification, 94, 97
 nature, 1
 phorbol esters and decrease in affinity of cell surface receptors, 83
Growth inhibition, TGF-$\beta$, 370
Growth inhibitor genes, 52–54
 agents mimicking, 54
 functions, 53–54, 354
Growth inhibitory sequences
 detection, 371–372
 mammalian cells, 375
Growth kinetics of H-ras-transformed NIH 3T3 cells, effect of cAMP, 82
Growth-regulated genes, human leukemias, expression, 165–167
$G_1$-specific block, cAMP, 78
GTP binding domain, 324–329
Guanine-nucleotide-dependent phospholipase C in v-fms transformants, 31–32

Ha-ras transformed Rat-1 cells, 112
Harvey- and Balb-murine sarcoma viruses, effects on mouse skin, 261–262
Heat-shock protein, inhibition by anti-sense RNAs, 314
HeLa cell(s)
 growth inhibition, protocol for detecting DNA sequences, 372
 kinetics of CAT synthesis, 255
 transfected with WI38 DNA, 370, 371
Hematopoietic disorders, c-fms locus relationship, 30
Hereditary cancers, 353
Heterologous receptor modulation, 142
H-GDP binding, mutant p21s, 328
Histone H3, ratio to growth-dependent genes, leukemias, 167
Histones, 160, 161, 164
Homeo box
 definition, 204
 regulatory role, 205
Homologies
 protein kinase C to other protein kinases, 51–52
 TGF-$\alpha$ and EGF, 5–6
Hormone-mediated gene activation pathways, 255
Hox-1 cluster, mouse 209–210
Hox-2 homeo box cluster, mouse, 208–209
Hox-3/m31, 210–211
H-ras gene-transfected cells, 20–21, 36
H-ras oncogene, transformation of NIH3T3 cell, 79
H-ras p21 oncogene, regulation of cAMP, 83
Human carcinoma cell lines, inhibition by TGF-$\beta$, 19
Hybrid cell lines, doubling times, 115
Hybridization
 in malignancy suppression, 112
 transformed CREF clone, 40–41
Hydroperoxides, action with inflammatory cells, 244
Hydrophobic segments, gP180 gag-fms, 28
Hypercalcemia, effects of EGF and TGF-$\alpha$, 9–10

Immortal cells
 evidence for multistep neoplastic transformation, 361
 susceptibility to transfection by different DNAs, 366
Immortalization, relation to promotion sensitivity, 136–137
Immortalization block, 355–356
Immune response, role of TGF-$\beta$, 22
Immunoprecipitation, fos-related proteins, 217
Inflammation promotion, 242
Inflammatory irritants, mechanism of tumor promotion, 239
Inhibin, 16

## Index

Inhibition of binding, Swiss 3T3 cells, PDBu, PDGF, medium from v-*sis*-transformed cells, 95
Initiated cells, response to phorbol esters, 187–188
Initiating effects, DMBA, 259
Initiation, alternative mechanisms, mouse skin tumors, 262–263
Initiation genes, inheritance, 137
Inositol phospholipids
  breakdown inhibition, reverse transformation, 73
  signal-induced degradation, 223
  turnover and signal transduction, 224
Integration, proliferation and differentiation, 180
Intercellular communication, 284
  blocked junctional, 287
Interferon
  as antioncogenes, 54
  combined effects, 154
Interleukin 1, 226

JB6 cells, types, 102
Junctional intercellular communication, blocked, and tumor promotion, 289

Keratinocyte cell lines, expression of mal sequences, 271
Keratinocyte progenitor cells, human, cellular differentiation and proliferation, 172
Keratinocytes, altered differentiation, 185–187
Kinase activities, separation by ion-exchange chromatography, 62, 64

Lanthanide induction, neoplastic transformation, JB6 $P^+$ and $P^-$ cells, 134
Lanthanides, and calcium-regulated events, 134–136
Leukemia, expression of growth-regulated genes, 165–167
Leukemia cells, RNA isolation, 162
Leukemic patients, identification of molecular groups, 167–168
Leukemogenesis, mechanism, 84
Lifespan extension in human BCNS cells, *pro* genes, 130–131
Lipocortin, 312
Lung carcinoma(s)
  ectopic hormones, 198
  human, insensitivity to differentiation inducers, 194
Lymphocyte activation, 226–227
Lymphoid neoplasia, aberrant c-*myc* expression, 337

Mal-1, gene product, 273
Mal-3 and mal-4 expression, mouse skin tumors, 269
Malignancy-induced hypercalcemia, 9–10
Malignancy reversal by activation of signal transduction pathways, 84
Malignant cells, selective growth advantage, 191
Malignant phenotype, suppression, 179
Malignant transformation
  role of inflammatory cells, 243
  role of TGF-α, 6–7
Mal overexpression, marker for malignancy, 273
Mal sequences, expression, mouse skin tumors, 268–270
MCA3/F cells, expression of mal sequences, 272
Membrane-associated receptors, pathways, 47
Mesenchymal stem cell differentiation and proliferation, factors mediating, 174
Metallothionein expression, 251
Metallothionein 11A transcription regulation, mechanism, 252–253
Metallothionein gene, cadmium resistance, 3T3 cells, 306–307
Methotrexate toxicity, TPA-enhanced resistance, 305
Methylation pattern, proto-oncogenes, relation to murine erythroleukemia cell differentiation, 287–288
Mezerein, incomplete tumor promoter, JB6, 128–129
$Mg^{2+}$ concentration, effect on kinase activities, 63
Mitogenic agents, 3T3 cells, 296
Mitogenic program, 48
Mitogen nonresponsive variants, selections, 296–297
Mitogen response
  nontransformed cells, 141–142
  3T3 cells, 296
Mo-en locus, 210
Molecular basis, EGF-nonresponsive phenotype, 3T3 cells, 297–302

Molecular characterization, TGF-$\beta$, 15–16
Molecular weight, heterogeneity, TGF-$\alpha$, 2
Monoclonal antibodies
    to EGF receptor, 153, 154–155
    to p21, 323
Morphology, CHO cells, effects of db-cAMP, 71
Mouse epidermal cells, model, 184–185
Mouse homeo boxes, isolation, 205–206
Mouse skin carcinogenesis model, two-stage, 193
Mouse skin tumors
    alternative mechanisms, initiation, 262–263
    carcinogen-specific mutation, 258–260
    expression of mal sequences, 268–270
    mal-1 and mal-2 expression, 269
mRNA
    inducible, 102
    WI38 cells, 164
m6 cluster, mouse, 206–208
Multistage carcinogenesis, 45
Multistep genetic process, neoplasia, 353, 361
Murine embryo Swiss 3T3 cells, model, 296
Murine erythroleukemia cell differentiation, methylation pattern of proto-oncogenes, 287–288
Murine homeo boxes, locations, 205
Murine homeo box genes, temporal expression, 207
Murine somatic development, effects of TGF-$\alpha$ and EGF, 9
Myelomonocytic growth factor, 7

Nasopharyngeal carcinoma, analog of TPA-transformed JB6 cells, 131–132
Negative regulatory element (NRE)
    CAT cells, orientation independence, 341
    human c-*myc* gene, 338
Neoplastic phenotype, genes inducing and maintaining, 131
Neoplastic transformation
    control of differentiation and proliferation, 176–178
    lanthanide induction, 134
    nonsequential, 137–138
Nerve growth factor, stimulation of c-*fos* expression, 218–219
*Neu* oncogene, 311–312
Neuroblastoma development, *Drosophila*, 53
Nickel ions, 196

NIH3T3 cells, reversal of H-*ras* oncogene transformation by cAMP, 79–83
Nitrosomethylurea (NMU), rat mammary carcinomas, 259–260
Normal human bronchial epithelial (NHBE) cells
    culture, 192
    effects of carcinogens in vitro, 196–197
Northern blot analysis
    poly (A) RNA preparations, JB6, 105
    somatic cell hybrids, 120, 121
Northern blots, total RNA, leukemic cells, 166
N-*ras* gene, activated, 278
N-*ras*-transformed PA-1 cells, tumorigenicity, 280
NRK cells, v-*sis*-transformed, 90, 91, 92
Nuclear "run-on" transformation, 106
Nuclear signals, c-*fos* and c-*myc*, 217

ODC activity, 3T3 cells, after TPA, 305
Oncogene combinations, 361
Oncogene methylation patterns, relation to TPA susceptibility, 291
Oncogene paradox, 353
Oncogenes
    growth-regulated, 161
    overexpressed human neoplasia, 165
    relation to growth factors and cell proliferation, 158
    specific point mutations, 277
Oncogene transfection, transformation of NHBE cells, 197
Oncogenic phenotype, 333
Oncogenic phenotype repression, E1a proteins, CREF cells, mechanism, 42
Oncogenic transformation, nature, 70
Ornithine decarboxylase, 161
Overexpression, growth-dependent genes, 165–166
Oxidant promoters
    poly ADP-ribosylation, chromosomal proteins, 245
    secondary reactions, 240
Oxidants with tumor-promoting properties, 240
Oxidant tumor promoters, mechanism, 239

PA-1 human terato cells, isolation, 278
Papilloma(s)
    formation, mouse skin, HaMSV initiation, 262

initiation with DMBA and TPA, 258
progression of virus-induced, 263–265
responsiveness to calcium differentiation stimulus, 186
Papilloma cell lines, response to phorbol esters, 187–188
Pathway of signal transduction, 52
PC12 pheochromocytoma cell line, 215–216
PDBu, comparison of binding block with v-*sis*-transformed cell medium and PDGF, 96
Peptides, effects on viral replication, 151
Peroxides, role in tumorigenesis, 243–244
Pertussis toxin, NHBE cells, 195, 198
P53, human leukemias, 166
Phagocytic leukocytes, role in tumorigenesis, 242–243
Phorbol ester binding sites, TPA-resistant and TPA-sensitive MELC, 285
Phorbol-ester-mediated tumor promotion, cellular basis, 187
Phorbol esters
  decrease in affinity of cell surface receptors for hormones or growth factors, 83
  limitations on study value, 226
  pleiotropic responses, 127
  and protein kinase C, 72
  role in activation of protein kinase C, 225–226
Phorbol ester tumor promoters, mechanism, 239
Phorbol-myristate-acetate (PMA)
  mechanism of tumor promotion, 239
  pro-oxidant action, 244–245
Phosphatidylinositol (PI) kinase
  activity, 57
  activity assay, 59–61
  metabolism, papilloma, 186
Phosphoinositide turnover, reverse transformation, 72–74
Phospholipase C, 31
  cellular enzyme, second messenger, 218–219
Phospholipid/calcium ion-dependent TPA binding, 3T3 cells, 303
Phosphoprotein distribution, CHO cells after cAMP and PMA, 75
Phosphorylation changes
  cAMP-mediated reverse transformation, *ras*-transformed NIH3T3 cells, 83
  PMA, 75
Phosphorylation protein, synergism of cAMP, phorbol esters and diacylglycerol, 74–77
Phosphotyrosine increase, transformation, 72
Photosensitization, 240
Platelet-derived growth factor (PDGF)
  binding after, 96
  comparison of binding block with v-*sis*-transformed cell medium and PDBu, 96
  EGF, synergism, 142
  generation of second messengers, 94
  as proliferation initiator, 141–142
  role in EGF binding, 92–93
Pleiotropic biological effects of cAMP, phorbol esters, and oncogenes, reconcilement, 83
Poly ADP-ribosylation of chromosomal proteins, 245
Polymorphism, variation with carcinogen type, 258–259
Polyoma T-antigen, role in transformation, 312
Polypeptide hormones, mechanism of tumor promotion, 239
Potassium ions, role in epidermal differentiation, 184
Precursors, human and rat TGF-$\alpha$, 2–5
Predifferentiation arrest state
  characteristics of cells, 175
  evidence for, 176
Primary human fibroblasts, 250
*Pro* genes, activities other than promotion sensitivity, 130–131
Proliferation, biological mechanisms regulating, 172–176
Promotion-relevant signal transductions, JB6 P$^-$ cells, 132
Promotion sensitivity
  existence of gene, 137
  function, 136–137
*Pro*-1, activated human, 131–132
*Pro*-1 and *pro*-2, genes specifying sensitivity to tumor promotion, 129–130
Pro-oxidant action, PMA, 244–245
Pro-oxidant genes, 239
Protein kinase C, 47
  activation, 224–225
    P$^+$ and P$^-$ cells, 133–134
    phorbol ester, in cAMP-induced growth inhibition, 78–79
    by TPA, 102, 302

activity, 3T3 cells, 303
depletion
  effect, 94
  method, 92
diacylglycerol pathway, 224
as mediator of changes in EGF receptor properties, 89–90
model, 50
possible role in action of EGF, 98
purification, 51
regulation of metallothionein and c-*fos* transcription, 252–253
role in cellular responses and feedback control, 226–227
role in epidermal differentiation, 185
role in TPA effects, 289
substrate, lanthanum sensitivity, 135
Protein kinases, 46
  cAMP dependent, 70
Protein phosphorylation
  by cAMP dependent protein kinase system, 70–71
  modulation of calcium-mediated processes, 229
Protein synthesis inhibition, 159
Protein synthesis inhibitors, effect on c-*fos* expression, 250
Protein synthesis requirement, loss of EGF binding capacity, 145
Proteolysis, role of TGF-β, 21–22
Proto-oncogenes, 45–46, 122–123
  definition and enumeration, 48–49
  expression, down-regulation, 124–125
  families, 49
  methylation state, 285
  normal growth, 313
  regulatory steps, cell growth, 215–216
Protoplast fusion, method, 279
Proviral DNA clones, transformation and GTP binding, 330
PtdIns turnover, 31
p21
  biochemical activities, 322
  expression in hybrids and parental cells, 114
  renaturing capability, 325–326
p21 *ras* proteins, function, 330
Purification, TGF-β, 15
Puromycin superinduction of c-*fos* expression, 252

Quiescence, growth inhibitory activity, 374

Rabies virus, 149
*Raf* oncogenes, lung carcinomas, 197
*Ras* cellular function, crucial roles of GTP binding domain, 329–331
*Ras* family, mutated oncogenes, 257
*Ras* gene function, $G_1$, 158–159
*Ras* genes
  characterization, 321–322
  initiation of carcinogenesis, 260–262
  reversal of effect, 81
*Ras* oncogene
  activation, skin carcinogenesis, 188–189
  NIH3T3 murine fibroblasts, 277
  transformation, mechanism, 282
*Ras* oncogene product p21, NHBE cells, 196
*Ras* proteins, effect, 81
Rat fibroblasts, morphological changes, TGF-α, 1
Rat TGF-α amino acid sequence, 3
Receptor interactions, 227–228
Receptor mechanism, role of inositol phospholipids, 223–224
Regulatory rules, E.B. Lewis, 203–204
Regulatory steps, cell growth control, proto-oncogene role, 215–216
Reovirus, 149
Restoration of normal properties to cancer cells, cAMP, 70
Restriction point
  cell division, 158
  protein characteristics, BALB/c-3T3 cells, 159
Retinoblastoma
  mechanism, 179
  paradigm, 353, 354
Retinoic acid, F9 teratocarcinoma variants responsive to, 311
Retinoids, as inhibitors of second-stage tumor promotion, 129
Retinoyl phorbol acetate (RPA)
  epidermal cell lines, 271
  JB6, 128–129
Retrovirally transformed cells, EGF binding sites, 7
Retrovirally transformed fibroblasts, 6
Reverse transformation, cAMP-mediated, 70
Rodent fibroblasts, transformed, TGF-α activity, 5–6
RNA blot analysis, EGF-receptor mRNA, 301

Sarcoma growth factor. *See* Transforming growth factors (TGFs)

Second messengers
    phosphatidylinositol (PtdIns)-derived, 31
    perturbation, 69
    stimulation of c-*fos* expression, 218–219
Senescence, tumor suppression in human cells, 355–356
Sensitivity, function specified by genes promoting, 136
Serum factor inhibiting epithelial cell growth, 194
Sequence homologies, murine homeo boxes, 206
Signal-gene interaction, models, 136
Signal transduction, 46–48
Signal transduction pathways, 47
    activation, mechanism for reversal of malignancy, 84–85
    cell-type specificity, 98
    c-*fos* expression, 219
*Sis* oncogene, 83
Skin carcinogenesis
    activation of *ras* oncogene, 188–189
    mouse, two-stage model, 267
Skin tumorigenesis, 257
Small cell lung carcinoma, malignant behavior, 197
Sodium, effects on vanadate-stimulated calcium influx in A431 cells, 236
Soft agar growth, NRK cells, enhancement by various factors, 17
Somatic cell hybrids, isolation, 120
Somatomedin C, effect on induction of 2ar, 108
S100 proteins, 168
Southern blot
    genomic, EGF-receptor gene, 301
    hybridization, 112–113
    papillomas, 263, 264
S phase, enzymes and DNA synthesis at onset, 160
Sp1 binding factors, 344
Sp1 binding sites, 345, 349
Stem cell differentiation, retinoblastoma, 179
Stem cell transformation, 172
Superoxide, clastogenic action, 241
Suppression/expression, anchorage-independent and tumorigenic phenotypes, hamster-hamster hybrids, 364
Suppression of malignancy, chromosome requirements, 112
Suppression of tumorigenicity, possible mechanism, 117

Suppressor genes, 116
    loss at late stages of carcinogenesis, 264–265
Synergism, cAMP and DAG, CHO cells, phosphoprotein, 76
Syrian hamster embryo cells
    characterization of TPA -sensitive and -resistant, 286
    progression from preneoplastic to neoplastic state, 360
    3-step model, neoplastic transformation, 366
    transformation system, advantages, 360
    variants, 284
Syrian hamster tumors, loss of chromosome, 15, 362–365

Target sites, cancer development, 184
Terminal differentiation, suppression by, 355
TGF-α
    biological activities vs. EGF, 7–11
    vs. EGF, potency, 10–11
    fetal synthesis, 6
    relationships to EGF and TGF-β, 14–15
    synthesis, cellular sources, 5–6
TGF-β
    activation, 20
    antagonism of differentiation induction by cAMP enhancers, 196
    binding after, 96
    discovery, 13
    as growth inhibitor, 14
    induction of terminal squamous differentiation, NHBE cells, 193
    inhibition of cell proliferation, 18–19
    in mediation of EGF binding inhibition, 93
    mechanism of growth stimulation, 17–18
    mitogenicity, 18
    potential roles in neoplasia and other diseases, 19–22
    properties, 15
    receptor, 16
3T3 fibroblasts, analysis of RNA, 108
3T3 quiescent fibroblasts, 106–108
Thymidine kinase
    assay for onset of DNA synthesis, 160
    as growth-regulated gene, 161
    inhibition by anti-sense RNAs, 314
Toxic agents, gene products, 250
TPA
    effects on gene expression, 102

exposure, 2ar induction, 107
and expression of mal sequences, 272
as hMT-IIA promoter regulator, 254
induction of terminal differentiation, 193–194
inhibition of differentiation, erythroleukemia cells, 284
mechanism for study, 52
and PDGF effect on EGF binding, 143–144
as *pro* gene inducer, 136
resistance, and diacylglycerols, 285–286
as tumor promoter, 47, 102
TPA-induced gene amplification, 3T3 cells, 305–307
TPA-nonresponsive phenotype, molecular basis 302–307
TPA-sensitive and -resistant clonal variants, Friend erythroleukemia cells, 290
TPA-sensitive and -resistant variant cells, OAG effects, 286
Transcription
    m6 region F9 stem and differentiating cells, 207
    pattern change, mal-3-related sequences, 273
    peak, mouse embryogenesis, 209
Transcriptional activity, C-Ha-*ras*, cki-*ras*, c-*myc*, c-*fos* oncogenes, 123
Transcriptional activity test, c-*myc* 5′ flanking sequences, 339
Transcriptional control
    c-*myc* expression, 337–338
    importance, 333
Transcriptional rate genes after UV and TPA, 250
Transcription level, oncogenes, 120
Transcripts, proto-oncogene, origins, 122–123
Transformation process, hybrids, 125
Transformation parameters, 114
Transformed foci, 52
Transforming factors, multiple pathways of EGF receptor modification, 94, 97
Transforming growth factor-α. *See* TGF-α
Transforming growth factor-β. *See* TGF-β
Transforming growth factors (TGFs)
    activity hybrids and parental cells, 115
    definition, 1, 96
    in induction of anchorage independence, 8
Transforming oncogenes of retroviruses (v-*onc* genes), generation, 27–28

Transglutaminase activity, marker for epidermal differentiation, 186
Translocation, MELC, 289
Tumor angiogenesis factor (TAF) regulation of production, 254–255
Tumorigenic cells, reduction after fusion with normal cells, 119
Tumorigenesis
mechanism, 116
role of N-*ras* oncogene, 280, 281
Tumorigenicity
    effect of WI38 × HeLa hybridization, 374
    hybrid cells, 113–114
    induced by v-Ha-*ras* plus v-*myc* oncogenes, 363
    recessive nature, 367
    suppression, 178–180
        by cell-cycle-dependent mechanisms, 177
Tumor necrosis factor, cell lines resistant to, 311
Tumor progression, 101–102
chromosomal changes, 263
Tumor promoters, 48
    classes, by mechanism, 239
    genetic variants for responses, 128
    induction of transient mal-related sequence expressions, 270–271
    role in activation of protein kinase C, 225–226
Tumor promotion, 101–102
initiated cells, 284
Tumor suppressor function, loss during chemically induced neoplastic progression, SHE cells, 364–366
Tumor suppressor genes, 179
evidence for existence, 353, 367
Tyrosine phosphorylation, 3T3 cells, 303
Tyrosine-specific protein kinase, 28–29
Tyrosyl kinase
    activity
        assay, 59–61
        EGF receptor, 233
    distinction, 66
    relationship to PI kinase activity, 57–58
Tyrosyl-specific protein kinase activity, 57

Ultraviolet irradiation and TPA, induction of overlapping gene products, 249–250

Vaccinia virus

preparation, 149
receptor, 149–150
replication, inhibition by EGF and peptides, 151–152
Vaccinia virus growth factor (VVGF), 5
Vanadate, 58
    effect on PI and tyrosyl phosphorylations, 63–65
    as phosphotyrosine phosphatase inhibitor, 234
    stimulation of calcium uptake, 234
Vanadate-stimulated calcium uptake, effects of depolarization, 237
Vascular tissue, effect of growth factors, 10
v-*fms*-coded glycoproteins, transmembrane orientation, 28
v-*fms* gene product, amino acid sequence, 28–29
v-*fms* oncogene, composition, 28
v-*fms* oncogene product, intracellular targets, 29
Viral promoters, 333–334
Viral transcription, E1A activation, E2 promoter, 334–335
Viral transformation, mechanisms, 30–31
Virus strains EGF receptor study, 150
v-*sis*-transformed cell medium, comparison of binding block with PDBu and PDGF, 96
v-*sis*-tranformed cell proteins, 89–90

WI38, cell culture, 162

XbaI polymorphism, NIH3T3 transformations, 260
Xeroderma cells, evidence for DNA damage, 252

Yin-Yang theory of cancer, 353–354